Applied Elementary Algebra

Math 092, Clark College

Charles P. McKeague

Katherine Yoshiwara

SECOND PRINTING — May 2018, Prealgebra Review chapter added

xyztextbooks

Applied Elementary Algebra

Math 092, Clark College

Charles P. McKeague

Katherine Yoshiwara

Publisher: XYZ Textbooks

Production Manager: Katherine Heistand Shields

Cover Image: © plfy

Printed in the United States of America

ISBN-13: 978-1-63098-191-4 / ISBN-10: 1-63098-191-5

For product information and technology assistance, contact us at
XYZ Textbooks, 1-877-745-3499

For permission to use material from this text or product,
e-mail: **info@mathtv.com**

XYZ Textbooks
1339 Marsh Street
San Luis Obispo, CA 93401
USA

For your course and learning solutions, visit **www.xyztextbooks.com**

Brief Contents

Contents

Preface

This edition was custom made as directed by the faculty of Clark College to meet the needs of their students studying Applied Elementary Algebra.

Description

Applied Elementary Algebra, Math 092, Clark College is an integration of XYZ Textbooks' popular introductory mathematics and algebra textbooks into a concise survey of the subject designed for a one-semester course covering Elementary Algebra and its applications. As with all XYZ titles, our emphasis on study skills, and our positive tone toward success, are reflected in the presentation.

You'll find this book to be rooted in the proven teaching methods that you come to expect from an XYZ Textbook, with some innovative features to support your students' success.

The materials for *Applied Elementary Algebra, Math 092, Clark College* are designed as an activity-based elementary algebra course. The textbook develops the concepts and skills of algebra in the context of modeling and problem solving. Each section includes embedded reading questions for students to check their understanding and exercises that reflect the material presented in the examples. The problem sets consist of two parts: skills practice problems that review prerequisite skills for the section and provide practice with new skills, and applications problems. The applications include a wide range of open-ended problems that emphasize mathematical modeling using tables of values, algebraic expressions, and graphs. Chapter reviews include a glossary and summary, as well as a set of review problems. Encourage your students to make regular use of these features—you'll find that they will gradually build a foundation of successful studying practices that will benefit them in all their future courses.

This textbook is more than just the book itself. We built this book to be the hub of a math "toolbox" of sorts. While most students still prefer to use the printed book in their studies, the eBook extends the reach of the book, giving students and instructors access to a wide array of supporting tools:

- The ebook includes free access to this textbook, plus over 20 other ebooks covering 8 math courses—great for remediation.
- It also includes free access to 10,000 MathTV videos, with 3-4 tutorials for every single example.
- Plus all of the accompanying worksheets and digital supplements for the book.

To Students: A Study Plan For Success

Before Class: Prepare

1. Read the textbook section slowly and carefully. Read one subsection at a time. Don't rush! Work on grasping each concept before going on. You may have to read a difficult passage two or three times before it begins to make sense.
2. Study each example. Read through the example step by step, and make sure you understand the reasoning behind each step. Read with a pencil in hand, and put a question mark beside any explanations that are unclear to you. Jot down in the margins any questions that occur to you.
3. Answer the reading questions and work the exercises as you go along. Check your answers to the exercises when you get to the end of the section.

During Class: Participate

4. Ask questions when you need help. Classtime is the best time to work on understanding new material. Try to get as many of your questions cleared up as possible before you leave class.
5. Look over the homework problems with your group. Exchange e-mail or phone numbers with your group so you can contact each other after class if you get stuck.

After Class: Practice

6. There are two types of homework problems at the end of each section in the textbook, skills practice and applications. Keep your homework problems in a notebook (a three-ring binder is best), so that you can use them to study for tests.
7. Write out the solutions to the skills practice problems neatly and completely, the same way you would write them on a test. If your class uses a computer homework system, you may be asked to enter your answers into the computer for grading.
8. Prepare the solutions to the application problems to hand in at your next class meeting. Follow your instructor's guidelines for written work.
9. Make any corrections necessary in previous Homework assignments. Finally, go back to Step 1 and prepare for the next day's lesson!

What to Do When You Need Help Outside of Class

In college courses it is not unusual for students to need extra help outside of class meetings; it's not possible to learn new ideas in one hour. It is very important that you take advantage of the resources available to you, and seek help as soon as you need it! Do not wait even one day before asking for help; it's very easy to get behind in a math class. Here are some strategies:

1. Go back to the textbook and look for similar examples, or an explanation of the relevant topic.
2. Contact your group members or study buddies and work together.
3. Email your instructor with your question. Try to be specific!
4. Visit your college's Math Lab or tutoring center and ask a tutor for help.
5. Visit your instructor during office hours. Come prepared with your homework notebook and specific questions.

The Secret of Success

In college most of your learning takes place outside the classroom. YOU are responsible for learning the material when it is assigned—don't wait! If you don't understand something, YOU must seek help right away! Practice is the key to learning skills: work problems every night until you have mastered each new topic.

Prealgebra Review

R

One of the most beautiful views in the world can be found from Skydeck 88 on the eighty-eighth floor of the Eureka Tower in Melbourne, Australia. Once you enter the building from the ground floor, it won't take long to get there. The elevators in Eureka Tower are the fastest in the southern hemisphere, traveling at over twenty miles per hour! When you arrive at Skydeck 88, you can stand in a glass cube that extends almost 10 feet out of the building, nearly 1,000 feet above the ground. The glass on the upper ten floors of the tower is plated in twenty-four carat gold, making the building also beautiful to look at from the ground.

Suppose you wanted to visit Skydeck 88 with some friends. The admission prices are listed in the table in Australian dollars.

	Skydeck Entry	Sun & Stars Entry (2 entries per day)
Adult	\$17.50	\$21.00
Child	\$9.00	\$12.50
Concession	\$11.00	\$14.50
Family of 4	\$39.90	\$49.50
Family of 3	\$29.50	\$39.50
Extra child	\$7.00	\$10.50

How much would it cost for you and two adult friends to purchase the Skydeck Entry and to pay the concession fee? How would you split this cost evenly between the three of you? How much would you pay if you treated your niece and nephew to the Sun & Stars Entry? Answering all of these questions requires that you know how to work with decimal numbers, which is one of the concepts we will review in this chapter.

R.1 Factors and Simplifying Fractions

OBJECTIVES

A Factor numbers into a product of primes.

B Simplify fractions by reducing to lowest terms.

KEY WORDS

prime number
composite number
divisor
factor
prime factor
fraction
numerator
denominator
lowest terms
equivalent fraction
improper fraction
proper fraction

©iStockphoto.com/kentarus

A Prime Numbers

Suppose you and a friend decide to split a medium-sized pizza for lunch. When the pizza is delivered, you find that it has been cut into eight equal pieces. If you eat four pieces, you have eaten $\frac{4}{8}$ of the pizza, but you also know that you have eaten $\frac{1}{2}$ of the pizza. The fraction $\frac{4}{8}$ is equivalent to the fraction $\frac{1}{2}$; that is, they both have the same value. The mathematical process we use to rewrite $\frac{4}{8}$ as $\frac{1}{2}$ is called *reducing to lowest terms*. Before we look at that process, we need to review some terms. Here is our first one.

DEFINITION prime number

A ***prime number*** is any whole number greater than 1 that has exactly two divisors—itself and 1. (A number is a divisor of another number if it divides it without a remainder.)

$$\text{Prime numbers} = \{2, 3, 5, 7, 11, 13, 17, 19, 23, 29, 31, 37, \dots\}$$

The list goes on indefinitely. Each number in the list has exactly two distinct divisors—itself and 1.

DEFINITION composite number

Any whole number greater than 1 that is not a prime number is called a ***composite number***. A composite number always has at least one divisor other than itself and 1.

EXAMPLE 1 Identify each of the numbers below as either a prime number or a composite number. For those that are composite, give two divisors other than the number itself or 1.

a. 43 **b.** 12

Solution **a.** 43 is a prime number, because the only numbers that divide it without a remainder are 43 and 1.

b. 12 is a composite number, because it can be written as $12 = 4 \cdot 3$, which means that 4 and 3 are divisors of 12. (These are not the only divisors of 12; other divisors are 1, 2, 6, and 12.)

Note You may have already noticed that the word *divisor* as we are using it here means the same as the word *factor.* A divisor and a factor of a number are the same thing. A number can't be a divisor of another number without also being a factor of it.

Practice Problems

1. Which of the numbers below are prime numbers, and which are composite? For those that are composite, give two divisors other than the number itself and 1.

   37, 39, 51, 59

Answers

1. Prime: 37, 59
   Composite: 39, 3 and 13; 51, 3 and 17

Every composite number can be written as the product of prime factors. Let's look at the composite number 108. One way we can write 108 is $2 \cdot 54$. The number 2 is a prime number, but 54 is not prime. Because 54 can be written as $2 \cdot 27$, we have

$$108 = 2 \cdot 54$$
$$= 2 \cdot 2 \cdot 27$$

The number 27 can be written as $3 \cdot 9$ or $3 \cdot 3 \cdot 3$ (because $9 = 3 \cdot 3$). Overall,

$$108 = 2 \cdot 54$$
$$108 = 2 \cdot 2 \cdot 27$$
$$108 = 2 \cdot 2 \cdot 3 \cdot 9$$
$$108 = 2 \cdot 2 \cdot 3 \cdot 3 \cdot 3$$

This last line is the number 108 written as the product of prime factors. We can use exponents to rewrite the last line.

$$108 = 2^2 \cdot 3^3$$

Note The numbers 2 and 54 are not the only divisors of 108. We could have begun with $3 \cdot 36$, $4 \cdot 27$, or even $6 \cdot 18$ and still end up with the same product of primes.

EXAMPLE 2 Factor 60 into a product of prime factors.

Solution We begin by writing 60 as $6 \cdot 10$ and continue factoring until all factors are prime numbers.

$$60 = 6 \cdot 10$$
$$= 2 \cdot 3 \cdot 2 \cdot 5$$
$$= 2^2 \cdot 3 \cdot 5$$

Notice that if we had started by writing 60 as $3 \cdot 20$, we would have achieved the same result. In other words, it doesn't matter which two factors you choose to begin, as long as you continue factoring until you have a product of primes.

$$60 = 3 \cdot 20$$
$$= 3 \cdot 2 \cdot 10$$
$$= 3 \cdot 2 \cdot 2 \cdot 5$$
$$= 2^2 \cdot 3 \cdot 5$$

2. Factor 90 into a product of prime factors.

Note There are some shortcuts to finding the divisors of a number. For instance, if a number ends in 0 or 5, then it is divisible by 5. If a number ends in an even digit (0, 2, 4, 6, or 8), then it is divisible by 2. A number is divisible by 3 if the sum of its digits is divisible by 3. For example, 921 is divisible by 3 because the sum of its digits is $9 + 2 + 1 = 12$, which is divisible by 3.

B Simplifying Fractions

As we proceed through Chapter 0, from time to time we will review some of the major concepts associated with fractions. To begin, here is the formal definition of a *fraction*:

DEFINITION fraction

If a and b are integers, then the expression

$$\frac{a}{b} \quad b \neq 0$$

is called a ***fraction*** or a rational number. The top number a is called the ***numerator***, and the bottom number b is called the ***denominator***. The restriction $b \neq 0$ keeps us from writing an expression that is undefined. (As you will see, division by zero is not allowed.)

We can use the method of factoring numbers into prime factors to help reduce fractions to *lowest terms*. Here is the definition for lowest terms:

Answers

2. $2 \cdot 3^2 \cdot 5$

3. Which of the following fractions are in lowest terms?

$\frac{1}{6}, \frac{2}{8}, \frac{15}{25}, \frac{9}{13}$

4. Reduce $\frac{12}{18}$ to lowest terms by dividing the numerator and the denominator by 6.

5. Reduce the fraction $\frac{15}{30}$ to lowest terms by first factoring the numerator and the denominator into prime factors and then dividing out the factors they have in common.

Note A common mistake when reducing to lowest terms is to write a fraction, such as

$\frac{12}{15}$ as $\frac{3+3+3+3}{3+3+3+3+3} \neq \frac{1}{3}$.

When finding an equivalent fraction, we can only divide out common factors, not common terms.

DEFINITION fraction in lowest terms

A *fraction* is said to be in *lowest terms* if the numerator and the denominator have no factors in common other than the number 1.

EXAMPLE 3 The fractions $\frac{1}{2}, \frac{1}{3}, \frac{2}{3}, \frac{1}{4}, \frac{3}{4}, \frac{1}{5}, \frac{2}{5}, \frac{3}{5}$, and $\frac{4}{5}$ are all in lowest terms, because in each case the numerator and the denominator have no factors other than 1 in common. That is, in each fraction, no number other than 1 divides both the numerator and the denominator exactly (without a remainder).

EXAMPLE 4 The fraction $\frac{6}{8}$ is not written in lowest terms, because the numerator and the denominator are both divisible by 2. To write $\frac{6}{8}$ in lowest terms, we divide both the numerator and the denominator by 2.

$$\frac{6}{8} = \frac{6 \div 2}{8 \div 2} = \frac{3}{4}$$

The fraction $\frac{3}{4}$ is in lowest terms, because 3 and 4 have no factors in common except the number 1.

Reducing a fraction to lowest terms is simply a matter of dividing the numerator and the denominator by all the factors they have in common to produce an *equivalent fraction* of the same value.

EXAMPLE 5 Reduce the fraction $\frac{12}{15}$ to lowest terms by first factoring the numerator and the denominator into prime factors and then dividing both the numerator and the denominator by the factor they have in common.

Solution The numerator and the denominator factor as follows:

$$12 = 2 \cdot 2 \cdot 3 \quad \text{and} \quad 15 = 3 \cdot 5$$

The factor they have in common is 3. We can divide both terms of the fraction by 3 to produce an equivalent fraction.

$$\frac{12}{15} = \frac{2 \cdot 2 \cdot 3}{3 \cdot 5} \qquad \text{Factor the numerator and the denominator completely.}$$

$$= \frac{2 \cdot 2 \cdot 3 \div 3}{3 \cdot 5 \div 3} \qquad \text{Divide by 3.}$$

$$= \frac{4}{5}$$

The fraction $\frac{4}{5}$ is equivalent to $\frac{12}{15}$ and is in lowest terms, because the numerator and the denominator have no factors other than 1 in common.

We can shorten the work involved in reducing fractions to lowest terms by using the slash to indicate division. For example, we can write the above problem this way:

$$\frac{12}{15} = \frac{2 \cdot 2 \cdot \cancel{3}}{\cancel{3} \cdot 5} = \frac{4}{5}$$

So long as we understand that the slashes through the 3s indicate that we have divided both the numerator and the denominator by 3, we can use this notation.

Answers

3. $\frac{1}{6}, \frac{9}{13}$

4. $\frac{2}{3}$

5. $\frac{1}{2}$

EXAMPLE 6 Laura is having a party. She puts 4 six-packs of diet soda in a cooler for her guests. At the end of the party, she finds that only 4 sodas have been consumed. What fraction of the sodas are left? Write your answer in lowest terms.

Solution She had 4 six-packs of soda, which is $4(6) = 24$ sodas. Only 4 were consumed at the party, so 20 are left. The fraction of sodas left is

$$\frac{20}{24}$$

Factoring 20 and 24 completely and then dividing out the factors they have in common gives us

$$\frac{20}{24} = \frac{2 \cdot 2 \cdot 5}{2 \cdot 2 \cdot 2 \cdot 3} = \frac{5}{6}$$

Note The slashes in Example 6 indicate that we have divided both the numerator and the denominator by $2 \cdot 2$, which is equal to 4. With some fractions it is apparent at the start what number divides the numerator and the denominator. For instance, you may have recognized that both 20 and 24 in Example 6 are divisible by 4. We can divide both terms by 4 without factoring first. This still guarantees that dividing both terms of the fraction by 4 will produce an equivalent fraction.

$$\frac{20}{24} = \frac{20 \div 4}{24 \div 4} = \frac{5}{6}$$

EXAMPLE 7 Reduce $\frac{6}{42}$ to lowest terms.

Solution We begin by factoring both terms. We then divide through by any factors common to both terms.

$$\frac{6}{42} = \frac{2 \cdot 3}{2 \cdot 3 \cdot 7} = \frac{1}{7}$$

We must be careful in a problem like this to remember that the slashes indicate division. They are used to indicate that we have divided both the numerator and the denominator by $2 \cdot 3 = 6$. The result of dividing the numerator 6 by $2 \cdot 3$ is 1. It is a very common mistake to call the numerator 0 instead of 1 or to leave the numerator out of the answer. Be careful not to make this mistake.

EXAMPLE 8 Reduce $\frac{4}{40}$ to lowest terms.

Solution
$$\frac{4}{40} = \frac{2 \cdot 2 \cdot 1}{2 \cdot 2 \cdot 2 \cdot 5} = \frac{1}{10}$$

All the fractions we have worked with so far have been *proper fractions.* The following example involves an *improper fraction.* Let's review a formal definition before working the problem.

DEFINITION improper fraction, proper fraction

A fraction in which the denominator is less than or equal to the numerator is called an ***improper fraction;*** whereas, a fraction in which the denominator is greater than the numerator is called a ***proper fraction.***

6. Reduce $\frac{30}{35}$ to lowest terms.

7. Reduce $\frac{8}{72}$ to lowest terms.

8. Reduce $\frac{5}{50}$ to lowest terms.

Answers

6. $\frac{6}{7}$
7. $\frac{1}{9}$
8. $\frac{1}{10}$

9. Reduce $\frac{120}{25}$ to lowest terms.

Note Unless instructed otherwise, it is perfectly acceptable to leave a final answer as an improper fraction as we have done in Example 9. However, for application problems, mixed numbers or decimals are preferred.

EXAMPLE 9 Reduce $\frac{105}{30}$ to lowest terms.

Solution
$$\frac{105}{30} = \frac{3 \cdot 5 \cdot 7}{2 \cdot 3 \cdot 5} = \frac{7}{2}$$

GETTING READY FOR CLASS

Each section of the book will end with some problems and questions like the ones below. They are for you to answer after you have read through the section but before you go to class. All of them require that you give written responses in complete sentences. Writing about mathematics is a valuable exercise. As with all problems in this course, approach these writing exercises with a positive point of view. You will get better at giving written responses to questions as the course progresses. Even if you never feel comfortable writing about mathematics, just attempting the process will increase your understanding and ability in this course.

After reading through the preceding section, respond in your own words and in complete sentences.

A. What is a prime number?

B. Why is the number 22 a composite number?

C. Factor 120 into a product of prime factors.

D. What is meant by the phrase "a fraction in lowest terms"?

Answers

9. $\frac{24}{5}$

Vocabulary Review

Choose the correct words to fill in the blanks below.

numerator denominator composite number lowest terms prime number equivalent

1. Any whole number greater than 1 that has exactly two divisors, 1 and the number itself, is called a ________________.
2. A ________________ is any whole number greater than 1 that is not a prime number.
3. The top number of a fraction is called the ________________.
4. The restriction $b \neq 0$ for the ________________ of a fraction keeps us from writing an expression that is undefined.
5. Fractions with the same value are said to be ________________.
6. A fraction is in ________________ if the numerator and the denominator have no factors in common other than the number 1.

Problems

A Identify each of the numbers below as either a prime number or a composite number. For those that are composite, give at least one divisor (factor) other than the number itself or the number 1.

1. 11
2. 23
3. 105
4. 41
5. 81
6. 50
7. 13
8. 219

Factor each of the following into a product of prime factors.

9. 12
10. 18
11. 81
12. 210
13. 215
14. 75
15. 15
16. 42

B Reduce each fraction to lowest terms.

17. $\frac{5}{10}$
18. $\frac{3}{6}$
19. $\frac{4}{6}$
20. $\frac{4}{10}$
21. $\frac{8}{10}$
22. $\frac{6}{10}$
23. $\frac{36}{20}$
24. $\frac{32}{12}$

25. $\frac{42}{66}$ **26.** $\frac{36}{60}$ **27.** $\frac{24}{40}$ **28.** $\frac{50}{75}$

29. $\frac{14}{98}$ **30.** $\frac{12}{84}$ **31.** $\frac{70}{90}$ **32.** $\frac{80}{90}$

33. $\frac{42}{30}$ **34.** $\frac{60}{36}$ **35.** $\frac{18}{90}$ **36.** $\frac{150}{210}$ **37.** $\frac{110}{70}$ **38.** $\frac{45}{75}$

39. $\frac{180}{108}$ **40.** $\frac{105}{30}$ **41.** $\frac{96}{108}$ **42.** $\frac{66}{84}$ **43.** $\frac{126}{165}$ **44.** $\frac{210}{462}$

45. $\frac{102}{114}$ **46.** $\frac{255}{285}$ **47.** $\frac{294}{693}$ **48.** $\frac{273}{385}$

49. Reduce each fraction to lowest terms.

a. $\frac{6}{51}$ **b.** $\frac{6}{52}$ **c.** $\frac{6}{54}$ **d.** $\frac{6}{56}$ **e.** $\frac{6}{57}$

50. Reduce each fraction to lowest terms.

a. $\frac{6}{42}$ **b.** $\frac{6}{44}$ **c.** $\frac{6}{45}$ **d.** $\frac{6}{46}$ **e.** $\frac{6}{48}$

51. Reduce each fraction to lowest terms.

a. $\frac{2}{90}$ **b.** $\frac{3}{90}$ **c.** $\frac{5}{90}$ **d.** $\frac{6}{90}$ **e.** $\frac{9}{90}$

52. Reduce each fraction to lowest terms.

a. $\frac{3}{105}$ **b.** $\frac{5}{105}$ **c.** $\frac{7}{105}$ **d.** $\frac{15}{105}$ **e.** $\frac{21}{105}$

53. The answer to each problem below is wrong. Give the correct answer.

a. $\frac{5}{15} = \frac{5}{3 \cdot 5} = \frac{0}{3}$ **b.** $\frac{5}{6} = \frac{3 + 2}{4 + 2} = \frac{3}{4}$ **c.** $\frac{6}{30} = \frac{2 \cdot 3}{2 \cdot 3 \cdot 5} = 5$

54. The answer to each problem below is wrong. Give the correct answer.

a. $\frac{10}{20} = \frac{7 + 3}{17 + 3} = \frac{7}{17}$ **b.** $\frac{9}{36} = \frac{3 \cdot 3}{2 \cdot 2 \cdot 3 \cdot 3} = \frac{0}{4}$ **c.** $\frac{4}{12} = \frac{2 \cdot 2}{2 \cdot 2 \cdot 3} = 3$

55. Which of the fractions $\frac{6}{8}, \frac{15}{20}, \frac{9}{16}, \frac{21}{28}$ does not reduce to $\frac{3}{4}$?

56. Which of the fractions $\frac{4}{9}, \frac{10}{15}, \frac{8}{12}, \frac{6}{12}$ do not reduce to $\frac{2}{3}$?

The number line below extends from 0 to 2, with the segment from 0 to 1 and the segment from 1 to 2 each divided into 8 equal parts. Locate each of the following numbers on this number line.

57. $\frac{1}{2}, \frac{2}{4}, \frac{4}{8}$, and $\frac{8}{16}$

58. $\frac{3}{2}, \frac{6}{4}, \frac{12}{8}$, and $\frac{24}{16}$

59. $\frac{5}{4}, \frac{10}{8}$, and $\frac{20}{16}$

60. $\frac{1}{4}, \frac{2}{8}$, and $\frac{4}{16}$

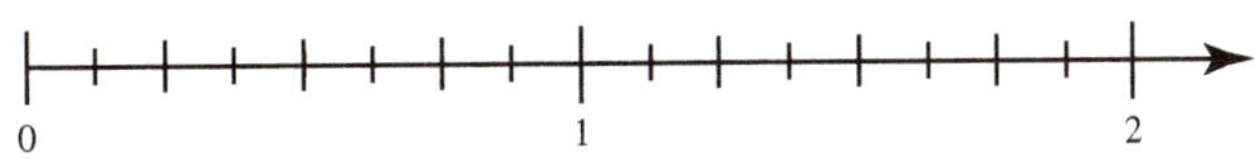

Applying the Concepts

61. Income A family's monthly income is $2,400, and they spend $600 each month on food. Write the amount they spend on food as a fraction of their monthly income in lowest terms.

62. Hours and Minutes There are 60 minutes in 1 hour. What fraction of an hour is 20 minutes? Write your answer in lowest terms.

63. Computers Use the chart to answer the questions about computers used by students.

a. Is the percentage of Apple and Dell computers less than or more than half of the computers used by students?

b. Is the percentage of Toshiba and HP computers used less than or more than one-third of computers used by students?

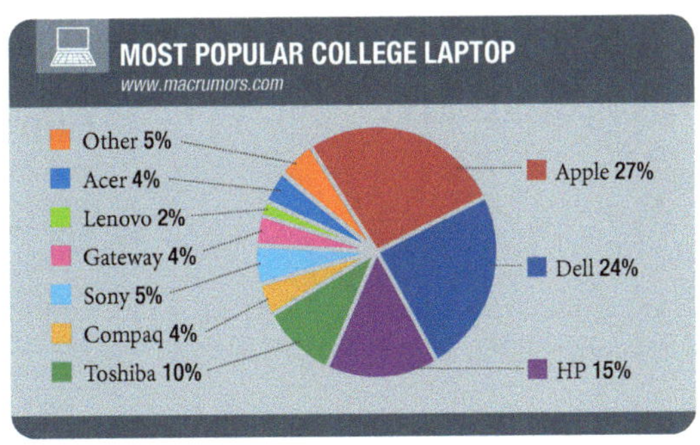

64. Students The table shows the number of male and female students enrolled at a small community college. What fraction of the students is male? Write your answer in lowest terms.

Community College Enrollment	
Men	1,400
Women	1,800

65. Final Exams Suppose 33 people took the final exam in a math class. If 11 people got an A on the final exam, what fraction of the students did not get an A on the exam? Write your answer in lowest terms.

66. Income Tax A person making \$21,000 a year pays \$3,000 in income tax. What fraction of the person's income is paid as income tax? Write your answer in lowest terms.

Nutrition The nutrition labels below are from two different kinds of cheese crackers. Use them to work Problems 67–72.

67. What fraction of the calories in Cheez-It crackers comes from fat?

68. What fraction of the calories in Goldfish crackers comes from fat?

69. For Cheez-It crackers, what fraction of the total fat is from saturated fat?

70. For Goldfish crackers, what fraction of the total fat is from saturated fat?

CHEEZ-IT CRACKERS

Nutrition Facts	
Serving Size 30 g. (About 27 crackers)	
Servings Per Container: 9	
Amount Per Serving	
Calories 150	Calories from fat 70
	% Daily Value*
Total Fat 8g	**12%**
Saturated Fat 2g	**10%**
Trans Fat 0g	
Polysaturated Fat 4g	
Monounsaturated Fat 2g	
Cholesterol 0mg	**0%**
Sodium 230mg	**10%**
Total Carbohydrate 17g	**6%**
Dietary Fiber less than 1g	**3%**
Sugars 0g	
Protein 3g	
Vitamin A 2% •	Vitamin C 0%
Calcium 4% •	Iron 6%

*Percent Daily Values are based on a 2,000 calorie diet

GOLDFISH CRACKERS

Nutrition Facts	
Serving Size 55 pieces	
Servings Per Container About 4	
Amount Per Serving	
Calories 140	Calories from fat 45
	% Daily Value*
Total Fat 5g	**8%**
Saturated Fat 1g	**5%**
Cholesterol 5mg	**2%**
Sodium 250mg	**10%**
Total Carbohydrate 20g	**7%**
Dietary Fiber 1g	**4%**
Sugars 1g	
Protein 4g	
Vitamin A 0% •	Vitamin C 0%
Calcium 4% •	Iron 2%

*Percent Daily Values are based on a 2,000 calorie diet

71. What fraction of the total carbohydrates in Cheez-It crackers is from sugar?

72. What fraction of the total carbohydrates in Goldfish crackers is from sugar?

73. Below are the 2011 prices for adult tickets at Disneyland based on the number of days planned for the visit. Complete the table to find the average cost per day for each type of ticket.

Duration	Total Cost	Cost per Day
1 day	$80	
2 days	$158	
3 days	$199	
4 days	$219	
5 days	$231	

Source: disneyland.disney.go.com

74. Suppose these are the average adult ticket prices for movies in the United States over the past 30 years.

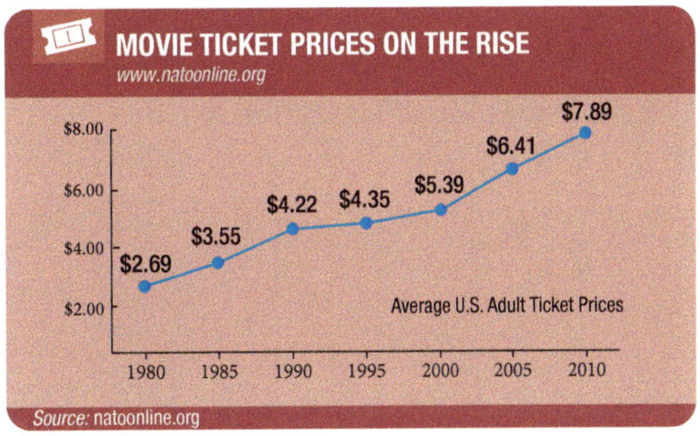

a. During which 5 year period did the prices go up the most? Which 5 year period did they go up the least?

b. Approximately what fraction represents the price in 1980 to the price in 2010?

75. In Universal City, Los Angeles, the adult price to see Harry Potter and the Deathly Hollows Part 2: IMAX 3D was $19.50. Refer to the chart in problem 74 to answer the following questions.

a. How many times more expensive is this than the average 2010 adult ticket price?

b. How many times more expensive is this than the average 1980 adult ticket price?

Estimating

76. Which of the following fractions is closest to the number $\frac{1}{2}$?

a. $\frac{5}{4}$ **b.** $\frac{5}{8}$ **c.** $\frac{5}{16}$

77. Which of the following fractions is closest to the number $\frac{1}{2}$?

a. $\frac{3}{8}$ **b.** $\frac{5}{8}$ **c.** $\frac{7}{8}$

78. Which of the following fractions is closest to the number $\frac{1}{2}$?

a. $\frac{3}{4}$ **b.** $\frac{5}{8}$ **c.** $\frac{9}{16}$

79. Which of the following fractions is closest to the number 1?

a. $\frac{3}{4}$ **b.** $\frac{7}{8}$ **c.** $\frac{15}{16}$

Calculator Problems

Use your calculator to help fill in the blanks for the following problems.

80. $\frac{5}{7} = \frac{\square}{574}$

81. $\frac{2}{3} = \frac{\square}{2{,}223}$

82. $\frac{8}{9} = \frac{728}{\square}$

83. $\frac{7}{8} = \frac{34{,}972}{\square}$

84. $\frac{3{,}787}{4{,}869} = \frac{\square}{9}$

85. $\frac{684}{836} = \frac{\square}{11}$

86. $\frac{262}{1{,}965} = \frac{\square}{15}$

87. $\frac{616}{1{,}672} = \frac{\square}{19}$

88. Ticket Prices The graphic shows the rise in Disneyland ticket prices since 1981. Which of the following is the closest fraction of the ticket price in 1981 to the ticket price in 2011?

a. $\frac{1}{5}$ **b.** $\frac{1}{7}$ **c.** $\frac{1}{10}$ **d.** $\frac{1}{12}$

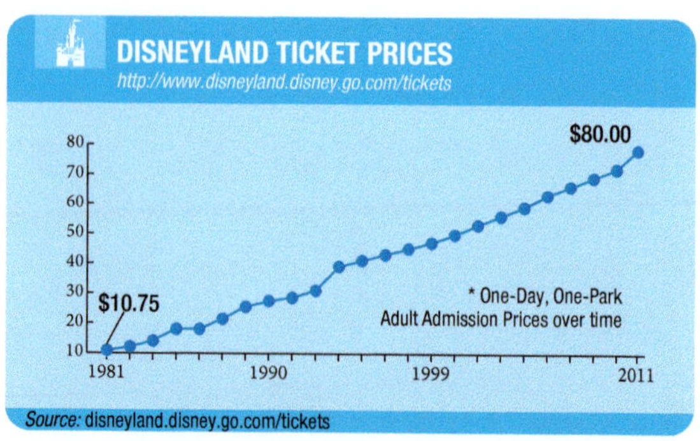

Find the Mistake

Each sentence below contains a mistake. Circle the mistake and write the correct word(s) or numbers(s) on the line provided.

1. A number is a divisor of another number if it divides it with a remainder. ____________________

2. 31 is a composite number because the only numbers that divide it without a remainder are 31 and 1. __________

3. Factoring 112 into the product of primes gives us $2 \cdot 7$. ____________________

4. To get the fraction $\frac{24}{36}$ in lowest terms, we divide the numerator and the denominator by 2. __________

Navigation Skills: Prepare, Study, Achieve

Studying is the key to success in this course. However, many students have never learned effective skills for studying. Study skills include but are not limited to the following:

- Work done on problems for practice and homework
- Amount of time spent studying
- Time of day and location for studying
- Management of distractions during study sessions
- Material chosen to review
- Order and process of review

Let's begin our discussion with the topic of homework. From the first day of class, we recommend you spend two hours on homework for every hour you are scheduled to attend class. Any less may drastically impact your success in this course. To help visualize this commitment, map out a weekly schedule that includes your classes, work shifts, extracurriculars, and any additional obligations. Fill in the hours you intend to devote to completing assignments and studying for this class. Post this schedule at home and keep a copy with your study materials to remind you of your commitment to success.

0.2 Operations with Fractions

OBJECTIVES

A Multiply fractions.

B Divide fractions.

C Add and subtract fractions with a common denominator.

D Find LCDs and solve problems involving fractions with different denominators.

KEY WORDS

multiplication

product

division

reciprocal

least common denominator

©iStockphoto.com/wojciech_gajda

Dock diving is a sport where dogs compete to see who can jump the highest or the farthest off the end of a dock into a body of water. Many of the dogs' handlers use a chase object to lure the dog as far as possible off the dock. The handler commands the dog to begin a running start and then throws the chase object off over the water. The dog leaps after the object, and once he splashes into the water, an event official measures his jump distance. The official measures the dog's final distance from the end of the dock to the point where the base of the dog's tail first hit the water.

Let's say that the top two dogs jumped 21 feet $10\frac{1}{4}$ inches and 21 feet $10\frac{7}{8}$ inches. How much farther did the second dog jump? To solve this problem, we need to know how to subtract $\frac{1}{4}$ from $\frac{7}{8}$. After reading this section, you will be more comfortable working problems involving fractions such as this. First, let's discuss a bit about the multiplication of fractions.

A Multiplication of Fractions

A recipe calls for $\frac{3}{4}$ cup of flour. If you are making only $\frac{1}{2}$ the recipe, how much flour do you use? This question can be answered by multiplying $\frac{1}{2}$ and $\frac{3}{4}$ because we need to find out how much is $\frac{1}{2}$ of $\frac{3}{4}$. Here is the problem written in symbols:

$$\frac{1}{2} \cdot \frac{3}{4} = \frac{3}{8}$$

As you can see from this example, to multiply two fractions, we multiply the numerators and then multiply the denominators. We begin this section with the rule for multiplication of fractions.

Note: As you may have covered in a previous math class, many operations in mathematics are represented in word problems as key words. In this case, the explanation used the word *of* to represent multiplication. We will revisit this and additional key words later in the book.

RULE Multiplication of Fractions

The product of two fractions is the fraction whose numerator is the product of the two numerators and whose denominator is the product of the two denominators. We can write this rule in symbols as follows:

If a, b, c, and d represent any numbers and b and d are not zero, then

$$\frac{a}{b} \cdot \frac{c}{d} = \frac{a \cdot c}{b \cdot d}$$

Practice Problems

1. Multiply: $\frac{2}{3} \cdot \frac{5}{9}$.

EXAMPLE 1 Multiply: $\frac{3}{5} \cdot \frac{2}{7}$.

Solution Using our rule for multiplication, we multiply the numerators and multiply the denominators.

$$\frac{3}{5} \cdot \frac{2}{7} = \frac{3 \cdot 2}{5 \cdot 7} = \frac{6}{35}$$

The product of $\frac{3}{5}$ and $\frac{2}{7}$ is the fraction $\frac{6}{35}$. The numerator 6 is the product of 3 and 2, and the denominator 35 is the product of 5 and 7.

2. Multiply: $\frac{2}{5} \cdot 7$.

EXAMPLE 2 Multiply: $\frac{3}{8} \cdot 5$.

Solution The number 5 can be written as $\frac{5}{1}$. That is, 5 can be considered a fraction with numerator 5 and denominator 1. Writing 5 this way enables us to apply the rule for multiplying fractions.

$$\frac{3}{8} \cdot 5 = \frac{3}{8} \cdot \frac{5}{1}$$
$$= \frac{3 \cdot 5}{8 \cdot 1} = \frac{15}{8}$$

3. Multiply: $\frac{1}{3} \cdot \frac{4}{5} \cdot \frac{1}{3}$.

EXAMPLE 3 Multiply: $\frac{1}{2} \cdot \frac{3}{4} \cdot \frac{1}{5}$.

Solution We find the product inside the parentheses first and then multiply the result by $\frac{1}{2}$.

$$\frac{1}{2} \cdot \frac{3}{4} \cdot \frac{1}{5} = \frac{1}{2} \cdot \frac{3 \cdot 1}{4 \cdot 5}$$
$$= \frac{1}{2} \cdot \frac{3}{20}$$
$$= \frac{1 \cdot 3}{2 \cdot 20} = \frac{3}{40}$$

The answers to all the examples so far in this section have been in lowest terms. Let's see what happens when we multiply two fractions to get a product that is not in lowest terms.

4. Multiply: $\frac{12}{25} \cdot \frac{5}{6}$.

EXAMPLE 4 Multiply: $\frac{15}{8} \cdot \frac{4}{9}$.

Solution Multiplying the numerators and multiplying the denominators, we have

$$\frac{15}{8} \cdot \frac{4}{9} = \frac{15 \cdot 4}{8 \cdot 9} = \frac{60}{72}$$

The product is $\frac{60}{72}$, which can be reduced to lowest terms by factoring 60 and 72 and then dividing out any factors they have in common.

$$\frac{60}{72} = \frac{2 \cdot 2 \cdot 3 \cdot 5}{2 \cdot 2 \cdot 2 \cdot 3 \cdot 3} = \frac{5}{6}$$

Answers

1. $\frac{10}{27}$
2. $\frac{14}{5}$
3. $\frac{4}{45}$
4. $\frac{2}{5}$

We can actually save ourselves some time by factoring before we multiply. Here's how it is done:

$$\frac{15}{8} \cdot \frac{4}{9} = \frac{15 \cdot 4}{8 \cdot 9}$$

$$= \frac{(3 \cdot 5) \cdot (2 \cdot 2)}{(2 \cdot 2 \cdot 2) \cdot (3 \cdot 3)}$$

$$= \frac{2 \cdot 2 \cdot 3 \cdot 5}{2 \cdot 2 \cdot 2 \cdot 3 \cdot 3} = \frac{5}{6}$$

The result is the same in both cases. Reducing to lowest terms before we actually multiply takes less time.

B Division of Fractions

Before we define division with fractions, we must first introduce the idea of *reciprocals*. Look at the following multiplication problems:

$$\frac{3}{4} \cdot \frac{4}{3} = \frac{12}{12} = 1 \qquad \frac{7}{8} \cdot \frac{8}{7} = \frac{56}{56} = 1$$

In each case the product is 1. Whenever the product of two numbers is 1, we say the two numbers are reciprocals.

DEFINITION reciprocal

Two numbers whose product is 1 are said to be ***reciprocals***. In symbols, the reciprocal of $\frac{a}{b}$ is $\frac{b}{a}$, because

$$\frac{a}{b} \cdot \frac{b}{a} = \frac{a \cdot b}{a \cdot b} = 1 \qquad (a \neq 0, b \neq 0)$$

Every number has a reciprocal except 0. The reason 0 does not have a reciprocal is because the product of any number with 0 is 0. It can never be 1. Reciprocals of whole numbers are fractions with 1 as the numerator. For example, the reciprocal of 5 is $\frac{1}{5}$, because

$$5 = \frac{1}{5} = \frac{5}{1} \cdot \frac{1}{5} = \frac{5}{5} = 1$$

Table 1 lists some numbers and their reciprocals.

Table 1

Number	Reciprocal	Reason
$\frac{3}{4}$	$\frac{4}{3}$	Because $\frac{3}{4} \cdot \frac{4}{3} = \frac{12}{12} = 1$
$\frac{9}{5}$	$\frac{5}{9}$	Because $\frac{9}{5} \cdot \frac{5}{9} = \frac{45}{45} = 1$
$\frac{1}{3}$	3	Because $\frac{1}{3} \cdot 3 = \frac{1}{3} \cdot \frac{3}{1} = \frac{3}{3} = 1$
7	$\frac{1}{7}$	Because $7 \cdot \frac{1}{7} = \frac{7}{1} \cdot \frac{1}{7} = \frac{7}{7} = 1$

Division with fractions is accomplished by using reciprocals. More specifically, we can define division by a fraction to be the same as multiplication by its reciprocal.

Note Defining division to be the same as multiplication by the reciprocal does make sense. If we divide 6 by 2, we get 3. On the other hand, if we multiply 6 by $\frac{1}{2}$ (the reciprocal of 2), we also get 3. Whether we divide by 2 or multiply by $\frac{1}{2}$, we get the same result.

RULE **Division of Fractions**

Division of fractions is the equivalent of multiplication by the reciprocal. We can write this rule in symbols as follows:

If a, b, c, and d represent any numbers and b and d are not zero, then

$$\frac{a}{b} \div \frac{c}{d} = \frac{a}{b} \cdot \frac{d}{c}$$

This definition states that dividing by the fraction $\frac{c}{d}$ is exactly the same as multiplying by its reciprocal $\frac{d}{c}$.

In some math classes, multiplication by the reciprocal is referred to as "inverting the divisor and multiplying" or "flip that guy and multiply." No matter how you say it, division by any number (except zero) is always equivalent to multiplication by its reciprocal. Here are additional examples that involve division by fractions.

EXAMPLE 5 Divide: $\frac{3}{8} \div \frac{9}{4}$.

Solution Dividing by $\frac{9}{4}$ is the same as multiplying by its reciprocal, which is $\frac{4}{9}$.

$$\begin{aligned}\frac{3}{8} \div \frac{9}{4} &= \frac{3}{8} \cdot \frac{4}{9} \\ &= \frac{3 \cdot 2 \cdot 2}{2 \cdot 2 \cdot 2 \cdot 3 \cdot 3} \\ &= \frac{1}{6}\end{aligned}$$

The quotient of $\frac{3}{8}$ and $\frac{9}{4}$ is $\frac{1}{6}$.

5. Divide: $\frac{5}{9} \div \frac{10}{3}$.

EXAMPLE 6 Divide: $\frac{2}{3} \div 2$.

Solution The reciprocal of 2 is $\frac{1}{2}$. Applying the definition for division of fractions, we have

$$\begin{aligned}\frac{2}{3} \div 2 &= \frac{2}{3} \cdot \frac{1}{2} \\ &= \frac{2 \cdot 1}{3 \cdot 2} \\ &= \frac{1}{3}\end{aligned}$$

6. Divide: $\frac{3}{4} \div 3$.

C Addition and Subtraction of Fractions (Common Denominator)

The solution to a problem that involves the addition of fractions is simply the sum of the numerators placed over the denominator both fractions share, also called the *common denominator*. In symbols, we have the following rule:

RULE **Addition and Subtraction of Fractions**

If a, b, and c are integers, and c is not equal to 0, then

$$\frac{a}{c} + \frac{b}{c} = \frac{a+b}{c}$$

This rule holds for subtraction as well. That is,

$$\frac{a}{c} - \frac{b}{c} = \frac{a-b}{c}$$

Answers

5. $\frac{1}{6}$
6. $\frac{1}{4}$
7. $\frac{2}{5}$
8. $\frac{a+8}{12}$

Find the sum or difference. Reduce all answers to lowest terms.

7. $\frac{3}{10} + \frac{1}{10}$

8. $\frac{a+5}{12} + \frac{3}{12}$

9. $\frac{8}{7} - \frac{5}{7}$

10. $\frac{5}{9} + \frac{8}{9} + \frac{5}{9}$

In Examples 7–10, find the sum or difference of the two fractions. (Add or subtract as indicated.) Reduce all answers to lowest terms. (Assume all variables represent nonzero numbers.)

EXAMPLE 7 $\frac{3}{8} + \frac{1}{8} = \frac{3+1}{8}$ Add numerators; keep the same denominator.

$= \frac{4}{8}$ The sum of 3 and 1 is 4.

$= \frac{1}{2}$ Reduce to lowest terms.

EXAMPLE 8 $\frac{a+5}{8} - \frac{3}{8} = \frac{a+5-3}{8}$ Combine numerators; keep the same denominator.

$= \frac{a+2}{8}$ The difference of 5 and 3 is 2.

Note This answer is in lowest terms because you cannot combine the variable *a* and the number 2 as they are not similar terms. We will discuss the concept of similar terms in the next chapter.

EXAMPLE 9 $\frac{9}{5} - \frac{3}{5} = \frac{9-3}{5}$ Subtract numerators; keep the same denominator.

$= \frac{6}{5}$ The difference of 9 and 3 is 6.

EXAMPLE 10 $\frac{3}{7} + \frac{2}{7} + \frac{9}{7} = \frac{3+2+9}{7}$

$= \frac{14}{7}$

$= 2$

As Examples 7–10 indicate, addition and subtraction are simple straightforward processes when all the fractions have the same denominator. We will now turn our attention to the process of adding fractions that have different denominators.

D Addition and Subtraction of Fractions (Different Denominators)

In order to get started, we need the following definition:

DEFINITION least common denominator

The *least common denominator* (LCD) for a set of denominators is the smallest number that is exactly divisible by each denominator. (Note that, in some books, the least common denominator is also called the least common multiple.)

In other words, all the denominators of the fractions involved in a problem must divide into the least common denominator exactly. That is, they divide it without leaving a remainder.

Answers

9. $\frac{3}{7}$

10. 2

11. 126

EXAMPLE 11 Find the LCD for the fractions $\frac{5}{12}$ and $\frac{7}{18}$.

Solution The least common denominator for the denominators 12 and 18 must be the smallest number divisible by both 12 and 18. We can factor 12 and 18 completely and then build the LCD from these factors. Factoring 12 and 18 completely gives us

$$12 = 2 \cdot 2 \cdot 3 \qquad 18 = 2 \cdot 3 \cdot 3$$

Now, if 12 is going to divide the LCD exactly, then the LCD must have factors of $2 \cdot 2 \cdot 3$. If 18 is to divide it exactly, it must have factors of $2 \cdot 3 \cdot 3$. We don't need to repeat the factors that 12 and 18 have in common.

$$12 = 2 \cdot 2 \cdot 3$$

12 divides the LCD.

$$\text{LCD} = 2 \cdot 2 \cdot 3 \cdot 3 = 36$$

18 divides the LCD.

$$18 = 2 \cdot 3 \cdot 3$$

The LCD for 12 and 18 is 36. It is the smallest number that is divisible by both 12 and 18; 12 divides it exactly three times, and 18 divides it exactly two times.

We can visualize the results in Example 11 with the diagram below. It shows that 36 is the smallest number that both 12 and 18 divide evenly. As you can see, 12 divides 36 exactly 3 times, and 18 divides 36 exactly 2 times.

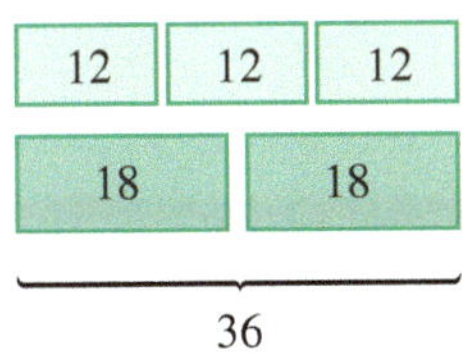

EXAMPLE 12 Add: $\frac{5}{12} + \frac{7}{18}$.

Solution We can add fractions only when they have the same denominators. In Example 11, we found the LCD for $\frac{5}{12}$ and $\frac{7}{18}$ to be 36. We change $\frac{5}{12}$ and $\frac{7}{18}$ to equivalent fractions that have 36 for a denominator. Recall from a previous math class that if the numerator and the denominator are multiplied by the same nonzero number, the resulting fraction is equivalent to the original fraction.

$$\frac{5}{12} = \frac{5 \cdot 3}{12 \cdot 3} = \frac{15}{36}$$

$$\frac{7}{18} = \frac{7 \cdot 2}{18 \cdot 2} = \frac{14}{36}$$

The fraction $\frac{15}{36}$ is equivalent to $\frac{5}{12}$, because it was obtained by multiplying both the numerator and the denominator by 3. Likewise, $\frac{14}{36}$ is equivalent to $\frac{7}{8}$, because it was obtained by multiplying the numerator and the denominator by 2. All we have left to do is to add numerators.

$$\frac{15}{36} + \frac{14}{36} = \frac{29}{36}$$

11. Find the LCD for the fractions $\frac{5}{18}$ and $\frac{3}{14}$.

Note The ability to find least common denominators is very important in mathematics. The discussion here is a detailed explanation of how to find an LCD.

12. Add: $\frac{5}{18} + \frac{3}{14}$.

Answers

12. $\frac{31}{63}$

13. 45

14. $\frac{22}{45}$

The sum of $\frac{5}{12}$ and $\frac{7}{18}$ is the fraction $\frac{29}{36}$. Let's write the complete problem again step by step.

$$\frac{5}{12} + \frac{7}{18} = \frac{5 \cdot 3}{12 \cdot 3} + \frac{7 \cdot 2}{18 \cdot 2}$$ Rewrite each fraction as an equivalent fraction with denominator 36.

$$= \frac{15}{36} + \frac{14}{36}$$

$$= \frac{29}{36}$$ Add numerators; keep the common denominator.

13. Find the LCD for $\frac{2}{9}$ and $\frac{4}{15}$.

EXAMPLE 13 Find the LCD for $\frac{3}{4}$ and $\frac{1}{6}$.

Solution We factor 4 and 6 into products of prime factors and build the LCD from these factors.

$$4 = 2 \cdot 2$$

$$\text{LCD} = 2 \cdot 2 \cdot 3 = 12$$

$$6 = 2 \cdot 3$$

The LCD is 12. Both denominators divide it exactly; 4 divides 12 exactly 3 times, and 6 divides 12 exactly 2 times.

14. Add: $\frac{2}{9} + \frac{4}{15}$.

Note We can visualize the work in Example 14 using circles and shading:

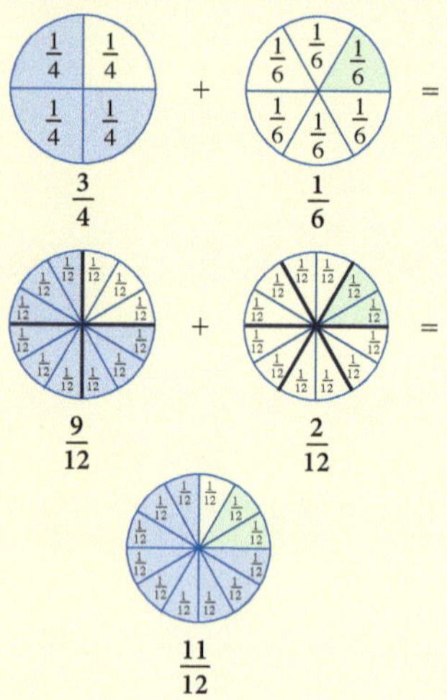

EXAMPLE 14 Add: $\frac{3}{4} + \frac{1}{6}$.

Solution In Example 13, we found that the LCD for these two fractions is 12. We begin by changing $\frac{3}{4}$ and $\frac{1}{6}$ to equivalent fractions with denominator 12.

$$\frac{3}{4} = \frac{3 \cdot 3}{4 \cdot 3} = \frac{9}{12}$$

$$\frac{1}{6} = \frac{1 \cdot 2}{6 \cdot 2} = \frac{2}{12}$$

The fraction $\frac{9}{12}$ is equal to the fraction $\frac{3}{4}$, because it was obtained by multiplying the numerator and the denominator of $\frac{3}{4}$ by 3. Likewise, $\frac{2}{12}$ is equivalent to $\frac{1}{6}$, because it was obtained by multiplying the numerator and the denominator of $\frac{1}{6}$ by 2. To complete the problem, we add numerators.

$$\frac{9}{12} + \frac{2}{12} = \frac{11}{12}$$

The sum of $\frac{3}{4}$ and $\frac{1}{6}$ is $\frac{11}{12}$. Here is how the complete problem looks:

$$\frac{3}{4} + \frac{1}{6} = \frac{3 \cdot 3}{4 \cdot 3} + \frac{1 \cdot 2}{6 \cdot 2}$$ Rewrite each fraction as an equivalent fraction with denominator 12.

$$= \frac{9}{12} + \frac{2}{12}$$

$$= \frac{11}{12}$$ Add numerators; keep the same denominator.

EXAMPLE 15 Subtract: $\frac{7}{15} - \frac{3}{10}$.

Solution Let's factor 15 and 10 completely and use these factors to build the LCD.

$$15 = 3 \cdot 5 \qquad 10 = 2 \cdot 5$$

$$\text{LCD} = 2 \cdot 3 \cdot 5 = 30$$

15 divides the LCD.

10 divides the LCD.

Changing to equivalent fractions and subtracting, we have

$$\frac{7}{15} - \frac{3}{10} = \frac{7 \cdot 2}{15 \cdot 2} - \frac{3 \cdot 3}{10 \cdot 3}$$ Rewrite as equivalent fractions with the LCD for the denominator.

$$= \frac{14}{30} - \frac{9}{30}$$

$$= \frac{5}{30}$$ Subtract numerators; keep the LCD.

$$= \frac{1}{6}$$ Reduce to lowest terms.

15. Subtract: $\frac{8}{25} - \frac{3}{20}$.

As a summary of what we have done so far, and as a guide to working other problems, we now list the steps involved in adding and subtracting fractions with different denominators.

HOW TO Add or Subtract Any Two Fractions

Step 1: Factor each denominator completely, and use the factors to build the LCD. (Remember, the LCD is the smallest number divisible by each of the denominators in the problem.)

Step 2: Rewrite each fraction as an equivalent fraction that has the LCD for its denominator. This is done by multiplying both the numerator and the denominator of the fraction in question by the appropriate whole number.

Step 3: Add or subtract the numerators of the fractions produced in Step 2. This is the numerator of the sum or difference. The common denominator of the sum or difference is the LCD.

Step 4: Reduce the fraction produced in Step 3 to lowest terms if it is not already in lowest terms.

The idea behind adding or subtracting fractions is really very simple. We can only add or subtract fractions that have the same denominators. If the fractions we are trying to add or subtract do not have the same denominators, we rewrite each of them as an equivalent fraction with the LCD for a denominator.

Here are some additional examples of sums and differences of fractions.

Answer

15. $\frac{17}{100}$

16. Subtract: $\frac{3}{4} - \frac{1}{5}$.

EXAMPLE 16 Subtract: $\frac{3}{5} - \frac{1}{6}$.

Solution The LCD for 5 and 6 is their product, 30. We begin by rewriting each fraction with this common denominator.

$$\begin{aligned}\frac{3}{5} - \frac{1}{6} &= \frac{3 \cdot 6}{5 \cdot 6} - \frac{1 \cdot 5}{6 \cdot 5}\\ &= \frac{18}{30} - \frac{5}{30}\\ &= \frac{13}{30}\end{aligned}$$

17. Add: $\frac{1}{9} + \frac{1}{4} + \frac{1}{6}$.

EXAMPLE 17 Add: $\frac{1}{6} + \frac{1}{8} + \frac{1}{4}$.

Solution We begin by factoring the denominators completely and building the LCD from the factors that result.

$$6 = 2 \cdot 3$$
$$8 = 2 \cdot 2 \cdot 2 \qquad \text{LCD} = 2 \cdot 2 \cdot 2 \cdot 3 = 24$$
$$4 = 2 \cdot 2$$

8 divides the LCD.

4 divides the LCD. 6 divides the LCD.

We then change to equivalent fractions and add as usual.

$$\frac{1}{6} + \frac{1}{8} + \frac{1}{4} = \frac{1 \cdot 4}{6 \cdot 4} = \frac{1 \cdot 3}{8 \cdot 3} + \frac{1 \cdot 6}{4 \cdot 6} = \frac{4}{24} + \frac{3}{24} + \frac{6}{24} = \frac{13}{24}$$

18. Subtract: $2 - \frac{3}{4}$.

EXAMPLE 18 Subtract: $3 - \frac{5}{6}$.

Solution The denominators are 1 (because $3 = \frac{3}{1}$) and 6. The smallest number divisible by both 1 and 6 is 6.

$$3 - \frac{5}{6} = \frac{3}{1} - \frac{5}{6} = \frac{3 \cdot 6}{1 \cdot 6} - \frac{5}{6} = \frac{18}{6} = \frac{5}{6} = \frac{13}{6}$$

GETTING READY FOR CLASS

After reading through the preceding section, respond in your own words and in complete sentences.

A. What is a reciprocal?

B. When adding two fractions with the same denominators, we always add their __________, but we never add their __________. Explain.

C. What does the abbreviation LCD stand for?

D. What is the first step when finding the LCD for the fractions $\frac{5}{12}$ and $\frac{7}{18}$?

Answers

16. $\frac{11}{20}$

17. $\frac{19}{36}$

18. $\frac{5}{4}$

EXERCISE SET R.2

Vocabulary Review

Choose the correct words to fill in the blanks below.

reciprocal factors least common denominator
multiplication division

1. The rule for the ______________ of fractions is given by $\frac{a}{b} \cdot \frac{c}{d} = \frac{a \cdot c}{b \cdot d}$.
2. The rule for the ______________ of fractions is given by $\frac{a}{b} \div \frac{c}{d} = \frac{a}{b} \cdot \frac{d}{c}$.
3. The ______________ of 4 is $\frac{1}{4}$.
4. The smallest number that is exactly divisible by each denominator in a set of denominators is called the ______________.
5. The first step to add any two fractions is to factor each denominator completely, and use the ______________ to build the least common denominator.

Problems

A Find each of the following products. (Multiply.)

1. $\frac{2}{3} \cdot \frac{4}{5}$
2. $\frac{5}{6} \cdot \frac{7}{4}$
3. $\frac{1}{2} \cdot \frac{7}{4}$
4. $\frac{3}{5} \cdot \frac{4}{7}$
5. $\frac{5}{3} \cdot \frac{3}{5}$
6. $\frac{4}{7} \cdot \frac{7}{4}$
7. $\frac{3}{4} \cdot 9$
8. $\frac{2}{3} \cdot 5$
9. $\frac{6}{7} \cdot \frac{7}{6}$
10. $\frac{2}{9} \cdot \frac{9}{2}$
11. $\frac{1}{2} \cdot \frac{1}{3} \cdot \frac{1}{4}$
12. $\frac{2}{3} \cdot \frac{4}{5} \cdot \frac{1}{3}$
13. $\frac{2}{5} \cdot \frac{3}{5} \cdot \frac{4}{5}$
14. $\frac{1}{4} \cdot \frac{3}{4} \cdot \frac{3}{4}$
15. $\frac{3}{2} \cdot \frac{5}{2} \cdot \frac{7}{2}$
16. $\frac{4}{3} \cdot \frac{5}{3} \cdot \frac{7}{3}$
17. $\frac{2}{3} \cdot \frac{2}{3} \cdot \frac{2}{3}$
18. $\frac{3}{4} \cdot \frac{3}{4} \cdot \frac{3}{4}$

Multiply each of the following. Be sure all answers are written in lowest terms.

19. $\frac{9}{20} \cdot \frac{4}{3}$

20. $\frac{135}{16} \cdot \frac{2}{45}$

21. $\frac{3}{4} \cdot 12$

22. $\frac{3}{4} \cdot 20$

23. $\frac{1}{3}(3)$

24. $\frac{1}{5}(5)$

25. $\frac{2}{5} \cdot 20$

26. $\frac{3}{5} \cdot 15$

27. $\frac{72}{35} \cdot \frac{55}{108} \cdot \frac{7}{110}$

28. $\frac{32}{27} \cdot \frac{72}{49} \cdot \frac{1}{40}$

B Find the quotient in each case by replacing the divisor by its reciprocal and multiplying.

29. $\frac{3}{4} \div \frac{1}{5}$

30. $\frac{1}{3} \div \frac{1}{2}$

31. $\frac{2}{3} \div \frac{1}{2}$

32. $\frac{5}{8} \div \frac{1}{4}$

33. $6 \div \frac{2}{3}$

34. $8 \div \frac{3}{4}$

35. $20 \div \frac{1}{10}$

36. $16 \div \frac{1}{8}$

37. $\frac{3}{4} \div 2$

38. $\frac{3}{5} \div 2$

39. $\frac{7}{8} \div \frac{7}{8}$

40. $\frac{4}{3} \div \frac{4}{3}$

41. $\frac{7}{8} \div \frac{8}{7}$

42. $\frac{4}{3} \div \frac{3}{4}$

43. $\frac{9}{16} \div \frac{3}{4}$

44. $\frac{25}{36} \div \frac{5}{6}$

45. $\frac{25}{46} \div \frac{40}{69}$

46. $\frac{25}{24} \div \frac{15}{36}$

47. $\frac{13}{28} \div \frac{39}{14}$

48. $\frac{28}{125} \div \frac{5}{2}$

49. $\frac{27}{196} \div \frac{9}{392}$

50. $\frac{16}{135} \div \frac{2}{45}$

51. $\frac{25}{18} \div 5$

52. $\frac{30}{27} \div 6$

53. $6 \div \frac{4}{3}$

54. $12 \div \frac{4}{3}$

55. $\frac{4}{3} \div 6$

56. $\frac{4}{3} \div 12$

C Find the following sums and differences, and reduce to lowest terms. (Add or subtract as indicated.)

57. $\frac{3}{6} + \frac{1}{6}$

58. $\frac{2}{5} + \frac{3}{5}$

59. $\frac{5}{8} - \frac{3}{8}$

60. $\frac{6}{7} - \frac{1}{7}$

61. $\frac{3}{4} - \frac{1}{4}$

62. $\frac{7}{9} - \frac{4}{9}$

63. $\frac{2}{3} - \frac{1}{3}$

64. $\frac{9}{8} - \frac{1}{8}$

65. $\frac{1}{4} + \frac{2}{4} + \frac{3}{4}$

66. $\frac{2}{5} + \frac{3}{5} + \frac{4}{5}$

67. $\frac{x+7}{2} - \frac{1}{2}$

68. $\frac{x+5}{4} - \frac{3}{4}$

69. $\frac{1}{10} + \frac{3}{10} + \frac{4}{10}$

70. $\frac{3}{20} + \frac{1}{20} + \frac{4}{20}$

71. $\frac{1}{3} + \frac{4}{3} + \frac{5}{3}$

72. $\frac{5}{4} + \frac{4}{4} + \frac{3}{4}$

D Find the LCD for each of the following; then use the methods developed in this section to add or subtract as indicated.

73. $\frac{4}{9} + \frac{1}{3}$

74. $\frac{1}{2} + \frac{1}{4}$

75. $2 + \frac{1}{3}$

76. $3 + \frac{1}{2}$

77. $\frac{3}{4} + 1$

78. $\frac{3}{4} + 2$

79. $\frac{1}{2} + \frac{2}{3}$

80. $\frac{1}{8} + \frac{3}{4}$

81. $\frac{1}{4} + \frac{1}{5}$

82. $\frac{1}{3} + \frac{1}{5}$

83. $\frac{1}{2} + \frac{1}{5}$

84. $\frac{1}{2} - \frac{1}{5}$

85. $\frac{5}{12} + \frac{3}{8}$

86. $\frac{9}{16} + \frac{7}{12}$

87. $\frac{8}{30} - \frac{1}{20}$

88. $\frac{9}{40} - \frac{1}{30}$

89. $\frac{3}{10} + \frac{1}{100}$

90. $\frac{9}{100} + \frac{7}{10}$

91. $\frac{10}{36} + \frac{9}{48}$

92. $\frac{12}{28} + \frac{9}{20}$

Find the Mistake

Each sentence below contains a mistake. Circle the mistake and write the correct word(s) or numbers(s) on the line provided.

1. To multiply the fractions $\frac{2}{8}$ and $\frac{3}{4}$, we put the product of 2 and 3 in the numerator over 8 in the denominator. ________________
2. The reciprocal of $\frac{5}{20}$ is $\frac{1}{5}$. ________________
3. To divide $\frac{7}{8}$ by $\frac{2}{5}$, multiply $\frac{2}{5}$ by $\frac{8}{7}$. ________________
4. The least common denominator for the fractions $\frac{3}{7}$ and $\frac{5}{6}$ is 13. ________________

R.3 Decimal Notation

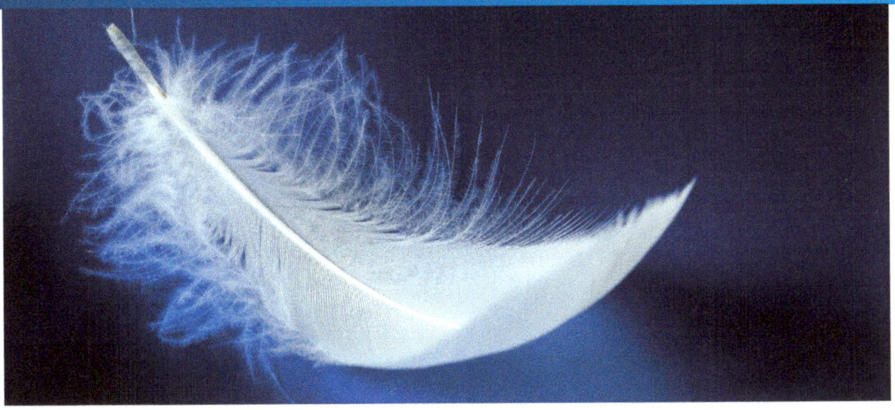

©iStockphoto.com/BrianAJackson

OBJECTIVES

A Write decimal numbers in expanded form.

B Round decimal numbers.

KEY WORDS

place value

expanded form

decimal point

decimal fraction

round

April 3rd is International Pillow Fight Day. In 2010, the 5th Pillow Fight NYC took place in Union Square, New York City. People from all over the city filled the square and wielded pillows. When the fight began, people lightly swung their pillows from side to side, hoping it would land on the nearest victim before getting hit themselves. The city's inaugural pillow fight took place in 2006 and had 500 participants. By 2010, the event garnered at least 5000 people.

In this section, we will learn about *place value* and its relation to decimals. Comparing the number of participants in the 2006 pillow fight to the 2010 pillow fight, we can see that the 5 moves from the hundreds place value to the thousands place value.

A Expanded Form

What is interesting and useful about decimals is their relationship to fractions and to powers of ten. The work we have done up to now—especially our work with fractions—can be used to develop the properties of decimal numbers.

Recall from a previous math class the idea of place value for the digits in a whole number. The name and the place value of each of the first seven columns in our number system are as follows:

Millions Column	Hundred Thousands Column	Ten Thousands Column	Thousands Column	Hundreds Column	Tens Column	Ones Column
1,000,000	100,000	10,000	1,000	100	10	1

As we move from right to left, we multiply by 10 each time. The value of each column is 10 times the value of the column on its right, with the rightmost column being 1. Up until now we have always looked at place value as increasing by a factor of 10 each time we move one column to the left.

Ten Thousands		Thousands		Hundreds		Tens		Ones
10,000	⟵ Multiply by 10	1,000	⟵ Multiply by 10	100	⟵ Multiply by 10	10	⟵ Multiply by 10	1

To understand the idea behind decimal numbers, we notice that moving in the opposite direction, from left to right, we *divide* by 10 each time.

Ten Thousands		Thousands		Hundreds		Tens		Ones
10,000	$\longrightarrow$ Divide by 10	1,000	$\longrightarrow$ Divide by 10	100	$\longrightarrow$ Divide by 10	10	$\longrightarrow$ Divide by 10	1

If we keep going to the right, the next column will have to be

$$1 \div 10 = \frac{1}{10} \qquad \text{Tenths}$$

The next one after that will be

$$\frac{1}{10} \div 10 = \frac{1}{10} \cdot \frac{1}{10} = \frac{1}{100} \qquad \text{Hundredths}$$

After that, we have

$$\frac{1}{100} \div 10 = \frac{1}{100} \cdot \frac{1}{10} = \frac{1}{1{,}000} \qquad \text{Thousandths}$$

We could continue this pattern as long as we wanted. We simply divide by 10 to move one column to the right. (And remember, dividing by 10 gives the same result as multiplying by $\frac{1}{10}$.)

To show where the ones column is, we use a *decimal point* between the ones column and the tenths column.

Note Because the digits to the right of the decimal point have fractional place values, numbers with digits to the right of the decimal point are called *decimal fractions.* In this book, we will also call them *decimal numbers,* or simply *decimals* for short.

Thousands	Hundreds	Tens	Ones		Tenths	Hundredths	Thousandths	Ten - Thousandths	Hundred - Thousandths
1,000	100	10	1	.	$\frac{1}{10}$	$\frac{1}{100}$	$\frac{1}{1{,}000}$	$\frac{1}{10{,}000}$	$\frac{1}{100{,}000}$
				Decimal point					

It is important to recognize the ones column, not the decimal point, can be thought of as the middle column, with columns larger than 1 to the left and columns smaller than 1 to the right. The first column to the right of the ones column is the tenths column, not the "oneths" column. The next column to the right is the hundredths column, the next is the thousandths column, and so on. The decimal point is always written between the ones column and the tenths column.

We can use the place value of decimal fractions to write them in *expanded form.*

Practice Problems

1. Write 785.462 in expanded form.

EXAMPLE 1 Write 423.576 in expanded form.

Solution $423.576 = 400 + 20 + 3 + \frac{5}{10} + \frac{7}{100} + \frac{6}{1{,}000}$

Answers

1. $700 + 80 + 5 + \frac{4}{10} + \frac{6}{100} + \frac{2}{1{,}000}$

EXAMPLE 2 Write each number in words.

a. 0.4 **b.** 0.04 **c.** 0.004

Solution

a. 0.4 is "four tenths."

b. 0.04 is "four hundredths."

c. 0.004 is "four thousandths."

When a decimal fraction contains digits to the left of the decimal point, we use the word "and" to indicate where the decimal point is when writing the number in words.

EXAMPLE 3 Write each number in words.

a. 5.4 **b.** 5.04 **c.** 5.004

Solution

a. 5.4 is "five and four tenths."

b. 5.04 is "five and four hundredths."

c. 5.004 is "five and four thousandths."

EXAMPLE 4 Write 3.64 in words.

Solution The number 3.64 is read "three and sixty-four hundredths." The place values of the digits are as follows:

3.64

3 ones 6 tenths 4 hundredths

We read the decimal part as "sixty-four hundredths" because

$$6 \text{ tenths} + 4 \text{ hundredths} = \frac{6}{10} + \frac{4}{100} = \frac{60}{100} + \frac{4}{100} = \frac{64}{100}$$

EXAMPLE 5 Write 25.4936 in words.

Solution Using the idea given in Example 4, we write 25.4936 in words as "twenty-five and four thousand, nine hundred thirty-six ten-thousandths."

EXAMPLE 6 Write the following numbers in order from largest to smallest.

0.03 0.3 0.032 0.006 0.003 0.2

Solution We know that the numbers with a digit greater than zero in the tenths place are greater than those with a zero in the tenths place, so our greatest numbers are 0.3 and 0.2.

$0.3 > 0.2$

Next, we look to the numbers with first digits greater than zero in the hundredths place.

$0.032 > 0.030$ so $0.032 > 0.03$

Finally, we look at numbers with the first digits greater than zero in the thousandths place.

$0.006 > 0.003$

Our complete list looks like this: 0.3 0.2 0.032 0.03 0.006 0.003

In order to understand addition and subtraction of decimals in the next section, we need to be able to convert decimal numbers to fractions or mixed numbers.

2. Write in words.

a. 0.06

b. 0.7

c. 0.008

3. Write each number in words.

a. 5.06

b. 4.7

c. 3.008

Note Sometimes we name decimal fractions by simply reading the digits from left to right and using the word "point" to indicate where the decimal point is. For example, using this method the number 5.04 is read "five point zero four." But for the purposes of this section, we will continue to use place values when naming decimal fractions.

4. Write 5.98 in words.

5. Write 305.406 in words.

6. Write the following numbers in order from largest to smallest.

0.01 0.001 0.012 0.2 0.1 0.12

Answers

2. a. Six hundredths **b.** Seven tenths **c.** Eight thousandths

3. a. Five and six hundredths **b.** Four and seven tenths **c.** Three and eight thousandths

4. Five and ninety eight hundredths

5. Three hundred five and four hundred six thousandths

6. 0.2 0.12 0.1 0.012 0.01 0.001

7. Write each number as a fraction or a mixed number. Do not reduce to lowest terms.

a. 0.06

b. 0.7

c. 0.008

Note Recall from a previous math class that a *mixed number* is the sum of a whole number and a proper fraction. In Example 6b, $3\frac{64}{100}$ is a mixed number.

EXAMPLE 7 Write each number as a fraction or a mixed number. Do not reduce to lowest terms.

a. 0.004 **b.** 3.64 **c.** 25.4936

Solution

a. Because 0.004 is 4 thousandths, we write

Three digits after the decimal point $\rightarrow 0.004 = \frac{4}{1{,}000}$ ← Three zeros

b. Looking over the work in Example 4, we can write

Two digits after the decimal point $\rightarrow 3.64 = 3\,\frac{64}{100}$ ← Two zeros

c. From the way in which we wrote 25.4936 in words in Example 5, we have

Four digits after the decimal point $\rightarrow 25.4936 = 25\,\frac{4936}{10{,}000}$ ← Four zeros

B Rounding Decimal Numbers

The rule for rounding decimal numbers is similar to the rule for rounding whole numbers. If the digit in the column to the right of the one we are rounding to is 5 or more, we add 1 to the digit in the column we are rounding to; otherwise, we leave it alone. We then replace all digits to the right of the column we are rounding to with zeros if they are to the left of the decimal point; otherwise, we simply delete them. Table 1 illustrates the procedure.

Table 1

	Round to the Nearest		
	Whole Number	Tenth	Hundredth
24.785	25	24.8	24.79
2.3914	2	2.4	2.39
0.98243	1	1.0	0.98
14.0942	14	14.1	14.09
0.545	1	0.5	0.55

8. Round 8,935.042 to the nearest hundred.

EXAMPLE 8 Round 9,235.492 to the nearest hundred.

Solution The number next to the hundreds column is 3, which is less than 5. We change all digits to the right to 0, and we can drop all digits to the right of the decimal point, so we write

9,200

9. Round 0.05067 to the nearest ten-thousandth.

EXAMPLE 9 Round 0.00346 to the nearest ten-thousandth.

Solution Because the number to the right of the ten-thousandths column is more than 5, we add 1 to the 4 and get

0.0035

Answers

7. a. $\frac{6}{100}$ **b.** $\frac{98}{100}$ **c.** $\frac{8}{1{,}000}$

8. 8,900

9. 0.0507

EXAMPLE 10 The bar chart below shows some ticket prices for a recent major league baseball season. Round each ticket price to the nearest dollar.

Solution Using our rule for rounding decimal numbers, we have the following results:

Least expensive: \$14.31 rounds to \$14.
League average: \$26.74 rounds to \$27.
Most expensive: \$52.32 rounds to \$52.

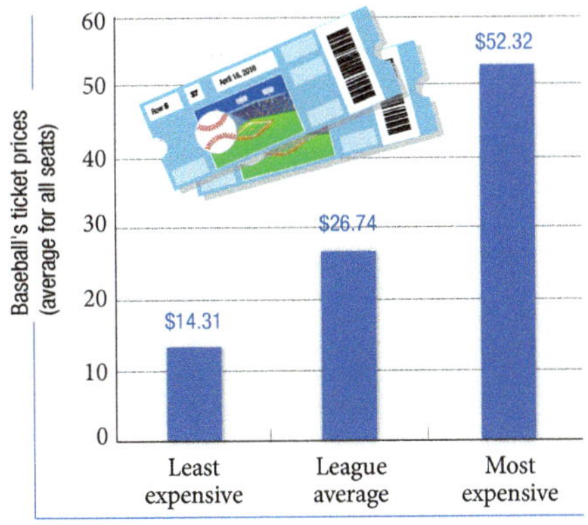

GETTING READY FOR CLASS

After reading through the preceding section, respond in your own words and in complete sentences.

A. Write 754.326 in expanded form.

B. Write $400 + 70 + 5 + \frac{1}{10} + \frac{3}{100} + \frac{7}{1,000}$ in decimal form.

C. Write seventy-two and three tenths in decimal form.

D. What is the rule for rounding decimal numbers?

10. Round each number in the bar chart to the nearest tenth of a dollar.

Answers

10. \$14.30, \$26.70, \$52.60

EXERCISE SET R.3

Vocabulary Review

Choose the correct words to fill in the blanks below.

hundredths tenths decimal point multiply divide

1. Moving from right to left in the place value columns, we ______________ by 10 each time.
2. Moving from left to right in the place value columns, we ______________ by 10 each time.
3. The symbol used between the ones column and the tenths column is called a ______________ ______________.
4. If we write 6.3 in words, we get "six and three ______________."
5. If we round 0.0456 to the nearest ______________ column, we get the answer 0.05.

Problems

A Write out the name of each number in words.

1. 0.3

2. 0.03

3. 0.015

4. 0.0015

5. 3.4

6. 2.04

7. 52.7

8. 46.8

Write each number as a fraction or a mixed number. Do not reduce your answers.

9. 405.36

10. 362.78

11. 9.009

12. 60.06

13. 1.234

14. 12.045

15. 0.00305

16. 2.00106

Give the place value of the 5 in each of the following numbers.

17. 458.327

18. 327.458

19. 29.52

20. 25.92

21. 0.00375

22. 0.00532

23. 275.01

24. 0.356

25. 539.76

26. 0.123456

27. 0.00045

28. 0.0000356

Write each of the following as a decimal number.

29. Fifty-five hundredths

30. Two hundred thirty-five ten-thousandths

31. Six and nine tenths

32. Forty-five thousand and six hundred twenty-one thousandths

33. Eleven and eleven hundredths

34. Twenty-six thousand, two hundred forty-five and sixteen hundredths

35. One hundred and two hundredths

36. Seventy-five and seventy-five hundred-thousandths

37. Three thousand and three thousandths

38. One thousand, one hundred eleven and one hundred eleven thousandths

B Complete the following table.

		Round the the Nearest			
	Number	Whole Number	Tenth	Hundredth	Thousandth
39.	47.5479	______	______	______	______
40.	100.9256	______	______	______	______
41.	0.8175	______	______	______	______
42.	29.9876	______	______	______	______
43.	0.1562	______	______	______	______
44.	128.9115	______	______	______	______
45.	2,789.3241	______	______	______	______
46.	0.8743	______	______	______	______
47.	99.9999	______	______	______	______
48.	71.7634	______	______	______	______

49. Write the following numbers in order from smallest to largest.

0.02 0.05 0.025 0.052 0.005 0.002

50. Write the following numbers in order from smallest to largest.

0.2 0.02 0.4 0.04 0.42 0.24

51. Which of the following numbers will round to 7.5?

7.451 7.449 7.54 7.56

52. Which of the following numbers will round to 3.2?

3.14999 3.24999 3.279 3.16111

Change each decimal to a fraction, and then reduce to lowest terms.

53. 0.25

54. 0.75

55. 0.35

56. 0.65

57. 0.125

58. 0.375

59. 0.625

60. 0.0625

61. 0.875

62. 0.1875

63. 0.012

64. 0.045

65. The following table lists the five tallest buildings in the world. Complete the chart by filing in each height in words.

Building	City, Country	Height (ft)	Height in Words
Burj Khalifa	Dubai, The United Arab Emirates	2,716	
Taipei 101	Taipei, Taiwan	1,667	
World Financial Center	Shangai, China	1,614	
Petronus Tower 1	Kuala Lunpur, Malaysia	1,483	
Petronus Tower 2	Kuala Lunpur, Malaysia	1,483	

66. The following table lists the famous "seven summits," the highest mountains on each of the seven continents. Complete the list by filling in the height in words of each mountain.

Summit	Continent	Height (ft)	Height in Words
Everest	Asia	29,035	
Aconcagua	South America	23,841	
McKinley	North America	20,320	
Kilimanjaro	Africa	19,340	
Elbrus	Europe	18,320	
Vinson Massif	Antartica	16,050	
Carstensz Pyramid	Australia	16,024	

Source: 7summits.com

Estimating

For each pair of numbers, choose the number that is closest to 10.

67. 9.9 and 9.99

68. 8.5 and 8.05

69. 10.5 and 10.05

70. 10.9 and 10.99

For each pair of numbers, choose the number that is closest to 0.

71. 0.5 and 0.05

72. 0.10 and 0.05

73. 0.01 and 0.02

74. 0.1 and 0.01

Find the Mistake

Each sentence below contains a mistake. Circle the mistake and write the correct word(s) or numbers(s) on the line provided.

1. The value of each column to the left of the decimal point is 10 times the value of the column on its left.

____

2. The fourth column to the right of the decimal point is called the thousandths column. ______________

3. 2,568 rounded to the nearest hundred is 2,560. ______________

4. 2.568 rounded to the nearest tenth is 2.5. ______________

0.4 Operations with Decimals

OBJECTIVES

A Add and subtract with decimals.

B Multiply with decimals.

C Divide with decimals.

KEY WORDS

column

fractional part

estimate

operation

remainder

At Noah's Ark Waterpark in Wisconsin, thrill seekers brave the country's first nearly-vertical loop waterslide. From a ten-story platform, a person enters the top of the waterslide's tube. Once the door of the tube closes, a countdown from ten begins. Then the floor drops from beneath the rider who plummets to the base of the loop, gaining as much speed as possible. Water sprays the inside of the tube to keep the ride smooth around the loop, and the rider comes shooting out happily on the other side.

The waterpark ran a summer special charging \$34.50 for an adult ticket and \$30.66 for a child's ticket. Suppose you bought tickets for two adults and two children. You would need to know how to perform operations with decimals in order to calculate how much you paid for the four tickets. In this section, we will do just that.

A Addition and Subtraction with Decimals

Let's begin with an example involving wages. Suppose you are earning \$8.50 an hour and you receive a raise of \$1.25 an hour. Your new hourly rate of pay is

$$\begin{array}{r} \$8.50 \\ +\ \$1.25 \\ \hline \$9.75 \end{array}$$

To add the two rates of pay, we align the decimal points, and then add in columns.

RULE Addition (or Subtraction) of Decimals

To add (or subtract) decimal numbers, we line up the decimal points and add (or subtract) as usual. The decimal point in the result is written directly below the decimal points in the problem.

We will use this rule for Examples 1–4.

EXAMPLE 1 Subtract: 39.812 − 14.236.

Solution We write the numbers vertically, with the decimal points lined up, and subtract as usual.

$$\begin{array}{r} 39.812 \\ -\ 14.236 \\ \hline 25.576 \end{array}$$

Practice Problems

1. Subtract: 78.674 − 23.431.

Answer

1. 55.243

2. Add: 16 + 0.033 + 4.6 + 0.08.

EXAMPLE 2 Add: 8 + 0.002 + 3.1 + 0.04.

Solution To make sure we keep the digits in the correct columns, we can write zeros to the right of the rightmost digits.

$$8 = 8.000$$
$$3.1 = 3.100$$
$$0.04 = 0.040$$

This doesn't change the value of any of the numbers, and it makes our task easier. Now we have

$$\begin{array}{r} 8.000 \\ 0.002 \\ 3.100 \\ +\ 0.040 \\ \hline 11.142 \end{array}$$

Note Writing the extra zeros here is really equivalent to finding a common denominator for the fractional parts of the original four numbers—now we have a thousandths column in all the numbers.

3. Subtract: 6.7 − 2.0563.

EXAMPLE 3 Subtract: 5.9 − 3.0814.

Solution In this case, it is very helpful to write 5.9 as 5.9000, since we will have to borrow in order to subtract.

$$\begin{array}{r} 5.9000 \\ -\ 3.0814 \\ \hline 2.8186 \end{array}$$

4. Subtract 5.89 from the sum of 7 and 3.567.

EXAMPLE 4 Subtract 3.09 from the sum of 9 and 5.472.

Solution Writing the problem in symbols, we have

$$\begin{aligned}(9 + 5.472) - 3.09 &= 14.472 - 3.09 \\ &= 11.382\end{aligned}$$

B Multiplication with Decimals

5. Change to a fractions and multiply.

0.4×0.6

EXAMPLE 5 Change each decimal to a fraction and multiply.

0.5×0.3

Solution Changing each decimal to a fraction and multiplying, we have

$$\begin{aligned}0.5 \times 0.3 &= \frac{5}{10} \times \frac{3}{10} && \text{Change to fractions.} \\ &= \frac{15}{100} && \text{Multiply numerators and multiply denominators.} \\ &= 0.15 && \text{Write the answer in decimal form.}\end{aligned}$$

The result is 0.15, which has two digits to the right of the decimal point.

Note To indicate multiplication, we are using a × sign here instead of a dot so we won't confuse the decimal points with the multiplication symbol.

Answers

2. 20.713
3. 4.6437
4. 4.677
5. 0.24

EXAMPLE 6 Multiply: 2.1×0.07.

Solution $2.1 \times 0.07 = 2\frac{1}{10} \times \frac{7}{100}$ *Change to fractions.*

$= \frac{21}{10} \times \frac{7}{100}$

$= \frac{147}{1{,}000}$ *Multiply numerators and denominators.*

$= 0.147$ *Write the answer as a decimal.*

Again, the digits in the answer come from multiplying $21 \times 7 = 147$. The decimal point is placed so that there are three digits to its right, because that is the total number of digits to the right of the decimal points in 2.1 and 0.07.

EXAMPLE 7 How many digits will be to the right of the decimal point in the following product?

2.987×24.82

Solution There are three digits to the right of the decimal point in 2.987 and two digits to the right in 24.82. Therefore, there will be $3 + 2 = 5$ digits to the right of the decimal point in their product.

EXAMPLE 8 Multiply: 3.05×4.36.

Solution We can set this up as if it were a multiplication problem with whole numbers. We multiply and then place the decimal point in the correct position in the answer.

```
    3.05   ⟵ Move decimal point 2 digits to the right.
  × 4.36   ⟵ Move decimal point 2 digits to the right, then multiply.
  ------
    1830
    915
  1220
  ------
  13.2980
```

The decimal point is placed so that there are $2 + 2 = 4$ digits to its right.

We summarize this discussion with a rule.

RULE Multiplication with Decimals

To multiply two decimal numbers:

1. Multiply as you would if the decimal points were not there.
2. Place the decimal point in the answer so that the number of digits to its right is equal to the total number of digits to the right of the decimal points in the original two numbers in the problem.

As you can see, multiplying decimal numbers is just like multiplying whole numbers, except that we must place the decimal point in the result in the correct position.

6. Change to fractions and multiply.

3.5×0.04

7. How many digits will be to the right of the decimal point in the following product?

3.706×55.88

8. Multiply: 4.03×5.22.

Answers

6. 0.14
7. 5
8. 21.0366

Estimating

Look back to Example 8. We could have placed the decimal point in the answer by rounding the two numbers to the nearest whole number and then multiplying them. Because 3.05 rounds to 3 and 4.36 rounds to 4, and the product of 3 and 4 is 12, we estimate that the answer to 3.05×4.36 will be close to 12. We then place the decimal point in the product 132980 between the 3 and the 2 in order to make it into a number close to 12.

9. Estimate the answer to each product.
 a. 82.3×5.8
 b. 37.5×178
 c. $(8.21)^2$

EXAMPLE 9 Estimate the answer to each of the following products.

a. 29.4×8.2 **b.** 68.5×172 **c.** $(6.32)^2$

Solution

a. Because 29.4 is approximately 30 and 8.2 is approximately 8, we estimate this product to be about $30 \times 8 = 240$. (If we were to multiply 29.4 and 8.2, we would find the product to be exactly 241.08.)

b. Rounding 68.5 to 70 and 172 to 170, we estimate this product to be $70 \times 170 = 11{,}900$. (The exact answer is 11,782.) Note here that we do not always round the numbers to the nearest whole number when making estimates. The idea is to round to numbers that will be easy to multiply.

c. Because 6.32 is approximately 6 and $6^2 = 36$, we estimate our answer to be close to 36. (The actual answer is 39.9424.)

C Division with Decimals

Suppose three friends go out for lunch and their total bill comes to \$18.75. If they decide to split the bill equally, how much does each person owe? To find out, they will have to divide 18.75 by 3. If you think about the dollars and cents separately, you will see that each person owes \$6.25. Therefore,

$$\$18.75 \div 3 = \$6.25$$

However, many division problems that involve decimals are not as intuitive as this one. Here are some examples that will help you find out how to do division with any combination of decimal numbers.

10. Divide: $4{,}626 \div 30$.

EXAMPLE 10 Divide: $5{,}974 \div 20$.

Solution

```
    298
 20)5974
    40
    197
    180
     174
     160
      14     Remainder of 14
```

Note We can estimate the answer to Example 10 by rounding 5,974 to 6,000 and dividing by 20.

$$\frac{6{,}000}{20} = 300$$

In the past, we have written this answer as $298\frac{14}{20}$ or, after reducing the fraction, $298\frac{7}{10}$. Because $\frac{7}{10}$ can be written as 0.7, we could also write our answer as 298.7. This last form of our answer is exactly the same result we obtain if we write 5,974 as 5,974.0 and continue the division until we have no remainder.

Answers

9. a. 480 b. 7,200 c. 64
10. 154.2

Here is how it looks:

$$\begin{array}{r} 298.7 \\ 20\overline{)5974.0} \\ \underline{40} \\ 197 \\ \underline{180} \\ 174 \\ \underline{160} \\ 14\,0 \\ \underline{14\,0} \\ 0 \end{array}$$

Let's try another division problem. This time one of the numbers in the original problem will be a decimal.

EXAMPLE 11 Divide: 34.8 ÷ 4.

11. Divide: 33.5 ÷ 5.

Solution We can use the ideas from Example 13 and divide as usual. The decimal point in the answer will be placed directly above the decimal point in the problem.

$$\begin{array}{r} 8.7 \\ 4\overline{)34.8} \\ \underline{32} \\ 2\,8 \\ \underline{2\,8} \\ 0 \end{array}$$

The answer is 8.7.

Check:

$$\begin{array}{r} 8.7 \\ \underline{\times 4} \\ 34.8 \end{array}$$

Note We never need to make a mistake with division, because we can always check our results with multiplication.

RULE Division with Decimals

To divide a decimal by a whole number, we do the usual long division as if there were no decimal point involved. The decimal point in the answer is placed directly above the decimal point in the problem.

Here are some more examples to illustrate the procedure:

EXAMPLE 12 Divide: 1,138.9 ÷ 35.

12. Divide: 1,138.5 ÷ 25.

Solution

$$\begin{array}{r} 32.54 \\ 35\overline{)1138.90} \\ \underline{105} \\ 88 \\ \underline{70} \\ 18\,9 \\ \underline{17\,5} \\ 1\,40 \\ \underline{1\,40} \\ 0 \end{array}$$

Hint: write 0 after the 9. It doesn't change the original number, but it gives us another digit to bring down.

The answer is 32.54.

Answers

11. 6.7

12. 45.54

Check:

$$\begin{array}{r} 32.54 \\ \times\ \ 35 \\ \hline 16270 \\ 97620 \\ \hline 1{,}138.90 \end{array}$$

Until now we have considered only division by whole numbers. Extending division to include division by decimal numbers is a matter of knowing what to do about the decimal point in the divisor.

13. Divide: 13.23 ÷ 4.2.

Note We do not always use the rules for rounding numbers to make estimates. For example, to estimate the answer to Example 13, 31.35 ÷ 3.8, we can get a rough estimate of the answer by reasoning that 3.8 is close to 4 and 31.35 is close to 32. Therefore, our answer will be approximately 32 ÷ 4 = 8.

EXAMPLE 13 Divide: 31.35 ÷ 3.8.

Solution In fraction form, this problem is equivalent to

$$\frac{31.35}{3.8}$$

If we want to write the divisor as a whole number, we can multiply the numerator and the denominator of this fraction by 10.

$$\frac{31.35 \times 10}{3.8 \times 10} = \frac{313.5}{38}$$

Since this fraction is equivalent to the original fraction, our original division problem is equivalent to

$$\begin{array}{r} 8.25 \\ 38\overline{)313.50} \\ \underline{304} \\ 95 \\ \underline{76} \\ 190 \\ \underline{190} \\ 0 \end{array}$$

Put 0 after the last digit

We can summarize division with decimal numbers by listing the following points, as illustrated by the previous examples.

Summary of Division with Decimals

1. We divide decimal numbers by the same process used to divide whole numbers. The decimal point in the answer is placed directly above the decimal point in the dividend.
2. We are free to write as many zeros after the last digit in a decimal number as we need.
3. If the divisor is a decimal, we can change it to a whole number by moving the decimal point to the right as many places as necessary, so long as we move the decimal point in the dividend the same number of places.

Answers

13. 3.15

EXAMPLE 14 Divide, and round the answer to the nearest hundredth.

$0.3778 \div 0.25$

Solution First, we move the decimal point two places to the right.

0.25. 0.37.78

Then we divide, using long division.

$$\begin{array}{r} 1.5112 \\ 25\overline{)37.7800} \\ \underline{25} \\ 12\,7 \\ \underline{12\,5} \\ 28 \\ \underline{25} \\ 30 \\ \underline{25} \\ 50 \\ \underline{50} \\ 0 \end{array}$$

Rounding to the nearest hundredth, we have 1.51. We actually did not need to have this many digits to round to the hundredths column. We could have stopped at the thousandths column and rounded off.

14. Divide, and round your answer to the nearest hundredth.
$0.4553 \div 0.32$

EXAMPLE 15 Divide, and round the answer to the nearest tenth.

$17 \div 0.03$

Solution Because we are rounding to the nearest tenth, we will continue dividing until we have a digit in the hundredths column. We don't have to go any further to round to the tenths column.

$$\begin{array}{r} 566.66 \\ 0.03.\overline{)17.00.00} \\ \underline{15} \\ 20 \\ \underline{18} \\ 20 \\ \underline{18} \\ 20 \\ \underline{18} \\ 20 \\ \underline{18} \\ 2 \end{array}$$

Rounding to the nearest tenth, we have 566.7.

15. Divide, and round to the nearest tenth.
$19 \div 0.06$

Answers

14. 1.423

15. 316.7

16. If the phone company in Example 19 charged $4.39 for a call, how long was the call?

EXAMPLE 16 A telephone company charges $0.43 for the first minute and then $0.33 for each additional minute for a long-distance call. If a long-distance call costs $3.07, how many minutes was the call?

Solution To solve this problem we need to find the number of additional minutes for the call. To do so, we first subtract the cost of the first minute from the total cost, and then we divide the result by the cost of each additional minute. Without showing the actual arithmetic involved, the solution looks like this:

$$\text{The number of additional minutes} = \frac{\overset{\substack{\text{Total cost}\\\text{of the call}\\\downarrow}}{3.07} - \overset{\substack{\text{Cost of the}\\\text{first minute}\\\downarrow}}{0.43}}{\underset{\substack{\uparrow\\\text{Cost of each additional minute}}}{0.33}} = \frac{2.64}{0.33} = 8$$

The call was 9 minutes long. (The number 8 is the number of additional minutes past the first minute.)

GETTING READY FOR CLASS

After reading through the preceding section, respond in your own words and in complete sentences.

A. Briefly explain the rule for adding and subtracting decimals.

B. Where do you place the decimal point in an answer to a multiplication problem that involves decimals? Explain.

C. The answer to the division problem in Example 13 is $298\frac{14}{20}$. In your own words, explain why we can also write this number as 298.7.

D. In Example 15, we place a 0 at the end of a number without changing the value of the number. Why is the placement of this 0 helpful?

Answer

16. 13 minutes

Vocabulary Review

Choose the correct words to fill in the blanks below.

divide multiplying add right above

1. To ______________ decimal numbers, we first line up the decimal points and then add as usual.
2. When ______________ decimal numbers, make sure to place the decimal point in the answer so that the number of digits to its right is equal to the total number of digits to the ______________ of the decimal points in the original two numbers.
3. To ______________ a decimal by a whole number, do the usual long division and then place the decimal point in the answer directly ______________ the decimal point in the problem.

Problems

A Find each of the following sums. (Add.)

1. $2.91 + 3.28$
2. $8.97 + 2.04$
3. $0.04 + 0.31 + 0.78$
4. $0.06 + 0.92 + 0.65$
5. $3.89 + 2.4$
6. $7.65 + 3.8$
7. $4.532 + 1.81 + 2.7$
8. $9.679 + 3.49 + 6.5$
9. $0.081 + 5 + 2.94$
10. $0.396 + 7 + 3.96$
11. $5.0003 + 6.78 + 0.004$
12. $27.0179 + 7.89 + 0.009$
13. $\begin{array}{r} 7.123 \\ 8.12 \\ +\ 9.1 \\ \hline \end{array}$
14. $\begin{array}{r} 5.432 \\ 4.32 \\ +\ 3.2 \\ \hline \end{array}$
15. $\begin{array}{r} 9.001 \\ 8.01 \\ +\ 7.1 \\ \hline \end{array}$
16. $\begin{array}{r} 6.003 \\ 5.02 \\ +\ 4.1 \\ \hline \end{array}$

17.
$$\begin{array}{r} 89.7854 \\ 3.4 \\ 65.65 \\ +\ 100.006 \\ \hline \end{array}$$

18.
$$\begin{array}{r} 57.4698 \\ 9.89 \\ 32.032 \\ +\ 572.0079 \\ \hline \end{array}$$

19.
$$\begin{array}{r} 543.21 \\ +\ 123.45 \\ \hline \end{array}$$

20.
$$\begin{array}{r} 987.654 \\ +\ 456.789 \\ \hline \end{array}$$

21.
$$\begin{array}{r} 14.2586 \\ 5.34 \\ 6.2 \\ +\ 12.293 \\ \hline \end{array}$$

22.
$$\begin{array}{r} 243.75 \\ +\ 582.346 \\ \hline \end{array}$$

Find each of the following differences. (Subtract.)

23. $99.34 - 88.23$

24. $47.69 - 36.58$

25. $5.97 - 2.4$

26. $9.87 - 1.04$

27. $6.3 - 2.08$

28. $7.5 - 3.04$

29. $149.37 - 28.96$

30. $796.45 - 32.68$

31. $45 - 0.067$

32. $48 - 0.075$

33. $8 - 0.327$

34. $12 - 0.962$

35. $765.432 - 234.567$

36. $654.321 - 123.456$

37. $10 - 1.942$

38. $10 - 5.0874$

Subtract each decimal.

39. $\begin{array}{r} 34.07 \\ -\ \ 6.18 \\ \hline \end{array}$

40. $\begin{array}{r} 25.008 \\ -\ \ 3.119 \\ \hline \end{array}$

41. $\begin{array}{r} 40.04 \\ -\ \ 4.40 \\ \hline \end{array}$

42. $\begin{array}{r} 50.05 \\ -\ \ 5.50 \\ \hline \end{array}$

43. $\begin{array}{r} 768.436 \\ -\ 356.998 \\ \hline \end{array}$

44. $\begin{array}{r} 495.237 \\ -\ 247.668 \\ \hline \end{array}$

Add and subtract as indicated.

45. $(7.8 - 4.3) + 2.5$

46. $(8.3 - 1.2) + 3.4$

47. $7.8 - (4.3 + 2.5)$

48. $8.3 - (1.2 + 3.4)$

49. $(9.7 - 5.2) - 1.4$

50. $(7.8 - 3.2) - 1.5$

51. $9.7 - (5.2 - 1.4)$

52. $7.8 - (3.2 - 1.5)$

53. Subtract 5 from the sum of 8.2 and 0.072.

54. Subtract 8 from the sum of 9.37 and 2.5.

55. What number is added to 0.035 to obtain 4.036?

56. What number is added to 0.043 to obtain 6.054?

B Find each of the following products. (Multiply.)

57. 0.7×0.4

58. 0.8×0.3

59. 0.07×0.4

60. 0.8×0.03

61. 0.03×0.09

62. 0.07×0.002

63. $2.6(0.3)$

64. $8.9(0.2)$

65. 0.9×0.88

66. 0.8×0.99

67. 3.12×0.005

68. 4.69×0.006

69. 4.003×6.07

70. 7.0001×3.04

71. $5(0.006)$

72. $7(0.005)$

73. 75.14×2.5

74. 963.8×0.24

75. 0.1×0.02

76. 0.3×0.02

77. $2.796(10)$

78. $97.531(100)$

79. 0.0043×100

80. $12.345 \times 1{,}000$

81. $49.94 \times 1{,}000$

82. $157.02 \times 10{,}000$

83. $987.654 \times 10{,}000$

84. $1.23 \times 100{,}000$

85. $1{,}000 \times 0.001$

86. $0.0001 \times 10{,}000$

A B Perform the following operations.

87. $2.1(3.5 - 2.6)$

88. $5.4(9.9 - 6.6)$

89. $0.05(0.02 + 0.03)$

90. $0.04(0.07 + 0.09)$

91. $2.02(0.03 + 2.5)$

92. $4.04(0.05 + 6.6)$

93. $(2.1 + 0.03)(3.4 + 0.05)$

94. $(9.2 + 0.01)(3.5 + 0.03)$

95. $(2.1 - 0.1)(2.1 + 0.1)$

96. $(9.6 - 0.5)(9.6 + 0.5)$

97. $3.08 - 0.2(5 + 0.03)$

98. $4.09 + 0.5(6 + 0.02)$

99. $4.23 - 5(0.04 + 0.09)$

100. $7.89 - 2(0.31 + 0.76)$

101. $2.5 + 10(4.3)^2$

102. $3.6 + 15(2.1)^2$

103. $100(1 + 0.08)^2$

104. $500(1 + 0.12)^2$

105. $(1.5)^2 + (2.5)^2 + (3.5)^2$

106. $(1.1)^2 + (2.1)^2 + (3.1)^2$

C Perform each of the following divisions.

107. $394 \div 20$

108. $486 \div 30$

109. $248 \div 40$

110. $372 \div 80$

111. $5\overline{)26}$

112. $8\overline{)36}$

113. $25\overline{)276}$

114. $50\overline{)276}$

115. $28.8 \div 6$

116. $15.5 \div 5$

117. $77.6 \div 8$

118. $31.48 \div 4$

119. $35\overline{)92.05}$

120. $26\overline{)146.38}$

121. $45\overline{)190.8}$

122. $55\overline{)342.1}$

123. $86.7 \div 34$

124. $411.4 \div 44$

125. $29.7 \div 22$

126. $488.4 \div 88$

127. $4.5\overline{)29.25}$

128. $3.3\overline{)21.978}$

129. $0.11\overline{)1.089}$

130. $0.75\overline{)2.40}$

131. $2.3\overline{)0.115}$

132. $6.6\overline{)0.198}$

133. $0.012\overline{)1.068}$

134. $0.052\overline{)0.23712}$

135. $1.1\overline{)2.42}$

136. $2.2\overline{)7.26}$

137. $0.15\overline{)1.8}$

138. $0.04\overline{)0.0064}$

Carry out each of the following divisions only so far as needed to round the results to the nearest hundredth.

139. $26\overline{)35}$

140. $18\overline{)47}$

141. $3.3\overline{)56}$

142. $4.4\overline{)75}$

143. $0.1234 \div 0.5$

144. $0.543 \div 2.1$

145. $19 \div 7$

146. $16 \div 6$

Applying the Concepts

147. Hot Air Balloon Since the pilot of a hot air balloon can only control the balloon's altitude, he relies on the winds for travel. To ride on the jet streams, a hot air balloon must rise as high as 12 kilometers. Convert this to miles by dividing by 1.61. Round your answer to the nearest tenth of a mile.

148. Hot Air Balloon December and January are the best times for traveling in a hot-air balloon because the jet streams in the Northern Hemisphere are the strongest. They reach speeds of 400 kilometers per hour. Convert this to miles per hour by dividing by 1.61. Round to the nearest whole number.

149. Wages If a woman earns $45.30 for working 6 hours, how much does she earn per hour?

150. Wages How many hours does a person making $7.65 per hour have to work in order to earn $191.25?

151. Gas Mileage If a car travels 336 miles on 15 gallons of gas, how far will the car travel on 1 gallon of gas?

152. Gas Mileage If a car travels 392 miles on 16 gallons of gas, how far will the car travel on 1 gallon of gas?

153. Wages Suppose a woman earns \$8.34 an hour for the first 40 hours she works in a week and then \$12.51 an hour in overtime pay for each additional hour she works in the same week. If she makes \$371.13 in one week, how many hours did she work overtime?

154. Wages Suppose a woman makes \$386.10 in one week. If she is paid \$7.80 an hour for the first 36 hours she works and then \$11.70 an hour in overtime pay for each additional hour she works in the same week, how many hours did she work overtime that week?

155. Phone Bill Suppose a telephone company charges \$0.41 for the first minute and then \$0.32 for each additional minute for a long-distance call. If a long-distance call costs \$2.33, how many minutes was the call?

156. Phone Bill Suppose a telephone company charges \$0.45 total for the first three minutes and then \$0.29 for each additional minute for a long-distance call. If a long-distance call costs \$2.77, how many minutes was the call?

157. The 2012 proposed federal budget contains \$2.627 trillion in revenue and \$3.729 trillion in spending. An alternative congressional plan contains \$2.533 trillion in revenue and \$3.529 trillion in spending. Calculate the deficits (revenue − spending) for each budget and compare.

158. The 2012 proposed federal budget contains \$241.6 billion in interest and \$925.2 billion in defense. Approximately how many times greater is the defense budget?

Find the Mistake

Each sentence below contains a mistake. Circle the mistake and write the correct word(s) or numbers(s) on the line provided.

1. To subtract 2.35 from 16.854, line up the right most digit of each number and subtract as usual.

   ____

2. The product of 7.19 and 189.4 will have 4 digits to the right of the decimal. ______________________________

3. To estimate the answer to the problem 6.79×28.6, we align the decimals and multiply as usual.

   ____

4. When dividing with decimal numbers, the decimal point in the answer is placed one digit to the right of the decimal in the dividend. ______________________________

Landmark Review: Checking Your Progress

This feature is intended as a review of key skills from the preceding sections in the chapter. Each review will give you the opportunity to utilize important concepts and check your progress as you practice problems of different types. Take note of any problems in this review that you find difficult. There is no better time than now to revisit and master those difficult concepts in preparation for any upcoming exams or subsequent chapters.

Identify each number as a prime or composite number. For those that are composite, factor into the product of prime factors.

1. 25

2. 19

3. 31

4. 21

Simplify each of the following expressions. Make sure each answer is written in lowest terms.

5. $\frac{2}{3} \cdot \frac{1}{2}$

6. $\frac{4}{5} \div \frac{3}{4}$

7. $\frac{3}{8} + \frac{5}{8}$

8. $\frac{5}{6} - \frac{1}{3}$

Write each number as a fraction or a mixed number. Do not reduce your answers.

9. 15.7

10. 0.0046

Write each number as a decimal.

11. $1\frac{4}{10}$

12. $11\frac{11}{100}$

Add or subtract each of the following.

13. $1.08 + 0.004 + 5$

14. $143.25 - 23.13$

Multiply or divide each of the following. Round to the nearest thousandth if necessary.

15. $2.3(0.3)$

16. $\begin{array}{r} 3.004 \\ \times\ 2.03 \\ \hline \end{array}$

17. $428 \div 12$

18. $532 \div 6$

0.5 Percents, Decimals, and Fractions

Image © Richard Ling, 2005

OBJECTIVES

A Write percents as fractions.

B Write percents as decimals.

C Write decimals as percents.

D Write percents as fractions.

E Write fractions as percents.

KEY WORDS

percent

pie chart

Scientists have discovered a toxin in the spit of a sea snail that works with greater effectiveness but in smaller dosages and without the addictive risk of the painkiller morphine. The marine cone snail dwells on the ocean floor. The snail shoots its harpoon-like teeth coated in toxic saliva into its prey to paralyze it. Researchers have discovered how to isolate the saliva's toxin and put it into a pill for humans in pain to ingest. A patient will feel the same pain-reducing effects with 1% of a dose of a popular neuropathic painkiller prescribed in hospitals. If the prescription for an adult of the popular painkiller starts at 300 milligrams, how many milligrams of the sea snail drug would be dosed? To answer this question, we need to review percents and how they interact with decimals and fractions.

A The Meaning of Percent

Percent means "per hundred." Writing a number as a percent is a way of comparing the number with 100. For example, the number 42% (the % symbol is read "percent") is the same as 42 one-hundredths; that is,

$$42\% = \frac{42}{100}$$

Percents are really fractions (or ratios) with denominator 100. Here are some examples that show the meaning of percent.

EXAMPLE 1

a. $50\% = \frac{50}{100}$

b. $75\% = \frac{75}{100}$

c. $25\% = \frac{25}{100}$

d. $33\% = \frac{33}{100}$

e. $6\% = \frac{6}{100}$

f. $160\% = \frac{160}{100}$

Practice Problems

1. Write each number as an equivalent fraction without the % symbol.

a. 40%

b. 80%

c. 15%

d. 37%

e. 8%

f. 150%

Answers

1. a. $\frac{40}{100}$ **b.** $\frac{80}{100}$ **c.** $\frac{15}{100}$ **d.** $\frac{37}{100}$ **e.** $\frac{8}{100}$ **f.** $\frac{150}{100}$

B Changing Percents to Decimals

To change a percent to a decimal number, we simply use the meaning of percent.

2. Change 25.2% to a decimal.

EXAMPLE 2 Change 35.2% to a decimal.

Solution We drop the % symbol and write 35.2 over 100.

$$35.2\% = \frac{35.2}{100}$$ Use the meaning of % to convert to a fraction with denominator 100.

$$= 0.352$$ Divide 35.2 by 100.

We see from Example 2 that 35.2% is the same as the decimal 0.352. The result is that the % symbol has been dropped and the decimal point has been moved two places to the *left*. Because % means "per hundred," we will always end up moving the decimal point two places to the left when we change percents to decimals. Because of this, we can write the following rule:

RULE Changing a Percent to a Decimal

To change a percent to a decimal, drop the % symbol and move the decimal point two places to the left.

Here are some examples to illustrate how to use this rule:

3. Change each percent to a decimal.

a. 40%

b. 80%

c. 15%

d. 5.6%

e. 4.86%

f. 0.6%

EXAMPLE 3

a. 25% = 0.25

b. 75% = 0.75

c. 50% = 0.50

d. 6.80% = 0.068

e. 3.62% = 0.0362

f. 0.4% = 0.004

Notice that the results in Example 3a, b, and c are consistent with the results in Example 1a, b, and c.

Notice in 3d and e that we put a 0 in front of the 6 so we can move the decimal point two places to the left.

In 3f, we put two 0's in front of the 4 in order to be able to move the decimal point two places to the left.

4. Write 0.58% as a decimal number.

EXAMPLE 4 Suppose a cortisone cream is 0.5% hydrocortisone. Writing this number as a decimal, we have

0.5% = 0.005

C Changing Decimals to Percents

Now we want to do the opposite of what we just did in Examples 2, 3 and 4. We want to change decimals to percents. We know that 42% written as a decimal is 0.42, which means that in order to change 0.42 back to a percent, we must move the decimal point two places to the *right* and use the % symbol.

0.42 = 42% Notice that we don't show the new decimal point if it is at the end of the number.

Answers

2. 0.252
3. **a.** 0.40 **b.** 0.80 **c.** 0.15 **d.** 0.056 **e.** 0.0486 **f.** 0.006
4. 0.0058

RULE **Changing a Decimal to a Percent**

To change a decimal to a percent, we move the decimal point two places to the right and use the % symbol.

Example 5 shows how we use this rule.

EXAMPLE 5

a. $0.27 = 27\%$

b. $4.89 = 489\%$

c. $0.2 = 0.20 = 20\%$ — Notice here that we put a 0 after the 2 so we can move the decimal point two places to the right.

d. $0.09 = 09\% = 9\%$ — Notice that we can drop the 0 at the left without changing the value of the number.

e. $25 = 25.00 = 2{,}500\%$ — Here, we put two 0's after the 5 so we can move the decimal point two places to the right.

EXAMPLE 6 A softball player has a batting average of 0.650. As a percent, this number is $0.650 = 65.0\%$.

As you can see from the examples above, percent is just a way of comparing numbers to 100. To multiply decimals by 100, we move the decimal point two places to the right. To divide by 100, we move the decimal point two places to the left. Because of this, it is a fairly simple procedure to change percents to decimals and decimals to percents.

D Changing Percents to Fractions

Recall from the beginning of this section our discussion of percents and how they are really fractions (or ratios) with denominator 100. We will now explore further how to change a percent to a fraction. To do so, drop the % symbol and write the original number over 100.

EXAMPLE 7 The pie chart shows who pays for college expenses. Change each percent to a fraction.

Solution In each case, we drop the percent symbol and write the number over 100. Then we reduce to lowest terms if possible.

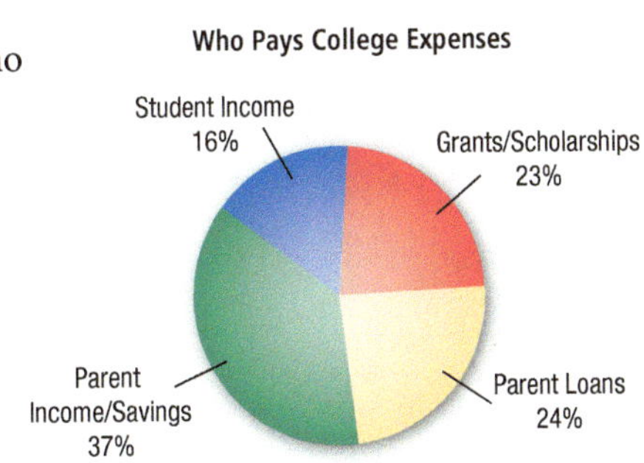

$$16\% = \frac{16}{100} = \frac{4}{25} \qquad 23\% = \frac{23}{100} \qquad 24\% = \frac{24}{100} = \frac{6}{25} \qquad 37\% = \frac{37}{100}$$

5. Change each decimal as a percent.
   a. 0.35
   b. 5.77
   c. 0.4
   d. 0.03
   e. 45

6. Write 0.69 as a percent.

7. Change 82% to a fraction in lowest terms.

Answers

5. **a.** 35% **b.** 577% **c.** 40% **d.** 3% **e.** 4,500%
6. 69%
7. $\frac{41}{50}$

8. Change 6.5% to a fraction in lowest terms.

EXAMPLE 8 Change 4.5% to a fraction in lowest terms.

Solution We begin by writing 4.5 over 100.

$$4.5\% = \frac{4.5}{100}$$

We now multiply the numerator and the denominator by 10 so the numerator will be a whole number.

$$\frac{4.5}{100} = \frac{4.5 \times 10}{100 \times 10} \quad \text{Multiply the numerator and the denominator by 10.}$$

$$= \frac{45}{1{,}000}$$

$$= \frac{9}{200} \quad \text{Reduce to lowest terms.}$$

The next example involves changing a mixed number into a complex fraction in order to reduce to lowest terms. A *complex fraction* contains fractions in the numerator and the denominator.

9. Change $42\frac{1}{2}\%$ to a fraction in lowest terms.

EXAMPLE 9 Change $32\frac{1}{2}\%$ to a fraction in lowest terms.

Solution Writing $32\frac{1}{2}\%$ over 100 produces a complex fraction. We change $32\frac{1}{2}$ to an improper fraction and simplify.

$$32\frac{1}{2}\% = \frac{32\frac{1}{2}}{100}$$

$$= \frac{\frac{65}{2}}{100} \quad \text{Change } 32\tfrac{1}{2} \text{ to the improper fraction } \tfrac{65}{2}.$$

$$= \frac{65}{2} \times = \frac{1}{100} \quad \text{Divide by 100 is the same as multiplying by } \tfrac{1}{100}.$$

$$= \frac{5 \cdot 13 \cdot 1}{2 \cdot 5 \cdot 20} \quad \text{Multiply.}$$

$$= \frac{13}{40} \quad \text{Reduce to lowest terms.}$$

Note that we could have changed our original mixed number to a decimal first and then changed to a fraction.

$$32\frac{1}{2}\% = 32.5\% = \frac{32.5}{100} = \frac{32.5 \times 10}{100 \times 10} = \frac{325}{1000} = \frac{5 \cdot 5 \cdot 13}{5 \cdot 5 \cdot 40} = \frac{13}{40}$$

The result is the same in both cases.

E Changing Fractions to Percents

To change a fraction to a percent, we can change the fraction to a decimal and then change the decimal to a percent.

10. Change $\frac{9}{10}$ to a percent.

EXAMPLE 10 Suppose the price your bookstore pays for your textbook is $\frac{7}{10}$ of the price you pay for your textbook. Write $\frac{7}{10}$ as a percent.

Solution We can change $\frac{7}{10}$ to a decimal by dividing 7 by 10.

$$\begin{array}{r} 0.7 \\ 10\overline{)7.0} \\ \underline{7\,0} \\ 0 \end{array}$$

Answers

8. $\frac{13}{200}$

9. $\frac{17}{40}$

10. 90%

We then change the decimal 0.7 to a percent by moving the decimal point two places to the right and using the % symbol.

$$0.7 = 70\%$$

You may noticed that we could have saved some time by simply writing $\frac{7}{10}$ as an equivalent fraction with denominator 100; that is,

$$\frac{7}{10} = \frac{7 \cdot 10}{10 \cdot 10} = \frac{70}{100} = 70\%$$

This is a good way to convert fractions like $\frac{7}{10}$ to percents. It works well for fractions with denominators of 2, 4, 5, 10, 20, 25, and 50, because they are easy to change to fractions with denominators of 100.

EXAMPLE 11 Change $\frac{3}{8}$ to a percent.

Solution We write $\frac{3}{8}$ as a decimal by dividing 3 by 8. We then change the decimal to a percent by moving the decimal point two places to the right and using the % symbol.

$$\frac{3}{8} = 0.375 = 37.5\%$$

$$\begin{array}{r} .375 \\ 8\overline{)3.000} \\ \underline{2\,4} \\ 60 \\ \underline{56} \\ 40 \\ \underline{40} \\ 0 \end{array}$$

11. Change $\frac{5}{8}$ to a percent.

EXAMPLE 12 Change $\frac{5}{12}$ to a percent.

Solution We begin by dividing 5 by 12.

$$\begin{array}{r} .4166 \\ 12\overline{)5.0000} \\ \underline{48} \\ 20 \\ \underline{12} \\ 80 \\ \underline{72} \\ 80 \\ \underline{72} \end{array}$$

Because the 6s repeat indefinitely, we can use mixed number notation to write

$$\frac{5}{12} = 0.41\overline{6} = 41\frac{2}{3}\%$$

Or rounding, we can write

$$\frac{5}{12} \approx 41.7\%$$ Round to the nearest tenth of a percent.

12. Change $\frac{7}{12}$ to a percent.

Note When rounding off, let's agree to round off to the nearest thousandth and then move the decimal point. Our answers in percent form will then be accurate to the nearest tenth of a percent, as in Example 12.

Answers

11. 62.5%
12. $58\frac{1}{3}\% = 58.3\%$
13. 375%

13. Change $3\frac{3}{4}$ to a percent.

EXAMPLE 13 Change $2\frac{1}{2}$ to a percent.

Solution We first change to a decimal and then to a percent.

$$2\frac{1}{2} = 2.5$$

$$= 250\%$$

Table 1 lists some of the most commonly used fractions and decimals and their equivalent percents.

Table 1

Fraction	Decimal	Percent
$\frac{1}{2}$	0.5	50%
$\frac{1}{4}$	0.25	25%
$\frac{3}{4}$	0.75	75%
$\frac{1}{3}$	$0.\overline{3}$	33.33%
$\frac{2}{3}$	$0.\overline{6}$	66.67%
$\frac{1}{5}$	0.2	20%
$\frac{2}{5}$	0.4	40%
$\frac{3}{5}$	0.6	60%
$\frac{4}{5}$	0.8	80%

GETTING READY FOR CLASS

After reading through the preceding section, respond in your own words and in complete sentences.

A. What is the relationship between the word *percent* and the number 100?

B. Explain in words how you would change 25% to a decimal.

C. Explain in words how you would change 25% to a fraction.

D. After reading this section, you know that $\frac{1}{2}$, 0.5, and 50% are equivalent. Show mathematically why this is true.

EXERCISE SET R.5

Vocabulary Review

Choose the correct words to fill in the blanks below.

right left percent fraction decimal denominator

1. Writing a number as a ________________ is a way of comparing the number with 100.
2. A percent is really a fraction with a ________________ of 100.
3. To change a percent to a decimal, drop the percent symbol and move the decimal point two places to the ________________.
4. To change a decimal to a percent, move the decimal point two places to the ________________ and use the percent symbol.
5. To change a percent to a ________________, drop the percent symbol and write the original number over 100.
6. To change a fraction to a percent, change the fraction to a decimal and then change the ________________ to a percent.

A Write each percent as a fraction with denominator 100.

1. 20% **2.** 40% **3.** 60% **4.** 80% **5.** 24% **6.** 48%

7. 65% **8.** 35% **9.** 48% **10.** 75% **11.** 10% **12.** 2%

B Change each percent to a decimal.

13. 23% **14.** 34% **15.** 92% **16.** 87% **17.** 9% **18.** 7%

19. 3.4% **20.** 5.8% **21.** 6.34% **22.** 7.25% **23.** 0.9% **24.** 0.6%

C Change each decimal to a percent.

25. 0.23 **26.** 0.34 **27.** 0.92 **28.** 0.87 **29.** 0.45 **30.** 0.54

31. 0.03 **32.** 0.04 **33.** 0.6 **34.** 0.9 **35.** 0.8 **36.** 0.5

37. 0.27 **38.** 0.62 **39.** 1.23 **40.** 2.34 **41.** 0.7 **42.** 0.07

D Change each percent to a fraction in lowest terms.

43. 60% **44.** 40% **45.** 75% **46.** 25% **47.** 4% **48.** 2%

49. 26.5% **50.** 34.2% **51.** 71.87% **52.** 63.6% **53.** 0.75% **54.** 0.45%

55. $6\frac{1}{4}\%$ **56.** $5\frac{1}{4}\%$ **57.** $33\frac{1}{3}\%$ **58.** $66\frac{2}{3}\%$ **59.** $2\frac{1}{2}\%$ **60.** $5\frac{1}{3}\%$

E Change each fraction or mixed number to a percent.

61. $\frac{1}{2}$ **62.** $\frac{1}{4}$ **63.** $\frac{3}{4}$ **64.** $\frac{2}{3}$ **65.** $\frac{1}{3}$ **66.** $\frac{1}{5}$

67. $\frac{4}{5}$ **68.** $\frac{1}{6}$ **69.** $\frac{7}{8}$ **70.** $\frac{1}{8}$ **71.** $\frac{7}{50}$ **72.** $\frac{9}{25}$

73. $3\frac{1}{4}$ **74.** $2\frac{1}{8}$ **75.** $1\frac{1}{2}$ **76.** $1\frac{3}{4}$ **77.** $2\frac{1}{2}$ **78.** $3\frac{3}{4}$

79. $\frac{21}{43}$ to the nearest tenth of a percent

80. $\frac{36}{49}$ to the nearest tenth of a percent

Calculator Problems

Use a calculator to write each fraction as a decimal, and then change the decimal to a percent. Round all answers to the nearest tenth of a percent.

81. $\frac{29}{37}$ **82.** $\frac{18}{83}$ **83.** $\frac{6}{51}$ **84.** $\frac{8}{95}$ **85.** $\frac{236}{327}$ **86.** $\frac{568}{732}$ **87.** $\frac{525}{739}$ **88.** $\frac{1024}{2575}$

Applying the Concepts

89. Physiology The human body is between 50% and 75% water. Write each of these percents as a decimal.

90. Alcohol Consumption In the United States, 2.7% of those over 15 years of age drink more than 6.3 ounces of alcohol per day. In France, the same figure is 9%. Write each of these percents as a decimal.

91. iPhones Use the chart to work the following problem.

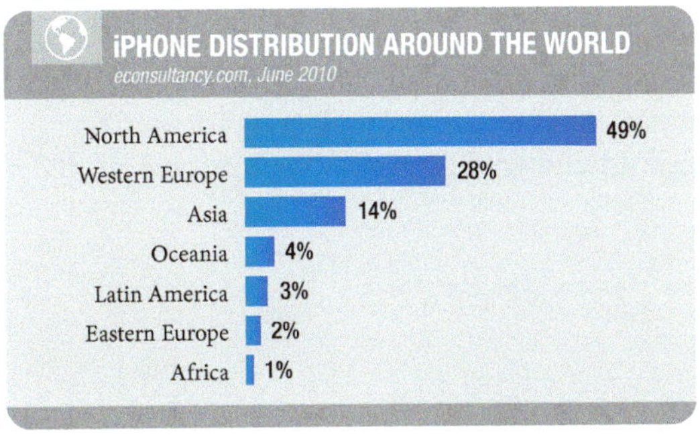

a. Convert each percent to a fraction.

b. Convert each percent to a decimal.

92. Commuting The pie chart below shows how we, as a nation, commute to work. Change each percent to a fraction in lowest terms.

How We Commute

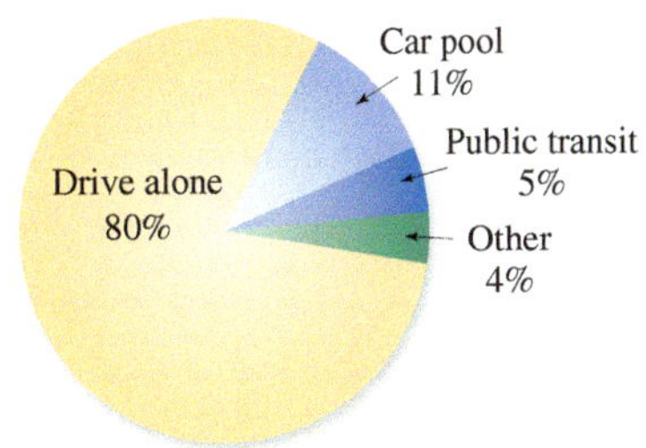

93. Nutrition Although, nutritionally, breakfast is the most important meal of the day, only $\frac{1}{5}$ of the people in the United States consistently eat breakfast. What percent of the population is this?

94. Children In School In Belgium, 96% of all children between 3 and 6 years of age go to school. In Sweden, the same figure is only 25%. In the United States, the figure is 60%. Write each of these percents as a fraction in lowest terms.

95. Student Enrollment Using the pie chart below, change each fraction to a percent.

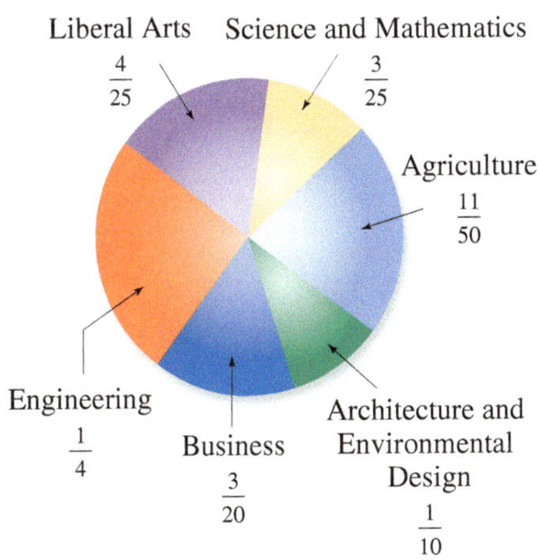

96. Mac Users The chart shows the increase in Mac users since 2003. Use the chart to answer the following questions.

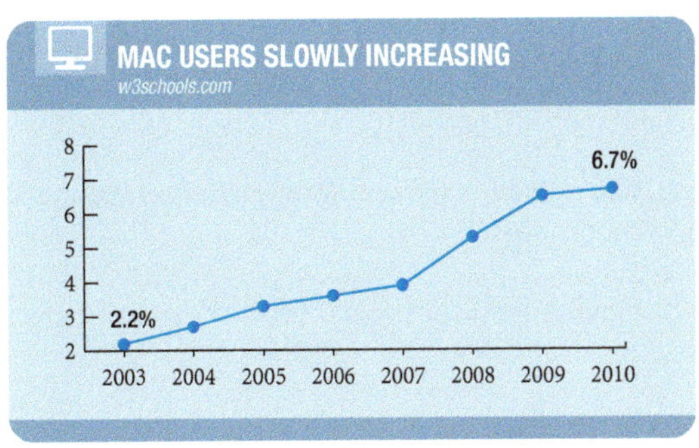

a. In 2006, estimate what fraction of people were Mac users.

b. During what year did the fraction of Mac users equal about $\frac{1}{20}$?

97. Number of Teachers The ratio of the number of teachers to the number of students in secondary schools in Japan is 1 to 17. In the United States, the ratio is 1 to 19. Write each of these ratios as a fraction and then as a percent. Round your answers to the nearest tenth.

98. Mountain Climbing Mountain climbing is generally known as a dangerous hobby or occupation. One way mountains are compared in terms of danger is their death to summit ratio, expressed as a percent. The famed Mount Everest, from 1922 to 2010, has had 5,101 summits and 219 deaths. Find the death to summit ratio as a percent. Round your answer to the nearest tenth.

99. Summit Deaths While climbing Mount Everest is certainly dangerous, there are far more dangerous climbs. Widely recognized by mountain climbers as the most dangerous summits are Ammapurna I in Nepal, with 130 summits and 63 deaths, and Gongjashan in China, with 22 summits and 16 deaths. Compare these two deadly mountains by finding their death to summit ratios as percents. Round your answer to the nearest percent.

100. National Defense In the proposed 2012 federal budget of \$3,728.7 billion, \$925.2 billion is proposed for defense. What percent of the federal budget is proposed for defense? Round your answer to the nearest percent.

101. Federal Budget In the proposed 2010 federal budget of \$3,728.7 billion, \$866.1 billion is proposed for health care, \$121.1 billion for education, and \$431.5 billion for welfare. What is the combined percent proposed for health, education, and welfare? Round your answer to the nearest percent.

102. Oxygen Ang Rita, a famous Mountain Climber from Nepal earned the nickname "The Snow Leopard" when he climbed Mount Everest 10 times without using supplemental oxygen. This has only been accomplished a total of 173 times. What percent of ascents without supplemental oxygen has Ang Rita participated in? Round your answer to the nearest tenth.

Find the Mistake

Each sentence below contains a mistake. Circle the mistake and write the correct word(s) or numbers(s) on the line provided.

1. The fraction $\frac{36}{100}$ given as a percent is 3.6%. ______________________

2. Changing the percent 57.4% to a decimal gives us 5.74. ______________________

3. Changing the decimal 0.1465 to a percent gives us 0.001465%. ______________________

4. To change a fraction to a percent, first change the fraction to its reciprocal and then change to a percent.

____

0.6 Exponents and Order of Operations

Image © sxc.hu, WR-Fife, 2010

OBJECTIVES

A Evaluate exponential expressions.

B Use order of operations to evaluate expressions.

KEY WORDS

base

exponent

order of operations

Each year, the coastal waters of Qingdao, China are infiltrated with a vibrant green algae bloom that covers more than 150 square miles. The algae thrive in high temperatures and in polluted waters caused by runoff from agriculture land and fish farms. The bloom sucks oxygen out of the water, threatening the local marine life. As it washes onshore and dries in the hot sun, it gives off a noxious odor that smells like rotten eggs. Soldiers and other volunteers work tirelessly to clear the water and beaches of the algae before the bloom drastically damages the local ecosystem.

The measurement used to describe the area covered by the algae was *square miles*. This term comes from the surface measurement made up of multiple squares (150 in the above example), each with a side length of one mile. Another notation used for square miles is mi^2, which is read "miles to the second power" or "miles squared." This second power is an exponent and is a shorthand way of writing repeated multiplication.

A Exponential Expressions

In the expression 2^3, the 2 is called the *base* and the 3 is called the *exponent*. The expression 2^3 is read "2 to the third power" or "2 cubed." The exponent 3 tells us to use the base 2 as a multiplication factor three times.

$2^3 = 2 \cdot 2 \cdot 2$ 2 is used as a factor three times.

We can simplify the expression by multiplication.

$$\begin{aligned} 2^3 &= 2 \cdot 2 \cdot 2 \\ &= 4 \cdot 2 \\ &= 8 \end{aligned}$$

The expression 2^3 is equal to the number 8. We can summarize this discussion with the following definition:

DEFINITION exponent

An ***exponent*** is a number that indicates how many times the base is to be used as a factor. Exponents indicate repeated multiplication.

For example, in the expression 5^2, 5 is the base and 2 is the exponent. The meaning of the expression is

$$5^2 = 5 \cdot 5 \quad \text{5 is used as a factor two times.}$$
$$= 25$$

The expression 5^2 is read "5 to the second power" or "5 squared."

Table 1 shows some more examples:

Table 1

Expression	Explanation
3^2	The base is 3, and the exponent is 2. The expression is read "3 to the second power" or "3 squared."
4^3	The base is 4, and the exponent is 3. The expression is read "4 to the third power" or "4 cubed."
2^4	The base is 2, and the exponent is 4. The expression is read "2 to the fourth power."

As you can see from these examples, a base raised to the second power is also said to be *squared*, and a base raised to the third power is also said to be *cubed*. These are the only two exponents (^{2} and ^{3}) that have special names. All other exponents are referred to only as "fourth powers," "fifth powers," "sixth powers," and so on.

The next examples show how we can simplify expressions involving exponents by using repeated multiplication.

Practice Problems

1. Simplify each of the following by using repeated multiplication.
   a. 5^2
   b. 9^2
   c. 2^3
   d. 1^3
   e. 2^5

EXAMPLE 1 Simplify each of the following by using repeated multiplication.

a. $3^2 = 3 \cdot 3 = 9$

b. $4^2 = 4 \cdot 4 = 16$

c. $3^3 = 3 \cdot 3 \cdot 3 = 9 \cdot 3 = 27$

d. $3^4 = 3 \cdot 3 \cdot 3 \cdot 3 = 9 \cdot 9 = 81$

e. $2^4 = 2 \cdot 2 \cdot 2 \cdot 2 = 4 \cdot 4 = 16$

USING TECHNOLOGY Calculators

Here is how we use a calculator to evaluate exponents, as we did in Example 1e:

Scientific Calculator: 2 [x^y] 4 [=]

Graphing Calculator: 2 [^] 4 [ENT] or 2 [x^y] 4 [ENT] (depending on the calculator)

Finally, we should consider what happens when the numbers 0 and 1 are used as exponents.

DEFINITION exponent 1

Any number with an ***exponent 1*** (raised to the first power) is itself. That is, if we let the letter a represent any number, then

$$a^1 = a$$

Answers

1. **a.** 25 **b.** 81 **c.** 8 **d.** 1 **e.** 32

To take care of the cases when 0 is used as an exponent, we must use the following definition:

DEFINITION exponent 0

Any number other than 0 that has an ***exponent 0*** (raised to the 0 power) is 1. That is, if a represents any nonzero number, then it is always true that

$$a^0 = 1$$

Any number raised to the first power is itself.

EXAMPLE 2

a. $5^1 = 5$

b. $9^1 = 9$

2. Simplify each of the following expressions.

a. 7^1

b. 4^1

Any nonzero number raised to the 0 power is 1.

EXAMPLE 3

a. $4^0 = 1$

b. $8^0 = 1$

3. Simplify each of the following expressions.

a. 9^0

b. 1^0

B Order of Operations

The symbols we use to specify operations, along with the symbols we use for grouping, ( ) and [ ], serve the same purpose in mathematics as punctuation marks in English. They may be called the punctuation marks of mathematics. Consider the following sentence:

Kendra said Tanner is tall.

It can have two different meanings, depending on how we punctuate it:

1. "Kendra," said Tanner, "is tall."
2. Kendra said, "Tanner is tall."

Without the punctuation marks, we don't know which meaning the sentence has. Now consider the following mathematical expression:

$$4 + 5 \cdot 2$$

It can have two different meanings, depending on the order in which we perform the operations. If we add 4 and 5 first, and then multiply 2, we get 18. On the other hand, if we multiply 5 and 2 first, and then add 4, we get 14. There seem to be two different answers. In mathematics, we want to avoid situations in which two different results are possible. Therefore, we follow the rule for order of operations.

Answers

2. a. 7 **b.** 4

3. a. 1 **b.** 1

RULE Order of Operations

When evaluating mathematical expressions, we will perform the operations in the following order:

1. If the expression contains grouping symbols, such as parentheses ( ), brackets [ ], or a fraction bar, then we perform the operations inside the grouping symbols, or above and below the fraction bar, first.
2. Then we evaluate, or simplify, any numbers with exponents.
3. Then we do all multiplications and divisions in order, starting at the left and moving right.
4. Finally, we do all additions and subtractions, from left to right.

According to our rule, the expression $4 + 5 \cdot 2$ would have to be evaluated by multiplying 5 and 2 first, and then adding 4. The correct answer—and the only answer—to this problem is 14.

$$4 + 5 \cdot 2 = 4 + 10 \quad \text{Multiply first.}$$
$$= 14 \quad \text{Then add.}$$

Here are some more examples that illustrate how we apply the rule for order of operations to simplify (or evaluate) expressions.

4. Simplify: $5 \cdot 7 - 3 \cdot 6$.

EXAMPLE 4 Simplify: $4 \cdot 8 - 2 \cdot 6$.

Solution We multiply first and then subtract.

$$4 \cdot 8 - 2 \cdot 6 = 32 - 12 \quad \text{Multiply first.}$$
$$= 20 \quad \text{Then subtract.}$$

5. Simplify: $7 + 3(6 + 4)$.

EXAMPLE 5 Simplify: $5 + 2(7 - 1)$.

Solution According to the rule for the order of operations, we must do what is inside the parentheses first.

$$5 + 2(7 - 1) = 5 + 2(6) \quad \text{Work inside parentheses first.}$$
$$= 5 + 12 \quad \text{Then multiply.}$$
$$= 17 \quad \text{Then add.}$$

6. Simplify: $6 \cdot 3^2 + 64 \div 2^4 - 2$.

EXAMPLE 6 Simplify: $9 \cdot 2^3 + 36 \div 3^2 - 8$.

Solution

$$9 \cdot 2^3 + 36 \div 3^2 - 8 = 9 \cdot 8 + 36 \div 9 - 8 \quad \text{Work exponents first.}$$
$$= 72 + 4 - 8 \quad \text{Then multiply and divide, left to right.}$$
$$= 76 - 8 \quad \text{Add and subtract, left to right.}$$
$$= 68$$

Answers

4. 17
5. 37
6. 56
7. 17
8. 0.1656

EXAMPLE 7 Simplify: $3 + 2[10 - 3(5 - 2)]$.

Solution The brackets, [ ], are used in the same way as parentheses. In a case like this, we move to the innermost grouping symbols first and begin simplifying.

$$\begin{aligned} 3 + 2[10 - 3(5 - 2)] &= 3 + 2[10 - 3(3)] \\ &= 3 + 2[10 - 9] \\ &= 3 + 2[1] \\ &= 3 + 2 \\ &= 5 \end{aligned}$$

7. Simplify: $5 + 3[24 - 5(6 - 2)]$.

We can use the rule for order of operations to simplify expressions involving decimal numbers and addition, subtraction, and multiplication.

EXAMPLE 8 Perform the indicated operations: $0.05(4.2 + 0.03)$.

Solution We begin by adding inside the parentheses.

$$\begin{aligned} 0.05(4.2 + 0.03) &= 0.05(4.23) && \text{Add.} \\ &= 0.2115 && \text{Multiply.} \end{aligned}$$

We could also have used the distributive property first, and the result would be unchanged. We will learn more about the distributive property in the next chapter, but for now, here is what the use of the property would look like:

$$\begin{aligned} 0.05(4.2 + 0.03) &= 0.05(4.2) + 0.05(0.03) && \text{Distributive property} \\ &= 0.210 + 0.0015 && \text{Multiply.} \\ &= 0.2115 && \text{Add.} \end{aligned}$$

8. Perform the indicated operations: $0.03(5.5 + 0.02)$

EXAMPLE 9 Simplify: $4.8 + 12(3.2)^2$.

Solution According to the rule for order of operations, we must first evaluate the number with an exponent, then multiply, and finally add.

$$\begin{aligned} 4.8 + 12(3.2)^2 &= 4.8 + 12(10.24) && (3.2)^2 = 10.24 \\ &= 4.8 + 122.88 && \text{Multiply.} \\ &= 127.68 && \text{Add.} \end{aligned}$$

9. Simplify: $5.7 + 14(2.4)^2$.

Answers

9. 86.34

10. How much will Sally make if she works 50 hours in one week?

EXAMPLE 10 Sally earns \$7.25 for each of the first 36 hours she works in one week and \$10.88 in overtime pay for each additional hour she works in the same week. How much money will she make if she works 42 hours in one week?

Solution The difference between 42 and 36 is 6 hours of overtime pay. The total amount of money she will make is

$$\underbrace{7.25(36)}_{\text{Pay for the first 36 hours}} + \underbrace{10.88(6)}_{\text{Pay for the next 6 hours}} = 261.00 + 65.28$$

$$= 326.28$$

She will make \$326.28 for working 42 hours in one week.

GETTING READY FOR CLASS

After reading through the preceding section, respond in your own words and in complete sentences.

A. What is an exponent?

B. In symbols, show that any number other than zero raised to the 0 power is 1.

C. Why is using the order of operations important to finding the correct solution to a problem?

D. Briefly list the steps of the order of operations.

Answers

10. \$413.32

EXERCISE SET R.6

Vocabulary Review

Choose the correct words to fill in the blanks below.

zero	first	base	product	difference
factor	sum	exponent	quotient	

1. For the exponential expression 5^3, the 5 is called the ______________ and the 3 is called the ______________.
2. An exponent is a number that indicates how many times the base is used as a ______________.
3. Any number raised to the ______________ power is the number itself.
4. Any non-zero number raised to the ______________ power is 1.
5. Twice the ______________ of 6 and 1 is written in symbols as $2(6 + 1)$.
6. 4 added to 5 times the ______________ of 3 and 2 is written in symbols as $5(3 - 2) + 4$.
7. 7 subtracted from the ______________ of 4 and 3 is written in symbols as $4 \cdot 3 - 7$.
8. The ______________ of 8 and 2 plus 9 is written in symbols as $8 \div 2 + 9$.

The following is a list of steps to perform operations when evaluating mathematical expressions. Number the steps in the correct order.

____ Do all multiplications and divisions in order, left to right.
____ Simplify any numbers with exponents.
____ Perform operations inside the grouping symbols, or above and below the fraction bar.
____ Do all additions and subtractions, in order left to right.

Problems

A For each of the following expressions, name the base and the exponent.

1. 4^5 **2.** 5^4 **3.** 3^6 **4.** 6^3 **5.** 8^2 **6.** 2^8

7. 9^1 **8.** 1^9 **9.** 4^0 **10.** 0^4 **11.** 4^8 **12.** 12^2

Use the definition of exponents to simplify each of the following expressions.

13. 6^2 **14.** 7^2 **15.** 2^3 **16.** 2^4 **17.** 1^4 **18.** 5^1

19. 9^0 **20.** 27^0 **21.** 9^2 **22.** 8^2 **23.** 10^1 **24.** 8^1

25. 12^1 **26.** 16^0 **27.** 45^0 **28.** 3^4 **29.** 3^3 **30.** 3^2

B Use the rule for the order of operations to simplify each expression.

31. $16 - 8 + 4$ **32.** $16 - 4 + 8$ **33.** $20 \div 2 \cdot 10$ **34.** $40 \div 4 \cdot 5$

35. $20 - 4 \cdot 4$ **36.** $30 - 10 \cdot 2$ **37.** $3 + 5 \cdot 8$ **38.** $7 + 4 \cdot 9$

39. $3 \cdot 6 - 2$ **40.** $5 \cdot 1 + 6$ **41.** $6 \cdot 2 + 9 \cdot 8$ **42.** $4 \cdot 5 + 9 \cdot 7$

43. $4 \cdot 5 - 3 \cdot 2$ **44.** $5 \cdot 6 - 4 \cdot 3$ **45.** $5^2 + 7^2$ **46.** $4^2 + 9^2$

47. $480 + 12(32)^2$ **48.** $360 + 14(27)^2$ **49.** $3 \cdot 2^3 + 5 \cdot 4^2$ **50.** $4 \cdot 3^2 + 5 \cdot 2^3$

51. $8 \cdot 10^2 - 6 \cdot 4^3$ **52.** $5 \cdot 11^2 - 3 \cdot 2^3$ **53.** $2(3 + 6 \cdot 5)$ **54.** $8(1 + 4 \cdot 2)$

55. $19 + 50 \div 5^2$

56. $9 + 8 \div 2^2$

57. $9 - 2(4 - 3)$

58. $15 - 6(9 - 7)$

59. $4 \cdot 3 + 2(5 - 3)$

60. $6 \cdot 8 + 3(4 - 1)$

61. $4[2(3) + 3(5)]$

62. $3[2(5) + 3(4)]$

63. $(7 - 3)(8 + 2)$

64. $(9 - 5)(9 + 5)$

65. $3(9 - 2) + 4(7 - 2)$

66. $7(4 - 2) - 2(5 - 3)$

67. $18 + 12 \div 4 - 3$

68. $20 + 16 \div 2 - 5$

69. $4(10^2) + 20 \div 4$

70. $3(4^2) + 10 \div 5$

71. $8 \cdot 2^4 + 25 \div 5 - 3^2$

72. $5 \cdot 3^4 + 16 \div 8 - 2^2$

73. $5 + 2[9 - 2(4 - 1)]$

74. $6 + 3[8 - 3(1 + 1)]$

75. $3 + 4[6 + 8(2 - 0)]$

76. $2 + 5[9 + 3(4 - 1)]$

77. $\dfrac{15 + 5(4)}{17 - 12}$

78. $\dfrac{20 + 6(2)}{11 - 7}$

Multiply each of the following. Be sure all answers are written in lowest terms.

79. $\left(\frac{3}{4}\right)^2 \cdot \frac{8}{9}$

80. $\left(\frac{5}{6}\right)^2 \cdot \frac{12}{15}$

81. $\left(\frac{1}{2}\right)^2 \cdot \left(\frac{3}{5}\right)^2$

82. $\left(\frac{3}{8}\right)^2 \cdot \left(\frac{4}{3}\right)^2$

83. $\left(\frac{1}{2}\right)^2 \cdot 8 + \left(\frac{1}{3}\right)^2 \cdot 9$

84. $\left(\frac{2}{3}\right)^2 \cdot 9 + \left(\frac{1}{2}\right)^2 \cdot 4$

Simplify each expression as much as possible.

85. $10 \div \left(\frac{1}{2}\right)^2$

86. $12 \div \left(\frac{1}{4}\right)^2$

87. $\frac{18}{35} \div \left(\frac{6}{7}\right)^2$

88. $\frac{48}{55} \div \left(\frac{8}{11}\right)^2$

89. $\frac{4}{5} \div \frac{1}{10} + 5$

90. $\frac{3}{8} \div \frac{1}{16} + 4$

91. $10 + \frac{11}{12} \div \frac{11}{24}$

92. $15 + \frac{13}{14} \div \frac{13}{42}$

93. $24 \div \left(\frac{2}{5}\right)^2 + 25 \div \left(\frac{5}{6}\right)^2$

94. $18 \div \left(\frac{3}{4}\right)^2 + 49 \div \left(\frac{7}{9}\right)^2$

95. $100 \div \left(\frac{5}{7}\right)^2 + 200 \div \left(\frac{2}{3}\right)^2$

96. $64 \div \left(\frac{8}{11}\right)^2 + 81 \div \left(\frac{9}{11}\right)^2$

Translate each English expression into an equivalent mathematical expression written in symbols. Then simplify.

97. 8 times the sum of 4 and 2

98. 3 times the difference of 6 and 1

99. Twice the sum of 10 and 3

100. 5 times the difference of 12 and 6

101. 4 added to 3 times the sum of 3 and 4

102. 25 added to 4 times the difference of 7 and 5

103. 9 subtracted from the quotient of 20 and 2

104. 7 added to the quotient of 6 and 2

105. The sum of 8 times 5 and 5 times 4

106. The difference of 10 times 5 and 6 times 2

Find the Mistake

Each sentence below contains a mistake. Circle the mistake and write the correct word(s) or numbers(s) on the line provided.

1. The expression 8^5 is read "5 to the 8th power." ____________________
2. In the expression 16^3, the 16 is used as a factor two times. ____________________
3. To simplify $9^2 + 4 - 6 \cdot 7$ using the order of operations, we first find the product of 6 and 7. ____________
4. To work the problem $5 + 6 - 2 + 18$ using the order of operations, we add and subtract right to left. ________
____

Chapter 1 Variables and Equations

Lesson 1.1 Variables

- Use letters to represent variables
- Make a table of values for an expression
- Find a pattern and write a rule
- Plot points from a table of values
- Read values from a graph

Lesson 1.2 Algebraic Expressions

- Write expressions
- Evaluate expressions
- Plot points from a table of values
- Use formulas to write and evaluate expressions

Lesson 1.3 Equations and Graphs

- Graph equations
- Solve equations graphically

Lesson 1.4 Solving Equations

- Solve linear equations with one operation
- Solve problems using formulas
- Write an equation to solve a problem

Lesson 1.5 Order of Operations

- Use the associative laws for addition and multiplication
- Follow the order of operations to simplify expressions

1.1 Variables

What Is a Variable?

A **variable** is a numerical quantity that changes over time or in different situations.

We can show the values of a variable in a table or a graph.

Example 1 Life expectancy is the average age to which people will live. The table at right shows that your life expectancy depends on the year of your birth. Thus, life expectancy is a variable.

a. To what age could people born in 1940 expect to live?

b. In what year did people's life expectancy reach 70 years of age?

Year Born	Life Expectancy
1900	49
1910	51
1920	58
1930	59
1940	63
1950	68
1960	70
1970	71
1980	74
1990	75

Solutions a. People born in 1940 lived to 63 years of age on average.

b. Life expectancy reached 70 years of age in 1960.

Example 2 The graph at right shows the average annual salaries of NFL football players, starting in 1940. The average salary of NFL football players is a variable.

a. Use the graph to estimate the average annual salary of NFL football players in 1975.

b. By how much did salaries increase from 1975 to 1980?

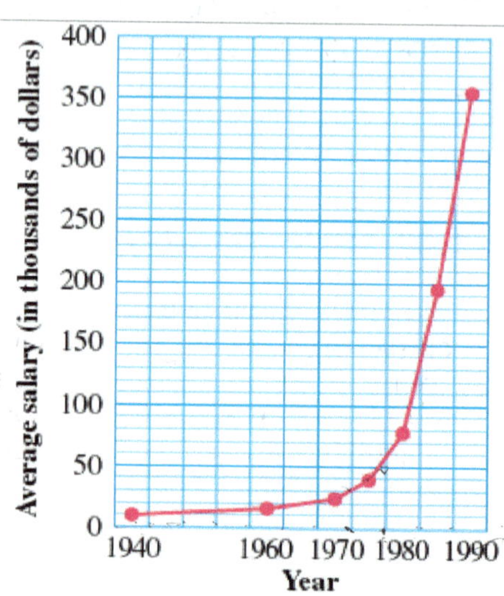

Solutions a. In 1975, the average NFL player's salary was $40,000.

b. In 1980, the average salary was $80,000, so salaries had increased from $40,000 to $80,000, or by

$$\$80,000 - \$40,000 = \$40,000$$

RQ1. A numerical quantity that changes over time or in different situations is called a Variable.

RQ2. We can show the values of a variable in a table or a Graph.

Look Closer: By displaying the values of a variable in a table or a graph, we sometimes see trends or patterns in those values. For example, we see that life expectancy has been increasing slowly over time, and that salaries of NFL football players have been increasing ever more rapidly since 1940.

Using Letters to Represent Variables

Imagine that you are traveling by airplane to another city. The table below shows the altitude of the airplane, in thousands of feet, at ten-minute intervals after take-off. The two quantities shown in the table, time and altitude, are both variables.

Time (min)	0	10	20	30	40	50	60	70	80	90	100	110
Altitude (1000 ft)	0	8	20	21	23	28	29	30	25	13	5	0

We often use a letter as a kind of shorthand to represent a variable. For this example, we use the following letters:

t stands for the **time** elapsed after take-off,
h stands for the plane's **altitude** at that time

We can get a better feel for the variables by plotting them on a graph. In the figure at right, the values of t are displayed on the horizontal scale, or **axis**, and the values of h are shown on the vertical axis. The graph shows how the values of h are related to the values of t.

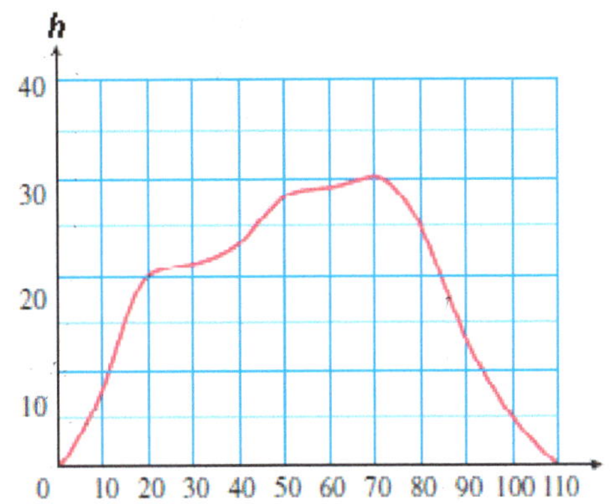

Example 3a. What is the value of h when $t = 85$? What do these values mean in this situation?
b. When is the plane descending?

Solutions a. The graph shows that h is approximately 20 when $t = 85$. This means that at 85 minutes into the flight, the plane's altitude is about 20,000 feet.
b. The plane is descending when its altitude is decreasing. On the graph, h begins to decrease at $t = 70$ and continues decreasing until $t = 110$. Thus, the plane is descending from the 70th minute until the end of the flight at 110 minutes.

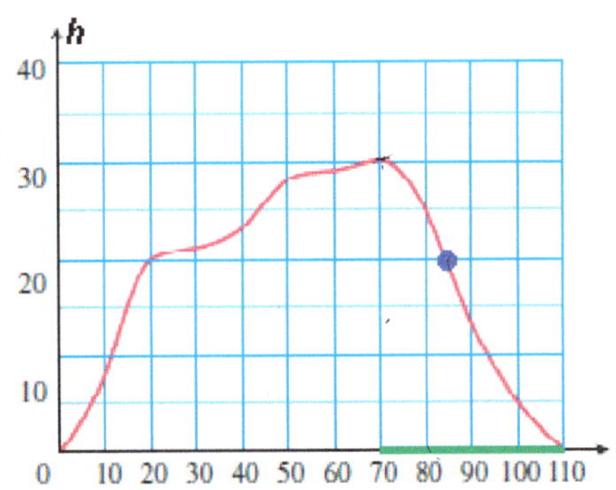

RQ3. We often use ________ as a shorthand to represent variable quantities.
RQ4. Which variable is displayed on the vertical axis in Example 3?

Look Ahead: It would be useful if we could discover a formula or rule to help us *predict* the values of an interesting variable.

Connections Between Variables

Sometimes there is a simple mathematical relationship between the values of two variables.

Example 4 Fernando plans to share an apartment with three other students next year. The table shows his share of the rent for apartments of various prices.

Rent	280	300	340	360	400	460	500
Fernando's share	70	75	85	90	100	?	?

115 125

a. Fill in the blanks in the table. Describe in words how you found the unknown values.

Divide the rent by 4

b. Write a mathematical sentence that gives Fernando's share of the total rent.

Fernando's share = **Rent ÷ 4**

c. The total rent and Fernando's share are variables. Let R stand for the total rent, and let F stand for Fernando's share. Using these letters, write a formula for Fernando's share of the rent.

$$F = R \div 4$$

d. Plot the points from the table and connect them with a smooth curve. Extend your line so that it reaches across the entire grid, not just the points that you plotted.

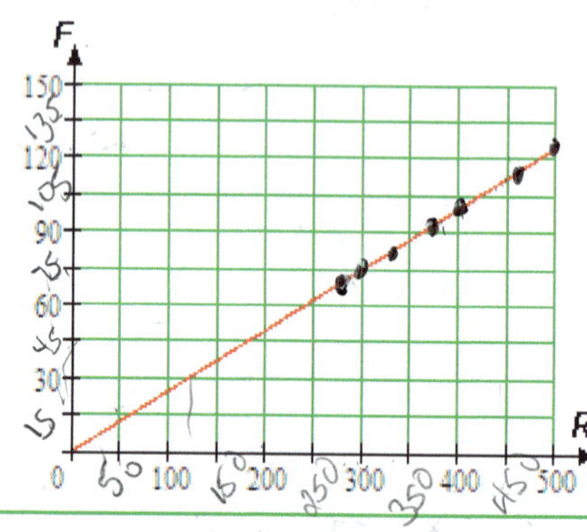

A formula relating two variables is a type of **equation**, and an equation is just a statement that two quantities are equal.

By studying the values in a table, we may be able to find a relationship between the values, and then write an equation relating the two variables.

Example 5 Write an equation that relates the two variables in the table.

x	5	10	15	20	25
y	2	7	12	17	22

Solution The two variables are x and y. Each column of the table shows a related pair of values. Notice that 2 is 3 less than 5, 7 is 3 less than 10, and so on. That is:

$$2 = 5 - 3, \quad 7 = 10 - 3,$$
$$12 = 15 - 3, \quad 17 = 20 - 3,$$

and so on. In each column, we can find y by subtracting 3 from x. Thus, $y = x - 3$.

RQ5. A statement that two quantities are equal is called an equation.

Skills Warm-Up

Choose the correct arithmetic operation (addition, subtraction, multiplication, or division) and explain how to answer the question.

1. An air conditioner keeps the inside temperature 16° cooler than the outside temperature. If the outside temperature is 90°, how can you find the inside temperature?

2. Tom's recipe for punch calls for three times as much fruit juice as soda. If he has half a gallon of soda, how can he find the amount of fruit juice he needs?

3. A clothesline should be 2 feet longer than the distance between the supporting poles so that it can be tied at each end. If the poles are 20 feet apart, how can you find the length of the clothesline?

4. The weight of a bridge is supported equally by 8 pillars. If the bridge weighs one million tons, how can you find the weight each pillar must support?

5. Katrin has 4 hours to complete both her math and geography homework. If her math assignment takes $2\frac{1}{2}$ hours, how can she calculate how long she has for geography?

6. The cost of leasing a compact car is 63% the cost of a luxury car. If the lease on the luxury car is $500 per month, how can you find the cost of leasing the compact car?

Answers: **1.** Subtract 16 from the outside temperature. **2.** Multiply the amount of soda by 3. **3.** Add 2 feet to the distance between the poles. **4.** Divide the weight of the bridge by 8. **5.** Subtract the time for math from 4 hours. **6.** Multiply the cost of the luxury car by 0.63.

Homework 1.1

Skills Practice

For Problems 1-4,

a. Find the pattern and fill in the table.

b. Write an equation for the second variable in terms of the first variable.

1.

m	g
2	5
3	6
5	8
10	13
12	
16	
18	
m	

2.

t	w
0	20
2	18
4	16
5	15
6	
10	
12	
t	

3.

b	x
0	0
2	1
4	2
5	2.5
6	
8	
9	
b	

4.

z	3	6	8	12	15	18	20	z
r	2	4	$\frac{16}{3}$	8				

Hint: What fraction can you multiply z by to get r?

In Problems 5-6, make your own table. Choose values for the first variable, and use the rule to find the values of the second variable.

5. $W = 1.2 \times n$

n					
W					

6. $M = \frac{3}{2} \times x$

x					
M					

For Problems 7-8, fill in the tables. What do you notice? Explain.

7a. $y = \frac{x}{8}$

x	y
4	
8	
10	
16	

b. $y = \frac{1}{8} \times x$

x	y
4	
8	
10	
16	

c. $y = 0.125 \times x$

x	y
4	
8	
10	
16	

8a. $y = \frac{3 \times x}{5}$

x	y
5	
10	
12	
1	

b. $y = \frac{3}{5} \times x$

x	y
5	
10	
12	
1	

c. $y = 0.6 \times x$

x	y
5	
10	
12	
1	

Applications

9. The temperature in Sunnyvale is usually 15° hotter than it is in Ridgecrest, which is at a higher elevation. Fill in the table.

Temperature in Ridgecrest	70	75	82	86	90	R
Calculation						
Temperature in Sunnyvale						

a. Explain in words how to find the temperature in Sunnyvale.

b. Write your explanation as a mathematical sentence:

Temp in Sunnyvale =

c. Let R stand for the temperature in Ridgecrest and S for the temperature in Sunnyvale. Write an equation for S in terms of R.

d. Plot the points from the table and connect them with a smooth curve.

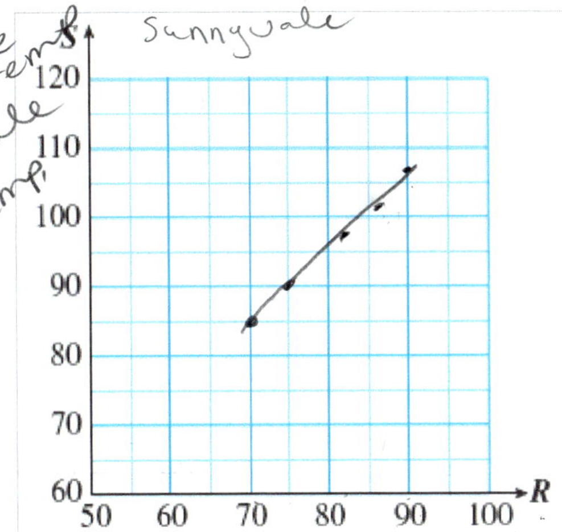

10. Jerome is driving from his home in White Falls to his parents' home in Castle Heights, 200 miles away. Fill in the table.

Miles driven	40	60	90	120	140	170	d
Calculation							
Miles remaining							

a. Explain in words how to find the number of miles Jerome has left to drive.

b. Write your explanation as a mathematical sentence:

Miles remaining =

c. Let d stand for the number of miles Jerome has driven and r for the number of miles that remain. Write an equation for r in terms of d.

d. Plot the points from the table and connect them with a smooth curve.

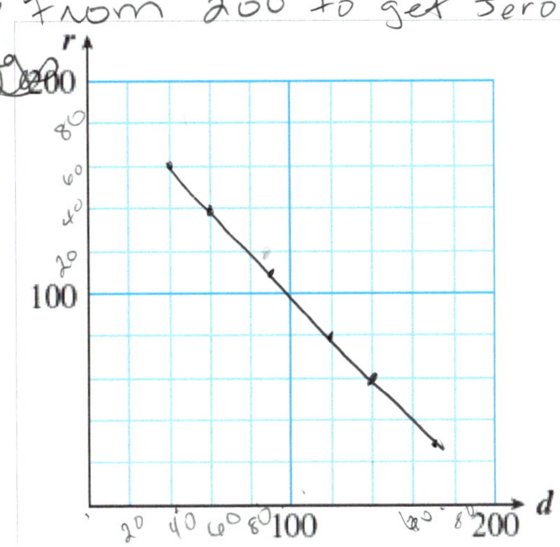

11. Milton goes to a restaurant with two friends, and they agree to split the bill equally. Fill in the table.

Total bill	15	30	45	60	75	81	b
Calculation							
Milton's share							

a. Explain in words how to find Milton's share of the bill.

b. Write your explanation as a mathematical sentence:

Milton's share =

c. Let b stand for the bill and s for Milton's share. Write an equation for s in terms of b.

d. Plot the points from the table and connect them with a smooth curve.

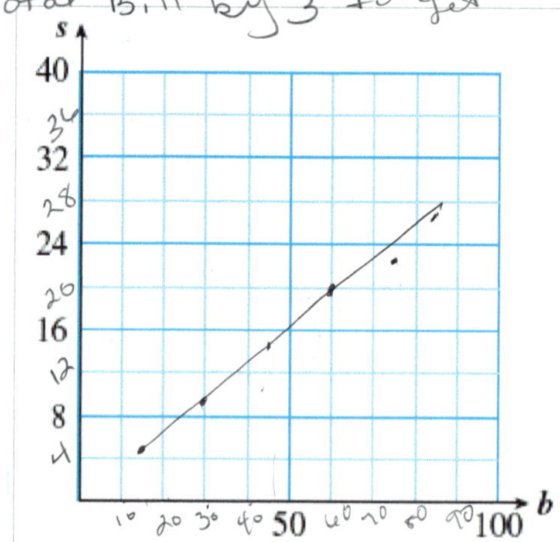

12. Nutrition experts tell us that no more than 30% of the calories we consume should come from fat. Fill in the table with the fat calories allowed daily for various calorie levels.

Total calories	1000	1500	2000	2500	3000	C
Calculation						
Fat calories						

a. Explain in words how to find the number of fat calories allowed.

b. Write your explanation as a mathematical sentence:

Fat calories =

c. Let C stand for the total number of calories and F for the number of fat calories. Write an equation for F in terms of C.

d. Plot the points from the table and connect them with a smooth curve.

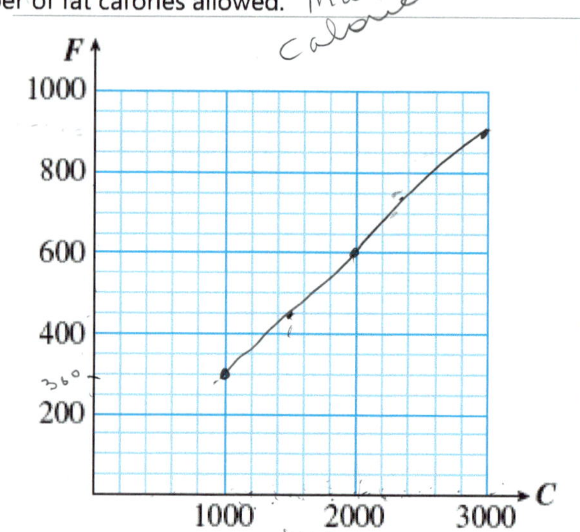

Use the graphs to answer the questions in Problems 13 and 14. You may have to estimate some of your answers.

13. Suppose you invest $2000 in a retirement account that pays 8% interest compounded continuously. The amount of money in the account each year is shown by the graph at left below.

a. What variable is displayed on the horizontal axis? The vertical axis?
b. How much money will be in the account 5 years from now?
c. When will the amount of money in the account exceed $6000?
d. How much will the account grow between the second and third years?
e. How much will the account grow between the twelfth and thirteenth years?

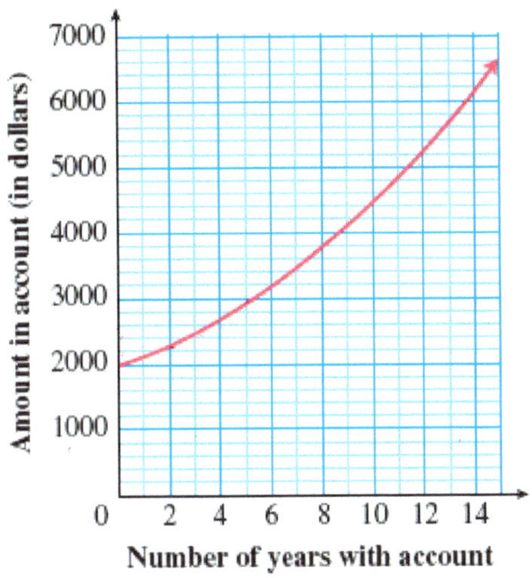

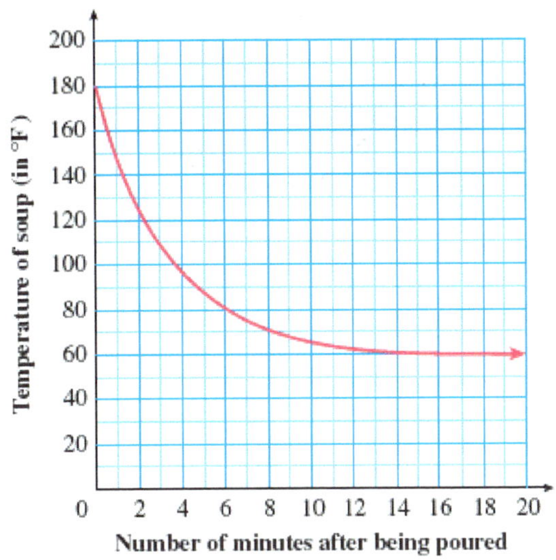

14. Wendy brings a Thermos of soup for her lunch on a hike in the mountains. When she pours the soup into a bowl, it begins to cool off. The graph above right shows the temperature of the soup after it is served.

a. What variable is displayed on the horizontal axis? The vertical axis?
b. How long does it take the soup to cool below 100°?
c. What is the temperature of the soup after 8 minutes?
d. How many degrees does the soup cool in the first 2 minutes?
e. After a while, the temperature of the soup levels off at the same as the outside temperature. What was the outside temperature that day?

15. A cantaloupe is dropped from a tall building. Its speed increases until it hits the ground. Which of the four graphs below best represents the speed of the cantaloupe? Explain why your choice is the best one.

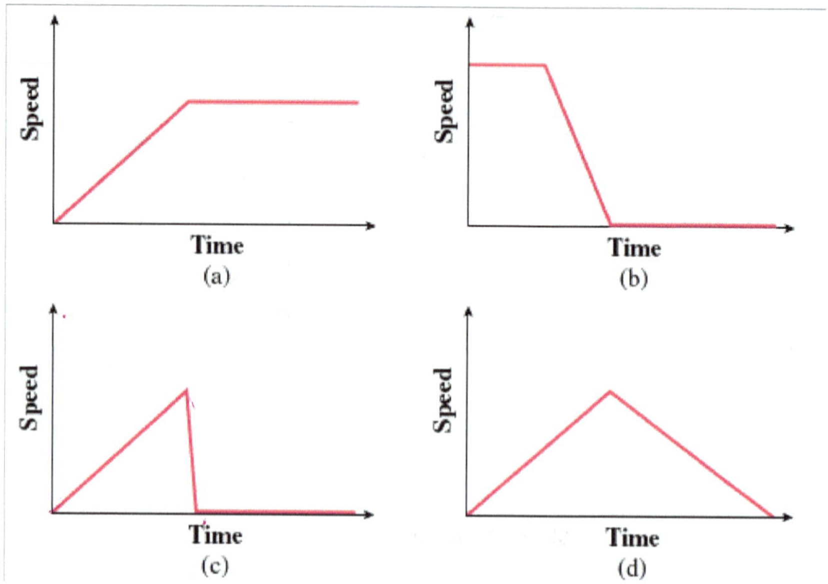

16. a. Match each of the following stories with the appropriate graph.

I. Delbert walks directly from home to school, then stays there.
II. Delbert walks towards school, but stops at coffee shop for a cappuccino, then he goes on to school and stays there.
III. Delbert walks toward school, but decides to return home, where he stays.

b. Write your own story for the remaining graph.

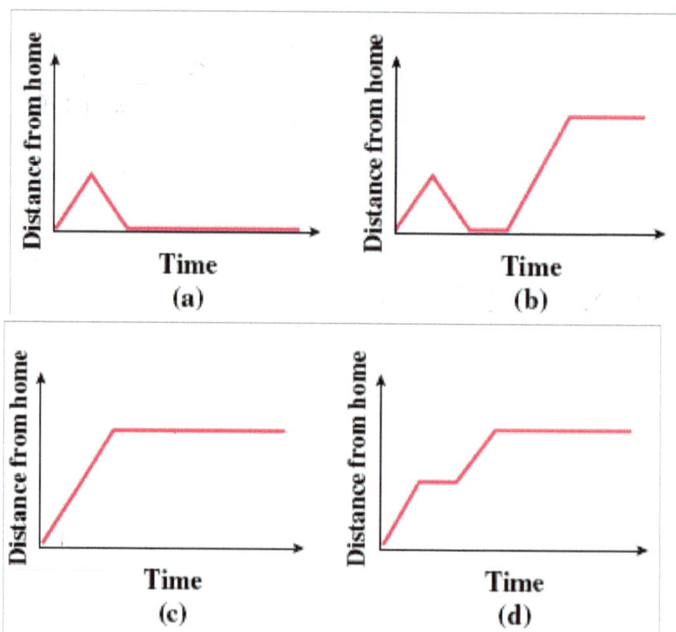

17. Plot each pair of values on the grid provided. Connect the points with a smooth curve. Which graph appears to be a straight line?

a.

t	0	1	2	5	6	8
v	16	14	12	6	4	0

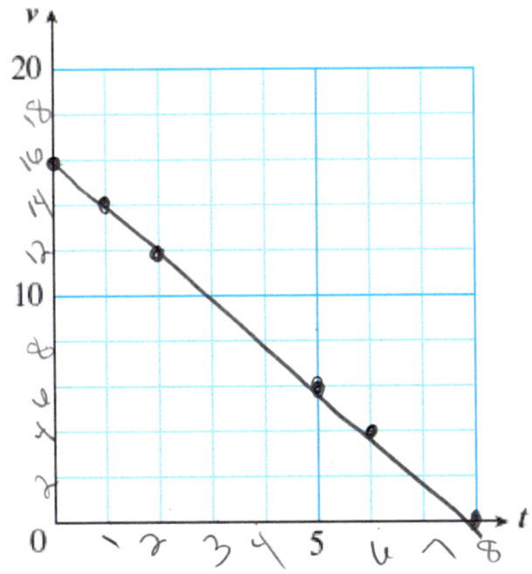

b.

w	0	1	2	4	6	10
Q	12	8	6	4	3	2

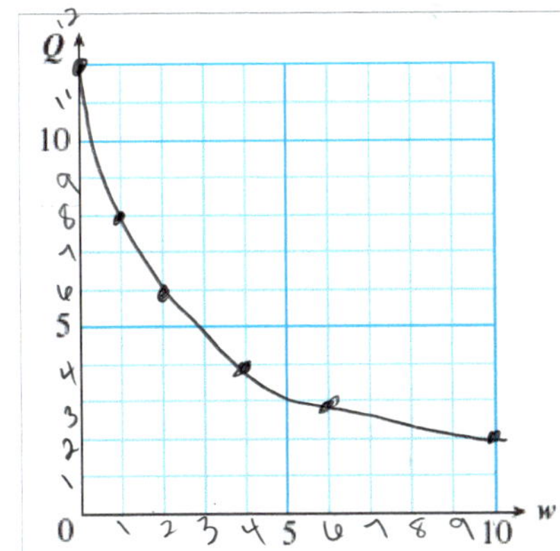

18. a. Read the graph at right to fill in the table.

x	0	10	30	40	60	70
y	70	60	40	30	10	0

b. Find a formula expressing y in terms of x.

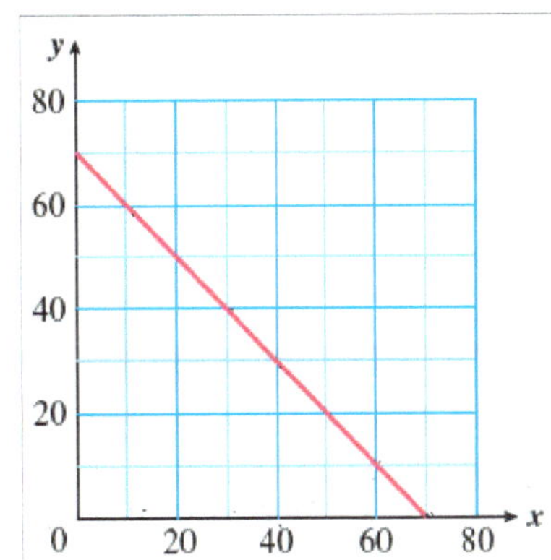

For Problems 19-20,

a. Write an equation for the second variable in terms of the first variable.

b. Fill in the table. Plot the points and connect them with a smooth curve.

19.

x	0	0.2	0.3	0.5	0.6	0.7
y	2	1.8	1.7	1.5	1.4	1.3

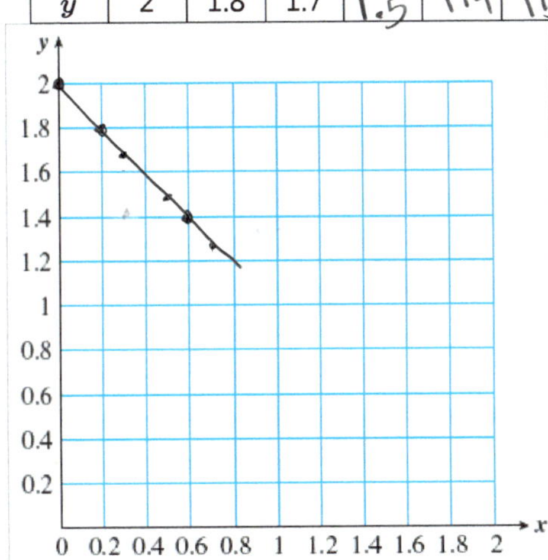

20.

x	1	2	3	4	6	8
y	120	60	40	30	20	15

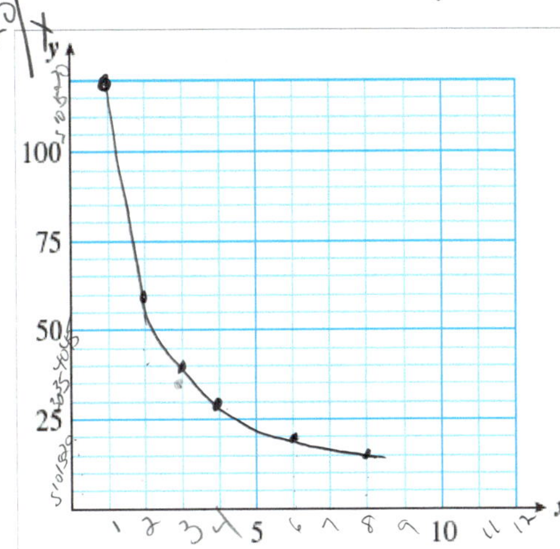

1.2 Algebraic Expressions

Writing an Algebraic Expression

> An **algebraic expression**, or simply an **expression**, is any meaningful combination of numbers, variables, and operation symbols.

For example,

$$4 \times g, \quad 3 \times c + p, \quad \text{and} \quad \frac{n-7}{w}$$

are algebraic expressions. An important part of algebra involves translating word phrases into algebraic expressions.

Example 1 Write an algebraic expression for each quantity.

a. 30% of the money invested in stocks

b. The cost of the dinner split three ways

Solutions a. The amount invested is unknown, so we choose a variable to represent it.

Amount invested in stocks: s

Next, we identify the operation described: "30% of" means 0.30 times:

The expression is $0.30s$

b. The cost of the dinner is unknown, so we choose a variable to represent it.

Cost of dinner: C

Next, we identify the operation described: "Split" means divided:

The expression is $\dfrac{C}{3}$

RQ1. A meaningful combination of numbers, variables and operation symbols is called an Expression

Sums

When we add two numbers a and b, the result is called the **sum** of a and b. We call the numbers a and b the **terms** of the sum.

terms: $\underbrace{2+3}_{\text{sum}}$ $\qquad$ terms: $\underbrace{x+5}_{\text{sum}}$

Example 2 Write sums to represent the following phrases.

a. Six more than x

b. Fifteen greater than r

Solutions a. $6+x$ or $x+6$

b. $15+r$ or $r+15$

The terms can be added in either order. We call this fact the **commutative law** for addition.

> **Commutative Law for Addition**
> If a and b are any numbers, then
> $$a + b = b + a$$

Products

When we multiply two numbers a and b, the result is called the **product** of a and b. The numbers a and b are the **factors** of the product.

Look Closer: In arithmetic we use the symbol $\times$ to denote multiplication. However, in algebra, $\times$ may be confused with the variable x. So, instead, we use a dot or parentheses for multiplication, like this:

$$\underbrace{2 \cdot 3}_{\text{product}} \qquad \underbrace{2(3)}_{\text{product}} \qquad \underbrace{(2)(3)}_{\text{product}}$$

(In each case, the numbers 2 and 3 are labelled "factors".)

If one of the factors is a variable, we can just write the factors side by side. For example,

ab means the product of a times b
$3x$ means the product of 3 times x

Example 3 Write products to represent the following phrases.

a. Ten times z **b.** Two-thirds of y **c.** The product of x and y

Solutions a. $10z$ **b.** $\frac{2}{3}y$ **c.** xy or yx

Look Closer: In Example 3a, $10z = 10 \cdot z = z \cdot 10$, because two numbers can be multiplied in either order to give the same answer. This is the **commutative law** for multiplication.

> **Commutative Law for Multiplication**
> If a and b are any numbers, then
> $$a \cdot b = b \cdot a$$

In algebra, we usually write products with the numeral first. Thus, we write $10z$ for "10 times z".

RQ2. When we add two numbers, the result is called the Sum and the two numbers are called the Terms.

RQ3. When we multiply two numbers, the result is called the product and the two numbers are called the factors.

Differences

When we subtract b from a the result is called the **difference** of a and b. As with addition, a and b are called **terms**.

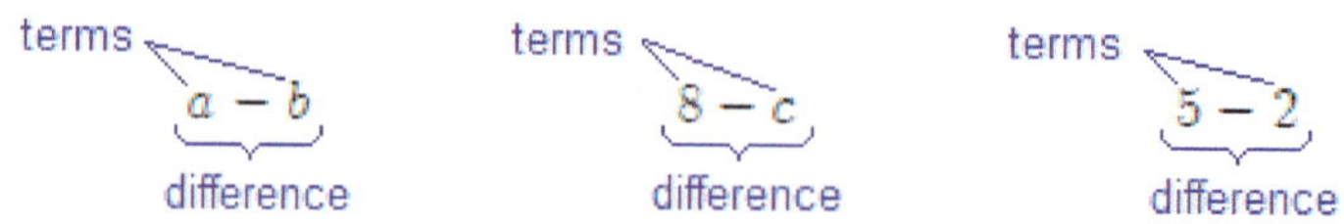

Caution! The difference $5-2$ is *not* the same as $2-5$. When we translate "a subtracted from b," the order of the terms is important.

$b-a$ means a subtracted from b
$x-7$ means 7 subtracted from x

The operation of subtraction is *not* commutative.

Quotients

When we divide a by b the result is called the **quotient** of a and b. We call a the **dividend** and b the **divisor**. In algebra we indicate division by using the division symbol, ÷, or a fraction bar.

Look Closer: The operation of division is **not** commutative. The order of the numbers in a quotient makes a difference. For example, $12 \div 3 = 4$ but $3 \div 12 = \frac{1}{4}$.

Example 4 Write an algebraic expression for each phrase.

a. z subtracted from 13 **b.** 25 divided by R

Solutions a. $13-z$ **b.** $\dfrac{25}{R}$

RQ4. Which operations obey the **commutative laws**? What do these laws say?
RQ5. When we subtract two numbers, the result is called the difference
RQ6. When we divide two numbers, the result is called the quotient

Evaluating an Algebraic Expression

An algebraic expression is a pattern or rule for different versions of the situation it describes. We can replace the variable by specific numbers to fit a particular situation.

Substituting a specific value into an expression and calculating the result is called **evaluating** the expression.

Example 5 Fernando has three roommates, and his share of the rent is $\frac{R}{4}$, where R is a variable representing the total rent on the apartment. If Fernando and his roommates find an apartment whose rent is \$540, we can find Fernando's share of the rent by replacing R by 540.

$$\frac{R}{4} = 540 \div 4 = 135$$

Fernando's share of the rent is \$135.

In Example 5, we evaluated the expression $\frac{R}{4}$ for $R = 540$.

Some Useful Algebraic Formulas

Here are some examples of common formulas. Formulas may involve several different variables.

1. To find the **distance** traveled by an object moving at a constant speed or rate for a specified time, we multiply the rate by the time.

distance = rate × time $\quad d = rt$

Example How far will a train moving at 80 miles per hour travel in three hours?

Solution Evaluate the formula with $r = 80$ and $t = 3$ to find

$$\begin{aligned} d &= rt \\ &= (80)(3) = 240 \end{aligned}$$

The train will travel 240 miles.

2. To find the **profit** earned by a company we subtract the costs from the revenue.

profit = revenue − cost $\quad P = R - C$

Example The owner of a sandwich shop spent \$800 last week for labor and supplies. She received \$1150 in revenue. What was her profit?

Solution Evaluate the formula with $R = 1150$ and $C = 800$ to find

$$\begin{aligned} P &= R - C \\ &= 1150 - 800 = 350 \end{aligned}$$

The owner's profit was \$350.

3. To find the **interest** earned on an investment we multiply the amount invested (the principal) by the interest rate and the length of the investment.

interest = principal × interest rate × time $\quad I = Prt$

Example How much interest will be earned on $800 invested for 3 years in a savings account that pays $5\frac{1}{2}\%$ interest?

Solution Evaluate the formula with $P = 800$, $r = 0.055$, and $t = 3$ to find

$$\begin{aligned} I &= Prt \\ &= 800 \times 0.055 \times 3 = 132 \end{aligned}$$

The account will earn $132 in interest.

4. To find a percentage or **part** of a given amount, we multiply the whole amount by the percentage rate.

$$\textbf{part} = \textbf{percentage rate} \times \textbf{whole} \qquad P = rW$$

Example How much ginger ale do you need to make 60 gallons of a fruit punch that is 20% ginger ale?

Solution Evaluate the formula with $r = 0.20$ and $W = 60$ to find

$$\begin{aligned} P &= rW \\ &= 0.20 \times 60 = 12 \end{aligned}$$

You would need 12 gallons of ginger ale.

5. To find the **average** of a collection of scores, we divide the sum of the scores by the number of scores.

$$\textbf{average score} = \frac{\textbf{sum of scores}}{\textbf{number of scores}} \qquad A = \frac{S}{n}$$

Example Bert's quiz scores in chemistry are 15, 16, 18, 18, and 12. What is his average score?

Solution Evaluate the formula with $S = 15 + 16 + 18 + 18 + 12$ and $n = 5$ to find

$$A = \frac{S}{n} = \frac{15 + 16 + 18 + 18 + 12}{5} = 15.8$$

Bert's average score is 15.8.

RQ7. What does it mean to evaluate an expression?

RQ8. How do we change a percent to a decimal fraction?

Look Ahead: You will also need to know the formulas for the area and perimeter of a rectangle. See the Skills Warm-Up to review these formulas.

Skills Warm-Up

Recall the formulas for the **area** and **perimeter** of a rectangle:

$$A = lw \qquad \text{and} \qquad P = 2l + 2w$$

1. Delbert's living room is 20 feet long and 12 feet wide. How much oak baseboard does he need to border the floor? (Don't worry about doorways.)
2. How much wood parquet tiling must he buy to cover the floor?

Which of the following phrases describe a perimeter, and which describe an area?

3. The distance you jog around the shore of a small lake
4. The amount of Astroturf needed for the new football field
5. The amount of grated cheese needed to cover a pizza
6. The number of tulip bulbs needed to border a patio

Find the perimeter and area of each figure. All angles are right angles.

7. 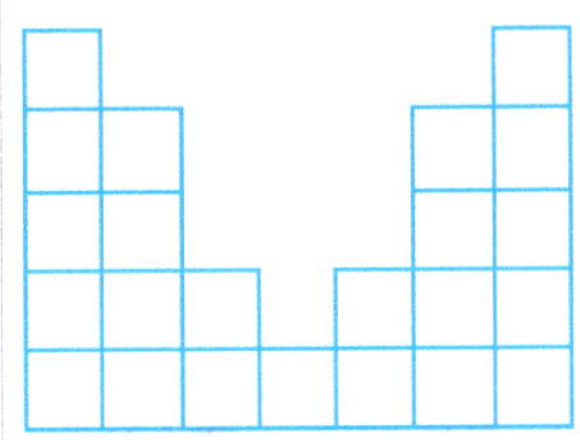

All squares are 1 centimeter on a side.

8. 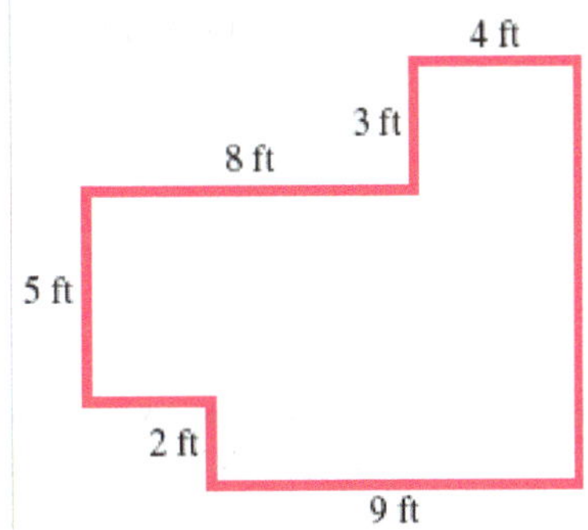

Answers: **1.** 64 ft **2.** 240 sq ft **3.** perimeter **4.** area **5.** area **6.** perimeter **7.** 32 cm; 23 sq cm **8.** 44 ft; 90 sq ft

Homework 1.2

Skills Practice

For Problems 1-10, write the word phrase as an algebraic expression.

1. Product of 4 and y
2. Twice b
3. 115% of g
4. t decreased by 5
5. The quotient of 7 and w
6. 4 divided into B
7. 20 more than T
8. 16 reduced by p
9. The ratio of 15 to M
10. The difference of R and 3.5

For Problems 11-14, choose the correct algebraic expression.

$n+6$	$n-6$	$6-n$
$6n$	$\frac{n}{6}$	$\frac{6}{n}$

11. Rashad is 6 years younger than Shelley. If Shelley is n years old, how old is Rashad?
12. Each package of sodas contains 6 cans. If Antoine bought n packages of sodas, how many cans did he buy?
13. Lizette and Patrick together own 6 cats. If Lizette owns n cats, how many cats does Patrick own?
14. Mitra divided 6 cupcakes among n children. How much of a cupcake did each child get?

For Problems 15-17, write an algebraic expression for the area or perimeter of the figure. Include units in your answers. The dimensions are given in inches.

15. 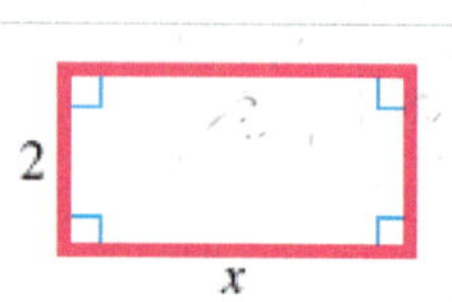

area =

16. 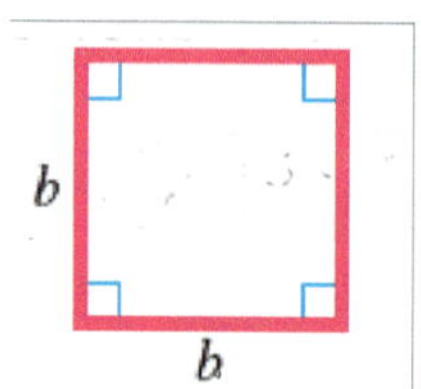

perimeter =

17. 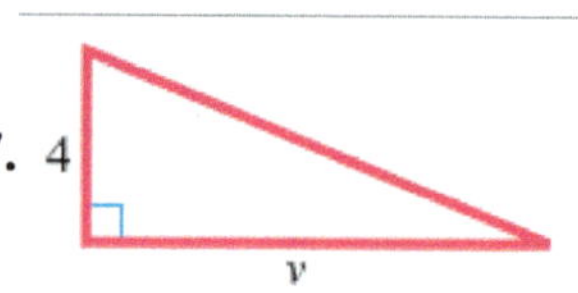

area =

18. a. Which of these expressions says "one-sixth of x" ?

$\frac{x}{6}$, $\frac{6}{x}$, $\frac{1}{6}x$, $\frac{1}{6x}$

b. Which of these expressions says "6% of x" ?

$0.6x$, $0.06x$, $\frac{6}{100}x$, $\frac{1}{6}x$

19. Francine is saving up to buy a car, and deposits $\frac{1}{5}$ of her spending money each month into a savings account.

a. If m stands for Francine's spending money, write an expression for the amount she saves.

b. Evaluate the expression to complete the table.

Spending money (dollars)	m	20	25	60	80
Amount saved (dollars)					

20. Delbert wants to enlarge his class photograph. Its height is $\frac{1}{4}$ of its width. The enlarged photo should have the same shape as the original.

a. If W stands for the width of the enlargement, write an expression for its height.

b. Evaluate the expression to complete the table.

Width of photo (inches)	W	6	10	24	30
Height of photo (inches)					

21. If the governor vetoes a bill passed by the State Assembly, $\frac{2}{3}$ of the members present must vote for the bill in order to overturn the veto.

a. If p stand for the number of Assembly members present, write an expression for the number of votes needed to overturn a veto.

b. Evaluate the expression to complete the table.

Members present	p	90	96	120	129
Votes needed					

22. Marla's investment club buys some stock. She will get $\frac{3}{5}$ of the dividends.

a. If D stands for the stock dividends, write an expression for Marla's share.

b. Evaluate the expression to complete the table.

Stock dividend	D	40	85	115	170
Marla's share					

Applications

For Problems 23-30, choose a variable for the unknown quantity and translate the phrase into an algebraic expression. Use the three steps you learned in the Lesson.

23. The product of the ticket price and \$15

24. Three times the cost of a light bulb

25. Three-fifths of the savings account balance

26. The price of the pizza divided by 6

27. The weight of the copper in ounces divided by 16

28. 9% of the school buses

29. \$16 less than the cost of the vaccine

30. The total cost of 32 identical computers

For Problems 31-32, write algebraic expressions in terms of x.

31. **a.** Daniel and Lara together made \$480. If Daniel made x dollars, how much did Lara make?

b. Alix spent \$500 on tuition and books. If she spent x dollars on books, how much was her tuition?

c. Thirty children signed up for summer camp. If x boys signed up, how many girls signed up?

32. **a.** Rona spent \$15 less than her sister on shoes. If Rona's sister spent x dollars, how much did Rona spend?

b. Phoenix had 12 fewer rain days than Boston last year. If Boston had x rain days, how many rain days did Phoenix have?

c. Jared scored 18 points lower on his second test than he scored on his first test. If he scored x points on the first test, what was his score on the second test?

For Problems 33-36, name the variable and write an algebraic expression.

33. Eggnog is 70% milk. Write an expression for the amount of milk in a container of eggnog.

34. Errol has saved \$1200 for his vacation this year. Write an expression for the average amount he can spend on each day of his vacation.

35. Garth received 432 fewer votes than his opponent in the election. Write an expression for the number of votes Garth received.

36. The cost of the conference was \$2000 over budget. Write an expression for the cost of the conference.

For Problems 37-40, use the formulas in this Lesson.

37. a. Write an equation for the distance d traveled in t hours by a small plane flying at 180 miles per hour.

b. How far will the plane fly in 2 hours? In $3\frac{1}{2}$ hours? In half a day?

38. a. A certain pesticide contains 0.02% by volume of a dangerous chemical. Write an equation for the amount of chemical C that enters the environment in terms of the number of gallons g of pesticide used.

b. How much of the chemical enters the environment if 400 gallons of the pesticide are used? 5000 gallons? 50,000 gallons?

39. a. BioTech budgets 8.5% of its revenue for research. Write an equation for the research budget B in terms of BioTech's revenue R.

b. What is the research budget if BioTech's revenue is \$100,000? \$500,000? \$2,000,000?

40. a. Hugo's Auto Shop paid \$4000 in expenses this month. Write an equation for their profit P in terms of their revenue R.

b. What was their profit if their revenue was \$10,000? \$6500? \$2500?

1.3 Equations and Graphs

Anatomy of a Graph

We use equations to express the relationship between two variabes. A graph is a way of visualizing an equation.

> A graph has two **axes**, horizontal and vertical, and the values for the variables are displayed along the axes. The first, or **input**, variable is displayed on the horizontal axis. The second, or **output**, variable is displayed on the vertical axis.

The graph itself shows how the two variables are related.

Example 1 The graph at right shows the relationship between Delbert's age, D, and Francine's age, F. Write an equation for Francine's age in terms of Delbert's age.

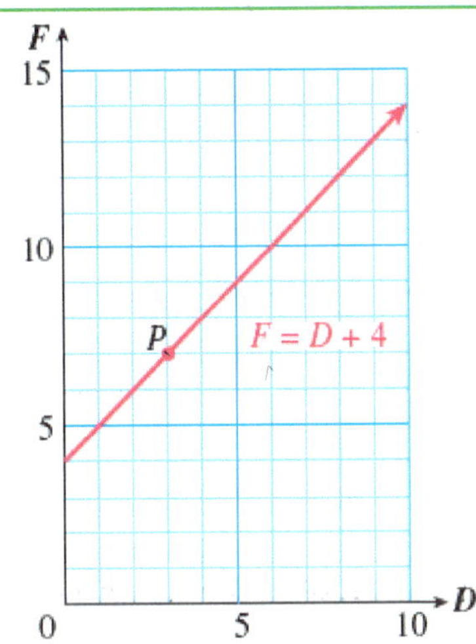

Solution In this graph, D is the input variable and F is the output variable. We make a table of values by reading points on the graph.

D	0	2	3	4	6	9
F	4	6	7	8	10	13

From the table, we see that the value of F is always 4 more than the value of D. Thus, Francine is exactly four years older than Delbert, or $F = D + 4$.

RQ1. A __graph__ is a way of visualizing an algebraic equation.

RQ2. The values of the variables are displayed on the __axes__.

RQ3. The input variable is located on the __horizontal__ axis.

Each point on a graph has two **coordinates**, which designate the position of the point. For example, the point labeled P in Example 1 has horizontal coordinate 3 and vertical coordinate 7.

> In Example 1 above, we write the coordinates of point P inside parentheses as an **ordered pair**: $(3,7)$. The order of the coordinates makes a difference. *We always list the horizontal coordinate first, then the vertical coordinate.*

RQ4. The position of a point on the graph is given by its __coordinates__

RQ5. The notation (x, y) is called __ordered pair__

RQ6. In an ordered pair, we always list the __horizontal__ coordinate first.

Look Closer: How does the graph illustrate the equation $F = D + 4$? The coordinates of each point on the graph are values for D and F that make the equation true. The coordinates of the point P, namely $D = 3$ and $F = 7$, represent the fact that when Delbert was 3 years old, Francine was 7 years old. If we substitute these values into our equation we get

$$F = D + 4$$
$$7 = 3 + 4$$

which is true.

An ordered pair that makes an equation true is called a **solution** of the equation. Each point on the graph represents a solution of the equation.

Exercise a. Locate on the graph each of the ordered pairs listed in the table above, and make a dot there. Label each point with its coordinates
b. By substituting its coordinates into the equation, verify that each point you labeled in part (a) represents a solution of the equation $F = D + 4$.

RQ7. An ordered pair whose coordinates make the equation true is called a Solution of the equation.

Graphing an Equation

The simplest way to draw a graph for an equation is to make a table of values and plot the points.

Example 2 Graph the equation $y = 8 - x$.

Solution In this equation, x is the input variable and y is the output variable. We choose several values for x and use the equation to find the corresponding values for y.

x	Calculation	y	(x, y)
0	$y = 8 - \mathbf{0} = 8$	8	$(0, 8)$
2	$y = 8 - \mathbf{2} = 6$	6	$(2, 6)$
5	$y = 8 - \mathbf{5} = 3$	3	$(5, 3)$
8	$y = 8 - \mathbf{8} = 0$	0	$(8, 0)$

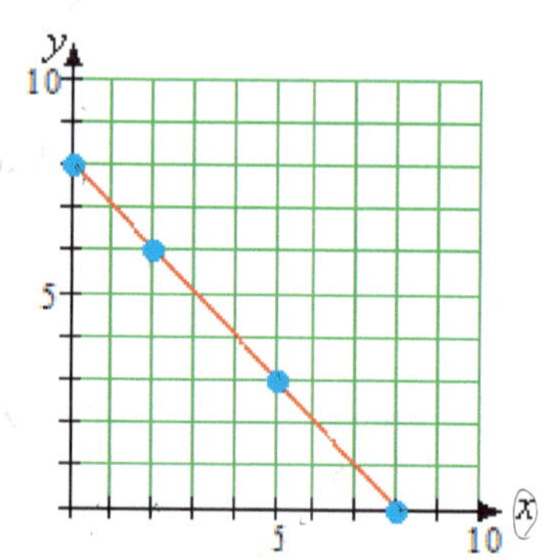

Each ordered pair (x, y) represents a point on the graph of the equation. We plot the points on the grid and connect them with a smooth curve, as shown at right.

Choosing Scales for the Axes

If the variables in an equation have very large (or very small) values, we must choose scales for the axes that fit these values.

Example 3 Graph the equation $y = 250 + x$.

Solution To graph this equation, we choose multiples of 50 for the x-values.

x	Calculation	y	(x, y)
0	$y = 250 + \mathbf{0} = 250$	250	$(0, 250)$
50	$y = 250 + \mathbf{50} = 300$	300	$(50, 300)$
100	$y = 250 + \mathbf{100} = 350$	350	$(100, 350)$
200	$y = 250 + \mathbf{200} = 450$	450	$(200, 450)$

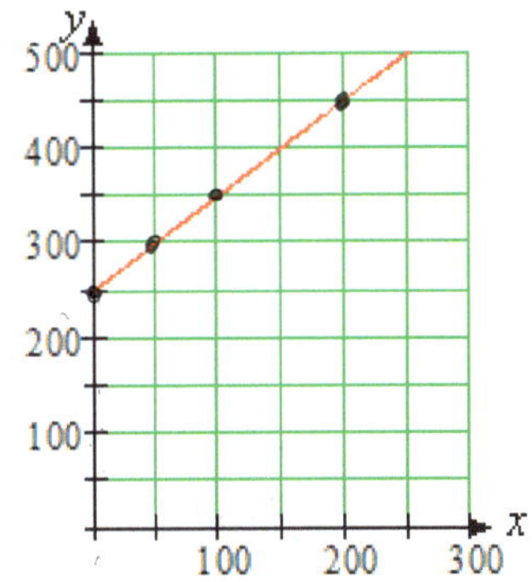

The largest y-value in the table is 450, so we scale the axis to a little larger than 450, say, 500. We plot the ordered pairs to obtain the graph shown at right.

Look Closer: How do we know what scales to use on the axes? For most graphs, it is best to have between ten and twenty tick marks on each axis, or the graph will be hard to read. We choose intervals of convenient size for the particular problem, such as 5, 10, 25, 100, or 1000. It is not necessary to use the same scale on both axes.

Caution! It is important that the scales on the axes be evenly spaced. Each tick mark must represent the same interval. The scales on the graph at right are incorrectly labeled.

Incorrect! →

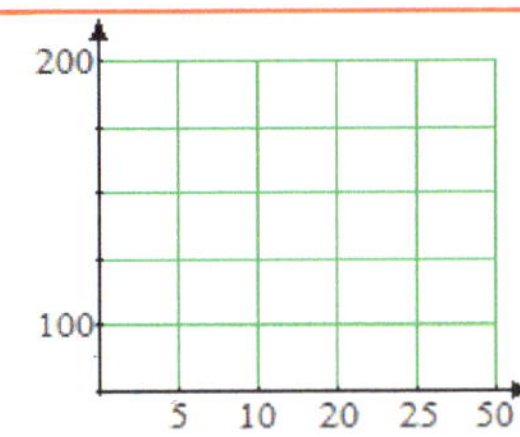

Solving Equations with Graphs

In Example 1 we showed a graph of the equation

$$F = D + 4$$

which gives Francine's age, F, in terms of Delbert's age, D. We can use the graph to answer two types of questions about the equation $F = D + 4$:

1. Given a value of D, find the corresponding value of F.
2. Given a value of F, find the corresponding value of D.

The first of these tasks is another way of **evaluating** the algebraic expression $D + 4$, and the second task is another way of **solving** an equation.

Example 4a. Use the graph of the equation $F = D + 4$ to evaluate the expression $D + 4$ for $D = 7$. Verify your answer algebraically.
b. Use the graph of the equation $F = D + 4$ to solve the equation $13 = D + 4$. Verify your answer algebraically.

Solutions a. We locate $D = 7$ on the horizontal axis, as shown at left below. Then we move vertically to point A on the graph with D-coordinate 7. Finally, we move horizontally from point A to the vertical axis to find the F-coordinate. The coordinates of point A are $(7, 11)$, which tells us that when $D = 7$, $F = 11$. To verify the answer algebraically, we **evaluate** the expression for $D = 7$: we substitute 7 for D into the equation:

$$F = D + 4 = 7 + 4 = 11$$

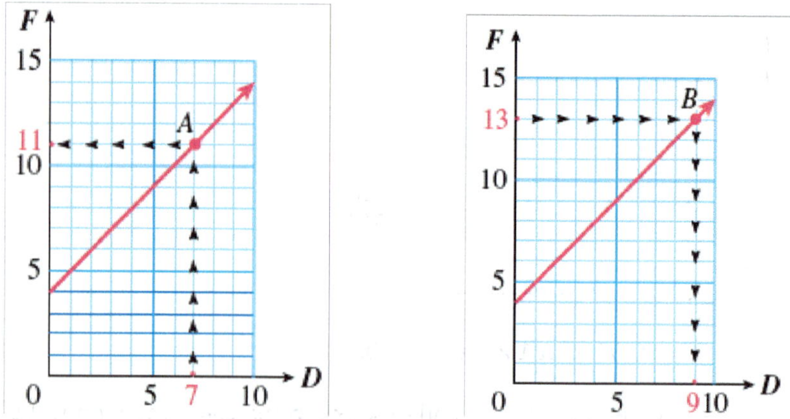

a. b.

b. We locate $F = 13$ on the vertical axis. We move horizontally to point B on the graph, with F-coordinate 13. From point B, we move vertically to the horizontal axis to find the D-coordinate. The coordinates of point B are $(9, 13)$, which tells us that when $F = 13$, $D = 9$. To verify the answer algebraically, we **solve** the equation when $F = 13$: we subtract 4 from both sides of the equation:

$$\begin{aligned} 13 &= D + 4 \\ \underline{-4} &\quad \underline{-4} \\ 9 &= D \end{aligned}$$

Subtract 4 from both sides.

Example 5 Here is a graph of

$$y = 1.5x$$

Use the graph to solve the equation $1.5x = 3.75$.

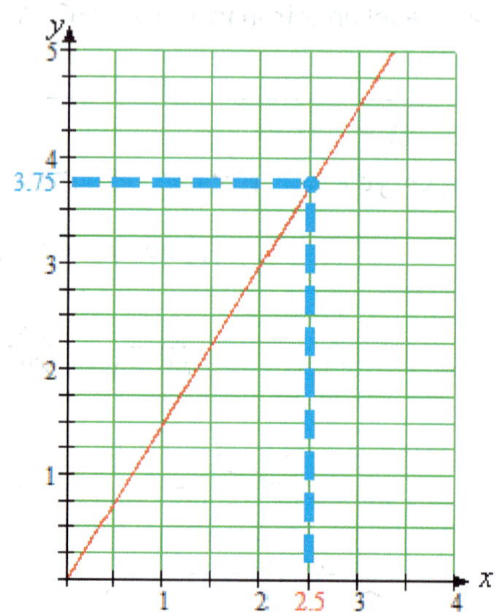

Solution By comparing the equation of the graph with the equation we want to solve, we see that y has been replaced by 3.75.

$$\begin{aligned} \boldsymbol{y} &= 1.5x \\ &\downarrow \\ \mathbf{3.75} &= 1.5x \end{aligned}$$

We locate 3.75 on the y-axis, then find the point on the graph with y-coordinate 3.75. The x-coordinate of this point is 2.5, so the ordered pair $(2.5, 3.75)$ is a solution of the equation $y = 1.5x$, and $x = 2.5$ is the solution of the equation $1.5x = 3.75$.

RQ8. The graph of an equation is a picture of its ____________.

RQ9. To solve an equation using a graph, we start on the _______ axis.

Skills Warm-Up

In Exercises 1 and 2, what value corresponds to each labeled point?

1. 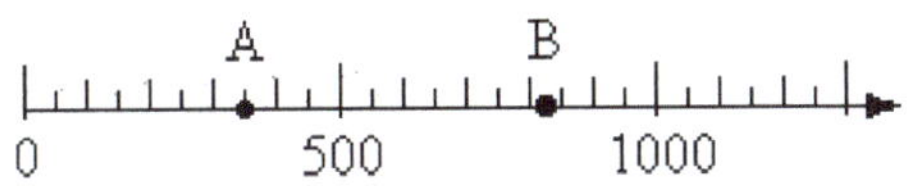

2. 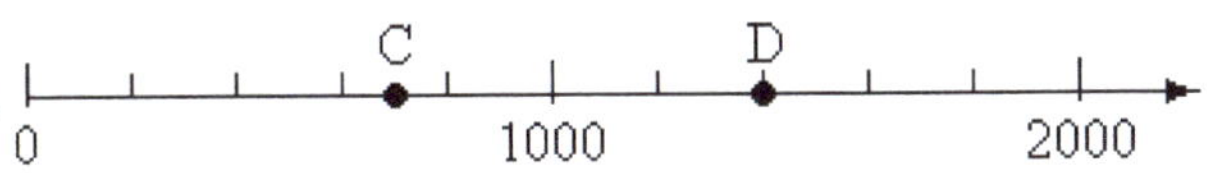

3. **a.** Label the axis. 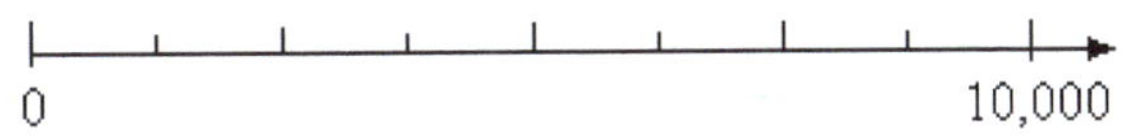

b. On the axis in part (a), plot 1400 and 8350.

4. **a.** Label the axis below with 16 tick marks (not counting zero), the highest being 800.

b. On the axis in part (a), plot 132 and 614.

5. **a.** Label an axis in increments of 40,000 from 0 to 600,000.

b. On your axis, plot 250,000 and 472,600.

6. **a.** Label an axis in increments of 0.5 from 0 to 5.

b. On your axis, plot 1.3 and 3.77.

Answers: **1.** A : 350; B : 825 **2.** 700; 1400

3. 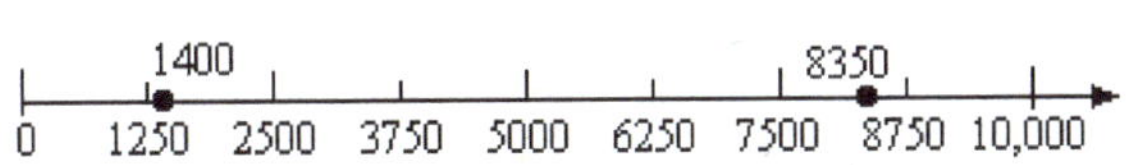

4. 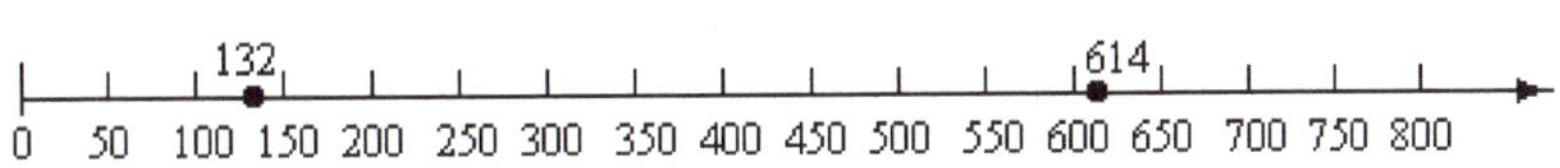

5. 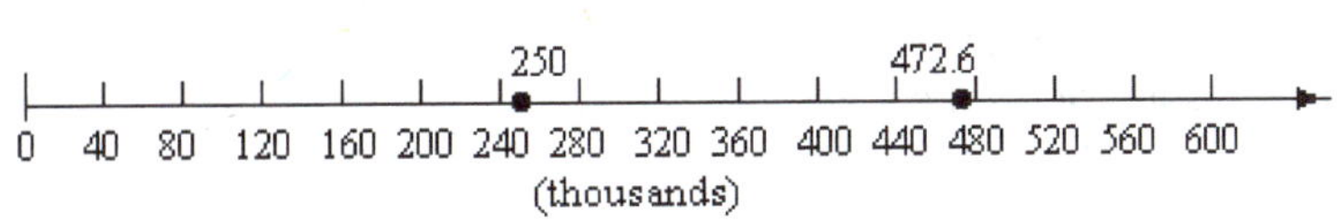

6. 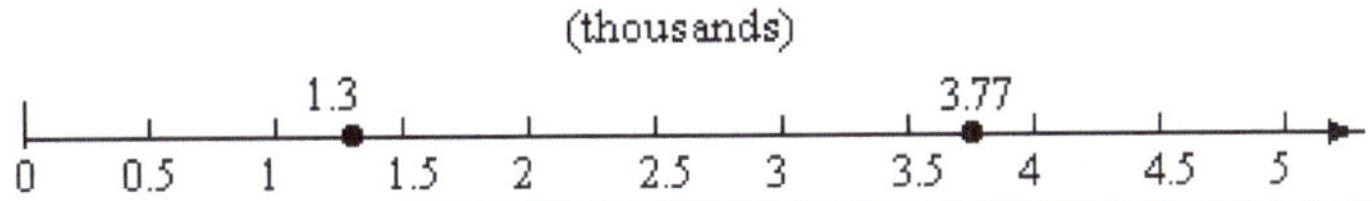

Homework 1.3

Skills Practice

For Problems 1-2, decide whether the ordered pairs are solutions of the equation whose graph is shown.

1. **a.** $(6.5, 3)$ **b.** $(0, 3.5)$ **c.** $(8, 2)$ **d.** $(4.5, 1)$

2. **a.** $(2, 6)$ **b.** $(4, 2)$ **c.** $(10, 0)$ **d.** $(11, 7)$

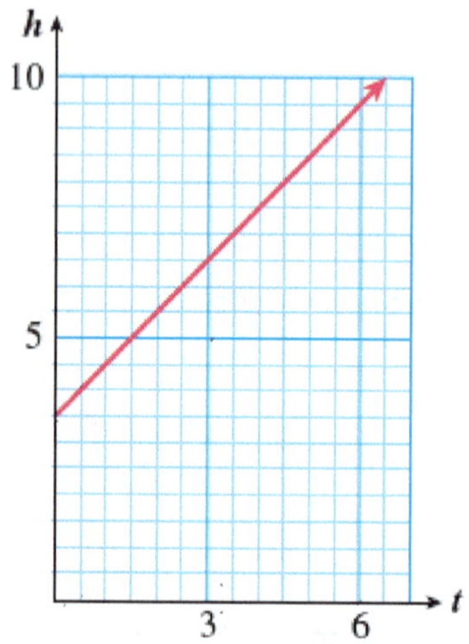

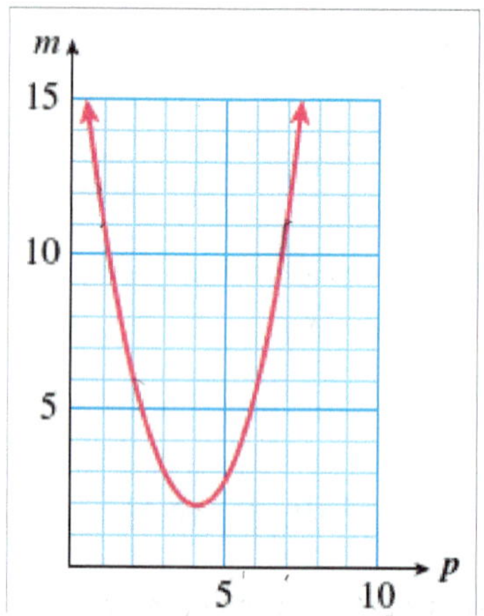

For Problems 3-6, decide whether the ordered pairs are solutions of the given equation.

3. $y = \frac{3}{4}x$
   a. $(8, 6)$ **b.** $(12, 16)$ **c.** $(2, 3)$ **d.** $(6, \frac{9}{2})$

4. $y = \frac{x}{2.5}$
   a. $(1, 2.5)$ **b.** $(25, 10)$ **c.** $(5, 2)$ **d.** $(8, 20)$

5. $w = z - 1.8$
   a. $(10, 8.8)$ **b.** $(6, 7.8)$ **c.** $(2, \frac{1}{5})$ **d.** $(9.2, 7.4)$

6. $w = 120 - z$
   a. $(0, 120)$ **b.** $(65, 55)$ **c.** $(150, 30)$ **d.** $(9.6, 2.4)$

For Problems 7-8, state the interval that each grid line represents on the horizontal and vertical axes.

7. 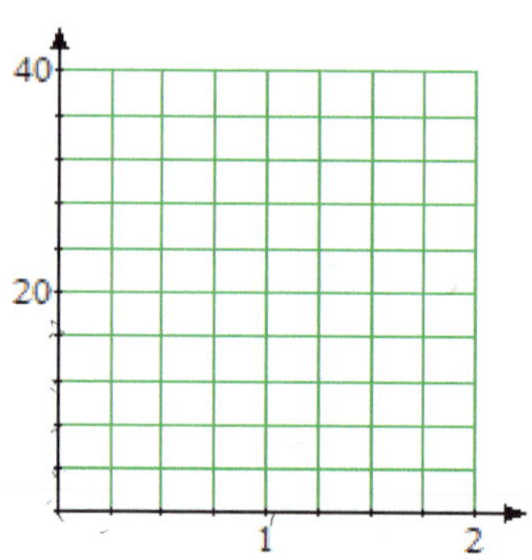

8. 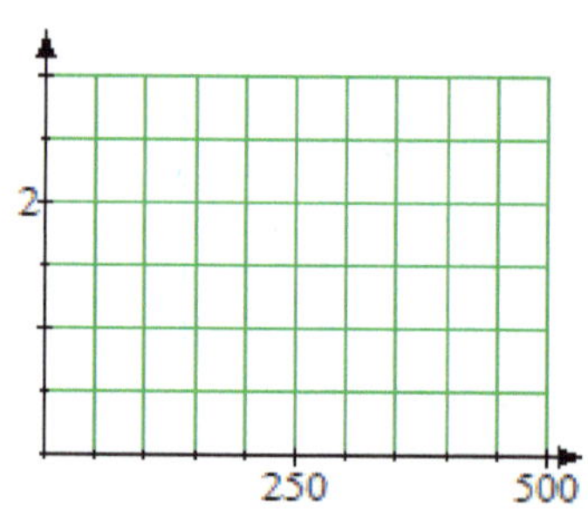

For the graphs in Problems 9-10:

a. Fill in the table with the correct coordinates. Choose your own points to complete the table.

b. Look for a pattern in your table, and write an equation for the second variable in terms of the first variable.

9.

x	0	2	5		10		
y	16			10			

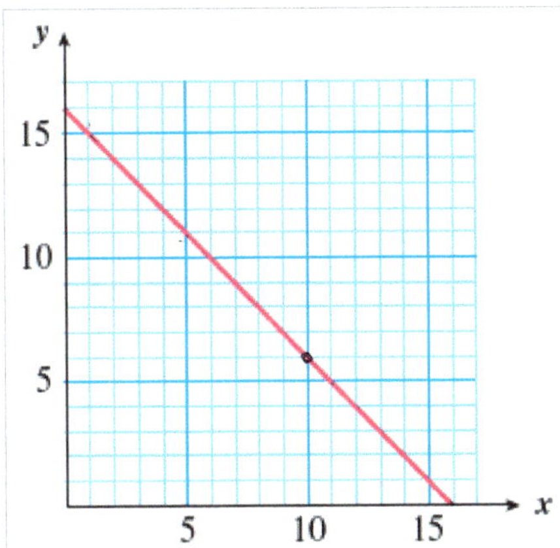

10.

x	0	1			7		
y			16	20			

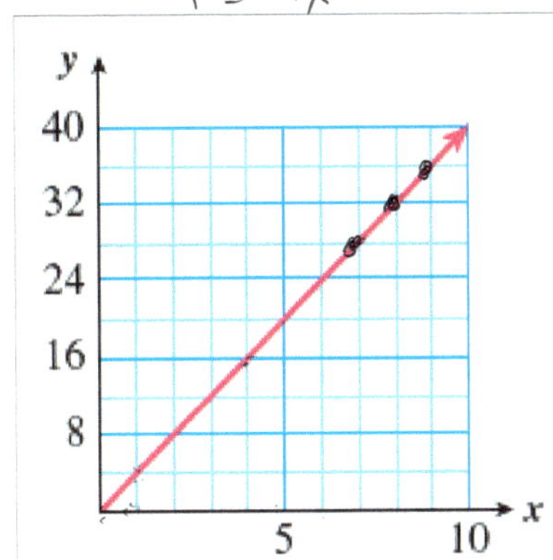

For Problems 11-14, choose the equation that describes the graph. (A graph might have more than one equation.)

a. $y = 150 - x$	**b.** $y = 15 - x$	**c.** $y = 0.2x$	**d.** $y = \frac{x}{60}$
e. $y = \frac{x}{5}$	**f.** $y = 15 + x$	**g.** $y = \frac{60}{x}$	**h.** $y = 5x$

11.

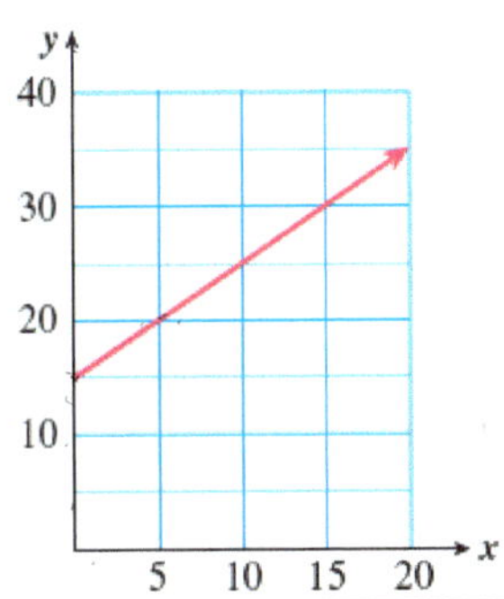

12.

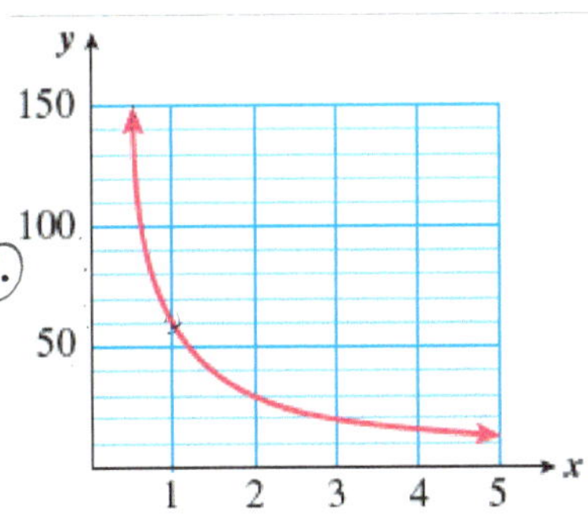

13.

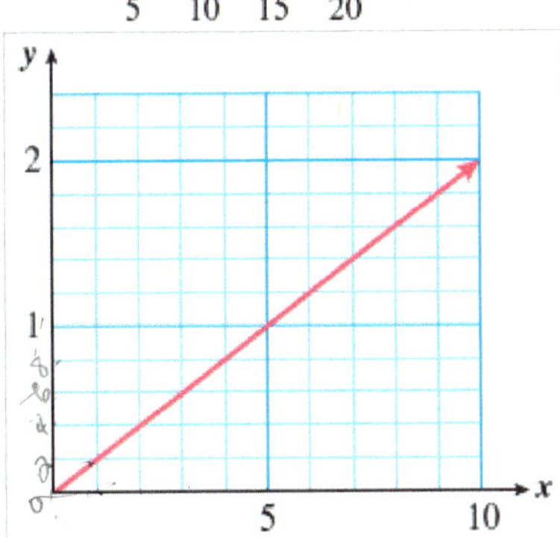

14.

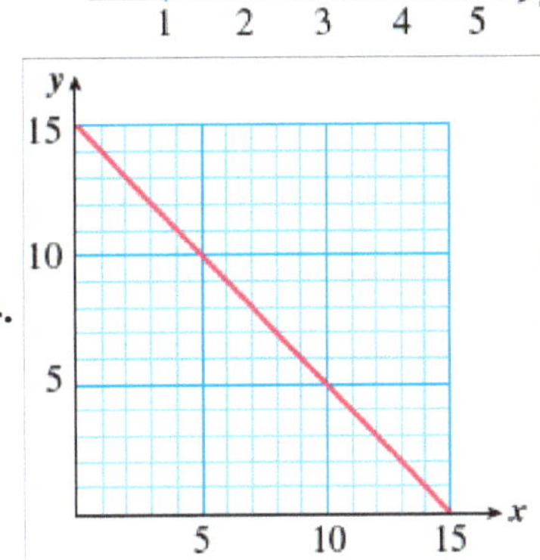

Applications

Use the three steps in the Lesson to make the graphs in Problems 15-17.

15. The college Bookstore charges a 25% markup over wholesale prices. If the wholesale price of a book is w dollars, the bookstore's price p is given by

$$p = 1.25\,w$$

a. How much does the bookstore charge for a book whose wholesale price is \$28? \$35

b. Evaluate the formula to find the bookstore's price for various books.

w	12	16	20	30	36	40
p	15	20	25	37.5	45	50

c. Plot the values from the table and graph the equation.

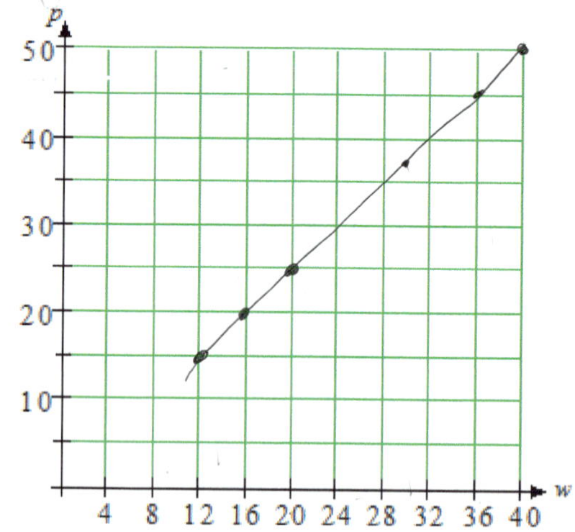

16. Uncle Ray's diet allows him to eat a total of 1000 calories for lunch and dinner. Let L stand for the number of calories in Uncle Ray's lunch, and D for the number of calories in his dinner. D = 1000 − L

a. Write an equation for D in terms of L.

b. Complete the table.

L	250	300	450	600
D	750	700	550	400

c. Label the scales on the axes and graph the equation. Use intervals of 100 on both axes.

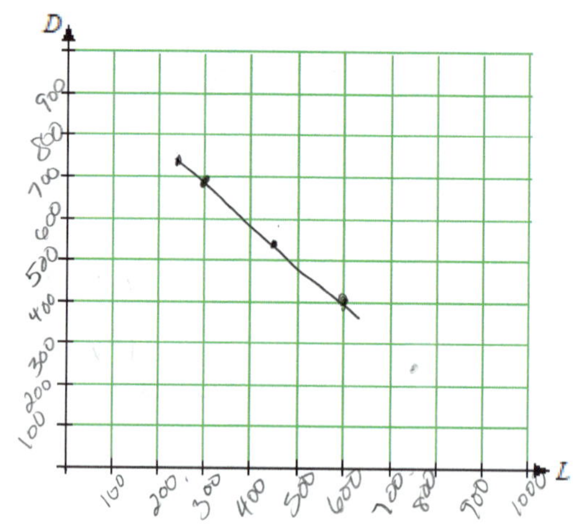

17. The Green Co-op plans to divide its profit of \$3600 from the sale of environmentally safe products among its members. Let m stand for the number of members, and S for the share of the profits each member will receive. S = 3600/m

a. Write an equation for S in terms of m.

b. Complete the table.

m	20	40	60	100
S	180	90	60	36

c. Label the scales on the axes and graph the equation. Use intervals of 20 on the vertical axis.

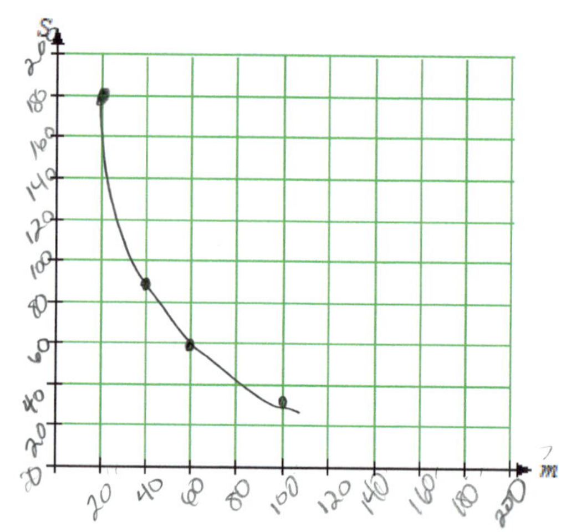

For Problems 18-19, solve the equation graphically by drawing on the graph. (You may have to estimate some of your solutions.) Then verify your answer by solving algebraically.

18. The graph of $k = p - 20$ is shown below. Use the graph to solve the equations:

a. $90 = p - 20$ **b.** $p - 20 = 40$

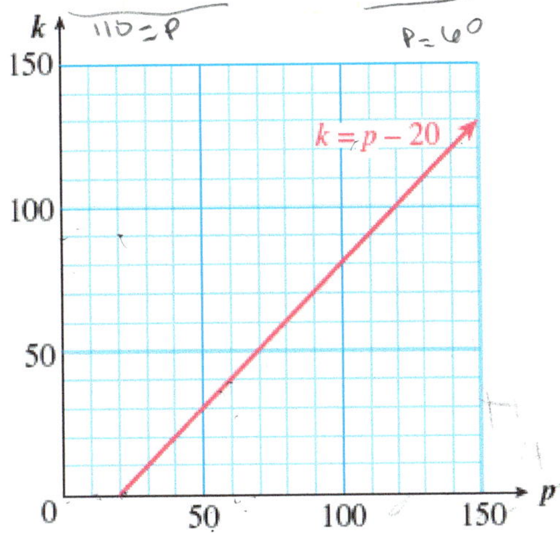

19. The graph of $d = 18g$ is shown below. Use the graph to solve the equations:

a. $250 = 18g$ **b.** $18g = 210$

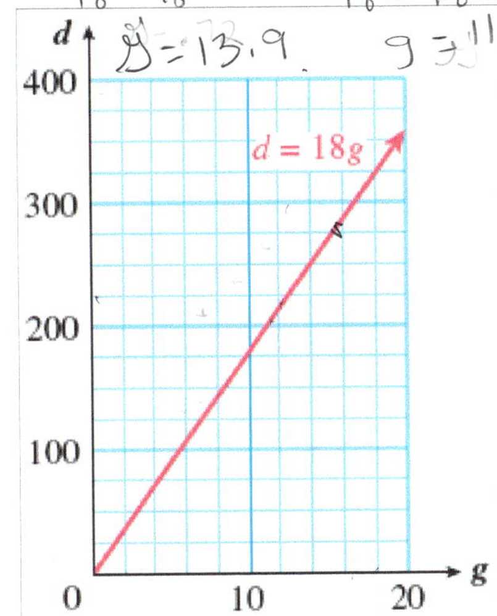

For Problems 20-21,

a. Give the coordinates of the indicated point.

b. Explain what those coordinates tell us about the problem situation.

20. The graph at left below gives the sales tax, T, in terms of the price p of an item, where the amounts are in dollars.

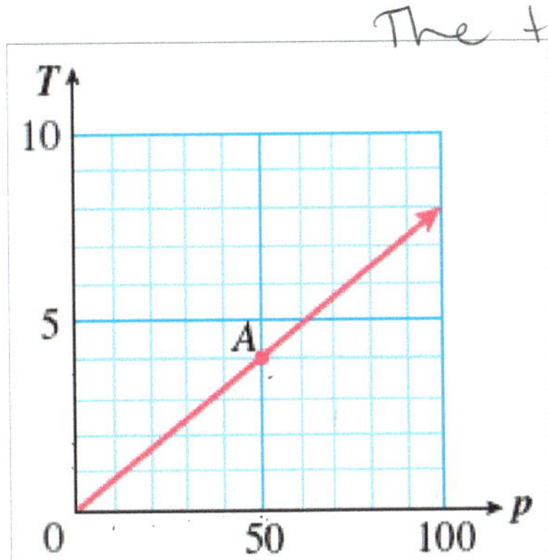

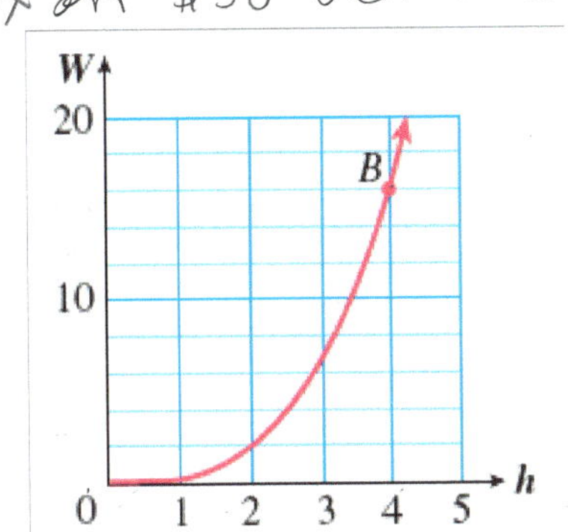

21. The graph at right above gives the weight, W, in ounces of a chocolate bunny that is h inches tall.

For Problems 22-25, use the formulas in Section 1.2.

22. a. Suppose your father loans you \$2000 to be repaid with 4% annual interest when you finish school. Write an equation for the amount of interest I you will owe after t years.

b. How much interest will you owe if you finish school in 2 years? In 4 years? In 7 years? Make a table showing your answers as ordered pairs.

c. Use your table to choose appropriate scales for the axes, then plot points and sketch a graph of the equation.

23. a. The Earth Alliance made \$6000 in revenue from selling tickets to Earth Day, an event for children. Write an equation for its profit P in terms of its costs.

b. What was their profit if their costs were \$800? \$1000? \$2500? Make a table showing your answers as ordered pairs.

c. Use your table to choose appropriate scales for the axes, then plot points and sketch a graph of the equation.

24. a. Edgar's great-aunt plans to put \$5000 in a trust for Edgar until he turns 21, three years from now. She has a choice of several different accounts. Write an equation for the amount of interest I the money will earn in 3 years if the account pays interest rate r.

b. How much interest will the \$5000 earn in an account that pays 8% interest? $10\frac{1}{4}$% interest? 12% interest? Make a table showing your answers as ordered pairs.

c. Use your table to choose appropriate scales for the axes, then plot points and sketch a graph of the equation.

25. a. Delbert keeps track of the total number of points he earns on homework assignments, each of which is worth 30 points. At the end of the semester he has 540 points. Write an equation for Delbert's average homework score S in terms of the number of assignments, n.

b. What is Delbert's average score if there were 20 assignments? 25 assignments? 30 assignments? Make a table showing your answers as ordered pairs.

c. Use your table to choose appropriate scales for the axes, then plot points and sketch a graph of the equation.

1.4 Solving Equations

What is an Equation?

An **equation** is a statement that two expressions are equal. It may involve one or more variables.

Example 1 Liz makes \$6 an hour as a tutor in the Math Lab. Her weekly earnings, w, depend on the number of hours she works, h. The equation relating these two variables is

$$w = 6h$$

If we know the value of h, we can evaluate the expression $6h$ to find Liz's earnings. For example, if $h = 7$, then

$$w = 6(7) = 42$$

Liz makes \$42 for 7 hours of work.

In Example 1 we used evaluation to find the value of w when we knew the value of h. What if we know the value of w and want to find h? Suppose Liz earned \$54 last week. How many hours did she work? To answer this question we substitute 54 for w, so that the equation becomes

$$54 = 6h$$

Now we have an equation in just one variable. We would like to find the value of h that makes this equation true.

A value of the variable that makes an equation true is called a **solution** of the equation, and the process of finding this value is called **solving** the equation.

Trial and Error

Sometimes we can solve an equation by trial and error. By trying different values of h, you will quickly see that the solution of the equation $54 = 6h$ is 9. We can use a table to help us solve an equation by trial and error.

Exercise One of Aunt Esther's Chocolate Dream cookies contains 42 calories, so d cookies contain c calories, and $c = 42d$. If Albert consumed 546 calories, how many cookies did he eat? Use trial and error to help you solve the equation $546 = 42d$.

d	5	6	7	8	9	10	11	12	13	14	15
c											

Look Ahead: Trial and error can be time-consuming, especially for complicated equations. In this section we investigate some algebraic methods for solving equations.

RQ1. What is an equation?
RQ2. What is a solution of an equation?

Inverse Operations

Instead of relying on trial and error to solve equations, we will use **inverse operations**. Try the Exercise below.

Exercise a. Choose any number for x: $x =$ ____________.
Multiply your number by 5: $5x =$ ____________.
Divide the result by 5: $\frac{5x}{5} =$ ____________.

Did you end up with your original number?
Try multiplying and dividing your number by another number besides 5. Do you still end up with your original number?

Multiplication and division are opposite or **inverse operations**, because each operation undoes the effects of the other.

Exercise b. Choose any number for x: $x =$ ____________.
Add 4 to your number: $x + 4 =$ ____________.
Subtract 4 from the result: $x + 4 - 4 =$ ____________.

Did you end up with your original number?
Try adding and subtracting another number besides 4. Do you still end up with your original number?

Addition and subtraction are opposite or **inverse operations**, because each operation undoes the effects of the other.

We see that the opposite operation for multiplication is division, and the opposite operation for addition is subtraction. Now we can solve the equation in Example 1.

Example 2 Solve the equation $54 = 6h$.

Solution We will transform this equation into a new one of the form $h = k$ (or $k = h$), where the number k is the solution. To do this, we must isolate the variable h on one side of the equals sign. Because h is multiplied by 6 in this equation, we divide both sides by 6, like this:

$$\frac{54}{6} = \frac{6h}{6} \quad \text{Divide both sides by 6.}$$
$$9 = h$$

The solution is 9. Liz worked for 9 hours.

In Example 2, we transformed the equation into a simpler one (namely, $h = 9$) which tells us the solution. We can use the following operations to transform an equation:

To solve an equation:
1. We can add or subtract the same number from both sides.
2. We can multiply or divide both sides by the same number, as long as that number is not zero.

Look Closer: Our method for solving equations works because of the **properties of equality**, one for each of the four arithmetic operations. They are listed below.

Addition and Subtraction Properties of Equality
If the same quantity is added to or subtracted from both sides of an equation, the solution is unchanged. In symbols,

If $a = b$, **then** $a + c = b + c$ **and** $a - c = b - c$

Multiplication and Division Properties of Equality
If both sides of an equation are multiplied or divided by the same nonzero quantity, the solution is unchanged. In symbols,

If $a = b$, **then** $ac = bc$ **and** $\frac{a}{c} = \frac{b}{c}$, $c \neq 0$

Applying any one of these properties to an equation produces an **equivalent equation**: one with the same solutions as the original equation.

Solving Equations

Now we can solve equations algebraically. We'll start with the simple Example from Section 1.3 about Delbert's and Francine's ages.

Example 3 Francine is exactly four years older than Delbert, so an equation relating their ages is $F = D + 4$. How old is Delbert when Francine is 19?

Solution We must solve the equation

$$19 = D + 4$$

We see that 4 has been added to the variable, D. To isolate D on one side of the equals sign, we subtract 4 from both sides of the equation, like this:

$$\begin{array}{rcl} 19 & = & D + 4 \\ \underline{-4} & & \underline{-4} \\ 15 & = & D \end{array} \qquad \text{Note that } D + 4 - 4 = D.$$

The solution to this equation is 15. You can check that substituting 15 for D does make the equation true.

Check: $19 = D + 4$ Substitute 15 for D.
$19 = 15 + 4$ True: the solution checks.

The last example is easy enough to solve without algebra, but the method will help us with harder problems. Here is a general strategy for solving such equations:

To solve an equation algebraically:

Step 1 Ask yourself what operation has been performed on the variable.

Step 2 Perform the opposite operation on *both* sides of the equation in order to isolate the variable.

RQ3. What is the opposite operation for subtraction? For multiplication?
RQ4. What are equivalent equations?

Problem Solving with Equations

Some problems can be solved by applying one of our familiar formulas.

Example 4 A saving's account at Al's Bank earns 5% annual interest. Jan's account earned \$42.50 in interest last year. What was her balance at the beginning of the year?

Solution First, we choose the appropriate formula. This is a problem about interest, so we use the interest formula: $I = Prt$
Next, we list the values of the variables. (Which variable is unknown?)

$$I = 42.50 \qquad r = 0.05$$
$$P = \text{unknown} \qquad t = 1$$

We substitute the values into the formula: $42.50 = P(0.05)(1)$
Then we solve the equation:

$$\frac{42.50}{0.05} = \frac{P(0.05)}{0.05} \qquad \text{Divide both sides by 0.05.}$$
$$850 = P$$

Finally, we answer the question in a sentence:

Jan's account balance was \$850 at the beginning of the year.

Many practical problems can be solved by first writing an equation that describes or **models** the problem, and then solving the equation. We will use three steps to solve applied problems.

Steps for solving an applied problem:

Step 1 Identify the unknown quantity and choose a variable to represent it.

Step 2 Find some quantity that can be described in two different ways, and write an equation using the variable to model the problem situation.

Step 3 Solve the equation and answer the question in the problem.

Example 5 Jerry needs an additional \$35 for airfare to New York. The ticket to New York costs \$293. How much money does Jerry have?

Solution We follow the three steps.

Step 1 The amount of money Jerry has is unknown.

Amount of money Jerry has: m

Step 2 The airfare to New York can be described in two different ways:

$$\underset{\text{amount Jerry has}}{m} + \underset{\text{amount he needs}}{35} = \underset{\text{airfare}}{293}$$

Step 3 We solve the equation and answer the question.

$$\begin{aligned} m + 35 &= 293 \\ \underline{-35} \quad & \underline{-35} \\ m &= 258 \end{aligned}$$

Subtract 35 from both sides.

Jerry has $258.

Caution! In Example 5, each side of the equation represents the airfare to New York, **not** the amount of money Jerry has, which is what we want to find. This is typical of applied problems: the equation does not usually represent the unknown quantity itself, but some related quantity.

Look Ahead: Although you may be able to solve the problems in this lesson with arithmetic, *the important thing is to learn to write the algebraic equation, or model,* for the problem. This skill will help you to solve problems that are too difficult to solve with arithmetic alone.

RQ5. What is the first step in solving an applied problem?
RQ6. True or false: One side of the equation that models a problem will be the unknown variable.

Skills Warm-Up

Choose the correct algebraic expression.

$m + 15$ $\qquad$ $m - 15$ $\qquad$ $15 - m$

1. Carol weighs 15 pounds less than Garth. If Garth weighs m pounds, how much does Carol weigh?
2. Amber and Beryl together planted 15 trees. If Amber planted m trees, how many trees did Beryl plant?
3. Fred earned $15 more this week than last week. If he earned m dollars last week, how much did he earn this week?
4. Meg bicycled 15 miles farther than Kwan. If Kwan rode m miles, how far did Meg ride?
5. There are 15 children in Amy's swim class. If there are m girls, how many are boys?
6. The sale price of a sweater is $15 less than the regular price. If the regular price is m dollars, what is the sale price?

Choose the correct algebraic expression.

$12p$ $\qquad$ $\dfrac{12}{p}$ $\qquad$ $\dfrac{p}{12}$

7. Julian earns $12 an hour. If he works for p hours, how much will he make?
8. Farmer Brown collected p eggs this morning. How many dozen is that?
9. Melissa bought 12 colored markers. If their total cost was p dollars, how much did each marker cost?

10. Rosalind is baby-sitting for p children. If she brings 12 puzzles, how many will each child get?

11. Hector has to read 12 chapters in his history text. If he has p days to complete the assignment, how many chapters should he read per day?

12. Roma swims 12 laps per day. After p days, how many laps has she swum?

Answers: **1.** $m-15$ **2.** $15-m$ **3.** $m+15$ **4.** $m+15$ **5.** $15-m$ **6.** $m-15$
7. $12p$ **8.** $\frac{p}{12}$ **9.** $\frac{p}{12}$ **10.** $\frac{12}{p}$ **11.** $\frac{12}{p}$ **12.** $12p$

Homework 1.4

Skills Practice

For Problems 1-2, fill in the table for each equation. Explain how you found the unknown values.

1. $q = 9 + t$

t	2	4			21	
q			15	18		39

2. $p = 5n$

n	0	2		5		
p			20		35	55

3. Decide whether the given value for the variable is a solution of the equation.

a. $x - 4 = 6;\quad x = 10$ **b.** $4y = 28;\quad y = 24$ **c.** $\dfrac{0}{z} = 0;\quad z = 19$

For Problems 4-12, solve, and check your solution.

4. $x - 3 = 11$ **5.** $10.6 = 7.8 + y$ **6.** $3y = 108$

7. $42 = 3.5b$ **8.** $2.6 = \dfrac{a}{1.5}$ **9.** $x - 4 = 0$

10. $34x = 212$ **11.** $6z = 20$ **12.** $9 = k + 9$

For Problems 13-16,

a. Choose the appropriate formula to write an equation.
b. Solve the equation and answer the question.

13. Clive loaned his brother some money to buy a new truck, and his brother agreed to repay the loan in 1 year with 3% interest. Clive earned $75 interest on the loan. How much did Clive loan his brother?

14. Andy's average homework score on eight assignments was 38.25. How many homework points did Andy earn altogether?

15. How long will it take a cyclist traveling at 13 miles an hour to cover 234 miles?

16. A roll of carpet contains 400 square feet of carpet. If the roll is 16 feet wide, how long is the piece of carpet?

For Problems 17-22, choose the appropriate equation.

$x + 7 = 26$	$7x = 26$	$\dfrac{x}{7} = 26$
$x - 7 = 26$	$\dfrac{x}{26} = 7$	$\dfrac{26}{x} = 7$

17. Sarah drove 7 miles farther to her high school reunion than Jenni drove. If Sarah drove 26 miles, how far did Jenni drive?

18. Lurline and Rozik live 26 miles apart. They meet at a theme park between their homes. If Lurline drove 7 miles to the park, how far did Rozik drive?

19. Doris is training for a triathlon. This week she averaged 26 miles per day on her bicycle. If she rode every day, what was her total mileage?

20. Glynnis jogged the same route every day this week for a total of 26 miles. How long is her route?

21. Astrid divided her supply of colored pencils among the 26 children in her class, and each child got 7 pencils. How many pencils does she have?

22. Ariel lost 7 of the beads on her necklace, and now there are 26. How many were there originally?

Applications

Follow the steps to solve Problems 23-26.

23. Lupe spent $24 at the Craft Fair. She now has $39 left. How much did she have before the Craft Fair?
    a. What are we asked to find? Choose a variable to represent it.
    b. Find two ways to express the amount of money Lupe had after the Craft Fair, and write an equation.
    c. Solve the equation and answer the question in the problem.

24. Danny weighs 32 pounds more than Brenda. If Danny weighs 157 pounds, how much does Brenda weigh?
    a. What are we asked to find? Choose a variable to represent it.
    b. Find two ways to express Danny's weight, and write an equation.
    c. Solve the equation and answer the question in the problem.

25. Miranda worked 20 hours this week and made $136. What is Miranda's hourly wage?
    a. What are we asked to find? Choose a variable to represent it.
    b. Find two ways to express Miranda's total earnings, and write an equation.
    c. Solve the equation and answer the question in the problem.

26. Struggling Students Gardening Service splits their profit equally among their eight members. If each member made $64 last week, what was the total profit?
    a. What are we asked to find? Choose a variable to represent it.
    b. Find two ways to express each member's share, and write an equation.
    c. Solve the equation and answer the question in the problem.

Write algebraic equations to solve Problems 27-29. Follow the three steps in the Lesson.

27. Martha paid $26 less for a suit at a discount store than her mother paid at a boutique for the same suit. If Martha paid $89 for the suit, how much did her mother pay?

28. Emily spends 40% of her monthly income on rent. If her rent is $360 a month, how much does Emily make?

29. After she wrote a check for $2378, Avril's bank account shows a balance of $1978. How much money was in Avril's account before the check cleared?

In Problems 30-32, we compare evaluating an expression and solving an equation.

30. A used car costs \$3400 less than the new version of the same model.

a. Choose two variables and write an equation for the cost of the used car in terms of the cost of the new car.

b. If the new car costs \$14,500, how much does the used car cost?

c. If the used car costs \$9200, how much does the new car cost?

31. The Dodgers won 60% of their games last season.

a. Choose two variables and write an equation for the number of games the Dodgers won in terms of the number of games they played.

b. If the Dodgers played 120 games, how many did they win?

c. If the Dodgers won 96 games, how many did they play?

32. Sunshine Industries manufactures beach umbrellas. Their profit on each umbrella is 18% of the selling price.

a. Choose two variables and write an equation for the profit on each umbrella in terms of its selling price.

b. If a beach umbrella sells for \$60, what is the profit?

c. If the profit on one umbrella is \$7.20, what is the selling price?

1.5 Order of Operations

Addition and Multiplication

If you add together three or more numbers, such as

$$2+5+8$$

it doesn't matter which addition you do first; you will get the same answer either way.

Example 1a. A sum of three or more terms can be added in any order. In the sums below, the parentheses tell us which part of the expression to simplify first.

$$(2+5)+8 \qquad \text{and} \qquad 2+(5+8)$$
$$=7+8=15 \qquad\qquad =2+13=15$$

b. Similarly, a product of three factors can be multiplied in any order. Thus

$$(3\cdot 2)\cdot 4 \qquad \text{and} \qquad 3\cdot(2\cdot 4)$$
$$=6\cdot 4=24 \qquad\qquad =3\cdot 8=24$$

These two facts illustrate the **associative laws** for addition and multiplication.

Associative Law for Addition

If a, b, and c are any numbers, then

$$(a+b)+c=a+(b+c)$$

Associative Law for Multiplication

If a, b, and c are any numbers, then

$$(a\cdot b)\cdot c=a\cdot(b\cdot c)$$

Subtraction and Division

What about a string of subtractions or a string of divisions, such as

$$20-8-5 \qquad \text{or} \qquad 36\div 6\div 2$$

In these calculations, we get different answers, depending on which operations we perform first, as you can see in Example 2.

Example 2a. Subtraction:

$$(20-8)-5=12-5=7 \qquad \text{but} \qquad 20-(8-5)=20-3=17$$

b. Division:

$$(36\div 6)\div 2=6\div 2=3 \qquad \text{but} \qquad 36\div(6\div 2)=36\div 3=12$$

The associative laws do **not** hold for subtraction or division.

So, if there are no parentheses in the expression, how do we know which operations to perform first?

1. In a string of additions and subtractions, we perform the operations **in order from left to right**.
2. Similarly, we perform multiplications and divisions **in order from left to right**.

Example 3 Simplify each expression.

a. $20 - 8 - 5$

b. $36 \div 6 \div 2$

Solutions Perform the operations in order from left to right.

a. $20 - 8 - 5 = 12 - 5 = 17$

b. $36 \div 6 \div 2 = 6 \div 2 = 3$

RQ1. What do the associative laws tell us?
RQ2. Which two operations are not associative?
RQ3. In what order should we perform a string of additions and subtractions?

Combined Operations

How should we simplify the expression $4 + 6 \cdot 2$?

If we do the addition first, we get

$$(4 + 6) \cdot 2 = 10 \cdot 2 = 20$$

If we do the multiplication first, we get

$$4 + (6 \cdot 2) = 4 + 12 = 16$$

Which one is correct? In order to avoid confusion, we make the following agreement.

Always perform multiplications and divisions before additions and subtractions.

Example 4 The correct way to simplify the expression $4 + 6 \cdot 2$ is

$4 + 6 \cdot 2$ Multiply first.
$= 4 + 12$ Then add.
$= 16$

Look Closer: In longer expressions, it can be helpful to group the expression into its terms before beginning. **Terms** are expressions separated by addition or subtraction symbols. We simplify each term before combining them.

Example 5 Simplify $6 + 2 \cdot 5 - 12 \div 3 \cdot 2$.

Solution We start by underlining each term of the expression separately. Then we simplify each term.

$\underline{6} + \underline{2 \cdot 5} - \underline{12 \div 3 \cdot 2}$ Perform multiplications and divisions from left to right.

$= 6 + 10 - 8$ Perform additions and subtractions from left to right.

$= 16 - 8 = 8$

Caution! 1. In Example 5, we do **not** start by adding $6+2$. We must perform multiplications before additions.
2. We simplify $12 \div 3 \cdot 2$ by performing the operations **from left to right**:

$$12 \div 3 \cdot 2 = 4 \cdot 2 = 8$$

Parentheses

What if the addition should come first in a particular calculation? In that case, we use parentheses to enclose the sum, like this:

$$(4+6) \cdot 2$$

Perform any operations inside parentheses first.

Example 6 Simplify.

a. $2+5(7-3)$ — Subtract inside parentheses.
$= 2+5(4)$ — Multiply.
$= 2+20 = 22$ — Add.

b. $6(10-2\cdot 4)\div 4$ — Multiply inside parentheses.
$= 6(10-8)\div 4$ — Subtract inside parentheses.
$= 6(2)\div 4$ — Multiply and divide in order from left to right.
$= 12 \div 4 = 3$

RQ4. Is it true that multiplications should be performed before divisions? Why or why not?
RQ5. What do parentheses tell us?
RQ6. If there are no parentheses, in what order should we perform multiplications and divisions?

Fraction Bars

Like parentheses, a fraction bar is a **grouping device**.

Expressions that appear above or below a fraction bar should be simplified first.

Example 7 Simplify $\dfrac{24+6}{12+6}$

Solution We begin by computing the sums above and below the fraction bar. Then we can reduce the fraction. Thus,

$$\frac{24+6}{12+6} = \frac{30}{18} = \frac{5}{3}$$

Caution! Do not be tempted to "cancel" the terms in Example 7. For example, it would be incorrect to write

$$\frac{24+6}{12+6} = \frac{\not{2}4+6}{\not{1}2+6} = \frac{2+6}{1+6} \quad \leftarrow \textbf{Incorrect!}$$

According to the order of operations, we must simplify the numerator and denominator first, before dividing.

RQ7. Besides parentheses, what other symbol serves as a grouping device?
RQ8. When adding fractions, which should you do first: reduce each fraction, or find an LCD?

Summary

Combining all our guidelines for simplifying expressions, we state the rules for the order of operations.

Order of Operations

1. First, perform any operations that appear inside parentheses, or above or below a fraction bar.
2. Next, perform all multiplications and divisions in order from left to right.
3. Finally, perform all additions and subtractions in order from left to right.

Algebraic Expressions

The order of operations applies to variables as well as constants. For example, the expression

$$5 + 2x$$

tells us to multiply x by 2, then add 5 to the result. Thus, to evaluate the expression for $x = 8$, we write

For $x = 8$, $\quad 5 + 2x = 5 + 2(8)$ **Multiply first.**
$\qquad\qquad = 5 + 16 = 21$

In Example 8, note how parentheses change the meaning of an expression.

Example 8 Write algebraic expressions for each of the following phrases. Then evaluate each phrase for $x = 2$ and $y = 6$.

a. Three times the sum of x and y **b.** The sum of $3x$ and y

Solutions **a.** $3(x + y)$
$3(2 + 6) = 3(8) = 24$

b. $3x + y$
$3(2) + 6 = 6 + 6 = 12$

Skills Warm-Up

Perform the operations on fractions as indicated.

1a. $\frac{5}{4}+\frac{3}{4}$ **b.** $\frac{5}{4}-\frac{3}{4}$ **c.** $\frac{5}{4}\cdot\frac{3}{4}$ **d.** $\frac{5}{4}\div\frac{3}{4}$

2a. $\frac{1}{2}+\frac{1}{6}$ **b.** $\frac{1}{2}-\frac{1}{6}$ **c.** $\frac{1}{2}\cdot\frac{1}{6}$ **d.** $\frac{1}{2}\div\frac{1}{6}$

3a. $\frac{2}{3}+\frac{1}{4}$ **b.** $\frac{2}{3}-\frac{1}{4}$ **c.** $\frac{2}{3}\cdot\frac{1}{4}$ **d.** $\frac{2}{3}\div\frac{1}{4}$

Answers: **1a.** 2 **b.** $\frac{1}{2}$ **c.** $\frac{15}{16}$ **d.** $\frac{5}{3}$ **2a.** $\frac{2}{3}$ **b.** $\frac{1}{3}$ **c.** $\frac{1}{12}$ **d.** 3
3a. $\frac{11}{12}$ **b.** $\frac{5}{12}$ **c.** $\frac{1}{6}$ **d.** $\frac{8}{3}$

Homework 1.5

Skills Practice

For Problems 1-12, simplify each expression by following the order of operations.

1. $2 + 4(3)$ **2.** $15 - \frac{3}{4}(16)$ **3.** $6 \div \frac{1}{4} \cdot 3$

4. $3 + 3(2 + 3)$ **5.** $\frac{1}{3} \cdot 12 - 3\left(\frac{5}{6}\right)$ **6.** $2 + 3 \cdot 8 - 6 + 3$

7. $\frac{3(8)}{12} - \frac{6+4}{5}$ **8.** $28 + 6 \div 2 - 2(5 + 3 \cdot 2)$

9. $3[3(3+2) - 8] - 17$ **10.** $\frac{3(3)+5}{6-2(2)} + \frac{2(5)-4}{9-4-2}$

11. $7[15 - 24 \div 2] - 9[5(4+2) - 4(6+1)]$ **12.** $20 - 4(9 - 3 \cdot 2) + 8 - 5 \cdot 3$

For Problems 13-16, use your calculator to simplify each expression. Round your answers to three decimal places if necessary.

13. $\frac{6.4+3.5}{3.6(3.2)}$ **14.** $\frac{14.6-6.8}{7.3+3.4}$

15. $\frac{26.2-9.1}{8.4 \div 7.7} + 5.1(6.9 - 1.6)$ **16.** $\frac{1728(847-603)}{216(98-38)} + 6(876 - 514)$

For Problems 17-22, evaluate each pair of expressions mentally.

17a. $8 + 2 \cdot 5$ **b.** $(8 + 2) \cdot 5$ **18a.** $\frac{24}{2+6}$ **b.** $\frac{24}{2} + 6$

19a. $(9 - 4) - 3$ **b.** $9 - (4 - 3)$ **20a.** $6 \cdot 8 - 6$ **b.** $6(8 - 6)$

21a. $\frac{36}{6(3)}$ **b.** $\frac{36}{6}(3)$ **22a.** $(30 - 5)(5)$ **b.** $30 - (5)(5)$

For Problems 23-24, fill in the table to evaluate the expression in two steps.

23. $5z - 3$

z	$5z$	$5z - 3$
2		
4		
5		

24. $2(12 + Q)$

Q	$12 + Q$	$2(12 + Q)$
0		
4		
8		

For Problems 25-30, evaluate for the given values.

25. $2y + x$ for $x = 8$ and $y = 9$

26. $4a + 3b$ for $a = 8$ and $b = 7$

27. $\frac{a}{b} - \frac{b}{a}$ for $a = 8$ and $b = 6$

28. $\frac{24-2x}{2+y} - \frac{4x+1}{3y}$ for $x = 4$ and $y = 6$

29. $\frac{a}{1-r}$ for $a = 6$ and $r = 0.2$

30. $mx + b$ for $m = \frac{3}{5}$, $x = \frac{2}{3}$, and $b = \frac{9}{10}$

Applications

31. Consider the expression $20 - 2 \cdot 8 + 1$. How would you change the expression if you wanted:

a. The subtraction performed first?
b. The addition performed first?
c. The multiplication performed first?

32. Write an algebraic expression for the following instructions:
"Multiply the sum of 5 and 7 by 4, then divide by the difference of 10 and 8."

33. Write two algebraic expressions for "the difference of 20 and 8, divided by the sum of 6 and 4 and 10":

a. Using a fraction bar
b. Using a division symbol

34. Without performing the calculations, write down the steps you would use to simplify the expression: $825 - 32(12) \div 4 + 2$

For problems 35-38, choose the correct algebraic expression for each English phrase.

$\frac{m}{12} - 3$	$\frac{m-3}{12}$
$\frac{12}{m-3}$	$3 - \frac{12}{m}$

35. Twelve divided by 3 less than m.
36. Three less than the quotient of m divided by 12.
37. The quotient of 3 less than m divided by 12.
38. Subtract from 3 the quotient of 12 and m.

39. Find and correct the error in the calculation:

$$\begin{aligned}&(5+4) - 3(8 - 3 \cdot 2)\\ &= 9 - 3(8-6)\\ &= 6(2) = 12\end{aligned}$$

40. Think of the region shown at right as a rectangle with a smaller rectangle removed, and write an expression for its area. Then simplify your expression to find the area. (The measurements given are in centimeters.)

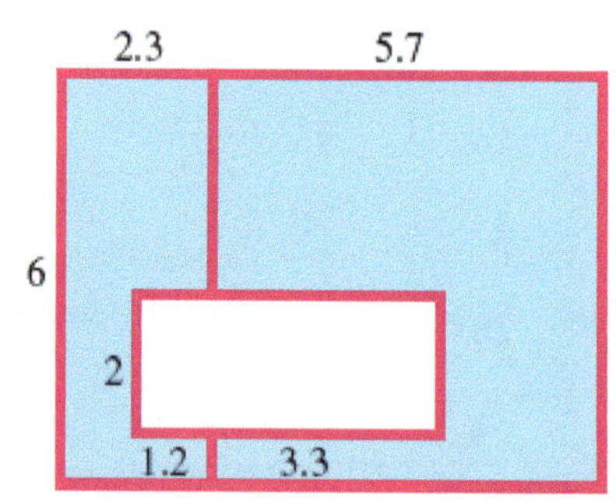

Chapter 1 Summary and Review

Lesson 1.1 Variables

- A **variable** is a numerical quantity that changes over time or in different situations.
- We can use a table, a graph, or an equation to compare the values of two related variables.
- A formula relating two variables is a type of **equation**, and an equation is just a statement that two quantities are equal.

Lesson 1.2 Algebraic Expressions

- An **algebraic expression** is any meaningful combination of numbers, variables, and operation symbols.

> **Commutative Laws**
>
> **Addition:** If a and b are any numbers, then
>
> $$a + b = b + a$$
>
> **Multiplication:** If a and b are any numbers, then
>
> $$a \cdot b = b \cdot a$$

- Subtraction and division are not commutative; the order of the numbers matters in a difference or a quotient.

> **To write an algebraic expression:**
>
> **Step 1** Identify the unknown quantity and write a short phrase to describe it.
>
> **Step 2** Choose a variable to represent the unknown quantity.
>
> **Step 3** Use mathematical symbols to represent the relationship.

- The process of substituting a specific value for the variable into an expression is called **evaluating** the expression.
- An **equation** is a mathematical statement that two algebraic expressions are equal. A **formula** is an equation that relates two or more variables.

Some Useful Algebraic Formulas	
1. distance = rate × time	$d = rt$
2. profit = revenue − cost	$P = R - C$
3. interest = principal × interest rate × time	$I = Prt$
4. part = percentage rate × whole	$P = rW$
5. average score = $\dfrac{\text{sum of scores}}{\text{number of scores}}$	$A = \dfrac{S}{n}$

Lesson 1.3 Equations and Graphs

- A graph is a way of visualizing an algebraic equation in two variables. The first or **input** variable is displayed on the horizontal axis and the second or **output** variable is displayed on the vertical axis.
- Each point on a graph has two **coordinates,** which designate the position of the point.

- An ordered pair of numbers that make an equation true is called a **solution** of the equation. Each point on the graph represents a solution of the equation.

> **To graph an equation:**
> 1. Make a table of values.
> 2. Choose scales for the axes.
> 3. Plot the points and connect them with a smooth curve.

- If the variables in an equation have very large (or very small) values, we must choose scales for the axes that fit these values.
- We can use a graph to evaluate an algebraic expression or to solve an equation.

Lesson 1.4 Solving Equations

- To solve an equation algebraically, we use **inverse**, or opposite, operations.

> **To solve an equation algebraically:**
>
> **Step 1** Ask yourself what operation has been performed on the variable.
>
> **Step 2** Perform the opposite operation on both sides of the equation. This should isolate the variable.

> **Operations for solving an equation:**
> 1. We can add or subtract the same number from both sides.
> 2. We can multiply or divide both sides by the same number, as long as that number is not zero.

> **Addition and Subtraction Properties of Equality**
>
> If the same quantity is added to or subtracted from both sides of an equation, the solution is unchanged. In symbols,
>
> **If** $a = b$, **then** $a + c = b + c$ **and** $a - c = b - c$
>
> **Multiplication and Division Properties of Equality**
>
> If both sides of an equation are multiplied or divided by the same nonzero quantity, the solution is unchanged. In symbols,
>
> **If** $a = b$, **then** $ac = bc$ **and** $\frac{a}{c} = \frac{b}{c}$, $c \neq 0$

- Many practical problems can be solved by first writing an equation that describes or **models** the problem, and then solving the equation.

> **Steps for solving an applied problem:**
>
> **Step 1** Identify the unknown quantity and choose a variable to represent it.
>
> **Step 2** Find some quantity that can be described in two different ways, and write an equation using the variable to model the problem situation.
>
> **Step 3** Solve the equation and answer the question in the problem.

Lesson 1.5 Order of Operations

Associative Laws

Addition: If a, b, and c are any numbers, then

$$(a+b)+c=a+(b+c)$$

Multiplication: If a, b, and c are any numbers, then

$$(a\cdot b)\cdot c=a\cdot(b\cdot c)$$

- The associative laws do *not* hold for subtraction or division.

Order of Operations

1. First, perform any operations that appear inside parentheses, or above or below a fraction bar.
2. Next, perform all multiplications and divisions in order from left to right.
3. Finally, perform all additions and subtractions in order from left to right.

- When using a calculator, you must enclose in parentheses any expression that appears above or below a fraction bar.

Review Questions

For Problems 1-10, use complete sentences to answer the questions.

1. What is a variable? Give an example.
2. What is the difference between an algebraic expression and an equation?
3. What is the difference between factors and terms?
4. **a.** Write as an algebraic expression "p subtracted from w."
   b. Evaluate your expression for $p=5.6$ and $w=56$.
5. **a.** Write as an algebraic expression "m divided by g."
   b. Evaluate your expression for $m=12$ and $g=192$.
6. State a formula from this chapter for each of the variables below.
   a. Distance: $d=$ **b.** Profit: $P=$
   c. Interest: $I=$ **d.** Percentage or part: $P=$
   e. Average value: $A=$
7. State formulas for the perimeter and area of a rectangle, and for the area and circumference of a circle.
8. When we construct the graph of an equation, which variable is displayed on which axis?
9. Write a paragraph explaining how the following terms are related to an equation in two variables: solution, coordinates, ordered pair.
10. State the order of operations for simplifying algebraic expressions.

Review Problems

1. Martha can travel 22 miles on a gallon of gasoline. Fill in the table with the distance Martha can drive on various amounts of gasoline.

Gallons of gas	3	5	8	10	11	12
Calculation						
Miles driven						

   a. Explain in words how to find the distance Martha can drive.
   b. Write your explanation as a mathematical sentence:
      Miles driven =
   c. Let g stand for the number of gallons of gas in Martha's car and m for the number of miles she can drive. Write an equation for m in terms of g.
   d. Use the grid below left to plot the values from your table. Connect the data points with a smooth curve.

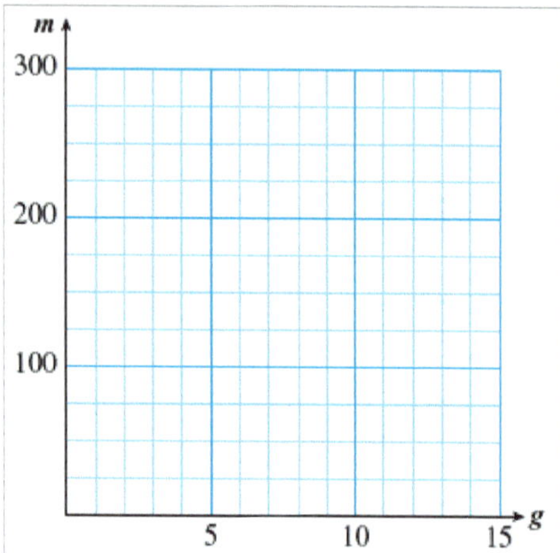

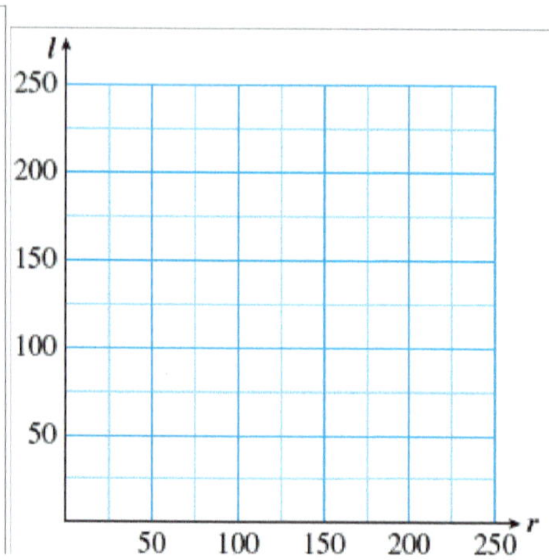

2. Mariel has a 200-page history assignment to read. Fill in the table with the number of pages Mariel has left to read after completing a certain number of pages

Pages read	20	50	85	110	135	180
Calculation						
Pages left						

   a. Explain in words how to find the number of pages left to read.
   b. Write your explanation as a mathematical sentence:
      Pages left =
   c. Let r stand for the number of pages Mariel has already read, and l for the number of pages she has left to read. Write an equation for l in terms of r.
   d. Use the grid above right to plot the values from your table. Connect the data points with a smooth curve.

3. Find the pattern and fill in the table. Then write an equation for the second variable in terms of the first.

x	0.5	1.0	1.5	2.0	4.0	6.0	7.5
y	0.125	0.25	0.375	0.5			

4. Use the graph to fill in the table. Then write an equation for the second variable in terms of the first.

x	y

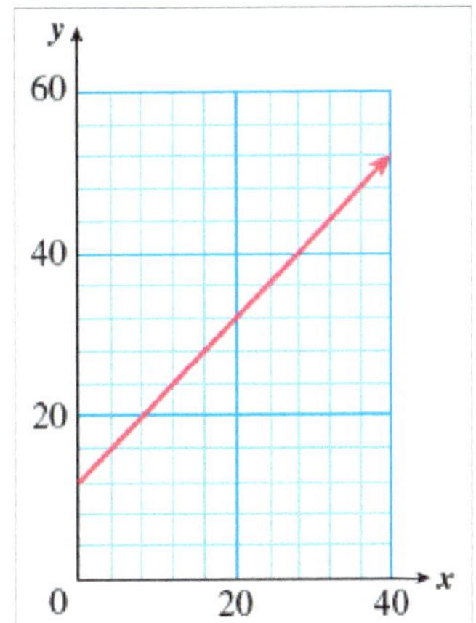

For Problems 5-12, write an algebraic expression for the phrase.

5. Five greater than z
6. 28% of t
7. Sixty dollars less than first-class air fare
8. The quotient of the volume of the sphere and 6
9. Six degrees hotter than yesterday's temperature
10. The cost of the gasoline divided three ways
11. 8% of the total bill
12. Five inches less than the height of the triangle

13. a. Write an expression for the distance traveled in t seconds by a car moving at 88 feet per second.
    b. How far will the car travel in 5 seconds? In half a second? In half a minute?

14. a. Write an expression for the distance you can travel in 3 hours at a speed of r miles per hour.
    b. How far can you travel in 3 hours on a bicycle at a speed of 6 miles per hour? On a motorboat at a speed of 20 miles per hour?

15. a. Write an expression for the amount of interest earned in t years by \$500 deposited in an account that pays 7% annual interest.
    b. How much interest will the account earn after 1 year? After 2 years? After 5 years?

16. a. Write an expression for the amount of interest earned in two years by \$1500 deposited in an account that pays interest rate r.
    b. How much interest will the money earn at $6\frac{1}{2}$% interest? At 8.3% interest?

17. Rachel has saved \$60 to help pay for her books this semester. Let b stand for the price of her books, and n for the amount she still needs.
    a. Write an equation for n in terms of b.
    b. Fill in the table.

b	100	120	150	180	200
n					

    c. Choose appropriate scales for the axes and graph your equation.

18. Matt wants to travel 360 miles to visit a friend over spring break. He is deciding whether to ride his bike or drive. If he travels at an average speed of r miles per hour, then the trip will take t hours, where $t = \frac{360}{r}$.

a. Fill in the table.

r	10	20	30	40	60	80	90
t							

b. Choose appropriate scales for the axes and graph your equation.

19. On a typical evening, there are twice as many non-smokers as smokers in a popular restaurant.

a. Write an equation for the number of non-smokers, n, in terms of the number of smokers, s.

b. Make a table of values and graph your equation.

20. Match each equation with its graph.

a. $y = 8 - x$ **b.** $y = x + 8$ **c.** $y = 8x$

I

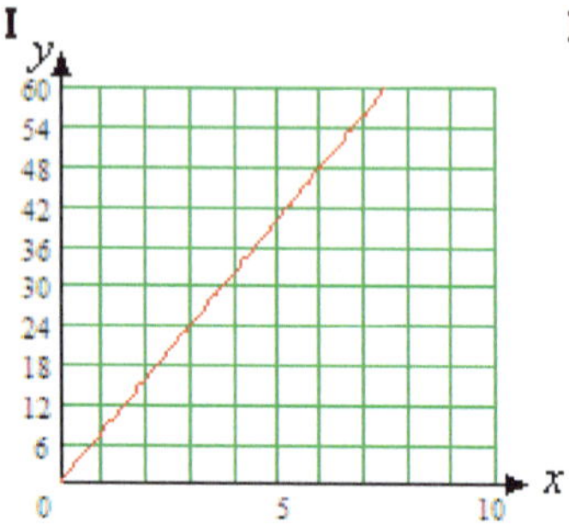

II

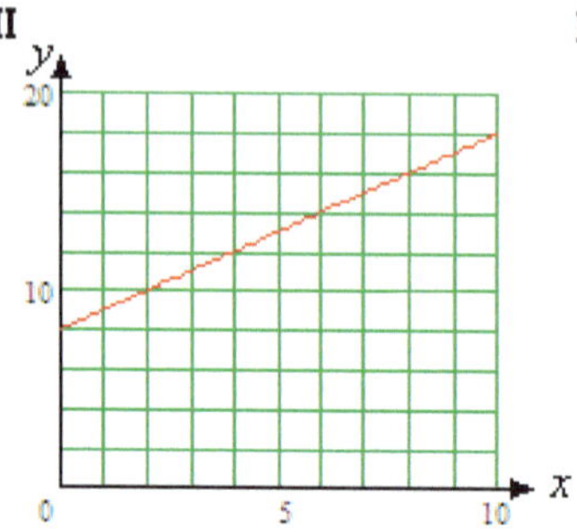

III

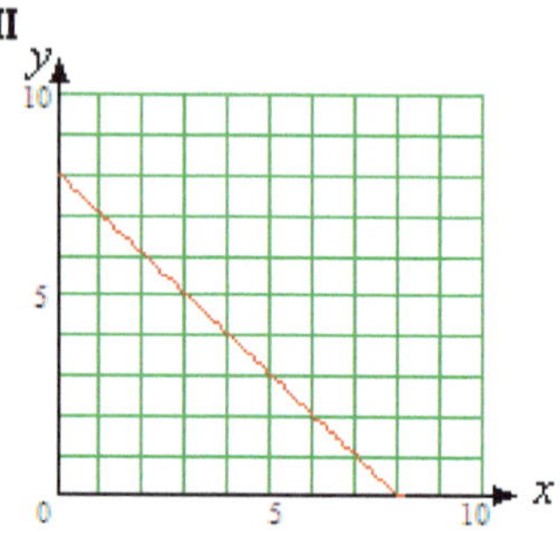

21. Decide which ordered pairs are solutions of the equation $y = \frac{5}{2}x$.

a. $(4, 10)$ **b.** $(5, 2)$

c. $(1.8, 4.5)$ **d.** $(3, 7.5)$

22. Decide which ordered pairs are solutions of the equation whose graph is shown.

a. $(5, 0)$ **b.** $(2, 8)$

c. $(8, 3)$ **d.** $(6, 8)$

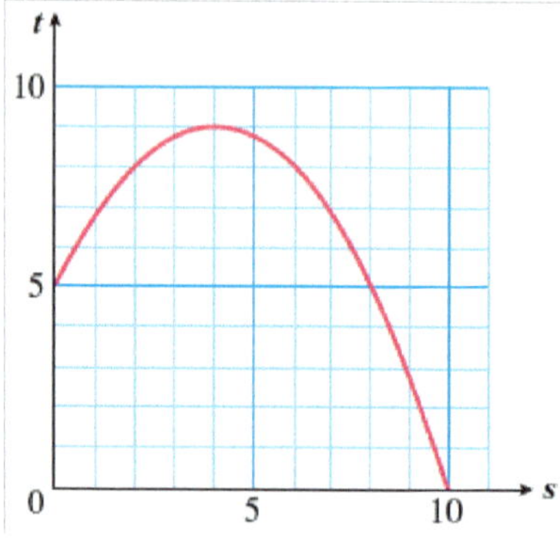

For Problems 23-24, decide whether the given values are solutions of the equation.

23. $\frac{t}{6} = 3$

a. $t = 18$ **b.** $t = 2$

24. $8z = 36.8$

a. $z = 28.8$ **b.** $z = 4.6$

For Problems 25-30, solve the equation.

25. $12 = 3 + y$

26. $2.7 = t - 1.8$

27. $12.2a = 4.88$

28. $8 = \dfrac{y}{4}$

29. $\dfrac{w}{2} = 0$

30. $9.5 = \dfrac{b}{0.6}$

For problems 31-32, solve the equation graphically.

31. The graph of $y = 1.25x$ is shown below left. Use the graph to solve each equation.

a. $1.25x = 40$ **b.** $30 = 1.25x$

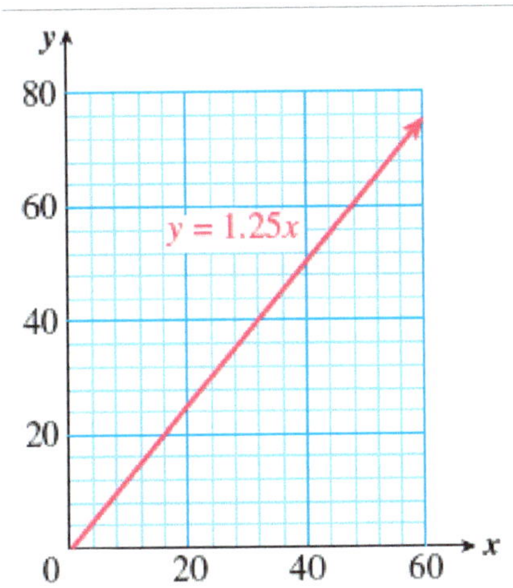

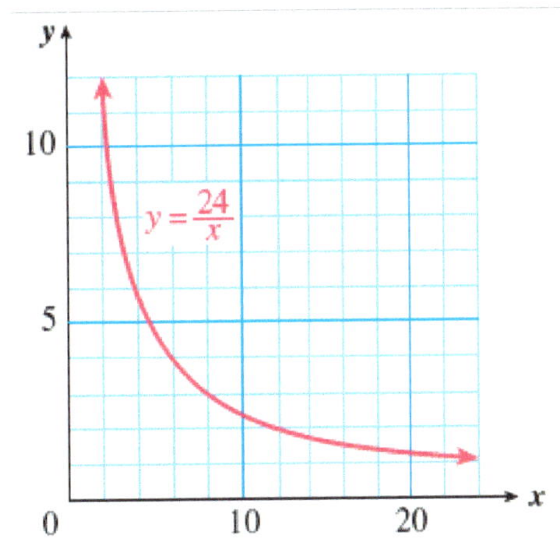

32. The graph of $y = \dfrac{24}{x}$ is shown above right. Use the graph to solve each equation.

a. $\dfrac{24}{x} = 6$ **b.** $1.5 = \dfrac{24}{x}$

For Problems 33-38, write and solve an equation to answer the question.

33. Staci invested some money in a T-bill account that pays $9\frac{1}{2}\%$ interest, and 1 year later the account had earned \$171 interest. How much did Staci deposit in the T-bill?

34. A jet flew 2800 miles at a speed of 560 miles per hour. How long did the trip take?

35. Fifty-three percent of the electorate voted for candidate Phil I. Buster. If candidate Buster received 106,000 votes, how many people voted?

36. Salim's test average was 3.2 points below the minimum grade for an A. If Salim's average was 89.3, what grade was required for an A?

37. Ivan has a new puppy whose adult weight will be about 85 pounds. When the ratio of the puppy's weight to its adult weight is 0.7, Ivan should take the puppy to the vet for shots. How much will the puppy weight when it is ready for shots?

38. If a rectangular garden plot is 12 feet wide, how long must it be to provide 180 square feet of gardening space?

For Problems 39-42, simplify by following the order of operations.

39. $6+18\div 6-3\cdot 2$

40. $\frac{2}{3}+\frac{1}{3}\cdot 4$

41. $36\div(9-3\cdot 2)\cdot 2-24\div 4\cdot 3$

42. $\frac{1.2(7.7)}{4.3-2.3}$

For Problems 43-36, write an algebraic expression for the phrase.

43. Four pages longer than half the length of the first draft
44. $100 short of twice last year's price
45. Three times the sum of the length and 5.6
46. 73% of the difference of Robin's weight and 100 pounds

For Problems 47-50, evaluate for the given values.

47. $mx+b$ for $m=\frac{1}{2},\ x=3,\ b=\frac{5}{2}$

48. $\frac{1}{m}-\frac{1}{n}$ for $m=4,\ n=6$

49. $\frac{3w+z}{z}$ for $w=8,\ z=6$

50. $2(l+w)$ for $l=\frac{1}{3},\ w=\frac{1}{6}$

51. The total amount of money in an account is given by $P(1+rt)$, where P is the initial investment, r is the interest rate, and t is the number of years the money has been in the account. How much money is in an account that earns 10% interest if $1000 was invested 3 years ago?

52. If an object is thrown downwards with initial speed v, then after t seconds it will fall a distance given in feet by $t(16t+v)$. How far will a penny fall in 2 seconds if it is thrown at a speed of 10 feet per second?

Chapter 2 Linear Equations

Lesson 2.1 Signed Numbers

- Perform arithmetic operations on signed numbers
- Write a subtraction fact as an equivalent addition fact
- Simplify quotients involving zero
- Use operations on signed numbers in applications

Lesson 2.2 Algebraic Expressions

- Write an expression with two operations
- Evaluate expressions using the order of operations
- Apply the order of operations to expressions involving signed numbers

Lesson 2.3 Graphs of Linear Equations

- Plot points on a Cartesian coordinate system
- Graph a linear equation in four quadrants
- Graph a linear model
- Use a graph to evaluate an expression

Lesson 2.4 Linear Equations and Inequalities

- Solve equations with two or more operations
- Write an equation to model an applied problem
- Solve an inequality

Lesson 2.5 Like Terms

- Decide whether two expressions are equivalent
- Combine like terms
- Write and simplify expressions using like terms
- Write an equation to model a situation
- Solve equations that involve like terms

2.1 Signed Numbers

Types of Numbers

Numbers greater than zero are called **positive numbers**, and numbers less than zero are **negative numbers**. We use a **number line** to illustrate relationships among numbers.

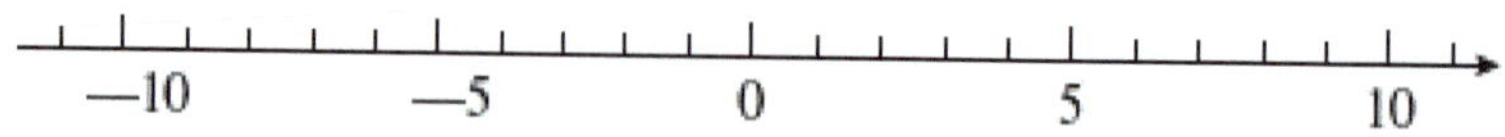

A Number Line

To the right of zero we mark the numbers 1, 2, 3, 4, ... at evenly spaced intervals. These numbers are called the **natural** or **counting numbers**. The **whole numbers** include the natural numbers and zero: 0, 1, 2, 3, ...

> The natural numbers, zero, and the negatives of the natural numbers are called the **integers**:
>
> $$\ldots, -3, -2, -1, 0, 1, 2, 3, \ldots$$

Fractions, such as $\frac{2}{3}$, $-\frac{5}{4}$, and 3.6, lie between the integers on the number line.

> **RQ1.** What are integers?

As we move from left to right on a number line, the numbers increase. In the figure below, the graph of −6 lies to the left of the graph of −2. Therefore, −6 is less than −2, or, equivalently, −2 is greater than −6.

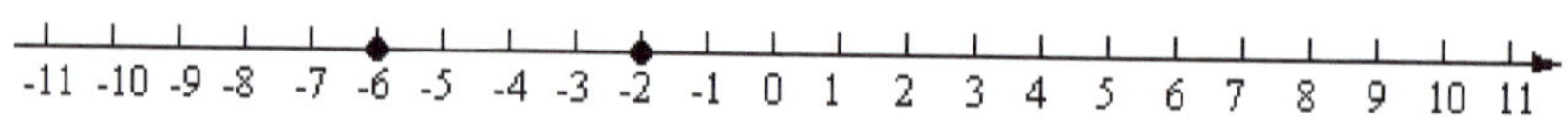

> We use special symbols to indicate order:
>
> $<$ means **is less than**
> $>$ means **is greater than**

For example,

$-6 < -2$ means −6 is less than −2
$-2 > -6$ means −2 is greater than −6

The small ends of the symbols $<$ and $>$ always point to the smaller number.

$3 < 5$ (points to smaller number) $5 > 3$ (points to smaller number)

> **Example 1a.** Which is the lower altitude, −81 feet or −94 feet?
> **b.** Express the relationship using one of the order symbols.
>
> **Solutions a.** Negative altitudes correspond to feet below sea level, and 94 feet is farther below sea level than 81 feet. Therefore, −91 feet is the lower altitude.
> **b.** $-94 < -81$, or $-81 > -94$.

Adding Two Numbers with the Same Sign

We'll use number lines to review operations on signed numbers.

Case 1: The sum of two positive numbers

Illustrate the sum $5+3$ on a number line.

1. Graph the first number, $+5$, as shown below.
2. Move 3 units in the positive direction, or to the right.
3. This brings us to the sum, which is $+8$. Thus, $5+3=8$.

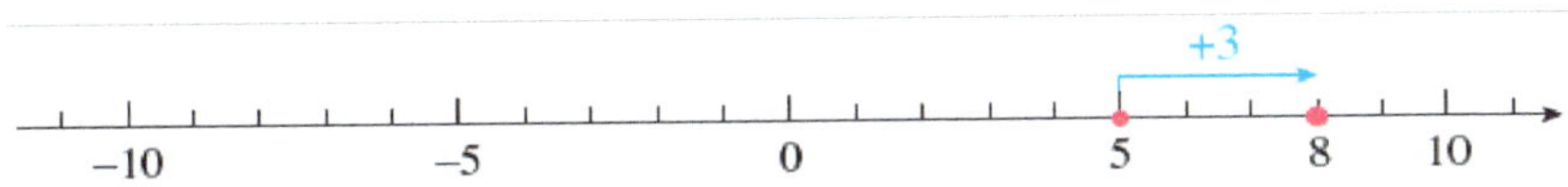

Case 2: The sum of two negative numbers

Think of negative numbers as representing debts. The sum of two debts is an even bigger debt, so we would expect the sum of two negative numbers to be a negative number. Thus, we guess that the sum $(-5)+(-3)$ is -8.

Illustrate the sum $(-5)+(-3)$ on a number line.

1. Graph -5, as shown below.
2. Move 3 units in the negative direction, or to the left.
3. This brings us to the sum, -8, as expected. Thus, $(-5)+(-3)=-8$.

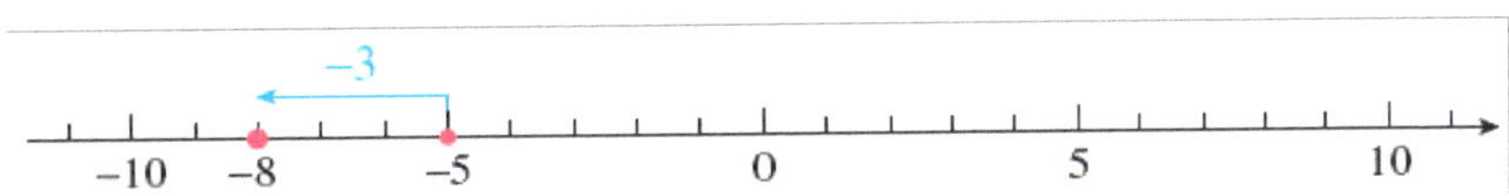

Look Closer: In the examples above, we see that:

> **The sum of two positive numbers is positive.**
> **The sum of two negative numbers is negative.**

Thus, the *sign* of the sum is the same as the sign of the two terms. To find the *value* of the sum, we add the **absolute values**, or unsigned parts, of the two terms.

Example 2 Add $(-4)+(-7)$

Solution The sum is negative. We add the absolute values of -4 and -7 (the numbers 4 and 7 without their signs) to get 11, then make the sum negative: $(-4)+(-7)=-11$ ■

Adding Two Numbers with Opposite Signs

Suppose you have debts of \$9, and assets of \$6. (Your "assets" are your possessions.) What is your net worth? We can model this situation by the sum

$$-9+(+6)$$

where debts are represented by negative numbers and assets by positive numbers. If you use your \$6 to pay off part of your debt, you will still owe \$3, so your net worth is $-\$3$. It makes sense that if your debts are greater than your assets, your net worth is negative.

Illustrate the sum $-9+(+6)$ on a number line.

1. Graph -9, as shown below.
2. Move 6 units in the positive direction, or to the right.
3. This brings us to the sum, -3.

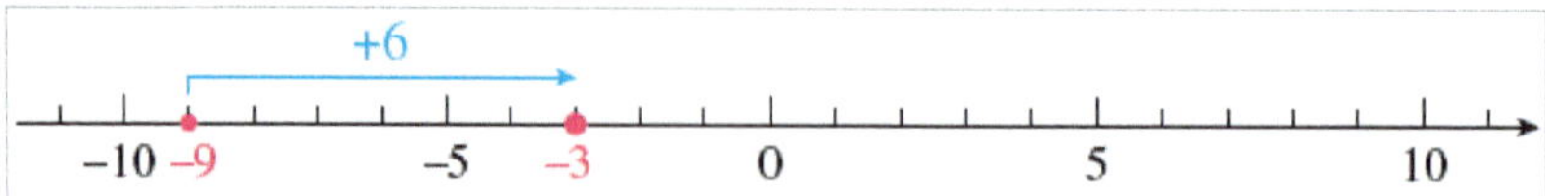

Look Closer: In the example above, we see that the sum has the same sign as the number with the larger absolute value. (Can you explain this in terms of debts and assets?) When the two numbers have opposite signs, we find the value of the sum by **subtracting the absolute values of the two numbers**.

Example 3a. Add $(-5)+(+9)$ **b.** Add $(+5)+(-9)$

Solutions a. The two numbers have opposite signs, so we subtract 5 from 9 to get 4, then make the answer positive, because $+9$ has a larger absolute value than -5. Thus, $(-5)+(+9)=4$.

b. We subtract 5 from 9 to get 4, then make the answer negative, because -9 has a larger absolute value than $+5$. Thus, $(+5)+(-9)=-4$.

RQ2. What is the absolute value of a number?

RQ3. When we add two numbers with opposite signs, we __________ their absolute values.

We now have two rules for adding signed numbers.

Rules for Adding Integers

1. To add two numbers with the **same** sign, add their absolute values. The sum has the same sign as the numbers.
2. To add two numbers with **opposite** signs, subtract their absolute values. The sum has the same sign as the number with the larger absolute value.

Subtracting a Positive Number

Let's compare the two problems:

$$9+(-5) \qquad \text{and} \qquad 9-(+5)$$

Illustrate the sum $9+(-5)$ on a number line.

1. Plot 9 on the number line.
2. Move 5 units in the negative direction, or to the left.
3. The sum is 4.

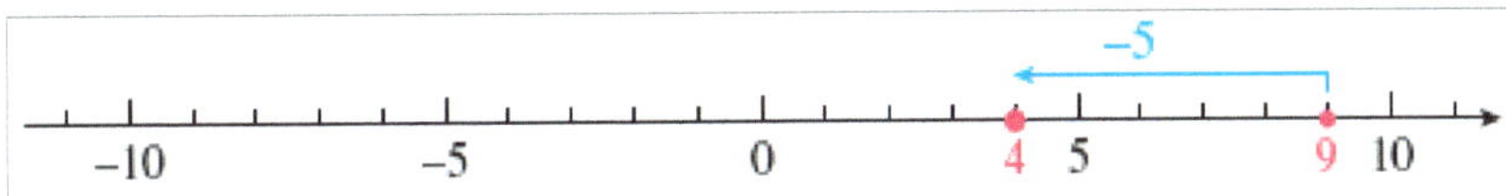

Illustrate the subtraction problem $9-(+5)$ on a number line.

1. Plot 9 on the number line
2. Move 5 units to the left, to indicate subtraction.
3. The result is 4. The picture is identical to the graph above.

Look Closer: We see that subtracting a positive number has the same result as adding the negative number with the same absolute value. Thus, **to subtract a positive number, we add its opposite**.

Subtracting a Negative Number

When we *add* a negative number we move to the *left* on the number line. There are only two directions we can move on a number line, so when we *subtract* a negative number we must move to the *right*.

Compare the graphs for adding –3 and subtracting –3.

Addition: $5+(-3)$ Move 3 units to the **left,** to get 2.

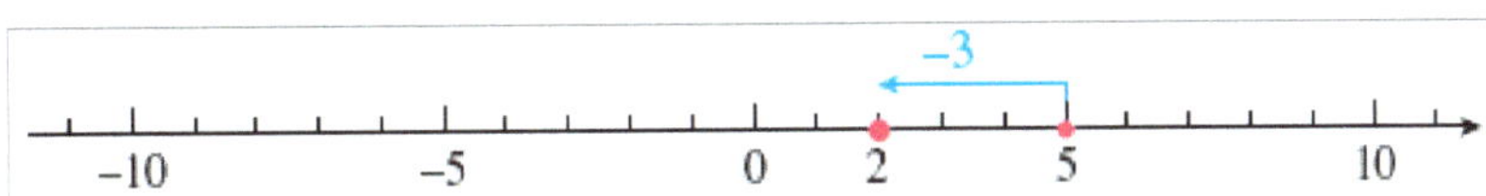

Subtraction: $5-(-3)$ Move 3 units to the **right,** to get 8.

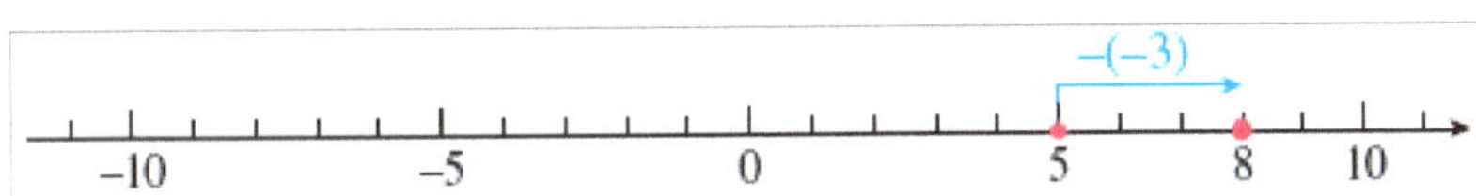

Look Closer: Notice that

$5-(-3)=8$ and $5+(+3)=8$

Subtracting a negative number has the same result as adding the positive number with the same absolute value. Thus, **to subtract a negative number, we add its opposite**.

RQ4. True or False: $-x$ always represents a negative number. Give an example to support your answer.

RQ5. Subtracting a negative number is the same as ______________ .

From the examples above, we see that to subtract any number, positive or negative, we add its opposite. We can do this in steps as follows.

Rule for Subtracting Integers

To subtract b from a:

1. Change the sign of b.
2. Change the subtraction to addition.
3. Proceed as in addition.

This rule tells us that **we can rewrite every subtraction problem as an addition problem by changing the sign of the second number**.

Example 4 Rewrite each subtraction problem as an addition, then compute the answer.

a. $3-(+10)=3+(-10)=-7$

change to addition / change sign

b. $-6-(+3)=-6+(-3)=-9$

change to addition / change sign

c. $2-(-6)=2+(+6)=8$

change to addition / change sign

d. $-7-(-4)=-7+(+4)=-3$

change to addition / change sign

Look Closer: At first it may be difficult to understand why subtracting a negative number is the same as adding a positive number. Try thinking in terms of debts and assets. For example, suppose you have \$25 in your wallet, and you have \$5 in your hand to pay off a debt you owe a friend. If your friend now cancels (or subtracts) the debt, you are actually \$5 richer:

$$25-(-5)=30$$

Caution! For simple subtractions we don't really need the subtraction rule. For example, we can subtract $11-5$ just as we did in arithmetic:

$$11-5=6$$

However, if the subtraction involves negative numbers, it is usually easier to use the subtraction rule. Thus,

$$11-(-5)=11+(+5)=16$$

Look Ahead: We can interpret the expression $7-9$ either as a subtraction or as an addition:

$$7-(+9) \qquad \text{or} \qquad 7+(-9)$$

Both give the same result, -2. Because addition is easier than subtraction, it is better to think of a long expression such as

$$7-9+12-6$$

as a *sum* of signed numbers. To do this, we add all the numbers together, and the $+$ or $-$ sign in front of each number tells us whether that number is positive or negative.

Remember that additions should be performed **in order from left to right**.

Example 5 Simplify $7-9+12-6$

Solution We add the signed numbers 7 and −9 and 12 and −6, in order from left to right.

$$7-9+12-6=-2+12-6 \quad \text{The sum of 7 and } -9 \text{ is } -2;$$
$$=10-6=4 \quad \text{the sum of } -2 \text{ and 12 is 10.}$$

Caution! The following calculation is incorrect:

$$16-9-4+3=16-5+3=11+3=14 \quad \leftarrow \text{Incorrect!}$$

We must perform additions and subtractions in order from left to right:

$$16-9-4+3=7-4+3=3+3=6$$

Multiplying Signed Numbers

Case 1: The product of numbers with opposite signs

We can think of multiplication as repeated addition. For example,

$3(2)$ means $2+2+2$ The sum of three 2's

Similarly,

$3(-2)$ means $-2+(-2)+(-2)$ The sum of three −2's

which is −6, so $3(-2)=-6$. Think in terms of money: if you owe three different people 2 dollars each, then you are 6 dollars in debt.

Because multiplication is commutative, it is also true that $-2(3)=-6$. This example illustrates the following fact.

The product of a positive number and a negative number is a negative number.

Example 6 You can use your calculator to verify each of the following products.

a. $(-4)(7)=-28$ **b.** $5(-2.6)=-13$

Case 2: The product of numbers with the same sign

You know that the product of two positive numbers is positive. It is also true that the product of two negative numbers is positive.

To understand why the product of two negative numbers is positive, you might think of canceling, or subtracting, several debts of the same amount. Or, you could observe the pattern in the lists of products below. As we move down the list, the product increases by equal amounts.

$3(-2)=-6$	$3(-5)=-15$	
$2(-2)=-4$	$2(-5)=-10$	
$1(-2)=-2$	$1(-5)=-5$	← Positive times negative is negative.
$0(-2)=0$	$0(-5)=0$	
$-1(-2)=2$	$-1(-5)=5$	← Negative times negative is positive.
$-2(-2)=4$	$-2(-5)=10$	
$-3(-2)=6$	$-3(-5)=15$	

Based on the pattern in these calculations, the following rule seems reasonable.

The product of two negative numbers is a positive number.

Example 7 You can use your calculator to verify the following products.

a. $(-6)(-3) = 18$ **b.** $(-2.1)(-3.4) = 7.14$

Now we have a pair of rules about the products of signed numbers.

Products of Signed Numbers

1. The product of two numbers with opposite signs is a negative number.
2. The product of two numbers with the same sign is a positive number.

Dividing Signed Numbers

Division is the inverse operation for multiplication. For example, we know that

$$\frac{12}{3} = \boxed{4} \quad \text{because} \quad 3 \cdot \boxed{4} = 12$$

Thus, every division fact can be converted into an equivalent multiplication fact. For example,

$$\text{The division fact } \frac{400}{16} = 25 \text{ is equivalent to } 16 \cdot 25 = 400$$

Note that the *numerator* in the division fact becomes the *product* in the multiplication fact. The same relationship between multiplication and division holds for negative numbers as well.

Example 8 Rewrite each division fact as a multiplication fact.

a. $\frac{-144}{64} = -2.25$ **b.** $-36 \div \frac{-3}{8} = 96$

Solutions The numerator in each division fact becomes the product in the multiplication fact.

a. $-144 = -2.25(64)$ **b.** $-36 = \frac{-3}{8}(96)$

Example 8 illustrates the rules for division of signed numbers.

Quotients of Signed Numbers

1. The quotient of two numbers with opposite signs is a negative number.
2. The quotient of two numbers with the same sign is a positive number.

RQ6. Which of these is a product?

$3 + (-4) \qquad -3 - 4 \qquad -3(-4) \qquad 3 - (-4)$

RQ7. Rewrite the division fact $\frac{48}{D} = Q$ as a multiplication fact.

Remember that a fraction can be thought of as a division. For example, $\frac{3}{4}$ means $3 \div 4$, or 0.75. What about negative fractions? Because the quotient of two numbers with unlike signs is negative, the fractions $\frac{-3}{4}$ and $\frac{3}{-4}$ both represent the negative number $-\frac{3}{4}$. On the other hand, $\frac{-3}{-4}$ is equal to the positive number $\frac{3}{4}$.

For many of the calculations in algebra, it is more convenient to write a negative fraction with the minus sign in the numerator, like this: $\frac{-3}{4}$.

> A negative fraction is written in **standard form** when the minus sign is in the numerator: $\frac{-a}{b}$.
>
> $$\frac{-a}{b} = \frac{a}{-b} = -\frac{a}{b}$$
>
> The standard form for a positive fraction has positive numerator and denominator.
>
> $$\frac{a}{b} = \frac{-a}{-b}$$

Quotients Involving Zero

What is the meaning of an expression such as $\frac{0}{2}$? We can rewrite this quotient as an equivalent multiplication fact.

$$\frac{0}{2} = \boxed{?} \quad \text{is equivalent to} \quad 2 \cdot \boxed{?} = 0$$

What number can we substitute for the question mark to make $2 \cdot \boxed{?} = 0$ a true statement? The only solution for this equation is 0. Thus, replacing the question marks by 0, we have

$$\frac{0}{2} = 0 \quad \text{because} \quad 2 \cdot 0 = 0$$

In general, **0 divided by any (non-zero) number is 0.**

Now consider a quotient with zero in the denominator, such as $\frac{2}{0}$.

$$\frac{2}{0} = \boxed{?} \quad \text{is equivalent to} \quad 0 \cdot \boxed{?} = 2$$

What number can we substitute for the question mark to make $0 \cdot \boxed{?} = 2$ a true statement? This equation has no solution, because 0 times any number is 0, not 2. Thus, there is no answer for the division problem $\frac{2}{0}$. We say that the quotient $\frac{2}{0}$ is **undefined**. In general, **we cannot divide any number by zero**.

> **Quotients Involving Zero**
>
> If a is any nonzero number, then
>
> $$\frac{0}{a} = 0 \quad \text{and} \quad \frac{a}{0} \text{ is undefined.}$$

Example 9a. $6 \div 0$ is undefined. **b.** $0 \div 8 = 0$

RQ8. Which of these is undefined? $\frac{-6}{0}$ or $\frac{0}{-6}$

Skills Warm-Up

1. Graph the numbers on the number line: $0,\ -3,\ 3,\ \frac{1}{2},\ \frac{-5}{3},\ 5\frac{1}{4}$

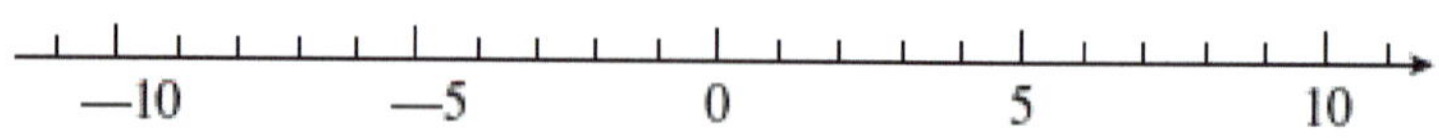

2. Graph the numbers on the number line: $-4.5,\ -2,\ -1.4,\ 0.6,\ 2$

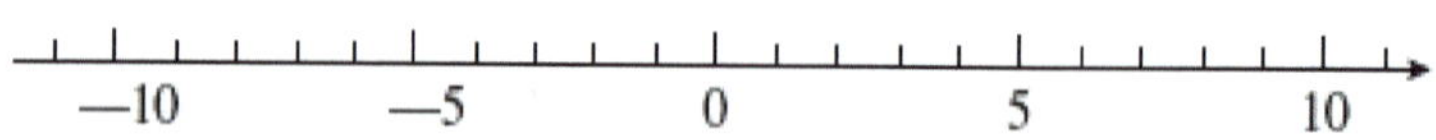

Replace the comma in each pair by the proper symbol, $<$, $>$, or $=$.

3. **a.** $0, -4$ **b.** $-5, -9$ **c.** $-2\frac{5}{8}, -2\frac{1}{8}$ **d.** $13.6, 13.66$

4. **a.** $3, -7$ **b.** $-1, -6$ **c.** $-3, -3\frac{3}{4}$ **d.** $-18.4, -19.6$

Answers: **3.a.** $>$ **b.** $>$ **c.** $<$ **d.** $<$ **4.a.** $>$ **b.** $>$ **c.** $>$ **d.** $>

Homework 2.1

Skills Practice

For Problems 1-4, add.

1. **a.** $5+(-3)$ **b.** $-5+3$ **c.** $-5+(-3)$

2. **a.** $-15+(-20)$ **b.** $15+(-20)$ **c.** $-15+20$

3. **a.** $-47+22$ **b.** $6.8+(-2.7)$ **c.** $-\frac{5}{6}+\frac{2}{3}$

4. **a.** $-13+(-36)$ **b.** $-2.5+4.9$ **c.** $-\frac{3}{4}+\left(-\frac{3}{8}\right)$

5. Rewrite each subtraction problem as an addition problem, and give the answer.

a. $4-8$ **b.** $3-(-9)$ **c.** $-8-(-6)$ **d.** $-6-5$

For Problems 6-8, add or subtract as indicated.

6. **a.** $12+(-6)$ **b.** $6-(-4)$ **c.** $-2-8$ **d.** $-7+9$

7. **a.** $-14-(-3)$ **b.** $-5+(-4)$ **c.** $-6-(-6)$ **d.** $-4-4$

8. **a.** $24-(-10)$ **b.** $18+(-12)$ **c.** $-16+14$ **d.** $-25-(-15)$

For Problems 9-10, compute each sum or difference in parts (a) and (b), and decide which is easier. Then decide how to simplify the expression in part (c).

9. **a.** $15-(+5)$ **b.** $15+(-5)$ **c.** $15-5$

10. **a.** $-6-(+2)$ **b.** $-6+(-2)$ **c.** $-6-2$

For Problems 11-12, compute mentally.

11. **a.** $2-7$ **b.** $-2-7$ **c.** $-2+7$ **d.** $2-(-7)$ **e.** $-2-(-7)$

12. **a.** $13-5$ **b.** $-13-5$ **c.** $-13+5$ **d.** $13-(-5)$ **e.** $-13-(-5)$

For Problems 13-14, add.

13. **a.** $-5+3+(-4)$ **b.** $-4+(-7)+(-7)$

14. **a.** $6+(-14)+12+(-17)$ **b.** $-35+(-5)+28+(-21)+13+(-14)$

15. Simplify by adding the signed numbers.

a. $2+5-8-1$ **b.** $-23+28-14+21$ **c.** $-34-52+68-21$

16. Simplify by following the rules for addition and subtraction.

a. $-6+5+(-3)-(-8)$ **b.** $-11-2-(-4)-(-3)$ **c.** $-14-(-16)-4+(-7)$

For problems 17-20, multiply or divide.

17. **a.** $(-8)(-4)$ **b.** $\frac{12}{-4}$ **c.** $-20\div(-5)$

18. **a.** $-6(-1)(3)$ **b.** $\frac{-8}{0}$ **c.** $(-5)(0)(6)$

19. **a.** $0\left(\frac{-7}{15}\right)$ **b.** $(-2)(-2)(-2)$ **c.** $-30\div 0$

20. a. $\left(\frac{1}{2}\right)\left(\frac{-3}{4}\right)$ **b.** $-0.1(-26)$ **c.** $\frac{-1}{2} \div \left(\frac{-3}{4}\right)$

For Problems 21-22, perform the indicated operations.

21. a. $-12 - 4$
b. $-12(-4)$
c. $-12 - (-4)$
d. $-12 + (-4)$
e. $-12 \div 4$

22. a. $-3 - 3$
b. $-3(-3)$
c. $-3 - (-3)$
d. $-3 + (-3)$
e. $-3 \div (-3)$

23. a. Find two numbers whose sum equals zero.
b. Find two numbers whose difference equals zero.
c. Find two numbers whose product equals zero.
d. Find two numbers whose quotient equals zero.

24. a. Find two numbers whose product is 1.
b. Find two numbers whose product is -1.
c. Find two numbers whose quotient is 1.
d. Find two numbers whose quotient is -1.

For Problems 25-30, solve.

25. $x - 9 = -4$ **26.** $-9z = 12$ **27.** $\frac{-a}{4} = 8$

28. $9 - x = 3$ **29.** $t + 5 = -8$ **30.** $-6b = -27$

31. a. What is the solution of the equation $-x = -3$?
b. What is the solution of the equation $-x = 6$?

32. Find and correct the error in the following calculation:

$$18 - 6 + 2 - 5 = 18 - 8 + 5 = 15 \quad \text{(Incorrect!)}$$

Applications

For problems 33-40, write and simplify an expression using negative numbers to answer the question.

33. Thelma's failing company is worth $-\$1000$. She decides to merge with Louise's company, whose net worth is \$1500. What is the combined worth of the two companies?

34. Emily is at an elevation of -87 feet when she begins climbing a mountain. What is her elevation after ascending 127 feet?

35. The temperature at noon was 2° and dropped 8° over the next 12 hours. What was the temperature at midnight?

36. Rocky lost \$450 betting on the horses this afternoon and then dropped \$245 at a poker game in the evening. What was the net change in his financial status for the day?

37. A tourist in California can travel from Death Valley, at an elevation of −282 feet, to Mt. Whitney, at 14,494 feet. What is his net change in elevation?

38. Despite being overdrawn by $24.20, Nelson writes a check for $11.20. What is his new balance?

39. Whitney is climbing down a sheer cliff. She has pitons spaced vertically every 6 meters. What is her net elevation change after descending to the eighth piton from the top?

40. The temperature on the ice planet Hoth dropped 115° in just 4 hours. What was the average temperature change per hour?

Choose the correct equation to model Problems 41-44.

$-12 + n = -5$	$-12 + n = 5$
$n + 12 = -5$	$n + 12 = 5$

41. Last night the temperature rose 12° and this morning the temperature is 5°. What was the temperature yesterday?

42. Marta's score was −12, but after the last hand her score rose to −5. How many points did she make on the last hand?

43. Bradley rode the elevator down 12 floors and emerged five floors below ground. What floor did he start on?

44. Orrin was $12 in debt, but he did some yard work this weekend and after paying his debt he has $5. How much did he make?

2.2 Expressions and Equations

Writing Algebraic Expressions

When we write an algebraic expression, we use the same operations on a variable that we would use to calculate with a specific number. Writing down the expression for a specific numerical value can help us write an algebraic expression.

Example 1 Alida keeps \$100 in cash from her weekly paycheck, and deposits 40% of the remainder in her savings account. If Alida's paycheck is p, write an expression for the amount she deposits in savings.

Solution How would we calculate Alida's deposit if we knew her paycheck? Suppose Alida's paycheck is \$500. First she subtracts \$100 from that amount to get $500 - 100$, and then she takes 40% of the remainder for savings:

$$0.40\,(500 - 100)$$ Perform operations inside parentheses first.

If her paycheck is p dollars, we perform the same operations on p instead of on 500. The expression is thus

$$0.40\,(p - 100)$$

RQ1. We write an expression with variables using the same operations we would use to calculate with a ________________ .

RQ2. In Example 1, how does Alida calculate the amount to deposit in savings? How do we write that in an algebraic expression?

Evaluating Algebraic Expressions

When we evaluate an algebraic expression, we must follow the order of operations.

Example 2 Four students bought concert tickets with their \$20 student-discount coupons.

a. If the regular price of a ticket is t dollars, write an expression for the total amount the four students paid.

b. How much did the students pay if the regular price of a ticket is \$38?

Solutions a. The discount price of one ticket is $t - 20$ dollars. So four tickets cost

$$4(t - 20) \text{ dollars}$$

b. We evaluate the expression in part (a) for $t = 38$.

$$\begin{aligned} 4(t - 20) &= 4(38 - 20) \\ &= 4(18) = 72 \end{aligned}$$ Simplify inside parentheses first.

The students paid \$72 for the tickets.

RQ3. Explain why we need parentheses in Example 2.

RQ4. How should we write "3 times the sum of x and 12"?

Negative Numbers

The order of operations applies to signed numbers. Operations inside parentheses or other grouping devices should be performed first.

Example 3 Simplify $4-3-[-6+(-5)-(-2)]$

Solution Perform the operations inside brackets first. Simplify each step by rewriting subtractions as equivalent additions.

$4-3-[-6+(-5)-(-2)]$	$-5-(-2)=-5+2$
$=4-3-[-6-5+2]$	$-6-5=-11;\ -11+2=-9$
$=4-3-[-9]$	Rewrite as an addition.
$=4-3+9=10$	Add from left to right.

RQ5. Which operation is performed first in the expression $6-4x$?
RQ6. What is the first step in evaluating the expression $2(18-x)$?

Caution! When using negative numbers, we must be careful to distinguish between products and sums. In Example 4, note how the parentheses and minus signs are used in each expression.

Example 4 Simplify each expression.

a. $3(-8)$ **b.** $3-(-8)$ **c.** $3-8$ **d.** $-3-8$

Solutions a. This expression is a product: $3(-8)=-24$

b. This is a subtraction. We follow the rule for subtraction by changing the sign of the second number and then adding: $3-(-8)=3+8=11$

c. This is an addition; the negative sign in front of 8 tells us that we are adding -8 to 3. Thus, $3-8=3+(-8)=-5$

d. This is also an addition: $-3-8=-3+(-8)=-11$

RQ7. What is wrong with this calculation: $4-3+9=4-12=-8$

Look Closer: When we evaluate an algebraic expression at a negative number, we enclose the negative numbers in parentheses. This will help prevent us from confusing multiplication with subtraction.

Example 5 Evaluate $2x-3xy$ for $x=-5$ and $y=-2$.

Solution We substitute -5 for x and -2 for y, then follow the order of operations.

$2x-3xy=2(-5)-3(-5)(-2)$	Do multiplications first.
$=-10-(-15)(-2)$	
$=-10-30=-40$	

RQ8. When we evaluate an expression at a negative number, we should ________ the negative number in ________________ .

Skills Warm-Up

Choose the correct algebraic expression for each of the following situations.

$n+12$	$12n$	$\frac{n}{12}$
$\frac{12}{n}$	$12-n$	$n-12$

1. Helen bought n packages of tulip bulbs. If each package contains 12 bulbs, how many bulbs did she buy?

2. Henry bought a package of n gladiolus bulbs, then bought 12 loose bulbs. How many bulbs did he buy?

3. Together Karen and Dave sold 12 tickets to the spring concert. If Karen sold n tickets, how many did Dave sell?

4. Together Karl and Diana collected n used books for the book sale. If Karl collected 12 books, how many did Diana collect?

5. Greta made n dollars last week. If she worked for 12 hours, how much did she make per hour?

6. Gert jogged for 12 minutes. If she jogged n miles, how many minutes does it take her to jog 1 mile?

Answers: **1.** $12n$ **2.** $n+12$ **3.** $12-n$ **4.** $n-12$ **5.** $\frac{n}{12}$ **6.** $\frac{12}{n}$

Homework 2.2

Skills Practice

1. The perimeter of a rectangle of length l and width w is given by $P = 2l + 2w$. Find the perimeter of a rectangular meeting hall with dimensions 8.5 meters by 6.4 meters.

2. The area of a trapezoid with bases B and b and height h is given by $A = \frac{1}{2}(B + b)h$. Find the area of a trapezoid whose bases are 9 centimeters and 7 centimeters and whose height is 3 centimeters.

3. If a company sells n items at a cost of c dollars each and sells then at a price of p dollars each, the company's profit is given by $P = n(p - c)$. Find the profit earned by a manufacturer of bicycle equipment by selling 300 bicycle helmets that cost $32 each to produce and sell for $50 apiece.

4. The temperature in degrees Celsius (°C) is given by $C = \dfrac{5}{9}(F - 32)$, where F stands for the temperature in degrees Fahrenheit (°F). Find normal body temperature in degrees Celsius if normal temperature is 98.6° Fahrenheit.

In Problems 5-8, simplify by following the order of operations.

5. $-18 - [8 - 12 - (-4)]$
6. $3 - (-6 + 2) + (-1 - 4)$
7. $-7 + [-8 - (-2)] - [6 + (-4)]$
8. $0 - [5 - (-1)] + [-6 - 3]$

9. Find each product.
   - a. $(-2)(3)(4)$
   - b. $(-2)(-3)(4)$
   - c. $(-2)(-3)(-4)$
   - d. $(-2)(-3)(4)(2)$
   - e. $(-2)(-3)(-4)(2)$
   - f. $(-2)(-3)(-4)(-2)$

10. Use your results from Problem 9 to complete the statements:
    - a. The product of an odd number of negative numbers is ________ .
    - b. The product of an even number of negative numbers is ________ .

In Problems 11-14, use the order of operations to simplify each expression.

11a. $-2(-3) - 4$ **b.** $5(-4) - 3(-6)$

12a. $(-4 - 3)(-4 + 3)$ **b.** $-3(8) - 6(-2) - 5(2)$

13a. $\dfrac{15}{-3} - \dfrac{4 - 8}{8 - 12}$ **b.** $\dfrac{4 - 2(-5)}{-4 + 3(-1)}$

14a. $12.6 - 0.32(0.25)(4.2 - 8.7)$ **b.** $(5.8 - 2.6)(-2.5)(-0.6 + 3)$

15. Simplify each expression.
    - a. $-3(-4)(-5)$
    - b. $-3(-4) - 5$
    - c. $-3(-4 - 5)$
    - d. $-3 - (-4 - 5)$
    - e. $-3 - (-4)(-5)$
    - f. $(-3 - 4)(-5)$

16. Simplify each expression.
    - a. $24 \div 6 - 2$
    - b. $24 \div (-6 - 2)$
    - c. $24 \div (-6) - 2$
    - d. $24 - 6 \div (-2)$
    - e. $24 \div (-6) \div (-2)$
    - f. $24 \div (-6 \div 2)$

In Problems 17-22, evaluate for the given values.

17. $15 - x - y$ for $x = -6,\ y = 8$

18. $p - (4 - m)$ for $p = -2,\ m = -6$

19. $12x - 3xy$ for $x = -3,\ y = 2$

20. $\dfrac{y-3}{x-4}$ for $x = -9,\ y = 2$

21. $\dfrac{1}{2}t(t-1)$ for $t = \dfrac{2}{3}$

22. $\dfrac{x-m}{s}$ for $x = 4,\ m = \dfrac{9}{4},\ s = \dfrac{3}{2}$

23. a. Evaluate each expression for $x = 5$. What do you notice?

(i) $\dfrac{-3}{4}x$ **(ii)** $\dfrac{-3x}{4}$ **(iii)** $-0.75\,x$

b. Does $\dfrac{-8}{5}x = \dfrac{-8}{5x}$? Support your answer with examples.

24. Which of the following are equivalent to $-\dfrac{4}{9}x$?

a. $\dfrac{4}{9}(-x)$ **b.** $\dfrac{-4x}{-9}$ **c.** $\dfrac{-4}{9x}$ **d.** $\dfrac{-4x}{9}$ **e.** $\dfrac{-4}{9}(-x)$

Write an algebraic expression for area or perimeter.

25.

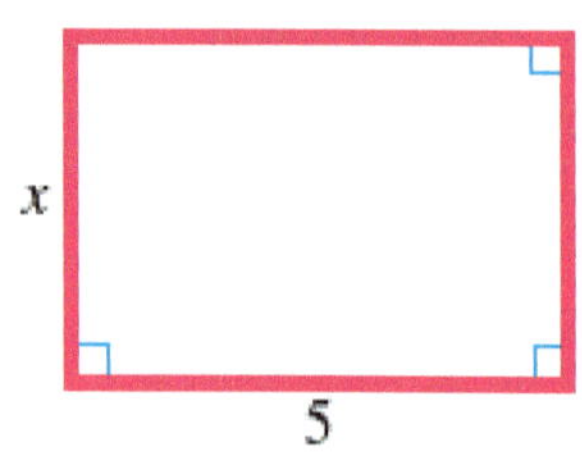

Perimeter =

26.

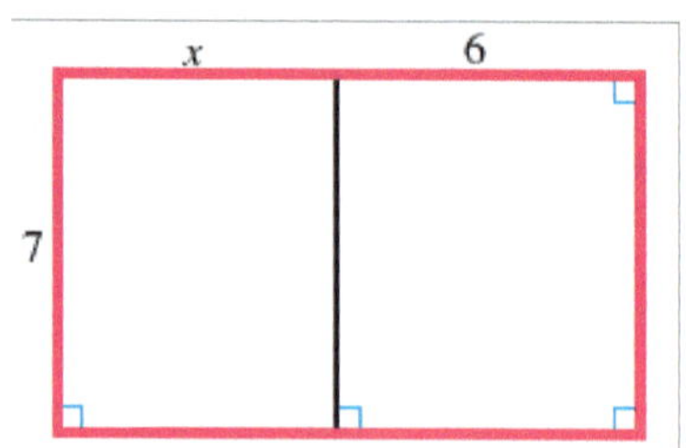

Area =

Applications

27. Salewa saved $5000 to live on while going to school full time. She spends $200 per week on living expenses.

a. How much of Salewa's savings will be left after 3 weeks? After 12 weeks?

b. Describe in words how you calculated your answers to part (a).

c. Fill in the table below.

Number of weeks	2	4	5	6	10	15	20
Calculation							
Savings left							

d. Write an algebraic expression for Salewa's savings after w weeks.

e. Write an equation that gives Salewa's savings, S, in terms of w.

28. To calculate how much state income tax she owes, Francine subtracts \$2000 from her income, and then takes 12% of the result.

a. What is Francine's state income tax if her income is \$8000? If her income is \$10,000?

b. Describe in words how you calculated your answers to part (a).

c. Fill in the table below.

Income	5000	7000	12,000	15,000	20,000	24,000	30,000
Calculation							
State Tax							

d. Write an algebraic expression for Francine's state income tax if her income is I.

e. Write an equation that gives Francine's state income tax, T, in terms of I.

For Problems 29-32,

a. Choose a variable for the unknown quantity and write an algebraic expression.

b. Evaluate the expression for the given values.

29. a. Three inches less than twice the width

b. The width is 13 inches.

30. a. Twenty dollars more than 40% of the principal

b. The principal is \$500.

31. a. \$8 times the number of children's tickets subtracted from 150

b. There are 83 children's tickets

32. a. One-third of \$50 less than the profit

b. The profit is \$500

In Problems 33-40, write an algebraic expression.

33. The oven temperature started at 65° and is rising at 30° per minute. Write an expression for the oven temperature after t minutes.

34. Luisa's parents have agreed to pay her tuition (\$800 per year) plus half her annual living expenses while she is in school. Write an expression for the amount her parents will pay if Luisa's annual expenses are a dollars.

35. Mildred canned 80 pints of tomatoes. She kept some for herslf and divided the rest equally among her four daughters. If Mildred kept M pints, write an expression for the number of pints she gave each daughter.

36. Moira's income is \$50 more than one-third of her mother's income. Write an expression for Moira's income if her mother's income is I.

37. Otis buys 200 pounds of dog food and uses 15 pounds a week for his dog Ralph. Write an expression for the amount of dog food Otis has left after w weeks.

38. Renee receives \$600 for appearing in a cola commercial, plus a residual of \$80 each time the commercial is aired. Write an expression for Renee's earnings if the commercial plays t times.

39. Digby bought scuba diving gear by making a \$50 down payment and arranging to pay the balance in ten equal installments. If the total cost of the gear is C dollars, write an expression for the amount of each installment.

40. Each passengers and three crew members on a small airplane is allowed 35 pounds of luggage. Write an expression for the weight of the luggage if there are x passengers.

In Problems 41-42,

a. Write an algebraic expression.

b. Make a table for the expression showing at least two positive values for the variable and two negative values.

41. The ratio of a number to 4 more than the number.

42. The sum of a and 5, times the difference of a and 5.

2.3 Graphs of Linear Equations

Graphing Equations

An equation of the form $y = ax + b$, where a and b are constants, is called a **linear equation** because its graph is a straight line.

We can graph a linear equation by evaluating the expression $ax + b$ at several values of x and then plotting points.

Example 1 At 6 am the temperature was 50°, and it has been rising by 4° every hour.

a. Write an equation for the temperature, T, after h hours.

b. Graph your equation.

Solutions a. We set $h = 0$ at 6 am. At that time, $T = 50$. Because the temperature is rising by 4° every hour, we multiply the number of hours by 4 and add the result to 50 to get the new temperature. Thus, our equation is $T = 50 + 4h$.

b. Step 1 We make a table of values for the equation. We choose some values for h, as shown in the first column of the table below. For each value of h, we evaluate the formula for T, and we record these values in the second column of the table.

h	T		Ordered pairs
0	50	$T = 50 + 4(\mathbf{0}) = 50 + 0 = 50$	$(0, 50)$
1	54	$T = 50 + 4(\mathbf{1}) = 50 + 4 = 54$	$(1, 54)$
3	62	$T = 50 + 4(\mathbf{3}) = 50 + 12 = 62$	$(3, 62)$
5	70	$T = 50 + 4(\mathbf{5}) = 50 + 20 = 70$	$(5, 70)$

Step 2 We use the values in the table to help us choose scales for the axes. For this graph, the x-values are less than 10, but the largest y-value is 70. We use increments of 1 on the horizontal or h-axis and increments of 10 on the vertical or T-axis.

Step 3 Each ordered pair gives us a point on the graph. We plot the points on the grid and connect them with a smooth curve. All the points should lie on a straight line.

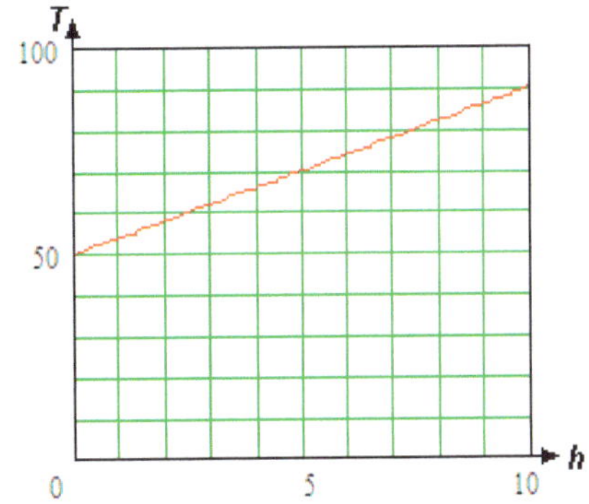

RQ1. A linear equation has the form ____________.

RQ2. We use __________ to help us choose scales for the axes.

Cartesian Coordinate System

Many graphs include negative values as well as positive values. To make such a graph, we construct a **Cartesian coordinate system**, named after the French mathematician and philosopher René Descartes.

We draw two perpendicular number lines for the horizontal and vertical axes, as shown below left. The two axes divide the plane into four **quadrants**, or regions, as shown at right. We often use x for the input variable and y for the output variable, so the horizontal axis is called the **x-axis** and the vertical axis is the **y-axis**. The axes intersect at the zero point of each number line. The coordinates of this intersection point, called the **origin**, are $(0, 0)$.

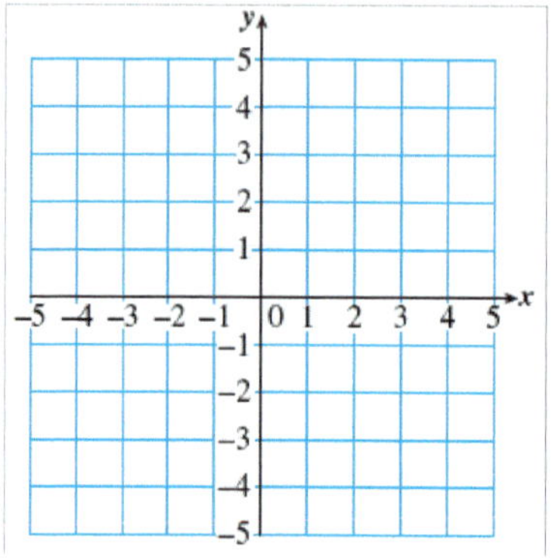

RQ3. What are the coordinates of the origin?
RQ4. True or False: The x-coordinate of any point on the y-axis is zero.

Example 2 The figure at right shows how to plot points on a Cartesian grid. Points in the upper right quadrant have both coordinates positive. Points with negative x-coordinates are plotted to the left of the y-axis, and points with negative y-coordinates are plotted below the x-axis. For example, the point $(-2, 1)$ is plotted 2 units to the left of the y-axis and 1 unit above the x-axis.

In which quadrant are both coordinates negative?

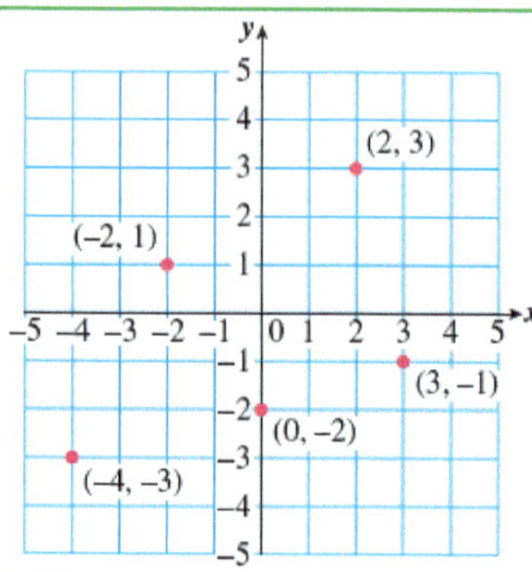

RQ5. True or False: The notation $(5, 3)$ represents two points, one on the x-axis and one on the y-axis.
RQ6. True or False: If both coordinates of a point are negative, the point is located in the lower left quadrant.

Now we can graph an equation that includes negative values.

Example 3 The temperature in Nome was -12°F at noon. It has been rising at a rate of 2°F per hour all day.

a. Write an equation for the temperature, T, after h hours.

b. Fill in the table and graph your equation. (Negative values of h represent hours before noon.)

h	-3	-2	-1	0	1	2	3
T							

Use your graph to answer the following questions:

c. What was the temperature 8 hours before noon?

d. When will the temperature reach $-4°$F?

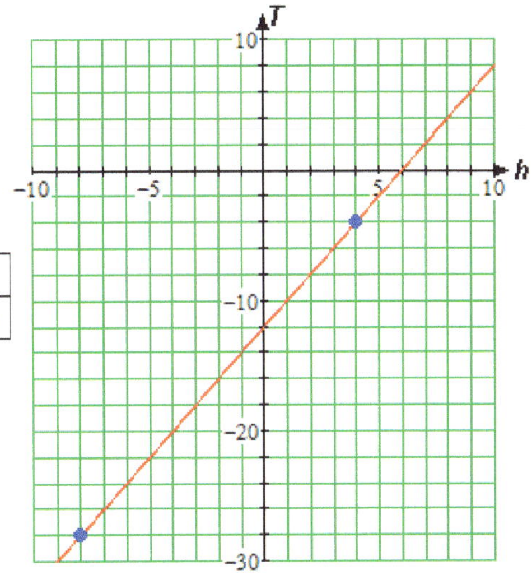

Solutions a. The temperature at h hours after noon is given by $T = -12 + 2h$.
b. We evaluate the expression for T at each of the h-values in the table.

h	−3	−2	−1	0	1	2	3
T	−18	−16	−14	−12	−10	−8	−6

c. We find the point on the graph with h-coordinate −8. Its T-coordinate is −28, so the temperature was −28°.
d. We find the point on the graph with T- coordinate −4. Its h-coordinate is 4, so it will be −4° at 4 pm.

Look Closer: We can also answer parts (c) and (d) of Example 3 using algebra. For part (c), we evaluate the expression for T, namely $-12 + 2h$, at $h = -8$. For part (d), we solve the equation $T = -12 + 2h$ with $T = -4$. Make sure that you understand both methods, graphical and algebraic.

Graphing Tip #1

If the coefficient of x is a fraction, we can make our work easier by choosing multiples of the denominator for the x-values. That way we won't have to work with fractions to find the y-values. In Example 4, the coefficient of x is $\frac{-2}{3}$, so we choose multiples of 3 for the x-values.

Example 4a. Graph $y = \dfrac{-2}{3}x - 4$
b. Use the graph to solve the equation $\dfrac{-2}{3}x - 4 = -2$.

Solutions a. We choose x-values that are multiples of 3 and make a table of values.

x	y
−9	2
−6	0
0	−4
3	−6
6	−8

$y = -\frac{2}{3}(\mathbf{-9}) - 4 = 2$
$y = -\frac{2}{3}(\mathbf{-6}) - 4 = 0$
$y = -\frac{2}{3}(\mathbf{0}) - 4 = -4$
$y = -\frac{2}{3}(\mathbf{3}) - 4 = -6$
$y = -\frac{2}{3}(\mathbf{6}) - 4 = -8$

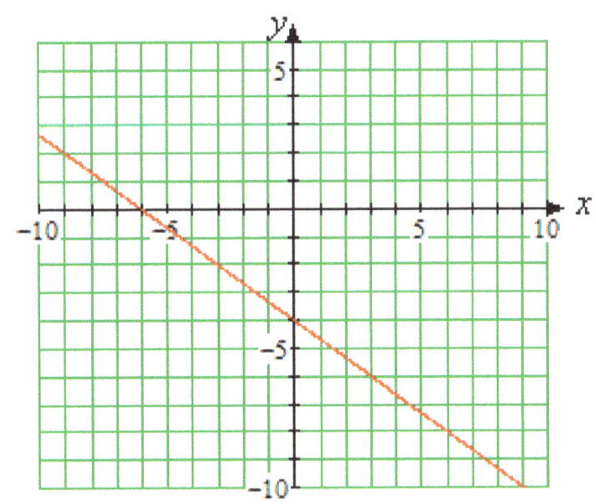

Then we plot the points and connect them with a straight line.
b. We see that y has been replaced by −2 in the equation of the graph. So we look for the point on the graph that has y-coordinate −2. This point, labeled P on the graph

at right, has x-coordinate -3. Because it lies on the graph, the point $P(-3, -2)$ is a solution of the equation $y = \frac{-2}{3}x - 4$.

When we substitute $x = -3$ and $y = -2$ into the equation, we get a true statement:

$$-2 = \frac{-2}{3}(-3) - 4$$

But this statement also tells us that -3 is a solution of the equation $-2 = \frac{-2}{3}x - 4$.

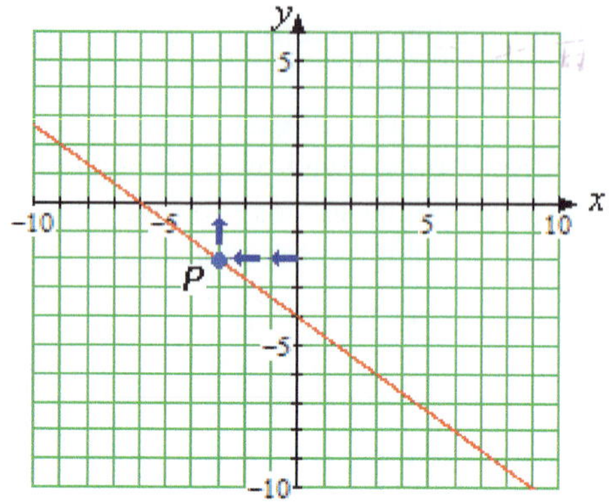

RQ7. What is the first step in graphing an equation?

RQ8. Which x-values would you choose to graph $y = \frac{5}{4}x - 3$?

Graphing Tip #2

When you graph a linear equation, you should extend the line far enough in both directions so that it will cross both the x-axis and the y-axis. The points where the graph crosses the axes are important for applications.

Look Closer: The graph of an equation is a picture of the solutions of the equation. Each point on the graph represents a solution. If a point lies on the graph, its coordinates make the equation true.

For example, the point $(-3, -2)$ lies on the graph in Example 4. When we substitute $x = -3$ and $y = -2$ into the equation $y = \frac{-2}{3}x - 4$, we get a true statement:

$$y = \frac{-2}{3}x - 4 \qquad \text{Substitute } x = -3 \text{ and } y = -2.$$
$$-2 = \frac{-2}{3}(-3) - 4$$
$$-2 = 2 - 4 \qquad \text{True.}$$

Skills Warm-Up

Choose the correct algebraic expression for each situation.

$2t + 12$	$12 - 2t$
$2(t + 12)$	$2t - 12$

1. Janine's history book has 12 chapters. If she studies 2 chapters a week, how many chapters will she have left after t weeks?
2. Arturo is 12 years older than twice the age of his nephew. If Arturo's nephew is t years old, how old is Arturo?
3. Rick made 12 fewer than twice as many phone calls as his roommate made this month. If Rick's roommate made t phone calls, how many calls did Rick make?
4. Every winter, the Civic Society knits mittens for the children of the county orphanage. This year there are 12 more children than last year. If there were t children last year, how many mittens will they need this year?

Answers: **1.** $12 - 2t$ **2.** $2t + 12$ **3.** $2t - 12$ **4.** $2(t + 12)$

Homework 2.3

Skills Practice

In Problems 1-2, the figure shown is the graph of an equation. Decide which of the given points are solutions of the equation.

1. **a.** $(2,0)$ **b.** $(3,2)$
   c. $(-1,-2)$ **d.** $(-3,-6)$

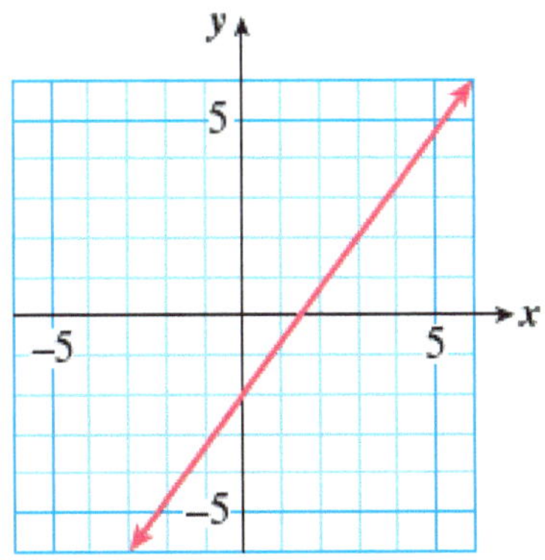

2. **a.** $(4,-2)$ **b.** $(4,4)$
   c. $(0,4)$ **d.** $(-1,-4)$

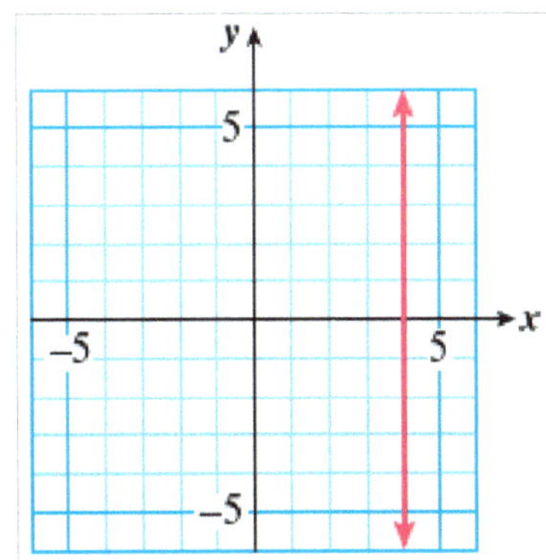

For Problems 3-5, give the coordinates of the labeled points.

3.

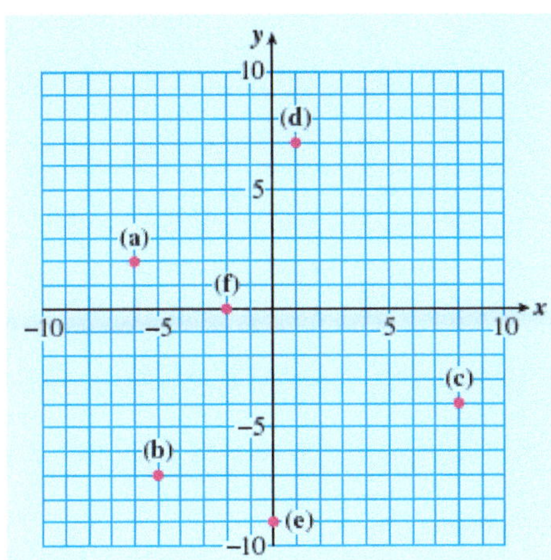

4.

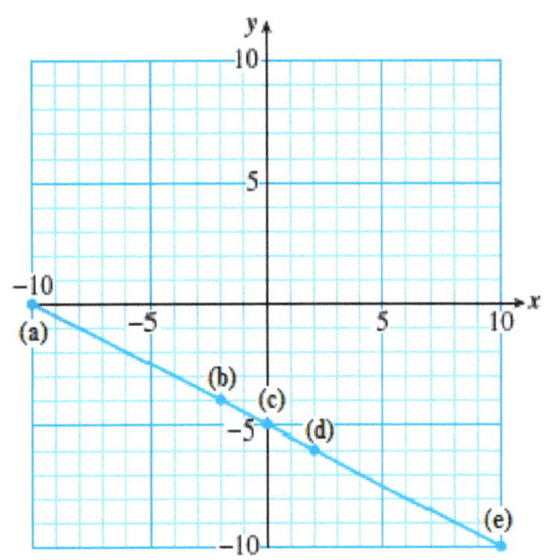

5.

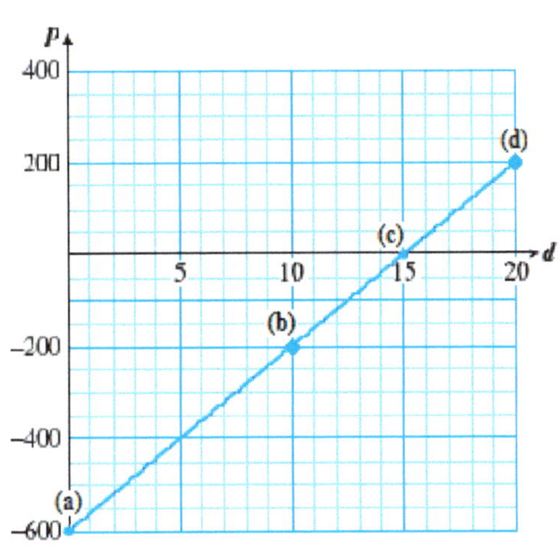

Applications

In Problems 6-7,

a. Write an equation relating the variables.
b. Complete the table of values.
c. Graph your equation on the grid.

6. Greta's math notebook has 100 pages, and she uses on average 6 pages per day for notes and homework. How many pages, P, will she have left after d days?

d	0	2	5	10	15
P					

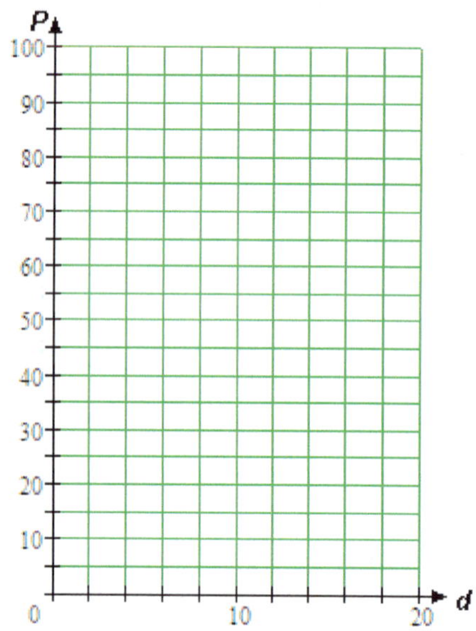

7. Delbert shares a house with four roommates. He pays \$200 rent per month, plus his share of the utilities. If the utilities cost U dollars, how much money, M, does Delbert owe?

U	20	40	80	100	200
M					

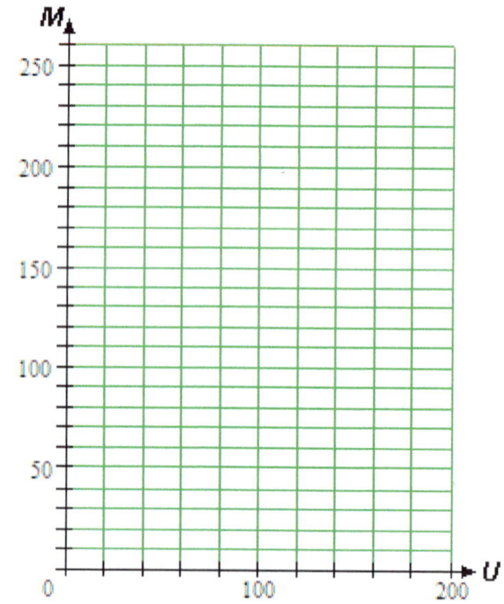

8. Darren inherited some money and has been spending it at the rate of \$100 per day. Right now he has \$2000 left.

a. Write an equation for Darren's balance, B, after d days.
b. Fill in the table. Negative values of d represent days in the past.

d	−15	−5	0	5	10	15	20
B							

c. Graph your equation, using the values in the table.
Use your graph to answer the following questions:
d. What was Darren's balance 15 days ago?
e. When will Darren's balance reach \$500?

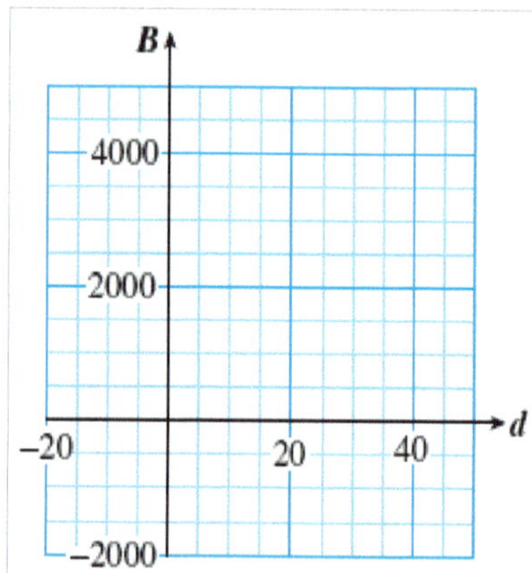

9. Gregory purchased stereo equipment on a monthly installment plan. After m months, Gregory still owes a balance of B dollars.
   a. What was the price of the stereo equipment?
   b. How much does Gregory pay each month?
   c. How much does Gregory owe after 6 months?
   d. How many monthly payments will Gregory make?
   e. Write an algebraic equation for B in terms of m.

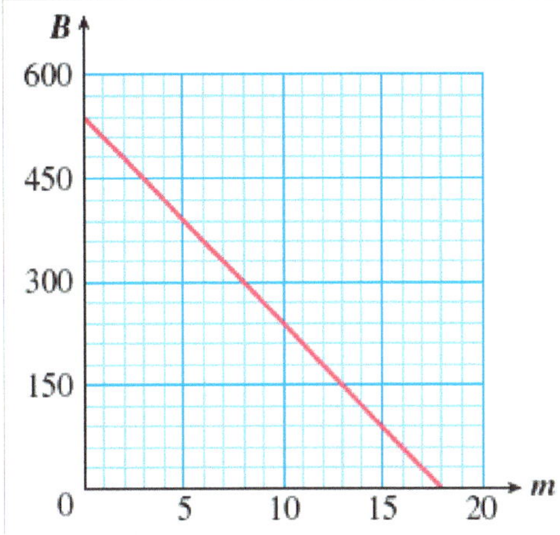

For part (c) of Problems 10-11, write an equation and use your graph to solve it.

10. On a 100-point test, Lori loses 5 points for each wrong answer.
   a. Write an equation for Lori's score, s, if she gives x wrong answers.
   b. Complete the table of values and graph your equation.

x	2	5	6	12
s				

   c. If Lori's score is 65, how many wrong answers did she give?

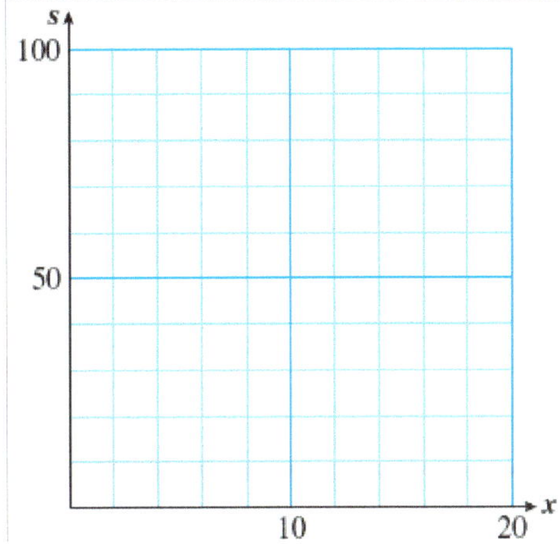

11. The water in Silver Pond is 10 feet deep, but the water level is dropping at a rate of $\frac{1}{2}$ inch per week.
   a. Write an equation for the depth, d, of the pond after w weeks.
   b. Complete the table of values and graph your equation.

w	24	36	84	120
d				

   c. How long will it take until the depth of the pond is 8 feet? (**Hint:** Convert all the units to inches.)

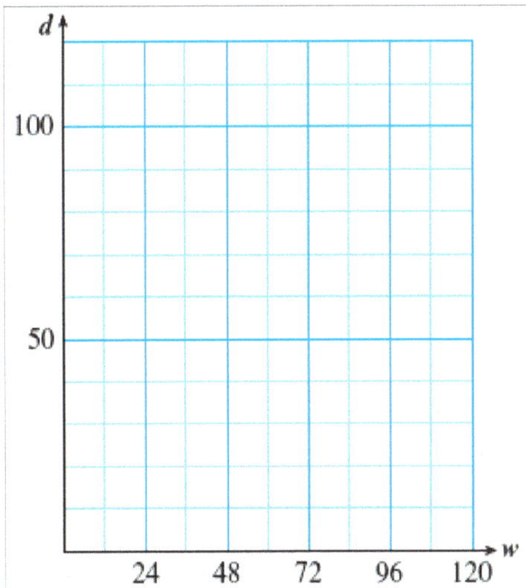

In Problems 12-15, choose four x-values and make a table of values, then graph the equation. Extend your line far enough that it crosses both axes.

12. $y = x + 3$

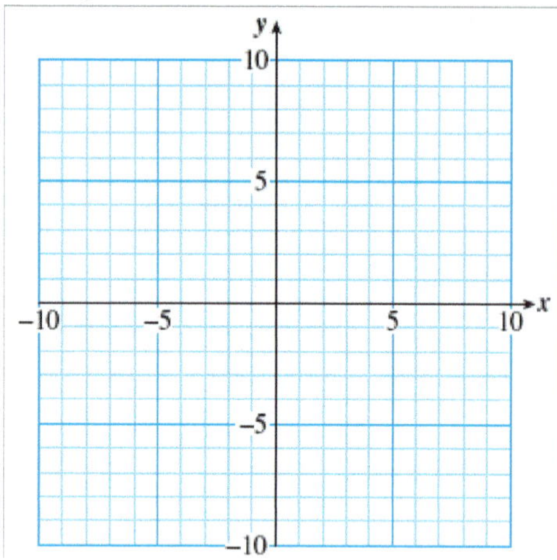

13. $y = 2x + 1$

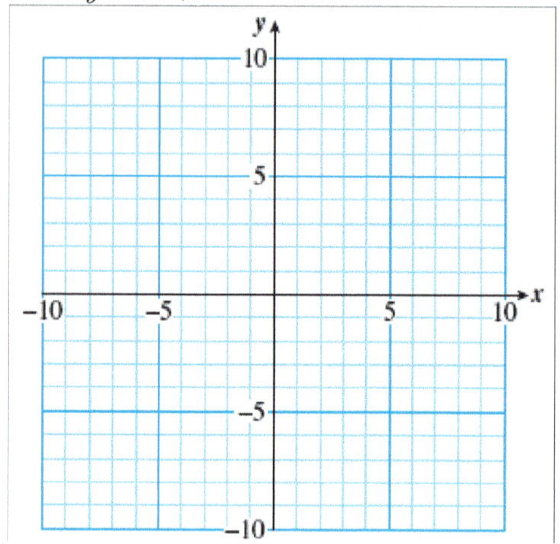

14. $y = -\frac{1}{2}x - 5$

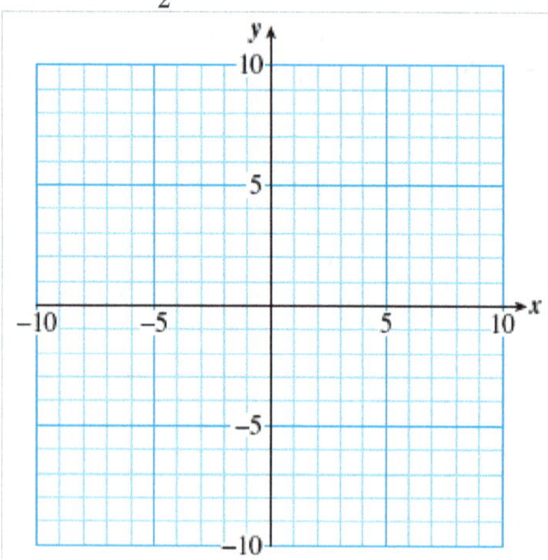

15. $y = \frac{5}{4}x - 4$

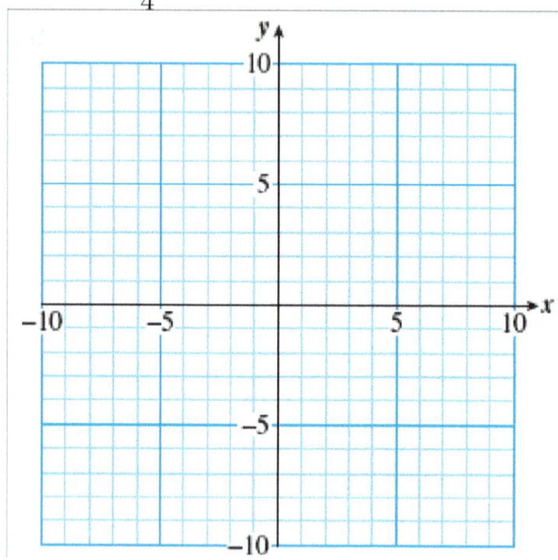

16. The graph of $y = 2x + 6$ is shown at right.

- **a.** Use the graph to evaluate the expression $2x + 6$ for $x = -5$.
- **b.** Find a point on the graph with $y = -4$. What is the x-value of the point?
- **c.** Verify that the coordinates of your point in part (b) satisfy the equation of the graph.
- **d.** Use the graph to find an x-value that produces a y-value of 8.
- **e.** Find two points on the graph for which $y > -4$. What are their x-values?

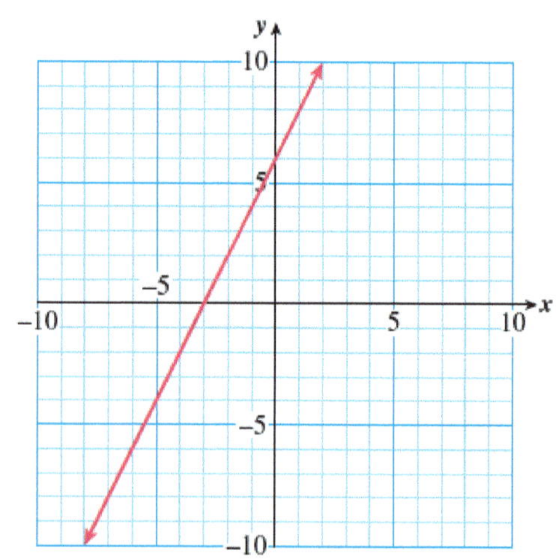

In Problems 17-18, use the graph to solve the equations. Show your solutions on the graph. Then check your solutions algebraically.

17.

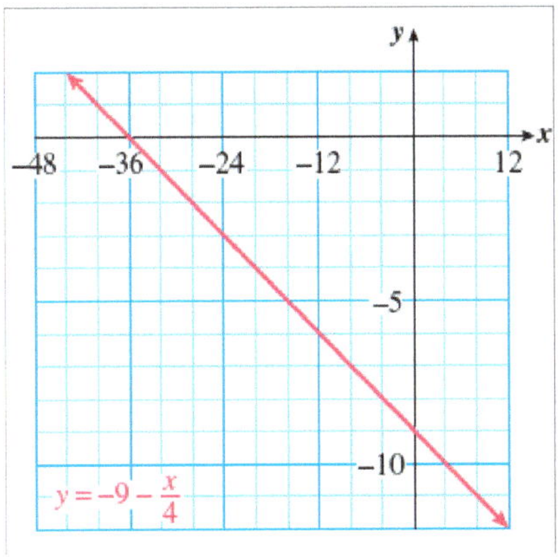

a. $-9 - \dfrac{x}{4} = -5$

b. $-9 - \dfrac{x}{4} = 1$

18.

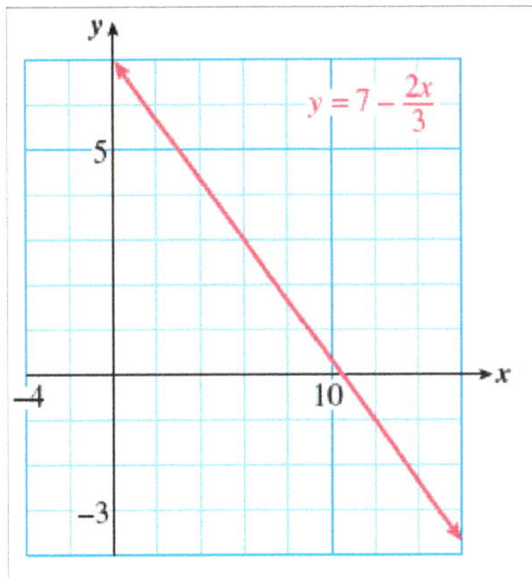

a. $7 - \dfrac{2x}{3} = 3$

b. $7 - \dfrac{2x}{3} = -1$

In Problems 19-20, use the graph to estimate the solution of the equation. Then use a calculator to help you verify the solution algebraically. Do you think your estimate was too big or too small?

19. $37.21 - 8.4t = 24.61$

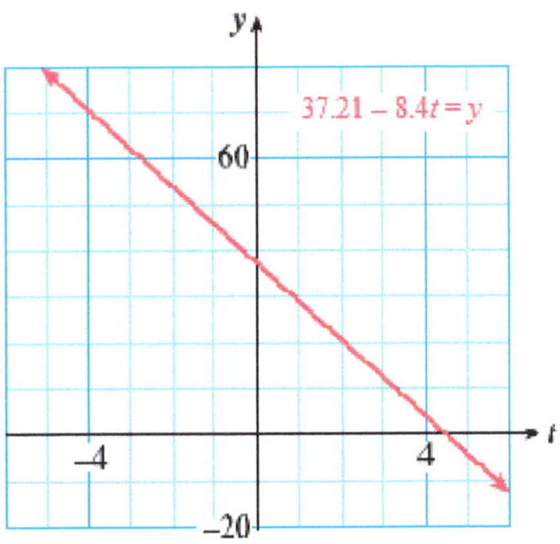

20. $-26.4 = -3.65 + 9.1x$

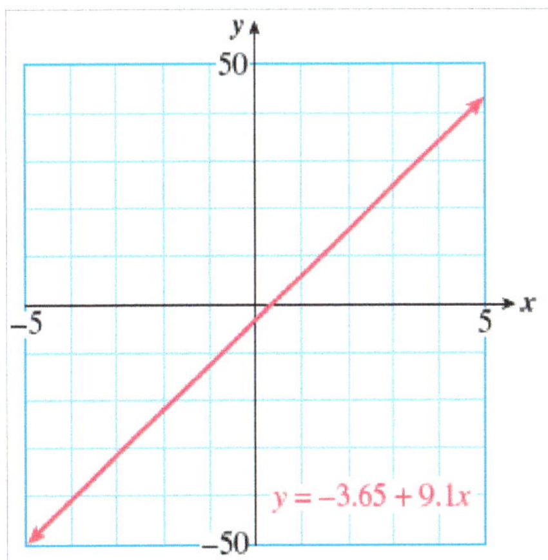

2.4 Linear Equations and Inequalities

Solving Equations

We solve an equation by isolating the variable on one side of the equation. If an equation involves two or more operations, we must undo those operations in reverse order.

Example 1 The expression $T = 50 + 15u$ describes college tuition consisting of a \$50 registration fee plus \$15 per unit. How many units can you take if you have \$290 saved for tuition?

Solution We can answer the question by solving the equation

$$290 = 50 + 15u$$

where we have substituted \$290 for the tuition, T. Think about the expression $50 + 15u$. How would you evaluate this expression if you were given a value for u? Following the order of operations, you would

1. first multiply by 15
2. then add 50

In order to solve the equation, we must *reverse these two steps* to undo the operations and isolate the variable. We first subtract 50 from both sides of the equation:

$$\begin{aligned} 290 &= 50 + 15u \\ -50 & \quad -50 \\ 240 &= 15u \end{aligned}$$ **Subtract 50 from both sides.**

This isolates the term that contains the variable, $15u$. Then we divide both sides of the equation by 15.

$$\frac{240}{15} = \frac{15u}{15}$$ **Divide both sides by 15.**

$$16 = u$$

You can enroll in 16 units. We can check the solution by substituting 16 for u in the original equation.

Check: $50 + 15u = 290$

$50 + 15(16) = 50 + 240 = 290$ **True.**

Because a true statement results, the solution checks.

Look Closer: In Example 1, notice how we reversed the operations used in the equation.

Operations performed on u	*Steps for solution*
1. Multiply by 15	1. Subtract 50
2. Add 50	2. Divide by 15

Strategy for solving equations.

To solve an equation that involves two or more operations, we undo those operations *in reverse order.*

Example 2 Solve $8 - 3x = -10$

Solution The left side of the equation has two terms: 8 and $-3x$. We want to isolate the term containing the variable, so we subtract 8 from both sides.

$$\begin{aligned} 8 - 3x &= -10 \\ -8 \quad &\quad -8 \\ -3x &= -18 \end{aligned}$$ Subtract 8 from both sides.

Next, we divide both sides by -3 to get

$$\frac{-3x}{-3} = \frac{-18}{-3}$$ Divide both sides by -3.

$$x = 6$$

Thus, the solution is 6.

Check: $8 - 3(6) = 8 - 18 = -10$

Caution! In Example 1, the term $-3x$ means "-3 times x," so we **divide** both sides by -3. Do not try to add 3 to both sides.

RQ1. If an equation involves more than one operation, how must we undo those operations?

RQ2. Delbert says he solved the equation $-5x = 15$ by adding 5 to both sides. What is wrong with his method?

Applied Problems

Problem solving often involves signed numbers, either in the equation that models the problem or in its solution, or both.

Example 3 The trout population in Clear Lake is decreasing by approximately 60 fish per year, and this year there are about 430 trout in the lake. If the population drops to 100, the Park Service will have to restock the lake.

a. Write an equation for the population P of trout x years from now.

b. When will the Park Service have to restock the lake?

c. Graph your equation for P, and illustrate your answer to part (b) on the graph.

Solutions a. The population starts this year at 430, and decreases by 60 for each following year. Thus, $P = 430 - 60x$.

b. We would like to find the value of x when $P = 100$. We substitute 100 for P, and solve the equation for x.

$$\begin{aligned} 100 &= \quad 430 - 60x \\ -430 &\quad -430 \\ -330 &= -60x \end{aligned}$$ Subtract 430 from both sides.

$$\frac{-330}{-60} = \frac{-60x}{-60}$$ Divide both sides by -60.

$$5.5 = x$$

The Park Service will have to restock the lake in five and a half years, if the population continues to decline at the current rate.

c. The figure shows the graph of the equation $P = 430 - 60x$. To solve

$$100 = 430 - 60x$$

we locate the point on the graph with $P = 100$, and read its x-coordinate, at about 5.5.

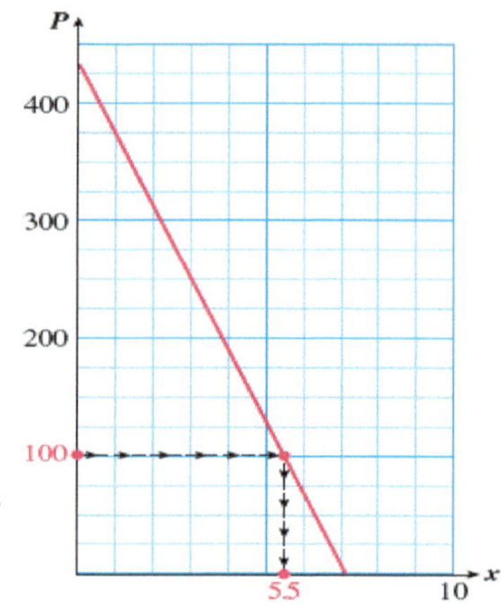

Solving Inequalities

A statement that uses one of the symbols $<$ or $>$ is called an **inequality**.

Examples of inequalities are

$$-1 > -3 \quad \text{and} \quad x < 2$$

Unlike the equations we have studied, which have at most one solution, an inequality can have infinitely many solutions. A solution of an equation or inequality is said to **satisfy** the equation or inequality.

Example 4 Solve the inequality $x < 2$.

Solution The solutions include $1, 0, -1, -2,$ and all the other negative integers, as well as fractions less than 2, such as $1\frac{3}{5}$, $\frac{2}{3}$, and $-\frac{17}{8}$. All of these values satisfy the inequality $x < 2$. In fact, *all* the numbers to the left of 2 on the number line are solutions of $x < 2$. Because we cannot list all these solutions, we often graph them on a number line as shown below.

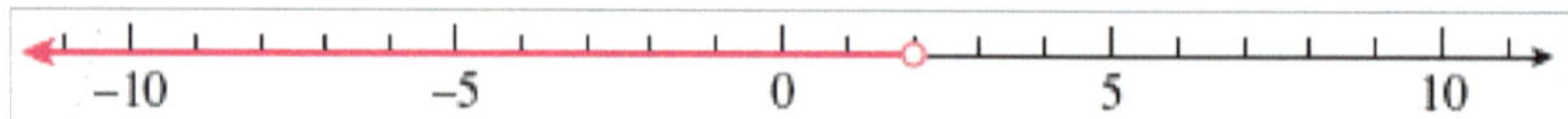

The open circle at 2 shows that 2 itself is not a solution (because 2 is not less than 2.)

Look Closer: An inequality that uses the symbol for less than, $<$, or greater than, $>$, is called a **strict inequality**. A **nonstrict inequality** uses one of the following symbols.

$\geq$ means **greater than or equal to**
$\leq$ means **less than or equal to**

For example, the graph of all solutions to the inequality

$$x \geq -2$$

is shown on the next page. We use a solid dot at -2 to show that $x = -2$ is included in the solutions.

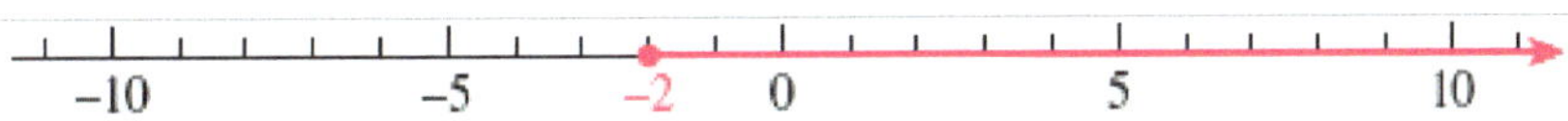

RQ3. A solution of an inequality is said to __________ the inequality.
RQ4. What is the difference between a strict and a nonstrict inequality?

The rules for solving inequalities are very similar to the rules for solving equations, with one important difference. In the Activities we will develop the following strategies for solving inequalities.

To Solve an Inequality:

1. We can add or subtract the same quantity on both sides.
2. We can multiply or divide both sides by the same positive number.
3. If we multiply or divide both sides by a **negative** number, we must **reverse** the direction of the inequality.

Example 5 Solve $-3x+1>7$, and graph the solutions on a number line.

Solution We isolate x on one side of the inequality.

$$-3x+1-1>7-1 \qquad \text{Subtract 1 from both sides.}$$
$$-3x>6$$
$$\frac{-3x}{-3}<\frac{6}{-3} \qquad \text{Divide both sides by } -3\text{; reverse the direction of the inequality.}$$
$$x<-2$$

The graph of the solutions is shown below.

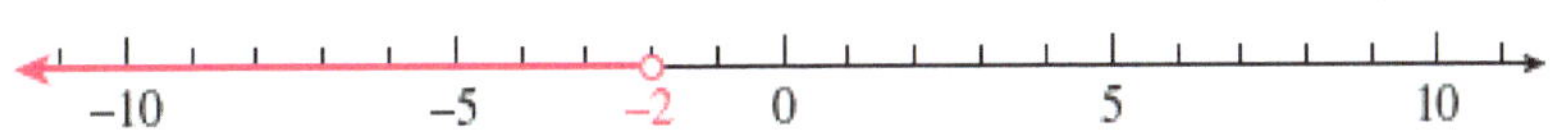

Compound Inequalities

An inequality in which the variable expression is bounded from above and from below is called a **compound inequality**. For example,

$$-3<2x-5\leq 6$$

is a compound inequality. To solve a compound inequality, we must perform the steps needed to isolate x on all three sides of the inequality.

Example 6 Solve $-3<2x-5\leq 6$

Solution To solve for x, we first add 5 on each side of the inequality symbols.

$$\begin{array}{rcccl} -3 & < & 2x-5 & \leq & 6 \\ +5 & & +5 & & +5 \\ \hline 2 & < & 2x & \leq & 11 \end{array}$$

Next, to solve $2<2x\leq 11$, we divide each side by 2.

$$\frac{2}{2} < \frac{2x}{2} \leq \frac{11}{2}$$
$$1 < x \leq \frac{11}{2}$$

The solution consists of all numbers greater than 1 but less than or equal to $\frac{11}{2}$. The graph of the solutions is shown below.

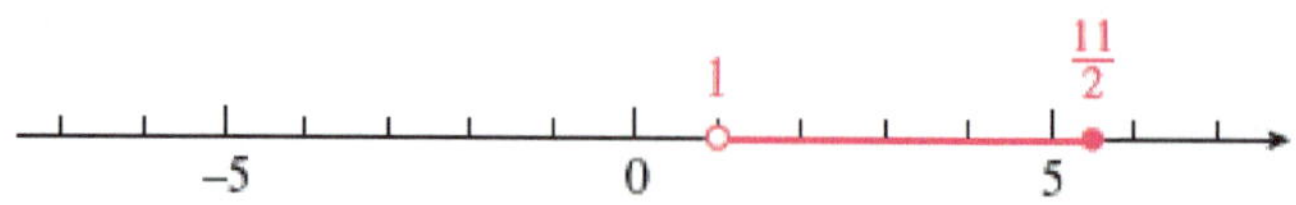

RQ5. When do we need to reverse the direction of an inequality?
RQ6. What is a compound inequality?

Skills Warm-Up

Solve each equation. Try to do so mentally (without using pencil and paper.)

1. $\frac{u}{3} = 6$ **2.** $7 = 3 + s$ **3.** $a - \frac{1}{3} = \frac{2}{3}$

4. $20 = 5m$ **5.** $\frac{1}{4}p = 8$ **6.** $7t = 5$

Fill in the tables. Then analyze the order of operations in your calculations:

Table 1

n	$3n$	$3n - 5$
2		
5		
		7
		22

Table 2

m	$\frac{m}{4}$	$\frac{m}{4} + 1$
8		
12		
		6
		2

7. Consider the equation $3n - 5 = p$. Look at Table 1 to help you answer the questions:
 a. Let $n = 2$. Explain how to find p in two steps.
 b. Let $p = 7$. Explain how to find n in two steps.

8. Consider the equation $\frac{m}{4} + 1 = h$. Look at Table 2 to help you answer the questions:
 a. Let $m = 8$. Explain how to find h in two steps.
 b. Let $h = 6$. Explain how to find m in two steps.

9. a. If you put on socks and then put on shoes, what operations are needed to reverse the process?
 b. You leave home and bicycle north for 3 miles and then east for 2 miles. Suddenly you notice that you have dropped your wallet. How should you retrace your steps?

Answers: **1.** 18 **2.** 4 **3.** 1 **4.** 4 **5.** 32 **6.** $\frac{5}{7}$

Homework 2.4

Skills Practice

Fill in the table in Problems 1-2. Use two steps for each row: Fill in the middle column first. Then use the table to help you answer the questions.

1.

x	$2x$	$2x+4$
3		
6		
		14
		20

Consider the equation $y = 2x + 4$.

a. Let $x = 3$. Explain how to find y in two steps.

b. Let $y = 14$. Explain how to find x in two steps.

2.

q	$q-3$	$5(q-3)$
3		
		10
4		
		20

Consider the equation $5(q-3) = R$.

a. Let $q = 3$. Explain how to find R in two steps.

b. Let $R = 10$. Explain how to find q in two steps.

For Problems 3-19, solve.

3. $6x - 13 = 5$

4. $\dfrac{2a}{5} = 8$

5. $\dfrac{x}{4} + 2 = 3$

6. $6x + 5 = 5$

7. $24 = 4(p - 7)$

8. $0 = \dfrac{5z}{7}$

9. $\dfrac{k+4}{5} = 9$

10. $\dfrac{2x}{3} - 5 = 7$

11. $7 = \dfrac{4b-3}{3}$

12. $3c - 7 = -13$

13. $-5 = -2 - 3t$

14. $-3(2h - 6) = -12$

15. $1 - \dfrac{b}{3} = -5$

16. $\dfrac{3y}{5} + 2 = -4$

17. $\dfrac{5x}{2} + 10 = 0$

18. $11.8w - 37.8 = 120.32$

19. $9.7(2.6 + v) = 58.2$

In Problems 20-28,

a. Solve each inequality algebraically.

b. Graph your solutions on a number line.

c. Give at least one value of the variable that is a solution, and one value that is not a solution.

20. $x + 10 \le -5$

21. $-3y < 15$

22. $\dfrac{x}{3} \le 4$

23. $2x + 3 > 7$

24. $-3x + 2 \le 11$

25. $-3 > \dfrac{2x}{3} + 1$

26. $-3 \le 3x \le 12$

27. $23 > 9 - 2b \ge 13$

28. $-8 \le \dfrac{5w+3}{4} < -3$

Applications

29. Delbert is thinking of a number. If he multiplies the number by 5 and then subtracts 6, the result is 29. What is the number?

In Problems 30-31, find the error in the "solution," then write a corrected solution.

30.
$$6 - 3x = -12$$
$$-3x = -6$$
$$x = 2$$

31.
$$-2 + \frac{2}{3}x = -4$$
$$-2 + 2x = -12$$
$$2x = -10$$
$$x = -5$$

In Problems 32-33, use the graphs to answer the questions.

32. Francine's new puppy weighed 3.8 pounds when she brought it home, and it should gain approximately 0.6 pound per week. The graph shows the puppy's weight, W, after t weeks.

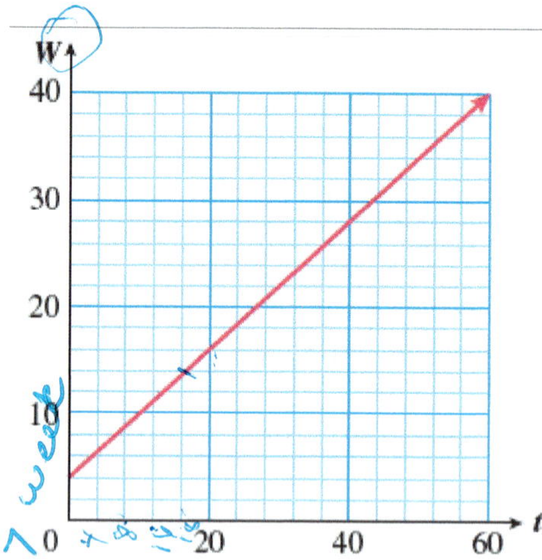

a. Estimate the puppy's weight after 8 weeks.

b. Estimate how long it will take for the puppy to reach 26 pounds.

c. Write an algebraic equation for W in terms of t.

d. Write and solve an equation to verify your answer to part (b).

33. Marvin's telephone company charges $13 a month plus $0.15 per minute. The graph shows Marvin's phone bill, P, if he talks for m minutes.

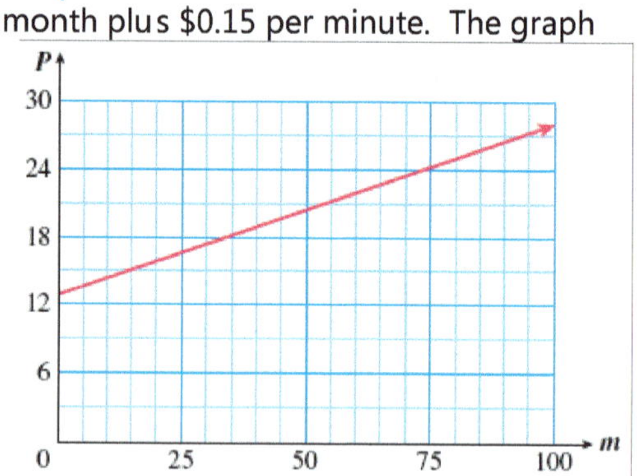

a. Estimate Marvin's phone bill if he talks for 45 minutes.

b. Estimate the number of minutes Marvin talked if his phone bill is $25.

c. Write an algebraic equation for P in terms of m.

d. Write and solve an equation to verify your answer to part (b).

In Problems 34-38, solve in three steps:

a. Choose a variable for the unknown quantity.

b. Write an equation that involves your variable.

c. Solve your equation, and answer the question.

34. Simona bought a car for $10,200. She paid $1200 down and will pay the rest in 36 monthly installments. How much is each installment?
(**Hint:** Write an equation about the total amount Simona pays.)

35. Every school day, each member of the cross-country team runs a certain distance d, which is assigned on the basis of ability, plus a half mile around the track. Greta runs 22.5 miles per week. What is her assigned distance?
(**Hint:** Write an equation about the total distance Greta runs in one school week.)

36. The Tree People planted new trees in an area that was burned by brush fires. 60% of the seeds sprouted, but gophers ate 38 of the new sprouts. That left 112 new saplings. How many seeds were planted?
(**Hint**: Write an equation about the number of seeds that survived.)

37. During a four-day warming trend, the temperature rose from $-6°$ to $26°$. What was the average change in temperature per day?

38. Eric is on a diet to reduce his current weight of 196 pounds to 162 pounds. If he loses 4 pounds per week, how long will it take him to reach his desired weight?

Use one of the formulas below to solve Problems 39-40.

Perimeter of a rectangle $P = 2l + 2w$

Area of a triangle $A = \frac{bh}{2}$

39. A farmer has 500 yards of fencing material to enclose a rectangular pasture. He would like the pasture to be 75 yards wide. How long will it be?

40. A triangular sail requires 12 square meters of fabric. If the base of the sail measures 4 meters, how tall is the sail?

In Problems 41-42, write and solve an equation to find the value of x.

41. 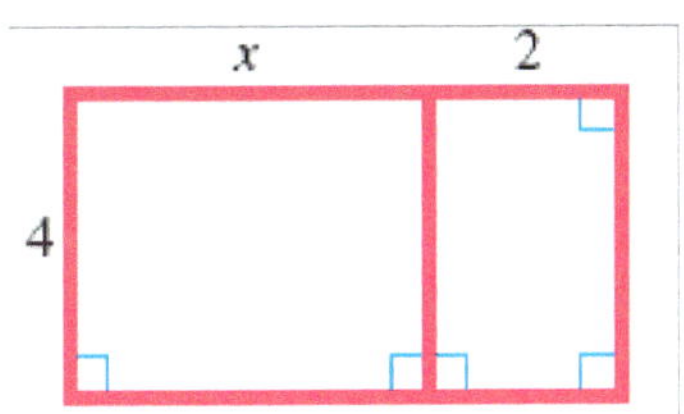

Area $= 36$

42. 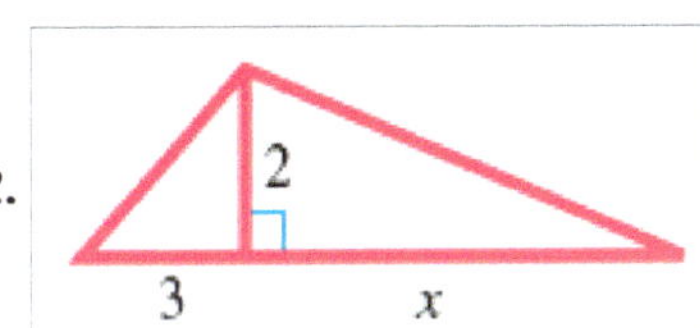

Area $= 20$

2.5 Like Terms

Equivalent Expressions

When we solve an equation such as

$$2 + 3x = 17$$

we find a value for the variable that makes the equation true. The solution to this equation is 5, because the two expressions $2 + 3x$ and 17 are equal if $x = 5$. If we use any other value for x, say $x = 9$ or $x = -1$, then $2 + 3x$ does not equal 17.

It is possible for two algebraic expressions to be equal for **all** values of the variable. For instance, the expression $2x + 3x$ and $5x$ will always give the same value when the same number is substituted for x in each expression. For example,

if $x = 3$, then $2x + 3x = 2(3) + 3(3) = 6 + 9 = 15$

and $5x = 5(3) = 15$

Both expressions equal 15 when $x = 3$.

Exercise Verify that $2x + 3x$ and $5x$ are equal when $x = -4$.

$2x + 3x =$ ________________

and $5x =$ ________________

Two algebraic expressions are **equivalent** if they name the same number for all values of the variable.

Thus, $2x + 3x$ and $5x$ are equivalent expressions, and we can replace one by the other if it will simplify our work.

Caution! Although two expressions may name the same number for *some* values of the variable, they are not necessarily equivalent. For example, to show that the expressions $2 + 3x$ and $5x$ are *not* equivalent, we only have to find a single value of x for which they are not equal. If we evaluate each expression for $x = 6$, we find:

$$2 + 3x = 2 + 3(6) = 20$$
$$5x = 5(6) = 30$$

Because they are not equal for $x = 6$, the expressions $2 + 3x$ and $5x$ are not equivalent. ■

RQ1. When are two expressions equivalent?

Like Terms

We saw above that $2x + 3x$ and $5x$ are equivalent expressions. This makes sense when we consider what these expressions mean: $2x$ means two x's added together, and $3x$ means three x's added together. Then

$$2x + 3x = (x + x) + (x + x + x) = 5x$$

We can add two x's to three x's and get five x's, just as we can add two pencils to three pencils, or two dollars to three dollars. In algebra this is called **combining like terms**.

Like terms are any terms that are exactly alike in their variable factors.

Example 1a. $2x$ and $3x$ are like terms
$-4a$ and $7a$ are like terms

because their variable factors are identical.

b.

$2x$ and $3y$ are not like terms
$2x$ and 3 are not like terms

because their variable factors are different.

The numerical factor in a term is called the **numerical coefficient**, or just the **coefficient** of the term.

Example 2a. In the expression $3xy$, the number 3 is the numerical coefficient.
b. In a term such as xy or b, the numerical coefficient is 1.
c. If a variable is preceded by a negative sign, the coefficient of the term is -1. For example,

$-x$ means $-1 \cdot x$
$-a$ means $-1 \cdot a$

Adding and Subtracting Like Terms

To add like terms:
Add the numerical coefficients of the terms. Do not change the variable factors of the terms.

Example 3 Add like terms.
a. $5n + 3n$ **b.** $-4y - 3y$ **c.** $-6st + 9st$

Solutions In each calculation, we do not change the variables we are adding; only the coefficient changes.
a. $5n + 3n = (5 + 3)n = 8n$ Add the coefficients, 5 and 3.
b. $-4y - 3y = (-4 - 3)y = -7y$ Add the coefficients, -4 and -3.
c. $-6st + 9st = (-6 + 9)st = 3st$ Add the coefficients, -6 and 9.

Look Ahead: In Example 3, we replaced each expression by a simpler but equivalent expression. In other words, $-7y$ is equivalent to $-4y - 3y$, and $3st$ is equivalent to $-6st + 9st$.

Replacing an expression by a simpler equivalent one is called **simplifying** the expression.

We subtract like terms in the same way that we add like terms.

To subtract like terms:
Subtract the numerical coefficients of the terms. Do not change the variable factors of the terms.

Example 4 $6a - (-8a) = [6 - (-8)]a = 14a$ *Subtract the coefficients.*

RQ2. What are like terms?
RQ3. How do we add or subtract like terms?

Caution! We cannot add or subtract unlike terms. Thus,

$2x + 3y$ cannot be simplified
$-6x + 4xy$ cannot be simplified
$6 - 2x$ cannot be simplified

Many expressions contain both like and unlike terms. In such expressions we can combine only the like terms.

Example 5 Simplify $-4a + 2 - 5 + 8a + 5a$

Solution We combine all the a-terms, and all the constant terms, like this:

$$-4a + 8a + 5a = 9a \quad \text{and} \quad 2 - 5 = -3$$

so

$$-4a + 2 - 5 + 8a + 5a = 9a - 3$$

Caution! In Example 5, $9a - 3 \neq 6a$, because $9a$ and -3 are not like terms.

Removing Parentheses

When we add signed numbers, we can remove parentheses that follow an addition symbol. For instance,

$$5 + (-3) = 5 - 3 = 2$$

We can also remove parentheses when we add like terms. Thus,

$$5x + (-3x) = 5x - 3x = 2x$$

Removing parentheses that follow a plus sign:
Parentheses preceded by a plus sign may be omitted, and each term within parentheses keeps its original sign.

Example 6 Simplify $(2x+3)+(5x-7)$

Solution Each set of parentheses is preceded by a plus sign, so we may remove the parentheses.

$$\begin{aligned}(2x+3)+(5x-7) &= 2x+3+5x-7 \quad \text{Combine like terms.}\\ &= 7x-4\end{aligned}$$

We must be careful when removing parentheses preceded by a **minus** sign. In the expression

$$(5x-2)-(7x-4)$$

the minus sign preceding $(7x-4)$ applies to *each term* inside the parentheses. Therefore, when we remove the parentheses we must change the sign of $7x$ from $+$ to $-$ *and* we must change the sign of -4 from $-$ to $+$. Thus,

$$\begin{aligned}(5x-2)-(7x-4) &= 5x-2-7x+4 \quad \text{Change each sign inside parentheses.}\\ &= 5x-7x-2+4 \quad \text{Combine like terms.}\\ &= -2x+2\end{aligned}$$

Removing parentheses that follow a minus sign:
If an expression inside parentheses is preceded by a minus sign, we **change the sign of each term within parentheses** and then omit the parentheses.

Example 7 Simplify $-5b-(3-2b)+(4b-6)$

Solution Before combining like terms, we must remove the parentheses. We change the signs of any terms inside parentheses preceded by a minus sign; we do *not* change the signs of terms inside parentheses preceded by a plus sign.

$$\begin{aligned}-5b-(3-2b)+(4b-6) &= -5b-3+2b+4b-6\\ &= -5b+2b+4b-3-6\\ &= b-9\end{aligned}$$

(signs changed: $-3+2b$; signs not changed: $+4b-6$)

RQ4. We can remove parentheses that follow ________________.
RQ5. What must you do if you remove parentheses that follow a minus sign?

Solving Equations and Inequalities

We can now solve equations in which two or more terms contain the variable.

Example 8 Solve the equation $2x+7=4x-3$

Solution We first get all terms containing the variable on one side of the equation. For this example, we will subtract $2x$ from both sides of the equation to get

$$\begin{array}{rrl} 2x+7= & 4x-3 & \text{Subtract } 2x \text{ from both sides.} \\ -2x & -2x & \\ \hline 7= & 2x-3 & \end{array}$$

Now the equation looks like ones we already know how to solve, and we proceed as usual to isolate the variable.

$$\begin{array}{rl} 7=2x-3 & \text{Add 3 to both sides.} \\ +3 \quad +3 & \\ 10=2x & \text{Divide both sides by 2.} \\ \frac{10}{2}=\frac{2x}{2} & \\ 5=x & \end{array}$$

The solution is 5.

Check: Does $2(\mathbf{5})+7=4(\mathbf{5})-3$? **True.**

If one side of an equation or inequality contains like terms, we should combine them before beginning to solve.

Example 9 Solve $2x-4x+14<3+5x-10$

Solution Combine like terms on each side of the inequality, to get

$$\begin{array}{rl} \mathbf{2x}-\mathbf{4x}+14<\mathbf{3}+5x-\mathbf{10} & 2x-4x=-2x \text{ and } 3-10=-7 \\ -2x+14<5x-7 & \end{array}$$

Now continue solving as usual: We want to get all the terms containing x on one side of the inequality, and all the constant terms on the other side.

$$\begin{array}{rrl} -2x+14< & 5x-7 & \text{Subtract } 5x \text{ from both sides.} \\ -5x & -5x & \\ \hline -7x+14< & -7 & \\ -14 & -14 & \text{Subtract 14 from both sides.} \\ \hline -7x< & -21 & \\ \frac{-7x}{-7}> & \frac{-21}{-7} & \text{Divide both sides by } -7\text{;} \\ & & \text{reverse the direction of the inequality.} \\ x>3 & & \end{array}$$

The solution is all x-values greater than 3. You can check the solution by substituting one x-value greater than 3 and one value less than 3 into the original inequality, for instance:

Check: $x=4:$ $2(\mathbf{4})-4(\mathbf{4})+14<3+5(\mathbf{4})-10$? **True:** $6<13$

$x=2:$ $2(\mathbf{2})-4(\mathbf{2})+14<3+5(\mathbf{2})-10$? **False:** $10 \not< 3$

RQ6. What should you do if one side of an equation or inequality contains like terms?

Here are our guidelines for solving linear equations.

Steps for Solving Linear Equations

1. Combine like terms on each side of the equation.
2. By adding or subtracting the same quantity on both sides of the equation, get all the variable terms on one side and all the constant terms on the other.
3. Divide both sides by the coefficient of the variable to obtain an equation of the form $x = a$.

Skills Warm-Up

Write algebraic expressions.

1. Getaway Tours offers a Caribbean cruise for \$2000 per person if 12 people sign up. For each additional person who signs up, the price per person is reduced by \$60. How much will you pay for a cruise if p additional people sign up?
2. Anita figures her taxes by subtracting a \$1200 deductible from her income, then taking 6% of the result. What is her tax bill if her income is I?
3. The area of a trapezoid is one-half the product of its height h and the sum of the two bases, a and b.
4. The ratio of a number k to 5 less than k.

Evaluate.

5. $3n(n-k)(n+k)$ for $n=-4$ and $k=-6$
6. $(t-1)(2t+3)$ for $t=\dfrac{2}{3}$

Answers: **1.** $2000-60p$ **2.** $0.06(I-1200)$ **3.** $\dfrac{1}{2}h(a+b)$
4. $\dfrac{k}{k-5}$ **5.** 240 **6.** $\dfrac{-13}{9}$

Homework 2.5

Skills Practice

In Problems 1-4, add or subtract like terms.

1. $-6x + 2x$
2. $-7.6a - 5.2a$
3. $3t - 4t + 2t$
4. $3bc - (-4bc) - 8bc$

In Problems 5-6, combine like terms.

5. $3 + 4y - (-8y) - 7$
6. $-2st - 2 + 5s - 6st - (-4s)$

In Problems 7-10, simplify and combine like terms.

7. $4x + (5x - 2)$
8. $22y - 34 - (16y - 24)$
9. $6a - 5 - 2a - (2a - 5)$
10. $(-2 - 3b) - (2b - 4)$

In Problems 11-16, solve each equation or inequality.

11. $4m - 3 = 2m + 5$
12. $15 - 9t = 33 - 5t$
13. $-6s = 3s$
14. $3x + 5 > 2x + 3$
15. $-8g + 35 = 9g - 13 + g$
16. $-15y + 5 - 2y - 4 \geq -12y + 21$
17. $-2 - 3w + 5w \leq 4w - 44$
18. $-5p - 4 + 3p = 2 + 7p + 3$

19. a. Evaluate the expression $-3y + 2 + 7y - 6y - 4y - 8$ for $y = 2.5$.
    b. Simplify the expression in part (a).
    c. Evaluate your answer to part (b) for $y = 2.5$. Check that you got the same answer for part (a).

20. Evaluate the expression for the given values of the variables.
    $4 - 3a + 6ab - (-8a) - 10 + 9ab + 2a$ for $a = -2,\ b = 5$

Applications

In Problems 21-22, choose a value for the variable and show that the following pairs of expressions are **not** equivalent.

21. $2 + 7x$ and $9x$
22. $-(a - 3)$ and $-a - 3$

Write and simplify an algebraic expression for the perimeter of the figure in Problems 23-24.

23. 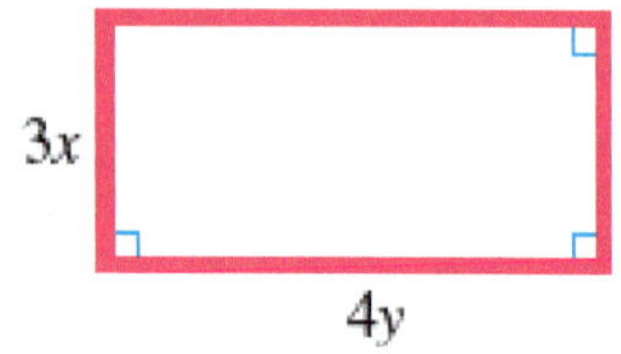

24. 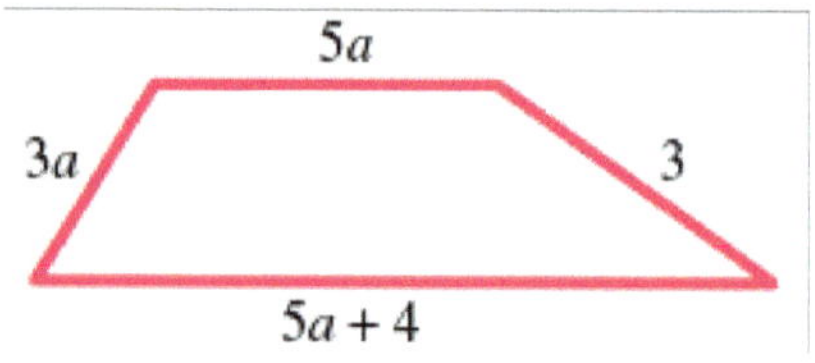

For Problems 25-29, write an algebraic expression or an equation to answer each question.

25. There are 8 more cats than dogs at the county shelter. Write expressions in terms of the number of dogs, d.
- **a.** The number of cats
- **b.** The total number of cats and dogs
- **c.** If there are 40 dogs and cats at the shelter, how many dogs are there?

26. For every smoker in the restaurant there are three nonsmokers. Write expressions in terms of x, the number of smokers.
- **a.** The number of nonsmokers
- **b.** The total number of people in the restaurant
- **c.** If there are 36 people in the restaurant, how many are smokers?

27. A tortoise and a hare are having a race. After t seconds, the tortoise has traveled $0.2t$ feet and the hare has traveled $10t$ feet.
- **a.** Express the distance between them in terms of t.
- **b.** How far apart are the tortoise and the hare after 10 seconds?
- **c.** When will they be 147 feet apart?

28. Delbert bought a high definition TV at a price of x dollars. The sales tax in his state is 8%.
- **a.** Write an expression for the tax on Delbert's TV.
- **b.** Write an expression for the price Delbert paid, including tax.
- **c.** Delbert paid $928.80 for his TV. What was the price before tax?

29. The cost of producing m stereos is $251m + 1355$ dollars, and each stereo sells for $847.
- **a.** What profit is made from producing and selling m stereos?
- **b.** How many stereos must be produced to make a profit of $16,525?

For Problems 30-34, write and solve an equation.

30. Last summer Kevin made 8 batches of barbecue sauce using one bottle of catsup. This summer he made only 5 batches and had 12 ounces of catsup left over. How much catsup does Kevin use in a batch of barbecue sauce?

31. Arch has two wooden beams of equal length. He makes 3 posts for mailboxes with one beam, and has 33 inches left. He makes another mailbox post with the other beam, and has 131 inches left. How long is each mailbox post?

32. Quentin bought 12 cans of cat food and a half-gallon of milk that cost $1.80. Tilda bought 8 cans of cat food and a $5 bag of kibble. They paid the same price at the check-out stand. How much does one can of cat food cost?

33. If Sarah drinks three glasses of milk a day she will exceed her minimum daily requirement for calcium by 70 milligrams. If she drinks only two glasses of milk she will still need another 220 milligrams of calcium. How many milligrams of calcium does a glass of milk contain?

34. On her last history test, Parisa answered 9 short-answer questions correctly and earned 15 points on the essay. Seana answered 10 short-answer questions correctly and got 23 points on the essay. Seana's score on the test was 14 points higher than Parisa's. How many points was each short-answer question worth?

Chapter 2 Summary and Review

Section 2.1 Signed Numbers

- The natural numbers, zero, and the negatives of the natural numbers are called the **integers.**
- We use special symbols to indicate order:

$<$ means **is less than**
$>$ means **is greater than**

- The **absolute value** of a number is the unsigned part of the number. It is never negative.

Rules for Operations on Integers
To add two integers:
1. To add two numbers with the same sign, add their absolute values. The sum has the same sign as the numbers.
2. To add two numbers with opposite signs, subtract their absolute values. The sum has the same sign as the number with the larger absolute value.

To subtract b from a:
1. Change the sign of b.
2. Change the subtraction to addition.
3. Proceed as in addition.

To multiply or divide two integers:
1. The product or quotient of two numbers with opposite signs is a negative number.
2. The product or quotient of two numbers with the same sign is a positive number.

Quotients Involving Zero
If a is any nonzero number, then

$$\frac{0}{a} = 0 \quad \text{and} \quad \frac{a}{0} \textbf{ is undefined}$$

Section 2.2 Expressions and Equations

- When we evaluate an algebraic expression, we follow the order of operations.
- When we evaluate an algebraic expression at a negative number, we enclose the negative numbers in parentheses.

Section 2.3 Graphs of Linear Equations

- An equation of the form $y = ax + b$, where a and b are constants, is called a **linear equation** because its graph is a straight line.
- We can graph a linear equation by evaluating the expression $ax + b$ at several values of x and then plotting points.
- We use a **Cartesian coordinate system** to make a graph that includes negative values of the variables.

- The graph of an equation is a picture of the solutions of the equation. Each point on the graph represents a solution.

Section 2.4 Linear Equations and Inequalities

- To solve an equation that involves two or more operations, we undo those operations *in reverse order*.
- A statement that uses one of the symbols $<$ or $>$ is called an **inequality**.

> **Rules for Solving Inequalities**
> 1. We can add or subtract the same quantity on both sides.
> 2. We can multiply or divide both sides by the same positive number.
> 3. If we multiply or divide both sides by a **negative** number, we must **reverse** the direction of the inequality.

- Each point on the graph of an equation represents a solution of the equation. To solve the equation for a particular y-value, we locate the corresponding point on the graph.
- To solve a **compound inequality**, we perform the steps needed to isolate x on all three sides of the inequality.

Section 2.5 Like Terms

- Two algebraic expressions are **equivalent** if they name the same number for *all* values of the variable.
- **Like terms** are any terms that are *exactly alike* in their variable factors. By adding or subtracting like terms we can replace one algebraic expression by a shorter or simpler one.
- The numerical factor in a term is called the **numerical coefficient**, or just the **coefficient** of the term.
- Replacing an expression by a simpler equivalent one is called **simplifying** the expression.

> **To add or subtract like terms:**
> Add or subtract the numerical coefficients of the terms. Do not change the variable factors of the terms.

> **Removing parentheses**
> 1. Parentheses following a plus sign may be omitted; each term within parentheses keeps its original sign.
> 2. To remove parentheses that follow a minus sign, we change the sign of each term within parentheses and then omit the parentheses and the minus sign.

> **Steps for Solving Linear Equations**
> 1. Combine like terms on each side of the equation.
> 2. By adding or subtracting the same quantity on both sides of the equation, get all the variable terms on one side and all the constant terms on the other.
> 3. Divide both sides by the coefficient of the variable to obtain an equation of the form $x = a$.

Review Questions

Use complete sentences to answer the questions in Problems 1-10.

1. Explain the terms natural numbers, whole numbers, and integers.
2. A classmate claims that the opposite of a number and the absolute value of a number are the same, and uses $x = -3$ as an example. Do you agree? Give examples of your own.
3. You have probably heard people say that "two negatives make a positive." For which of the four arithmetic operations is this statement
   a. always true?
   b. always false?
   c. sometimes true and sometimes false?
   Make up an example for each case.
4. In a Cartesian coordinate system, the axes divide the plane into four ________ . The point $(0, 0)$ is called the ________ .
5. Explain how to add or subtract like terms.
6. What is wrong with this statement? $8x - (2x - 3) = 6x - 3$
7. When solving an inequality, we must remember to ________ if we ________ both sides by ________ .
8. Explain how to use a graph to solve an equation.
9. Give an example of a quotient that is undefined.
10. Explain why $-\frac{a}{b}$ is not equal to $\frac{-a}{-b}$.

Review Problems

For Problems 1-6, choose the appropriate equation.

a. $5x - 8 = 30$	**b.** $\frac{x}{5} - 30 = 8$	**c.** $\frac{x - 30}{5} = 8$
d. $5x + 8 = 30$	**e.** $\frac{x}{5} - 8 = 30$	**f.** $\frac{x + 30}{5} = 8$

1. Ilciar has earned a total of 30 points on the first four quizzes in his biology class. What must he earn on the fifth quiz to end up with an average of 8?
2. Jocelyn ordered five exotic plants from a nursery. She paid a total of \$30, including an \$8 shipping fee. How much did she pay for each plant?
3. The five members of the chess team pitched in to buy new equipment. They used \$30 from their treasury, and each member donated \$8. How much did the new equipment cost?
4. Hemman bought 5 tapes on sale, and he cashed in a gift certificate for \$8. He then owed the clerk \$30. How much was each tape?
5. Nirusha and four other people won the office baseball pool. After spending \$30 of her share, Nirusha had \$8 left. What was the total amount in the pool?
6. One-fifth of the members at Sportslife Health Club signed up for a yoga class. Eight of them dropped out, leaving 30 in the class. How many members are in the Club?

For Problems 7-10, solve.

7. $3x - 4 = 1$
8. $1.2 + 0.4z = 3.2$
9. $\frac{7v}{8} - 3 = 4$
10. $13 = \frac{2}{7}x + 13$

For Problems 11-12, write an equation relating the variables, make a table of values, and graph your equation on the grid provided.

11. A new computer station for a graphics design firm costs \$2000 and depreciates in value \$200 every year. Write and graph an equation that gives the value, V, of the station after t years. Use the grid below left.

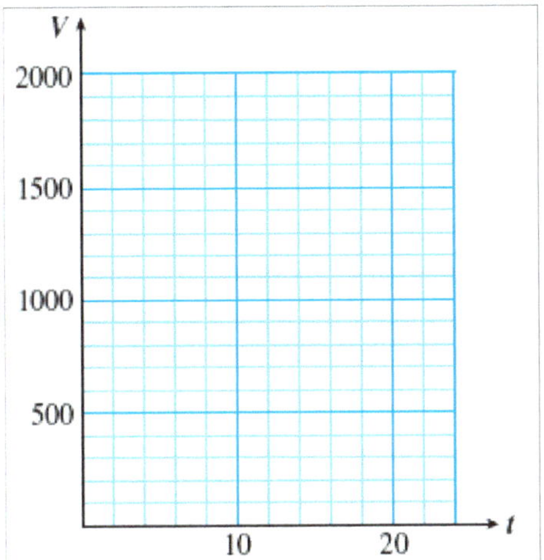

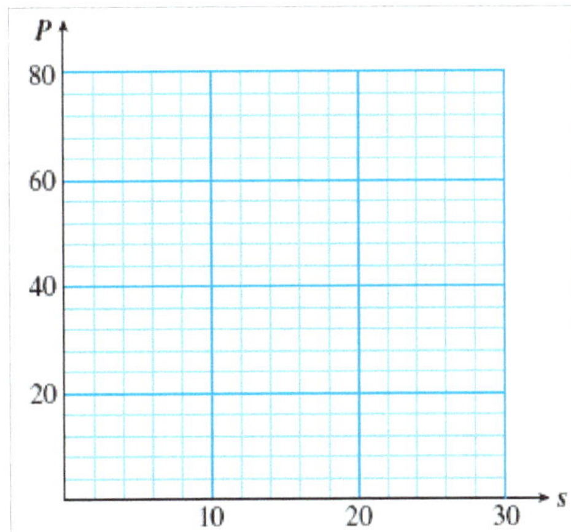

12. Ranwa teaches aerobic dance. Her gym pays her \$5 for an hour-long class, plus \$2 for each paying student in the class. Write and graph an equation that gives Ranwa's pay, P, if there are s students in the class. Use the grid above right.

For Problems 13-14, write and solve an equation to answer the question.

13. Customers at Sunny Orchard can pick their own fruit. There is a \$5.75 entrance fee, and the fruit costs \$4.50 a bushel. Bahn spent \$19.25 at the orchard. How many bushels of fruit did he pick?

14. The perimeter of a rectangle is the sum of twice its width and twice its length. If 150 meters of fence enclose a rectangular yard of length 45 meters, what is the width of the yard?

15. If $x = -3$, evaluate each expression.

a. $-x$ **b.** $x - 3$ **c.** $3 - x$ **d.** $-(-x)$

16. Graph each set of numbers on a separate number line.

a. $1.4,\ 2\frac{1}{2},\ -2\frac{1}{3},\ -3.5$ **b.** $-15, -5, 10, 25$

For Problems 17-20, replace the comma by the proper symbol; $>$, $<$, or $=$.

17. $-2, -3$ **18.** $-2.02, -2.1$

19. $2 - (-5), -7$ **20.** $-6(\frac{-1}{3}), -2$

For Problems 21-28, simplify.

21. $28 - 14 - 9 + 15$ **22.** $11 - 14 + (-24) - (-18)$

23. $12 - [6 - (-2) - 5]$ **24.** $-2 + [-3 - (-14) + 6]$

25. $5 - (-4)3 - 7(-2)$ **26.** $5 - (-4)(3 - 7)(-2)$

27. $\dfrac{6(-3)-8}{-4(-3-5)}$

28. $\dfrac{6(-2)-8(-9)}{5(2-7)}$

For Problems 29-32, evaluate.

29. $2-ab-3a,$ for $a=-5,\ b=-4$

30. $(8-6xy)xy,$ for $x=-2,\ y=2$

31. $\dfrac{-3-y}{4-x},$ for $x=-1,\ y=2$

32. $\dfrac{5}{9}(F-32)+273,$ for $F=-22$

33. If the overnight low in Lone Pine was -4°F, what was the temperature after it had warmed by 10°F?

34. The winter temperature in the city is typically 3°F warmer than in the adjoining suburb. If the temperature in the suburb is -7°F, what is the temperature in the city?

35. Jordan's clothing company is worth 280,000 dollars and Asher's is worth $-180{,}000$ dollars. How much more is Jordan's company worth than Asher's?

36. A certain arsenic compound has a melting point of -8.5°F. If the melting point is reduced by 1.1°F, what is the new melting point?

37. In his first football game, Bo rushes for three consecutive losses of 4 yards each. What is his total net yardage for the three plays?

38. In testing a military aircraft's handling at low altitude, the pilot runs a series of flights over a set course, each at an altitude 150 feet lower than the previous flight. What is the net change in altitude between the first and the sixth flights?

For Problems 39-42, find three solutions for the inequality.

39. $x-2<5$

40. $4x\geq -12$

41. $-9\leq -3x$

42. $-15>5+x$

43. The temperature in Maple Grove was 18°F at noon, and it has been dropping ever since at a rate of 3°F per hour.

a. Fill in the table. T stands for the temperature h hours after noon. Negative values of h represent hours before noon.

h	-4	-2	0	1	3	5	8
T							

b. Write an equation for the temperature, T, after h hours.

c. Graph your equation on the grid.

d. What was the temperature at 10 am?

e. When will the temperature reach -15°F?

f. How much did the temperature drop between 3 pm and 9 pm?

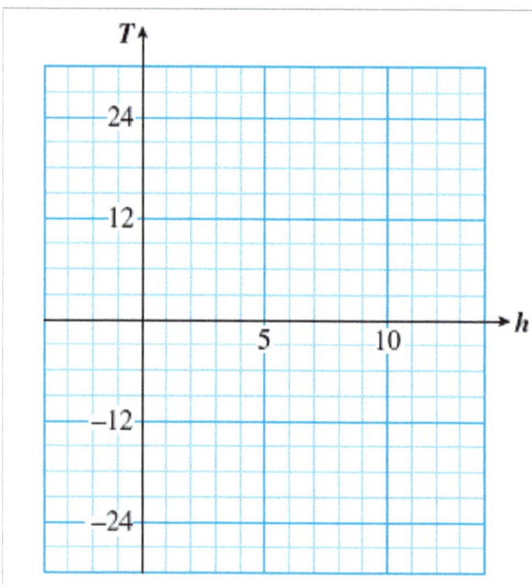

44. Graph each pair of points on the grid, then find the distance between them.

a. $A(-2,-4)$ and $B(-7,-4)$

b. $P(6,-2)$ and $Q(6,8)$

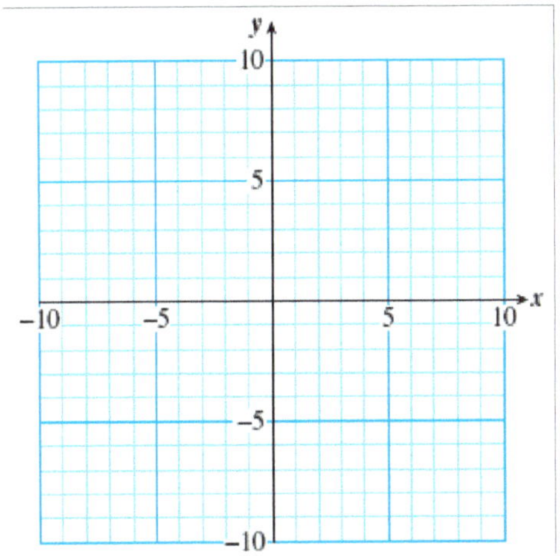

For Problems 45-46, make a table of values and graph each equation.

45. $y = -2x + 7$

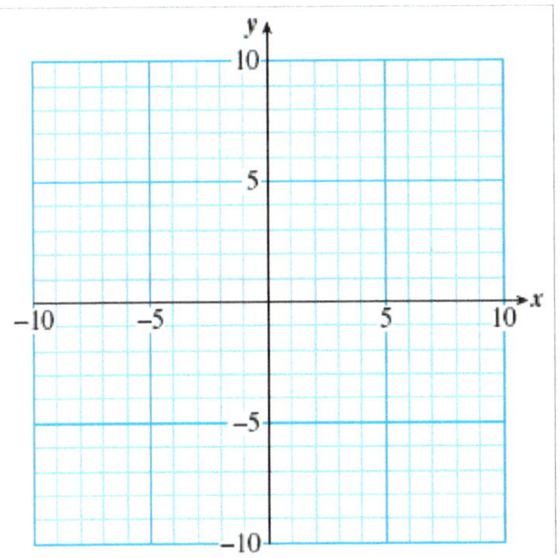

46. $y = \frac{4}{3}x - 2$

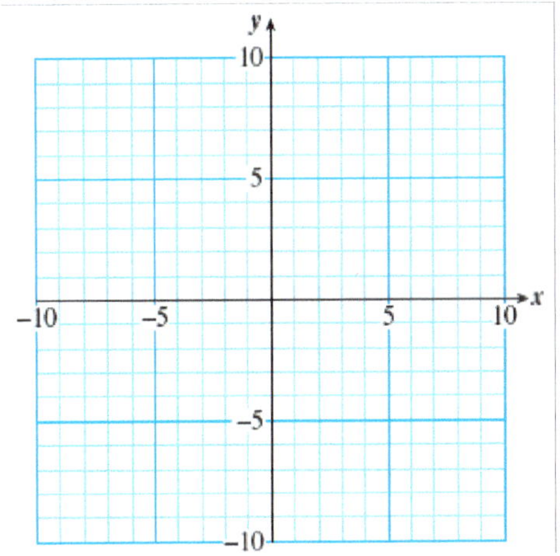

For Problems 47-48, simplify the expression.

47. $(4m + 2n) - (2m - 5n)$

48. $(-5c - 6) + (-11c + 15)$

For problems 49-54, solve.

49. $4z - 6 = -10$

50. $3 - 5x = -17$

51. $-1 = \frac{5w}{3} + 4$

52. $4 - \frac{2z}{5} = 8$

53. $3h - 2 = 5h + 10$

54. $7 - 9w = w + 7$

For Problems 55-56, use the graph to solve the equation. Estimate your solutions if necessary.

55. a. $-3x+9=24$
b. $-3x+9=3$

56. a. $24x-1800=-1250$
b. $24x-1800=500$

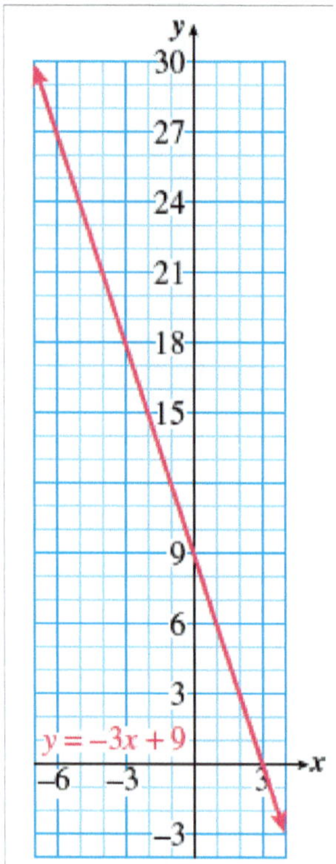

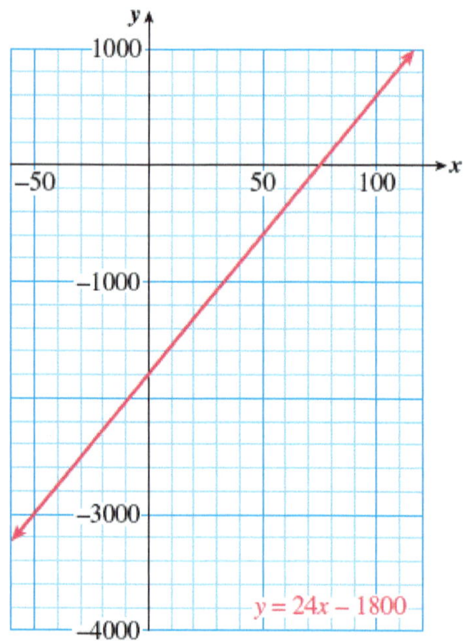

57. The length of a rectangle is 3 times its width.
a. If the width of a rectangle is x, what is its length?
b. Express the perimeter of the rectangle in terms of x.
c. Suppose the perimeter of the rectangle is 48 centimeters. Find the dimensions of the rectangle.

58. In a city council election the winner received 132 votes more than her opponent.
a. If the winner received y votes, how many did her opponent receive?
b. Write an expression for the total number of votes cast for the two candidates.
c. If 12,822 votes were cast, how many did each candidate receive?

For Problems 59-64, solve the inequality algebraically, and graph the solutions on a number line.

59. $2-3z \le -7$

60. $\dfrac{t}{-3} - 1.7 > 2.8$

61. $3k-13 < 5+6k$

62. $12a-28 < -18+2a$

63. $-9 < 5-2n \le -1$

64. $15 \ge -6+3m \ge -6$

Chapter 3 Graphs of Linear Equations

Lesson 3.1 Intercepts

- Find the intercepts of a line from its equation
- Graph a line by the intercept method
- Interpret the intercepts in context

Lesson 3.2 Ratio and Proportion

- Compare two quantities with a ratio or a rate
- Solve a proportion
- Write a proportion to model a problem
- Recognize proportional variables

Lesson 3.3 Slope

- Compute the slope of a line
- Interpret the slope as a rate of change

Lesson 3.4 Slope-Intercept Form

- Write an equation in slope-intercept form
- Identify the slope and y-intercept of a line from its equation
- Graph a linear equation by the slope-intercept method
- Interpret the slope and y-intercept in context

Lesson 3.5 Properties of Lines

- Find equations for parallel or perpendicular lines
- Find equations for horizontal or vertical lines
- Use the two-point formula to compute slope

3.1 Intercepts

Intercepts of a Line

The **intercepts** of a line are the points where the graph crosses the axes.

It is easy to recognize the intercepts of a line on a good graph. Here is a graph of the line

$$y = \frac{1}{2}x + 4$$

We can see that its x-intercept is the point $(-8, 0)$, and its y-intercept is $(0, 4)$.

If we have an equation for a line we can find the intercepts of its graph algebraically. In Example 1 we'll use algebra to find the intercepts of this line.

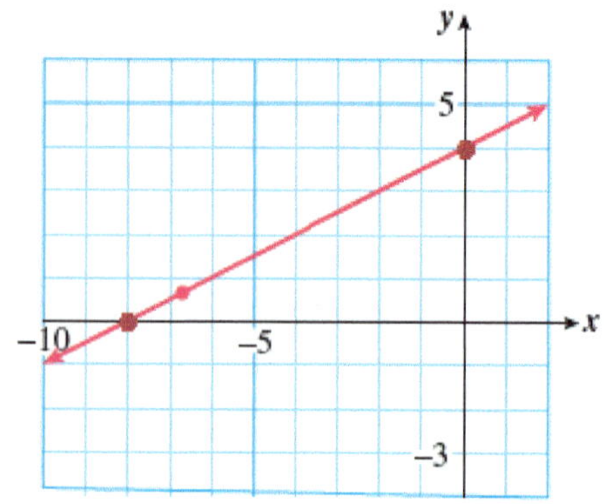

Look Closer: Look at the intercepts on the graph above. Because the y-intercept of a graph lies on the y-axis, its x-coordinate must be zero. And because the x-intercept lies on the x-axis, its y-coordinate must be zero.

Example 1 Use algebra to find the intercepts of the graph of $y = \frac{1}{2}x + 4$.

Solution To find the y-intercept, we substitute 0 for x in the equation and solve for y.

$$y = \frac{1}{2}(0) + 4 = 4$$

The y-intercept is $(0, 4)$, as we saw above. To find the x-intercept we substitute 0 for y and solve the equation for x.

$$\begin{aligned} 0 &= \frac{1}{2}x + 4 && \text{Subtract 4 from both sides.} \\ -4 &= \frac{1}{2}x && \text{Multiply both sides by 2.} \\ -8 &= x \end{aligned}$$

The x-intercept is $(-8, 0)$, as expected.

To find the x-intercept of a graph:
Substitute 0 for y in the equation and solve for x.

To find the y-intercept of a graph:
Substitute 0 for x in the equation and solve for y.

RQ1. What are the intercepts of a line?
RQ2. How do we find the x-intercept of the graph of an equation?

The Intercept Method of Graphing

We can use the x- and y-intercepts to graph a linear equation quickly. Instead of choosing several different values of x to find points on the graph, we find the two intercepts and fill in the short table shown at right. By finding the missing values for x and y we are finding the intercepts of the graph.

x	y
0	
	0

Example 2 Graph the equation $3x + 2y = 7$ by the intercept method.

Solution First, we find the x- and y-intercepts of the graph.

To find the y-intercept, we substitute 0 for x and solve for y.
The y-intercept is the point $(0, 3\frac{1}{2})$.

$$3(0) + 2y = 7 \quad \text{Simplify the left side.}$$
$$2y = 7 \quad \text{Divide both sides by 2.}$$
$$y = \frac{7}{2} = 3\frac{1}{2}$$

To find the x-intercept, we substitute 0 for y and solve for x.
The x-intercept is the point $(2\frac{1}{3}, 0)$.

$$3x + 2(0) = 7 \quad \text{Simplify the left side.}$$
$$3x = 7 \quad \text{Divide both sides by 3.}$$
$$x = \frac{7}{3} = 2\frac{1}{3}$$

Here is a table showing the two intercepts. We plot the intercepts and connect them with a straight line to obtain the graph below.

x	y
0	$3\frac{1}{2}$
$2\frac{1}{3}$	0

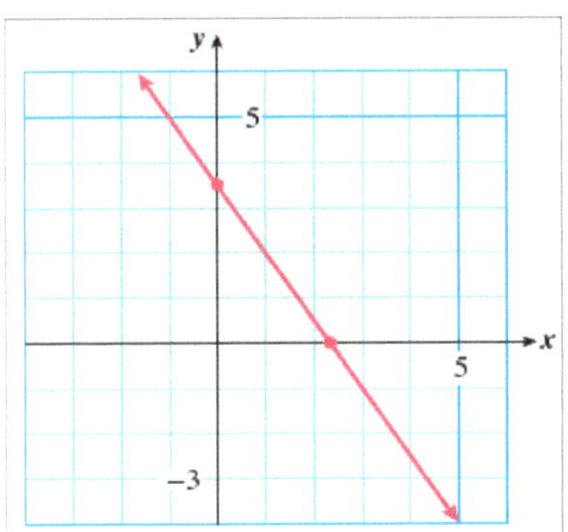

It is a good idea to find a third point as a check. We choose $x = 1$ and solve for y.

$$3(1) + 2y = 7 \quad \text{Subtract 3 from both sides.}$$
$$2y = 4 \quad \text{Divide by 2.}$$
$$y = 2$$

You can check that the point $(1, 2)$ lies on the graph.

The **intercept method** for graphing a line is often faster and more efficient than making a table and plotting points.

To Graph a Linear Equation Using the Intercept Method:

1. Find the x- and y-intercepts of the graph.
2. Draw the line through the two intercepts.
3. Find a third point on the graph as a check. (Choose any convenient value for x and solve for y.)

RQ3. How do we graph a line by the intercept method?
RQ4. Make a table that you can use with the intercept method.
RQ5. What is the y-intercept of the line $3x + 2y = 7$?

Interpreting the Intercepts

The intercepts of a graph give us valuable information about a problem. They often represent the starting or ending values for a particular variable.

Example 3 The temperature in Nome was $-12°$ at noon and has been rising at a rate of $2°$ per hour all day. Find the intercepts of the graph. What do they tell us about the temperature in Nome?

Solution An equation for the temperature T at time h, where h represents the number of hours from noon, is

$$T = -12 + 2h$$

The graph of the equation is shown below. To find the T-intercept, we set $h = 0$ and solve for T.

$$T = -12 + 2(0) = -12$$

The T-intercept is $(0, -12)$. This point tells us that when $h = 0,\ T = -12$, or the temperature at noon was $-12°$. To find the h-intercept, we set $T = 0$ and solve for h.

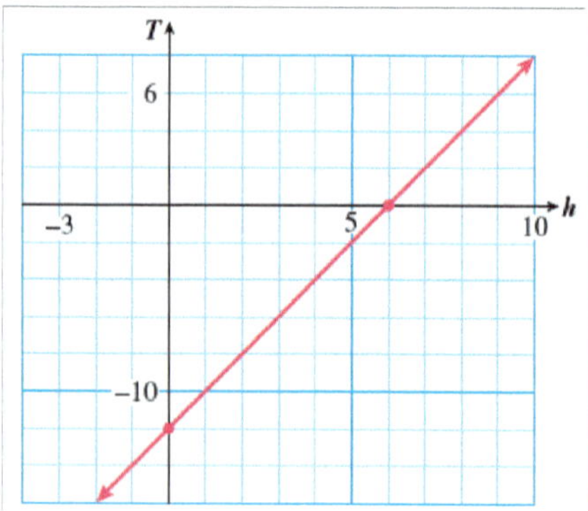

$0 = -12 + 2h$ **Add 12 to both sides.**

$12 = 2h$ **Divide by 2.**

$6 = h$

The h-intercept is the point $(6, 0)$. This point tells us that when $h = 6,\ T = 0$, or the temperature will reach zero degrees at six hours after noon, or 6 pm.

RQ6. The intercepts of a graph often represent the ________ or ________ values for a particular variable.

RQ7. What does the h-intercept tell us in Example 3?

Skills Warm-Up

Choose the correct algebraic expression for each of the following situations.

$5x - 8 = 30$	$\frac{x}{5} - 30 = 8$	$\frac{x-30}{5} = 8$
$5x + 8 = 30$	$\frac{x}{5} + 30 = 8$	$\frac{x+30}{5} = 8$

1. Ilciar has earned a total of 30 points on the first four quizzes in his biology class. What must he earn on the fifth quiz to end up with an average of 8?
2. Jocelyn ordered five exotic plants from a nursery. She paid a total of \$30, including an \$8 shipping fee. How much did she pay for each plant?
3. To buy new equipment, the five members of the chess club used \$30 from the treasury, and each member donated \$8. How much did the equipment cost?

4. Hemman bought 5 tapes on sale, and he cashed in a gift certificate for $8. He then owed the clerk $30. How much was each tape?
5. Nirusha and four other people won the office baseball pool. After spending $30 of her share, Nirusha had $8 left. What was the total amount in the pool?

Answers: **1.** $\dfrac{x+30}{5}=8$ **2.** $5x+8=30$ **3.** $\dfrac{x-30}{5}=8$ **4.** $5x-8=30$ **5.** $\dfrac{x}{5}-30=8$

Homework 3.1

Skills Practice

■ For Problems 1-6,

a. Find the x- and y-intercepts of each line.

b. Use the intercept method to graph the line.

1. $2x + 4y = 8$

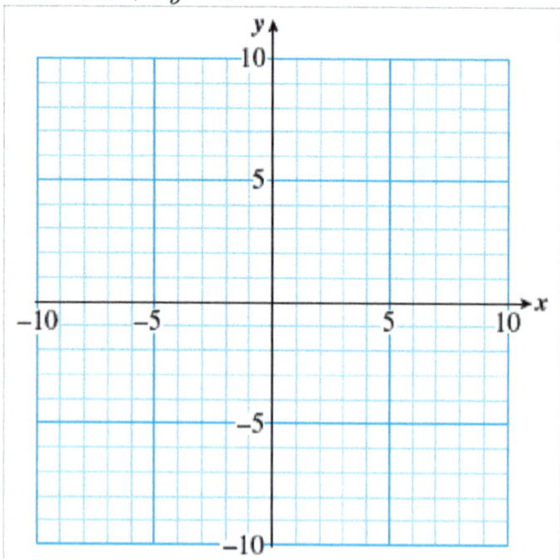

2. $x + 2y + 10 = 0$

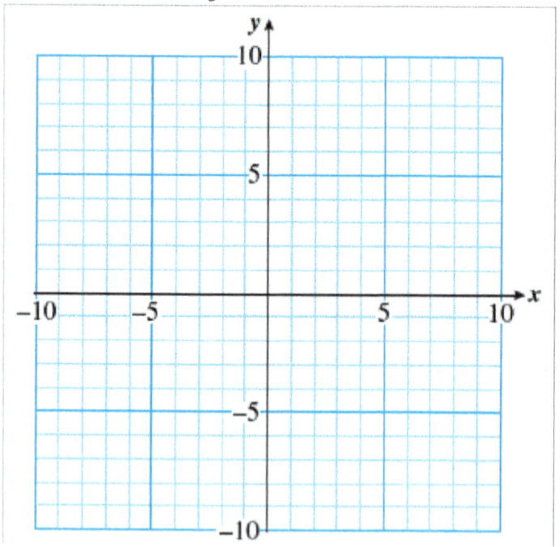

3. $2x = 14 + 7y$

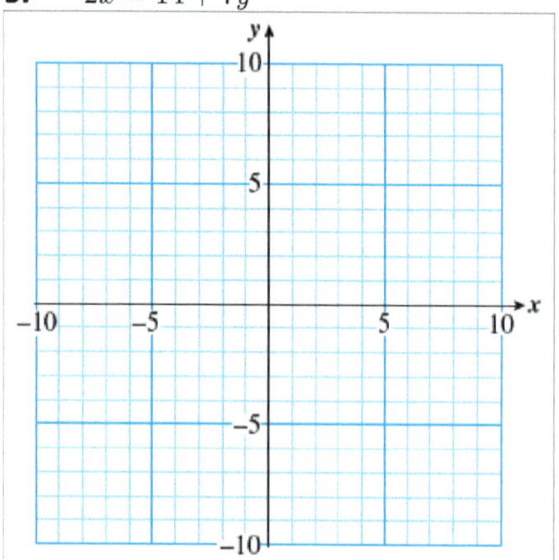

4. $y = -4x + 8$

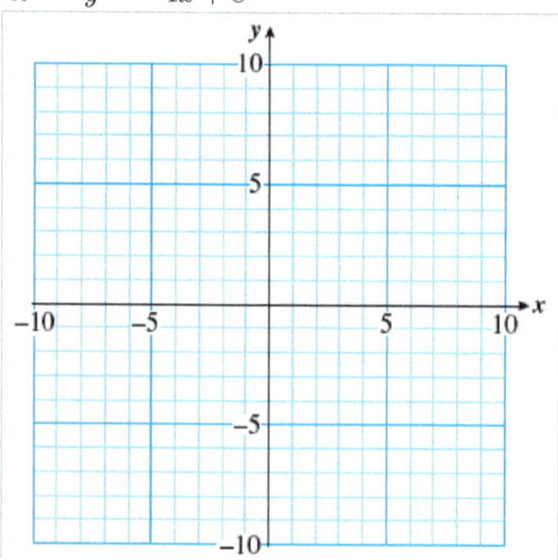

5. $\dfrac{x}{20} + \dfrac{y}{30} = 1$

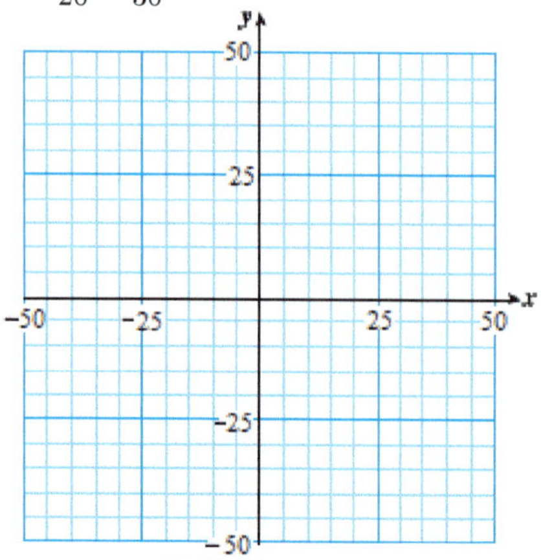

6. $3x - 2y = 120$

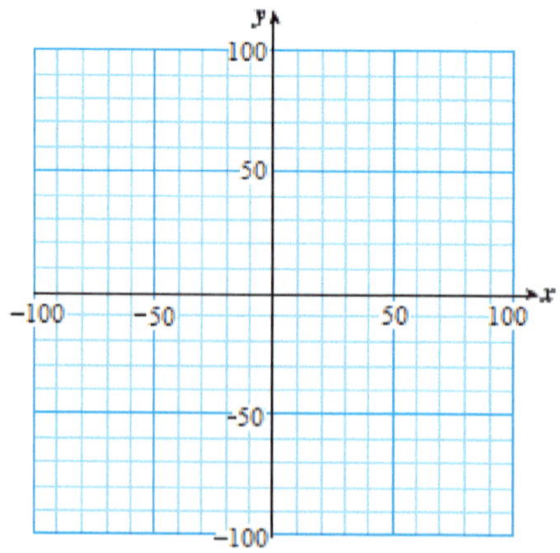

For Problems 7-12, match each equation with its graph. (More than one equation may describe the same graph.)

7. $2x + 3y = 12$ **8.** $2x - 3y = 12$ **9.** $3x - 2y = 12$

10. $-3y - 2x = 12$ **11.** $\frac{x}{6} - \frac{y}{4} = 1$ **12.** $\frac{x}{4} - \frac{y}{6} = 1$

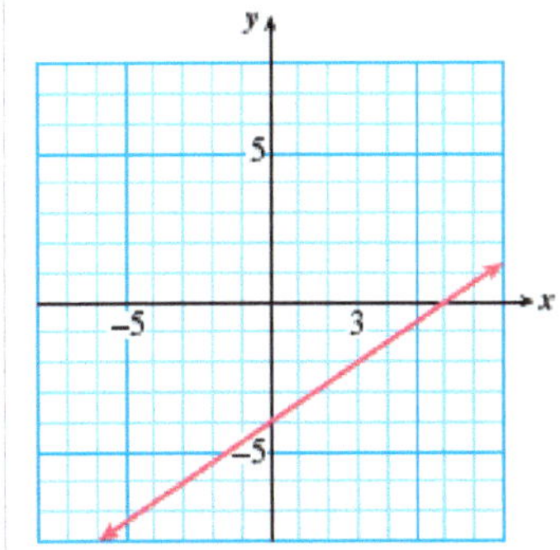

a.

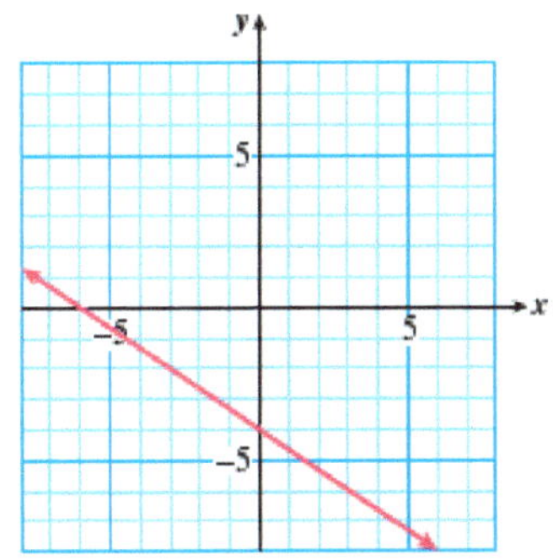

b.

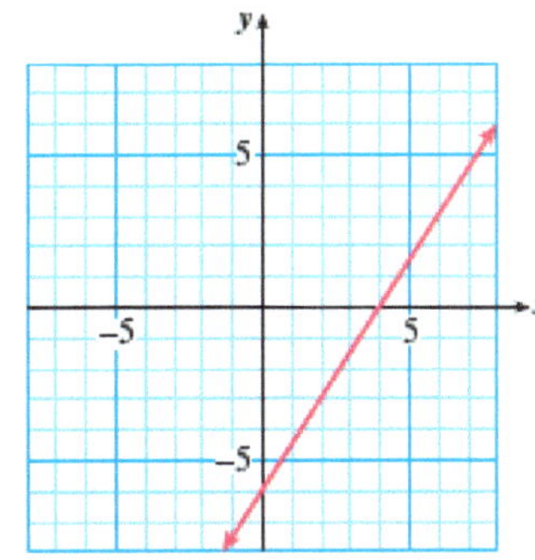

c.

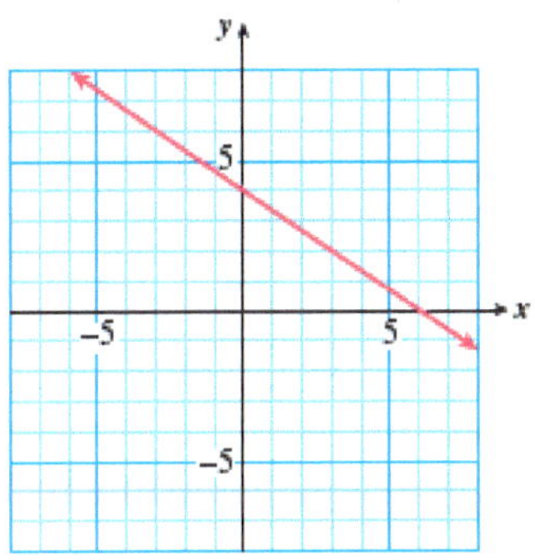

d.

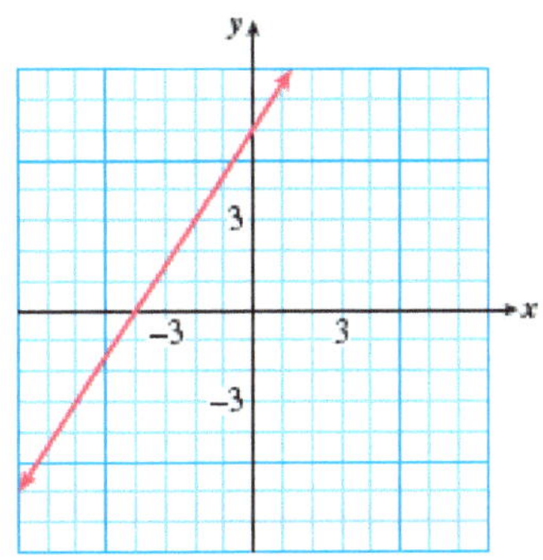

e.

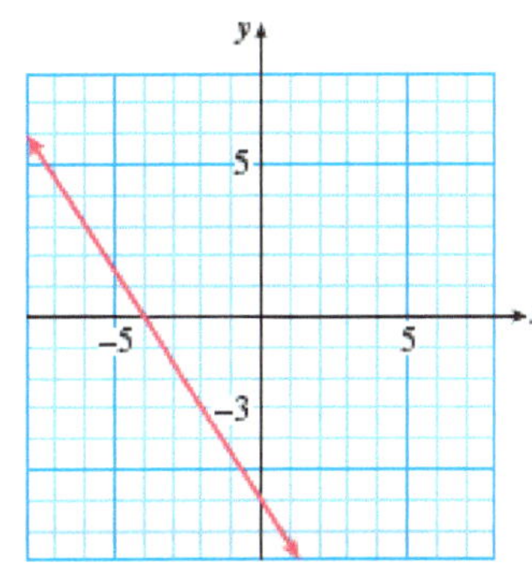

f.

For Problems 13-18, solve each pair of equations. In part (b) of each problem, your answer will involve the constant k.

13. a. $-2x = 6$
b. $-2x = k$

14. a. $x + 4 = 7$
b. $x + k = 7$

15. a. $x - 5 = 9$
b. $x - 5 = k$

16. a. $2x + 3 = 8$
b. $2x + k = 8$

17. a. $15 - 4x = 3$
b. $15 - 4x = k$

18. a. $9 + 3x = -1$
b. $9 + kx = -1$

Applications

Use the graph to answer the questions.

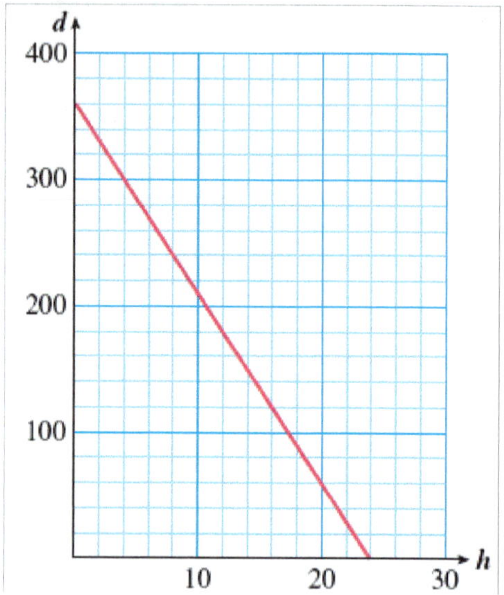

19. During spring break, Francine took the train to San Francisco and then bicycled home. The graph below left shows Francine's distance d from home, in miles, after cycling for h hours.

a. How far is it from San Francisco to Francine's home?

b. How many hours did Francine cycle to get home?

c. After cycling for 12 hours, how far is Francine from home?

d. How far did Francine cycle in the first four hours?

For Problems 20-22,

a. Find the intercepts of each linear equation.

b. Use the intercept method to graph the line.

c. Explain what the intercepts mean in terms of the problem situation.

20. The amount of home heating oil (in gallons) in the Olsons' tank is given by the equation $G = 200 - 15w$, where w is the number of weeks since they turned on the furnace.

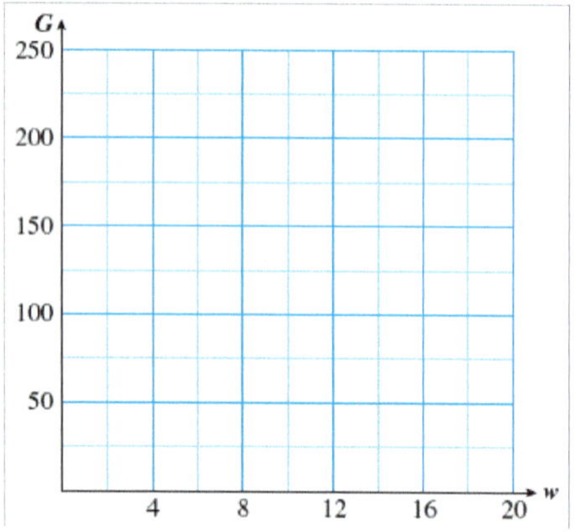

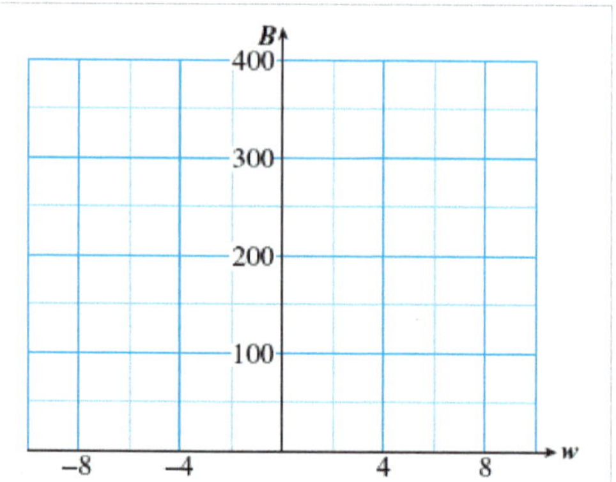

21. Dana joined a savings plan some weeks ago. Her bank balance is growing each week according to the formula $B = 225 + 25w$, where $w = 0$ represents this week.

22. Delbert bought some equipment and went into the dog-grooming business. His profit is increasing according to the equation $P = -600 + 40d$, where d is the number of dogs he has groomed.

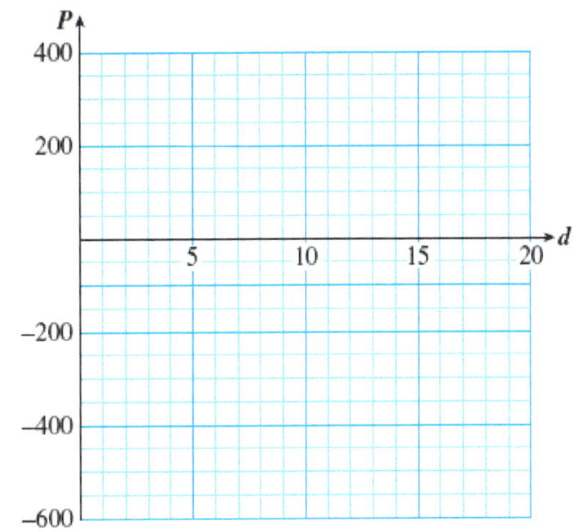

23. a. At what point does the graph of $2x - 3y = 25$ cross the x-axis?

b. At what point does the graph of $1.4x + 3.6y = -18$ cross the y-axis?

24. The x-intercept of a line is positive and its y-intercept is negative. Is the line increasing or decreasing? Sketch a possible example of such a line.

3.2 Ratio and Proportion

Ratios

A **ratio** is a type of quotient used to compare two numerical quantities.

For example, suppose the ratio of pupils to computers at a local elementary school is 10 to 3. This means that for every 10 pupils in the school there are 3 computers. We often display a ratio as a fraction, like this:

$$\frac{\text{Number of pupils}}{\text{Number of computers}} = \frac{10}{3}$$

Caution! This ratio does *not* mean that there are exactly 10 pupils and 3 computers at the school. There could be 40 pupils and 12 computers or 200 pupils and 60 computers (or many other combinations) because

$$\frac{40}{12} = \frac{200}{60} = \frac{10}{3}$$

Example 1 In a survey of 100 of his classmates, Greg found that 68 students supported a strict gun control law, while 32 students did not. What is the ratio of survey respondents who support gun control to those opposed?

Solution The ratio of those supporting gun control to those opposed is 68 to 32, or $\frac{68}{32}$. This ratio reduces to $\frac{17}{8}$, or 17 to 8.

Caution! When we write a ratio, it is important to notice which number is the numerator, and which is the denominator. When we write "the ratio of a to b" as a fraction, a is the numerator and b is the denominator.

In Example 1, we could also compute the ratio of opponents to supporters of the gun control law. In that case, the answer would be $\frac{32}{68}$ or $\frac{8}{17}$, the reciprocal of the answer to Example 1.

Ratio

A **ratio** is a type of quotient used to compare two numerical quantities. The **ratio of a to b** is written $\frac{a}{b}$.

Look Closer: A ratio can be expressed as a decimal instead of a common fraction. Because $\frac{17}{8}$ is equal to 2.125, we can write the ratio in Example 1 as 2.125 to 1. We usually express a ratio as a decimal fraction if the numbers being compared are decimal numbers.

Example 2 A small circuit board measures 4.2 centimeters long by 2.4 centimeters wide. What is the ratio of its length to its width?

Solution The ratio of length to width is $\frac{4.2 \text{ cm}}{2.4 \text{ cm}}$, or 1.75, as you can verify with your calculator. (Can you express this ratio as a common fraction?)

RQ1. What is a ratio?
RQ2. When do we express a ratio as a decimal fraction?

Rates

A **rate** is a ratio that compares two quantities with different units.

You are already familiar with many rates. If you say that your average speed was 50 miles per hour, or that apples cost 89 cents per pound, you are using rates. Because *per* usually indicates division,

50 miles per hour means $\frac{50 \text{ miles}}{1 \text{ hour}}$

89 cents per pound means $\frac{89 \text{ cents}}{1 \text{ pound}}$

Example 3 Sarah traveled 390 miles on 15 gallons of gas. Express her rate of fuel consumption as a ratio, and then as a rate.

Solution As a ratio, Sarah's rate of fuel consumption was $\frac{390 \text{ miles}}{15 \text{ gallons}}$. We simplify by dividing the denominator into the numerator, to get $\frac{26 \text{ miles}}{1 \text{ gallon}}$, or 26 miles per gallon.

Proportions

A **proportion** is a statement that two ratios are equal.

Thus, a proportion is an equation in which both sides are ratios. Here are some proportions:

$$\frac{6}{9} = \frac{8}{12} \quad \text{and} \quad \frac{3.2}{8} = \frac{1}{2.5}$$

A proportion may involve variables. How can we solve a proportion such as this one?

$$\frac{7}{5} = \frac{x}{6}$$

Because x is divided by 6, we could multiply both sides of the equation by 6 to clear the fraction. But we can clear both fractions at the same time if we multiply by the LCD (Lowest Common Denominator) of the two fractions. In this case, the LCD is $5 \cdot 6 = 30$. Multiplying both sides by 30 gives us

$$30\left(\frac{7}{5}\right) = \left(\frac{x}{6}\right)30$$
$$42 = 5x$$

Now we can divide both sides by 5 to get the answer,

$$x = \frac{42}{5} = 8.4$$

Look Ahead: There is a short-cut we can use that avoids calculating an LCD. We can obtain the equation $42 = 5x$ by *cross-multiplying.*

$$\frac{7}{5} = \frac{x}{6}$$
$$42 = 5x$$

Then we complete the solution as before.

Property of Proportions

We can clear the fractions from a proportion by **cross-multiplying**.

If $\dfrac{a}{b} = \dfrac{c}{d}$, then $ad = bc$.

Example 4 Solve $\dfrac{2.4}{1.5} = \dfrac{8.4}{x}$

Solution We apply the property of proportions and cross-multiply to get

$$2.4x = 1.5(8.4) \quad \text{Divide both sides by 2.4.}$$
$$x = \frac{1.5(8.4)}{2.4} \quad \text{Simplify the right side.}$$
$$x = 5.25$$

We can check the solution by substituting $x = 5.25$ into the original proportion.

Caution! Cross-mutiplying works **only** for solving proportions! Do not try to use cross-multiplying on other types of equations, or for other operations on fractions. In particular, do not use cross-multiplying when multiplying fractions or adding fractions.

RQ3. What is a rate?
RQ4. What is a proportion?
RQ5. What is the short-cut for solving a proportion called?

Proportional Variables

Suppose that grape juice costs 80 cents per quart, and complete the table.

Number of Quarts	Total Price (cents)	$\frac{\text{Total Price}}{\text{Number of Quarts}}$
1	80	$\frac{80}{1} = 80$
2	160	$\frac{160}{2} =$ 80
3	240	240/3 = 80
4	320	320/4 = 80

Did you find that the ratio $\frac{\text{total price}}{\text{number of quarts}}$, or *price per quart*, is the same for each row of the table? This agrees with common sense: The *price per quart* of grape juice is the same no matter how many quarts you buy.

Two variables are **proportional** if their ratio is always the same.

Example 5 One quart of grape juice costs 80 cents. The price of grape juice is proportional to the number of quarts you buy. How much grape juice can you buy for \$10.00?

Solution Let x represent the number of quarts you can buy for \$10.00, or 1000 cents. Because the variables *total price* and *number of quarts* are proportional, we know that their ratio is constant. Thus, the ratio

$$\frac{1000 \text{ cents}}{x \text{ quarts}} \quad \text{is equal to the ratio} \quad \frac{80 \text{ cents}}{1 \text{ quart}}$$

$$\frac{1000}{x} = \frac{80}{1}$$

We solve the proportion by cross-multiplying to get

$$1000\,(1) = 80x \qquad \text{Divide both sides by 80.}$$

$$x = \frac{1000}{80} = 12.5$$

You can buy 12.5 quarts of grape juice for \$10.00. (Or, if you can only buy quart bottles, then you can buy 12 quarts and have 40 cents left over.)

Caution! When writing a proportion, we must be careful that *both* ratios have the *same units* in their numerators, and the same units in their denominators. In the example above, it would *not* be correct to equate $\frac{1000}{x}$ and $\frac{1}{80}$, because the ratios do not have the same units.

For the grape juice example, the ratio $\dfrac{\text{total price}}{\text{number of quarts}}$ was always equal to 80. The number 80 is called the **constant of proportionality**.

Look Closer: If we let P stand for the price of the grape juice and q stand for the number of quarts purchased, we have the equation

$$\frac{P}{q} = 80$$

or

$$P = 80q$$

The graph of this equation is a straight line that passes through the origin, as shown at right.

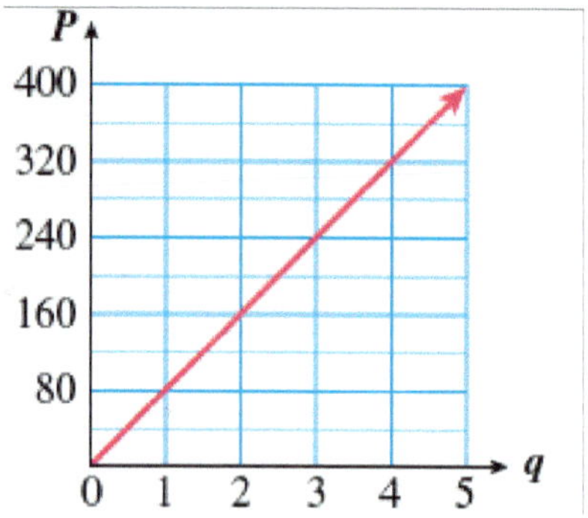

Proportional Variables

If two variables are proportional, they are related by the equation

$$y = kx$$

where k is the **constant of proportionality**.

RQ6. What are proportional variables?
RQ7. If x and y are proportional, what equation do they satisfy?

Skills Warm-Up

Convert each common fraction to a decimal, and each decimal to a common fraction.

1. $\frac{3}{8}$ **2.** $\frac{7}{5}$ **3.** $\frac{4}{3}$ **4.** $\frac{5}{12}$

5. 0.04 **6.** $0.\overline{6}$ **7.** 1.875 **8.** 2.2

Answers: **1.** 0.375 **2.** 1.4 **3.** $1.\overline{3}$ **4.** $0.41\overline{6}$ **5.** $\frac{1}{25}$ **6.** $\frac{2}{3}$ **7.** $\frac{15}{8}$ **8.** $\frac{11}{5}$

Homework 3.2

Skills Practice

For Problems 1-4, write ratios or rates.

1. In a survey of 300 employees at a large company, 125 used public transportation to commute to work. What is the ratio of employees who use public transportation to those who do not?

2. Erica tutored in the math lab for 6.5 hours last week and made $56.68. What is Erica's rate of pay in dollars per hour?

3. The instructions for mixing a rose fertilizer call for $\frac{3}{4}$ cup of potash and $1\frac{3}{8}$ cups of nitrogen. What is the ratio of nitrogen to potash?

4. An orange contains 0.14 milligram of thiamin and 0.6 milligram of niacin. What is the ratio of niacin to thiamin?

5. Which of the following expressions are proportions?

a. $\frac{3}{x} = \frac{8}{15}$ **b.** $\frac{v}{9} - \frac{12}{5}$ **c.** $\frac{w}{6} + \frac{2}{5} = \frac{3}{2}$ **d.** $\frac{2.6}{m} = \frac{m}{1.6}$

For Problems 6-9, solve the proportion.

6. $\frac{x}{16} = \frac{9}{24}$ **7.** $\frac{182}{65} = \frac{21}{w}$ **8.** $\frac{7}{b} = \frac{5}{9}$ **9.** $\frac{7/3}{3} = \frac{z}{9/2}$

For Problems 10-12, decide whether the two variables are proportional.

10.

Time	Distance
1	45
2	90
4	180
5	225

11.

Length	Area
3	9
4	16
8	64
10	100

12.

Rate	Time
20	40
40	20
50	16
80	10

For Problems 13-16, decide whether the two quantities are proportional.

13. A 9-inch pizza costs $6, and a 12-inch pizza costs $10.
14. A commuter train travels 10 miles in 20 minutes, and 15 miles in 30 minutes.
15. It takes $4\frac{1}{2}$ cups of flour to make two loaves of bread, and 18 cups to make eight loaves.
16. An 18-foot sailboat sleeps four, and a 32-foot sailboat sleeps six.

Applications

For Problems 17-22,
a. Write a proportion.
b. Solve the proportion and answer the question.

17. If 3 pounds of coffee makes 225 cups, how many pounds of coffee is needed to make 3000 cups of coffee?

18. Gunther's car uses 32 liters of gas to travel 184 kilometers. How many liters will he need to travel 575 kilometers?

19. On a map of Fairfield County, $\frac{3}{4}$ centimeter represents a distance of 10 kilometers. If Eastlake and Kenwood are 6 centimeters apart on the map, what is the actual distance between the two towns?

20. In a survey of 1200 voters, 863 favored the construction of a light rail system. If 8000 people vote, about how many will vote for the light rail system?

21. If 1 inch equals 2.54 centimeters, what is the length in inches of a wire 35 centimeters long?

22. A cinnamon bread recipe calls for $1\frac{1}{4}$ tablespoons of cinnamon and 5 cups of flour. How much cinnamon would be needed with 8 cups of flour?

For Problems 23-24,

a. Find the constant of proportionality and write an equation relating the variables.
b. Make a table of values and graph the equation.
c. Use your equation to answer the question.

23. Megan's wages, w, are proportional to the time, h, she works. Megan made \$60 for 8 hours work as a lab assistant. How long will it take her to make \$500?

h	w

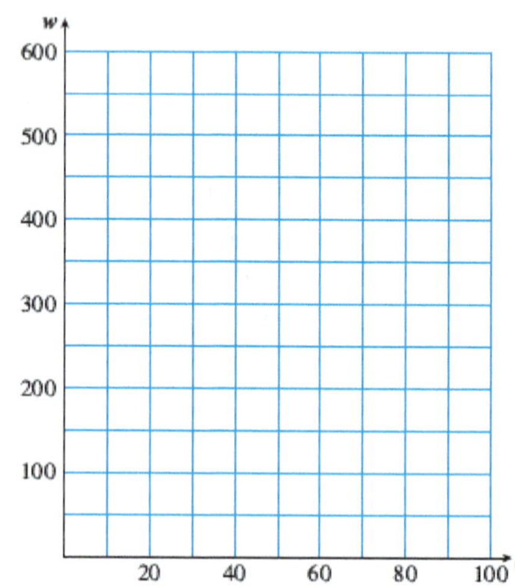

24. The amount of pure copper, c, produced by Copperfield Mine is proportional to the amount of ore extracted, g. A new lode produced 24 grams of copper from 800 grams of ore. How many grams of copper will one kilogram of ore yield?

g	c

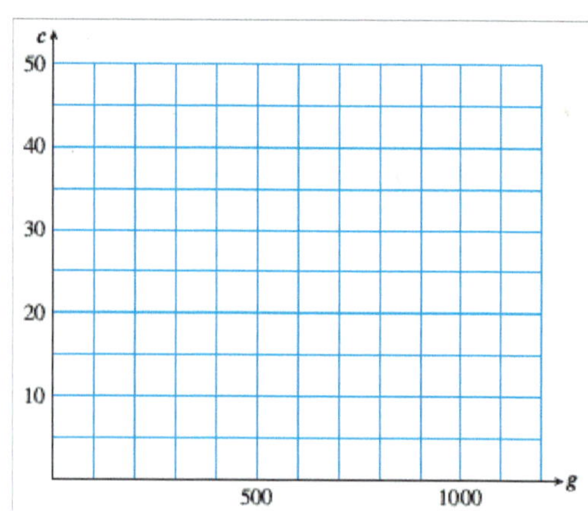

25. Trish's real-estate commission is proportional to the amount of the sale. She made \$3200 commission on a sale of \$80,000.

a. Write an equation relating Trish's sales and her commission.

b. What would her commission be on the sale of a \$200,000 property?

26. Everett can bicycle 16 miles in 2 hours and 24 miles in 3 hours. The distance d that Everett travels is proportional to the time t that he bicycles.

a. Write an equation for d in terms of t.

b. Graph your equation.

c. Complete the tables, and use them to answer the question below.

t	d
2	
3	
5	

t	d
4	
6	
10	

What happens to the distance Everett travels when he doubles the time he bicycles?

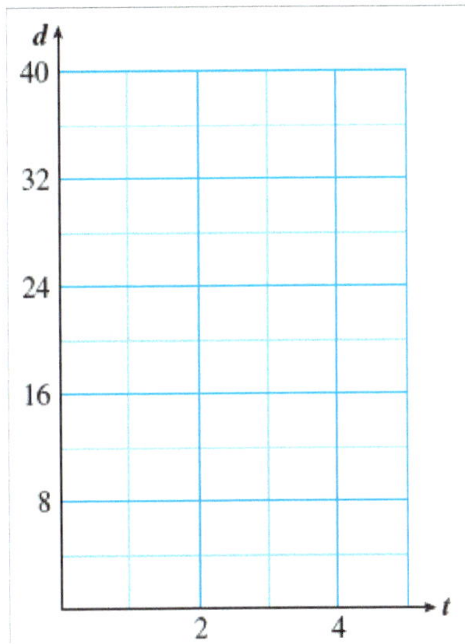

27. The sales tax on \$12 is 72 cents, and the sales tax on \$15 is 90 cents. The tax T is proportional to the price P.

a. Write an equation for T in terms of P.

b. Graph your equation.

c. Complete the tables, and use them to answer the question below.

P	T
12	
20	
30	

P	T
6	
10	
15	

What happens to the sales tax when the price is cut in half?

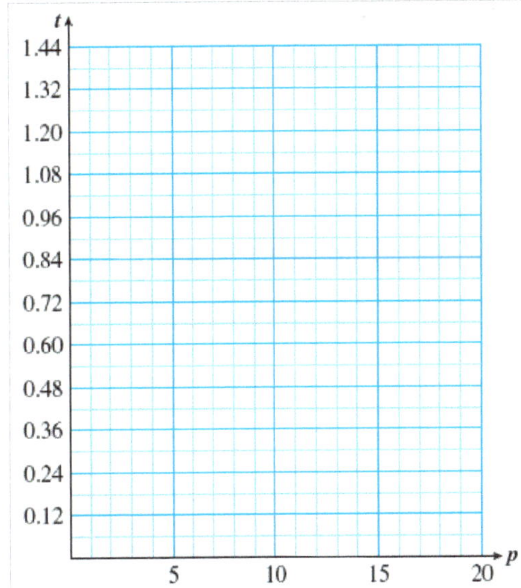

28. The cost of tuition at Walden College is given by the formula $T = 500 + 40u$, where u is the number of units you take.

a. Fill in the tables.

Units	Tuition
3	620
5	700
8	820

Units	Tuition
6	740
10	900
16	1,140

b. Does doubling the number of units you take double your tuition? No

c. Is T proportional to u? NO

29. a. State a formula for the circumference of a circle in terms of its radius.

b. Complete the table.

r	2	5	7	10
C				

c. Graph your equation, using the values in your table.

d. Is the circumference of a circle proportional to its radius?

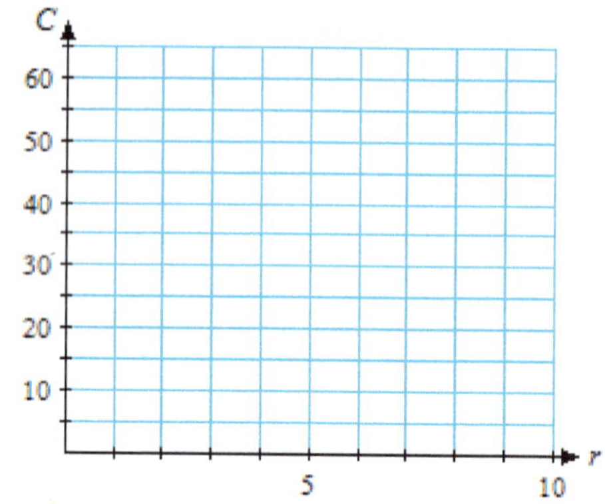

30. a. State a formula for the area of a square in terms of the length of its side.

b. Complete the table.

s	2	5	6	8
A				

c. Graph your equation, using the values in your table.

d. Is the area of a square proportional to the length of its side?

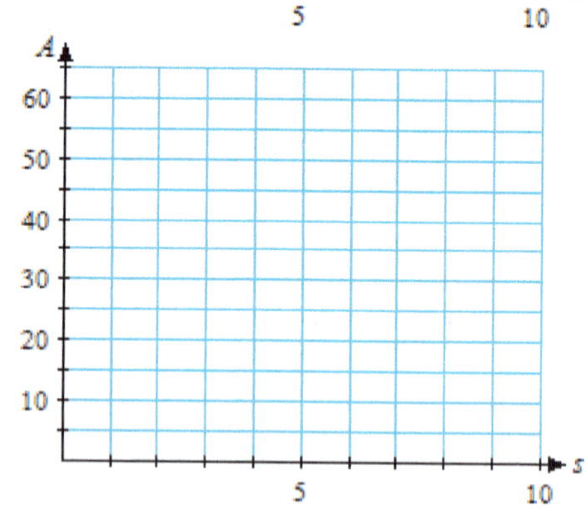

31. The following recipe for Cadet Mess Sloppy Joes comes from the West Point Officers Wives' Club Cookbook.

For 1000 servings:

310 lb. ground beef
100 lb. ground turkey
$5\frac{1}{2}$ cans mushrooms
4 lb. green peppers, sliced
18 cans tomato puree
12 cans tomato catsup
1 qt. cider vinegar
2 cans diced red peppers
20 lb. onions, diced
3 lb. chili powder
1 lb. garlic powder
$3\frac{1}{2}$ lb. sugar
1 qt. light flavor glow
28 oz. salt and pepper

Cook off turkey and beef in oven; place in pots. Braise off chopped peppers and diced onions (turkey will absorb liquid). Add stock to turkey to moisten well before adding other products. Stir together. Heat till piping hot. Serve on rolls.

a. What is the ratio of tomato puree to tomato catsup? How many servings of sloppy joes can you make if you have only 5 cans of tomato catsup? How many cans of tomato puree will you need?

b. How much chili powder (in ounces) will you need to serve sloppy joes to 50 people? How many pounds of ground beef?

c. The cookbook also lists the ingredients needed for 4500 servings. How many cans of mushrooms are required for 4500 servings? How many pounds of onions?

3.3 Slope

Rate of Change

A **rate of change** is a type of ratio that measures how one variable changes with respect to another.

Example 1 In order to fire a particular kind of pottery, the pieces must first be cured by raising the temperature slowly and evenly. Sonia checks the temperature in the drying oven at ten-minute intervals, and records the following data.

Time, x	0	10	20	30	40	50	60
Temperature, y	70	74	78	82	86	90	94

The heat in the oven should not increase any faster than 0.5 degree per minute. Is the temperature in the oven within the safe limits?

Solution A graph of the data is shown at right. Sonia calculates the rate at which the temperature is rising by finding the following ratio:

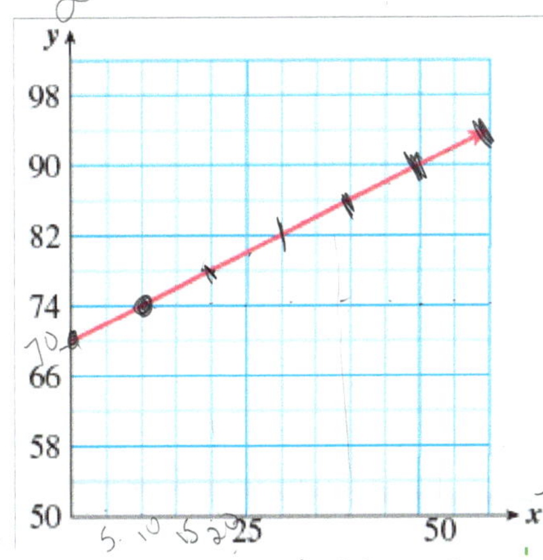

$$\frac{\text{change in temperature}}{\text{change in time}}$$

For example, over the first 10 minutes, the temperature rises from 70 degrees to 74 degrees, so

$$\frac{\text{change in temperature}}{\text{change in time}} = \frac{4 \text{ degrees}}{10 \text{ minutes}}$$

or 0.4 degree per minute. This is less than the maximum rate recommended for curing the pottery.

You can check that over each ten-minute interval the temperature again rises by four degrees, so it appears that the oven is heating up at an acceptable rate.

RQ1. What is a rate of change?
RQ2. What are the units of the rate of change in Example 1?

Slope

We introduce some new notation to use when calculating a rate of change.

The Greek letter Δ ("delta") is used in mathematics to indicate **change**.

In Example 1, we used the variable x to represent time and y to represent the temperature, so we denote the ratio $\frac{\text{change in temperature}}{\text{change in time}}$ by $\frac{\Delta y}{\Delta x}$. With this notation, we calculate the rate of change of temperature between the data points $(20, 78)$ and $(50, 90)$ as follows:

$$\frac{\Delta y}{\Delta x} = \frac{12 \text{ degrees}}{30 \text{ minutes}} = 0.4 \text{ degrees per minute}$$

We can illustrate the rate of change on a graph of the data, as shown at right. We move from the point $(20, 78)$ to the point $(50, 90)$ by moving horizontally a distance of $\Delta x = 30$ and then vertically a distance of $\Delta y = 12$.

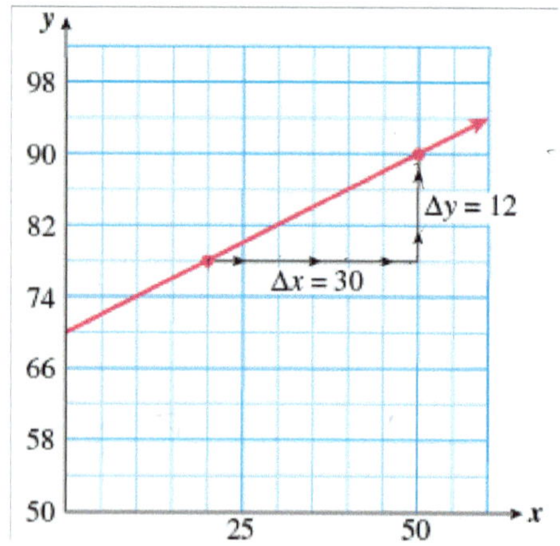

The rate of change of one variable with respect to another is so important in applications that the ratio $\frac{\Delta y}{\Delta x}$ is given a name; it is called **slope**, and is usually denoted by the letter m.

Slope

The **slope** of a line is defined by the ratio

$$\textbf{slope} = \frac{\textbf{change in } y\textbf{-coordinate}}{\textbf{change in } x\textbf{-coordinate}}$$

as we move from one point to another on the line. In symbols,

$$m = \frac{\Delta y}{\Delta x}$$

Look Closer: The slope of a line measures how fast the y-coordinate changes as we increase the x-coordinate of points on the line. More specifically, when we move one unit in the x-direction, how many units should we move in the y-direction to get back to the line?

Example 2 Use the points A and B to compute the slope of the line shown.

Solution The point A has coordinates $(1, 3)$, and B has coordinates $(5, 6)$. As we move along the line from $A(1, 3)$ to $B(5, 6)$, the y-coordinate changes by 3 units, and the x-coordinate changes by 4 units. The slope of the line is thus

$$\frac{\Delta y}{\Delta x} = \frac{\text{change in } y\text{-coordinate}}{\text{change in } x\text{-coordinate}} = \frac{3}{4}$$

The slope tells us that if we start at any point on the line and move 1 unit in the x-direction, we must move $\frac{3}{4}$ unit in the y-direction to return to the line.

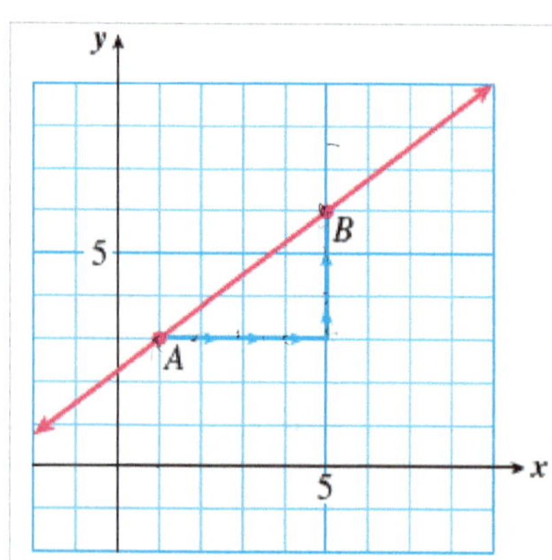

Caution! Note the difference between the statements $y = 3$ and $\Delta y = 3$; they are not the same. When we discuss a graph, $y = 3$ means that the y-coordinate of a particular point is 3, but $\Delta y = 3$ means that the y-coordinate changes by 3 units when we move from one point to another.

RQ3. What does Δ mean in mathematics?
RQ4. How do we indicate Δx on a graph?
RQ5. What is the name of the ratio $\frac{\Delta y}{\Delta x}$, and what letter is used to represent it?

Meaning of Slope

In Example 1, we graphed the temperature of a pottery oven over time. We calculated the slope of the graph as

$$\frac{\Delta y}{\Delta x} = 0.4 \text{ degrees per minute}$$

The slope gives us the rate of change of the temperature with respect to time: the temperature is increasing at a rate of 0.4 degrees per minute.

The **slope** of a line measures the **rate of change** of y with respect to x.

In different situations, this rate might be interpreted as a rate of growth or a speed. The slope of a graph can give us valuable information about the variables involved.

Example 3 The graph shows the distance traveled by a driver for a cross-country trucking firm in terms of the number of hours she has been on the road.

a. Compute the slope of the graph.
b. What is the meaning of the slope for this problem?

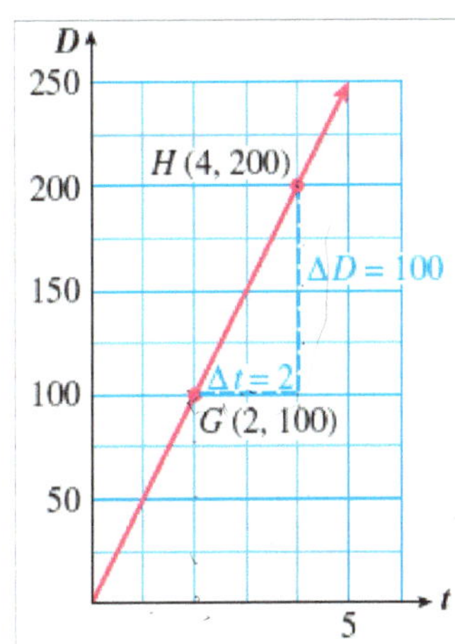

Solutions a. Choose any two points on the line, say $G(2, 100)$ and $H(4, 200)$ shown in the figure. As we move from G to H we find

$$m = \frac{\Delta D}{\Delta t} = \frac{100}{2} = 50$$

The slope of the line is 50.

b. The best way to understand the slope is to include units in the calculation.

$$\frac{\Delta D}{\Delta t} \text{ means } \frac{\text{change in distance}}{\text{change in time}} \quad \text{or} \quad \frac{\Delta D}{\Delta t} = \frac{100 \text{ miles}}{2 \text{ hours}} = 50 \text{ miles per hour}$$

The slope represents the trucker's average speed or velocity.

Caution! In Example 3, we refer to a point by a capital letter and the coordinates of the point, like this: $H(4, 200)$. This means that $t = 4$ and $D = 200$ at the point H. Do not confuse the coordinates (t, D) of a particular point with the values of Δt and ΔD obtained by moving from one point to a second point.

RQ6. What does the slope of a line measure?
RQ7. What does the slope of the line measure in Example 3?

Geometrical Meaning of Slope

Suppose we graph two lines with positive slope on the same coordinate system. If we move along the lines from left to right, then the line with the larger slope will be steeper. This makes sense if we think of the slope as a rate of change: The line whose y-coordinate is increasing faster with respect to x is the steeper line.

Look Closer: You can verify the slope given for each line in Figure (a) by computing $\frac{\Delta y}{\Delta x}$. For each unit you increase in the x-direction, the steepest line increases 2 units in the y-direction, the middle line increases 1 unit in the y-direction, and the flattest line increases only $\frac{1}{3}$ unit.

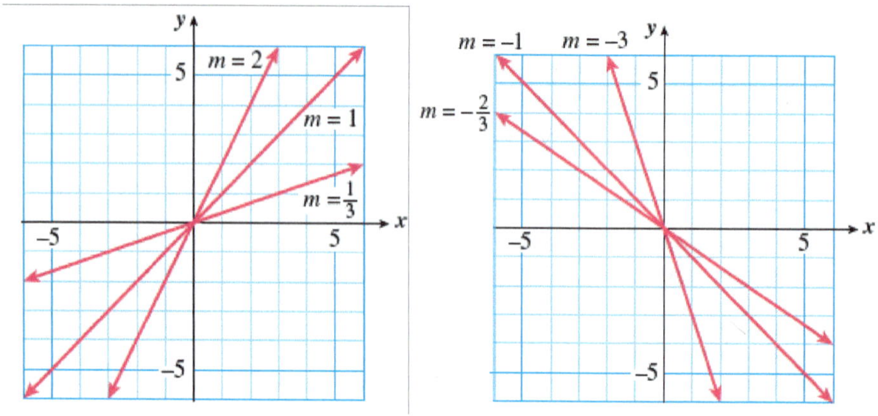

a. b.

Figure (b) shows several lines with negative slopes. These lines slant downwards or decrease as we move from left to right. The more negative the slope, the more sharply the line decreases. For both increasing and decreasing graphs, the larger the absolute value of the slope, the steeper the graph.

Caution! Slopes measure the relative steepness of two lines only if they are graphed on axes with the same scales. Changing the scale on either the x-axis or the y-axis can greatly alter the appearance of a graph.

RQ8. What sort of lines have negative slopes?

Skills Warm-Up

Write a rate for each of the following situations, including units.

1. Zack's average speed, if he drove 426 miles in 9 hours.
2. Zelda's average speed, if she ran 6.6 miles in 55 minutes.
3. The rate at which water flows through a pipe, if a 400-gallon storage tank fills in 20 minutes.
4. A baby whale's rate of growth, if it gains 3000 pounds in its first 40 days of life.
5. Earnest's rate of pay, if he earns $344 for a 40-hour week.
6. Meg's rate of pay, if she charges $90 to type a 40-page paper.

Answers: **1.** $47.\overline{3}$ miles per hour **2.** 0.12 miles per minute **3.** 20 gallons per minute **4.** 75 pounds per day **5.** 8.60 dollars per hour **6.** 2.25 dollars per page

$\frac{426}{9} = 47.3$ $\frac{6.6}{55} = .12$ $\frac{400}{20} = 20$

$\frac{3000}{40} = 75$ $\frac{344}{40} = 8.60$ $\frac{90}{40} =$ \$ 2.25

Homework 3.3

Skills Practice

For Problems 1-2, find the slope of each line segment.

1a.

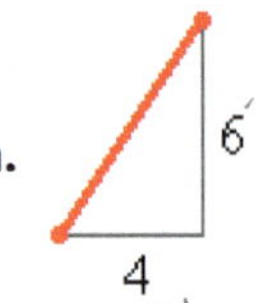

b.

2a.

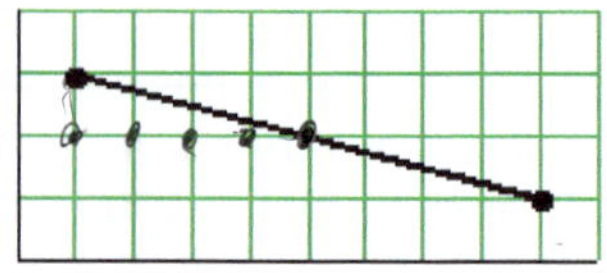

b.

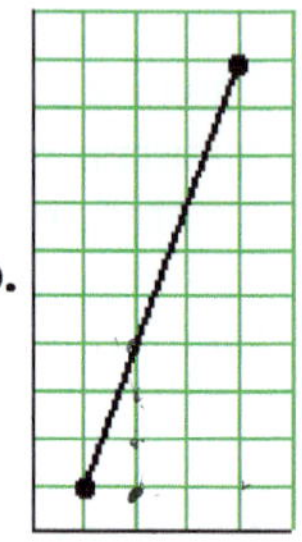

3. Choose two points from the table and compute the slope of the line.

x	0	2	6	8
y	−30	0	60	90

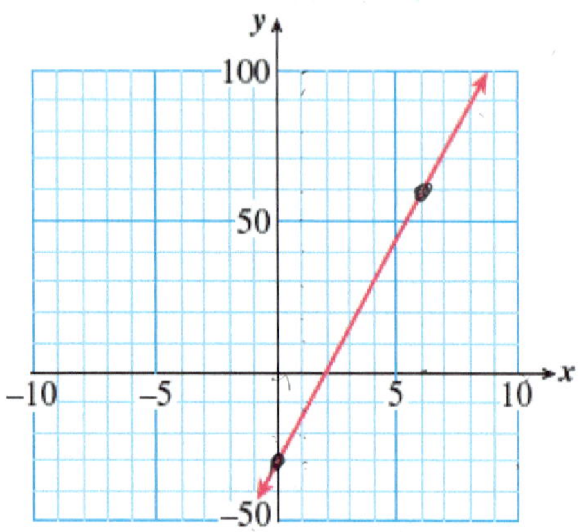

4. Graph the line and compute its slope: $y = -12x + 32$

x	−2	0	3	4
y				

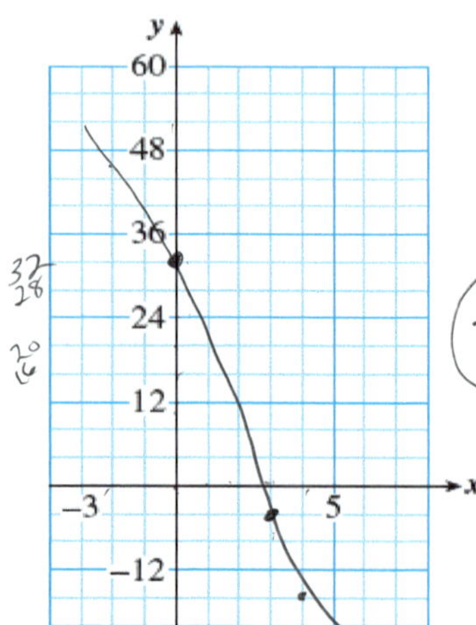

For Problems 5-6, find the slope of the line. Illustrate Δx and Δy on the graph.

5. $x + 2y = 6$

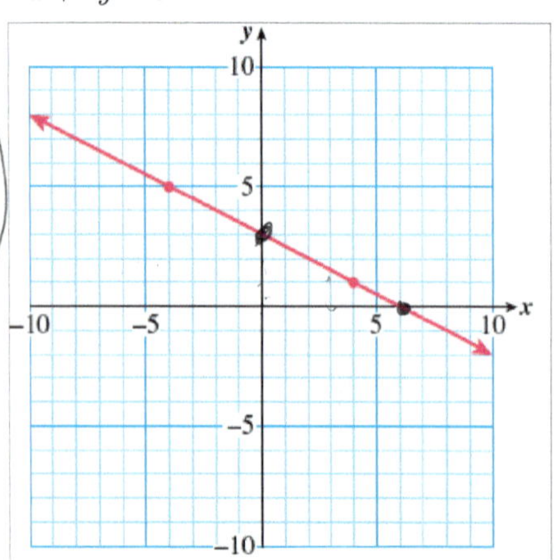

6. $3x - 2y = 0$

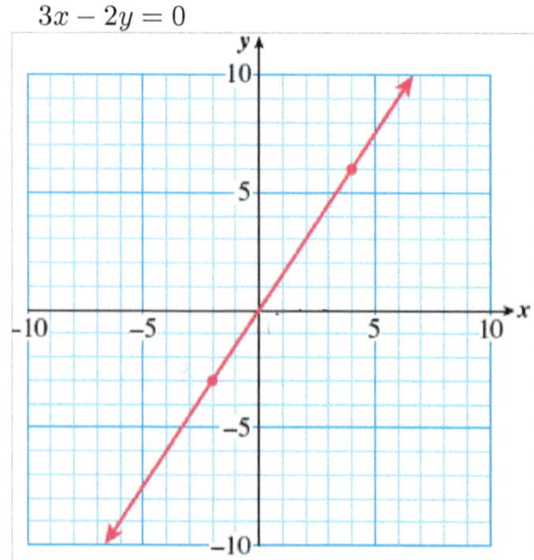

For Problems 7-10,

a. Find the intercepts of each line.
b. Graph the line on the grid provided. Use the intercept method.
c. Use the intercepts to calculate the slope of the line.
d. Calculate the slope again using the suggested points on the line.

7. $2x + 3y = 12$
$(-3, 6)$ and $(3, 2)$

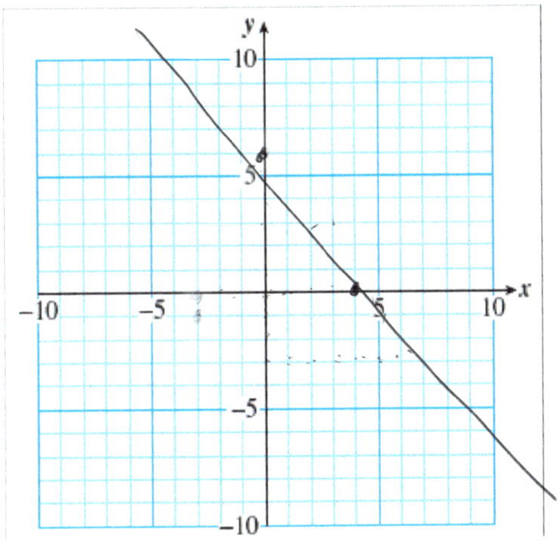

8. $5x - 2y = 10$
$(-2, -10)$ and $(4, 5)$

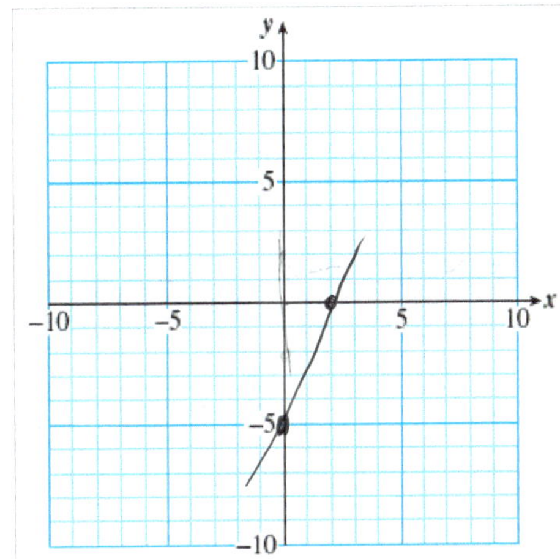

9. $x + y = 5$
$(-3, 8)$ and $(8, -3)$

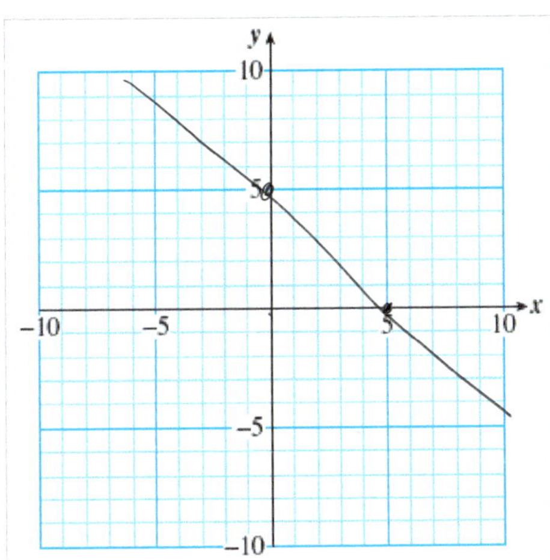

10. $x - 2y = 4$
$(6, 1)$ and $(-4, -4)$

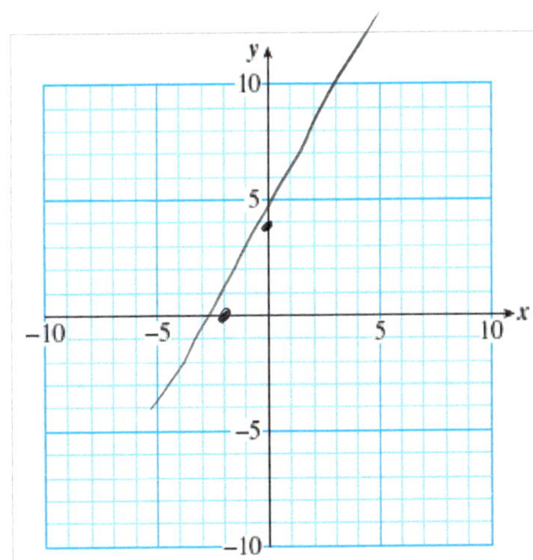

11. A line contains the points $(0, 0)$ and $(3, 2)$. What is its slope?
12. A line contains the points $(0, 0)$ and $(-30, 50)$. What is its slope?
13. Which line is steeper: one with slope $\frac{3}{5}$ or one with slope $\frac{5}{3}$?
14. Which line is decreasing: one with slope $\frac{1}{4}$ or one with slope -2?

15. The line shown in Figure (a) below has slope $\frac{5}{2}$. If $\Delta x = 7$, find Δy.

a.

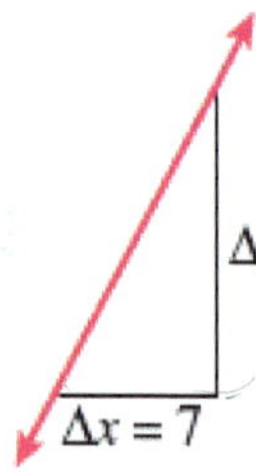

b.

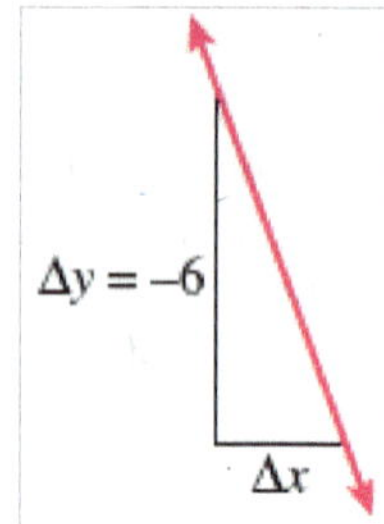

16. The line shown in Figure (b) above has slope -4. If $\Delta y = -6$, find Δx.

Applications

For Problems 17-18,

a. Compute the slope of the graph, including units.
b. Interpret the slope as a rate; what does it tell you about the problem?

17. Audrey can drive 150 miles on 6 gallons of gas, and 225 miles on 9 gallons of gas. Write an equation for the distance, d, that Audrey can drive on g gallons of gas.

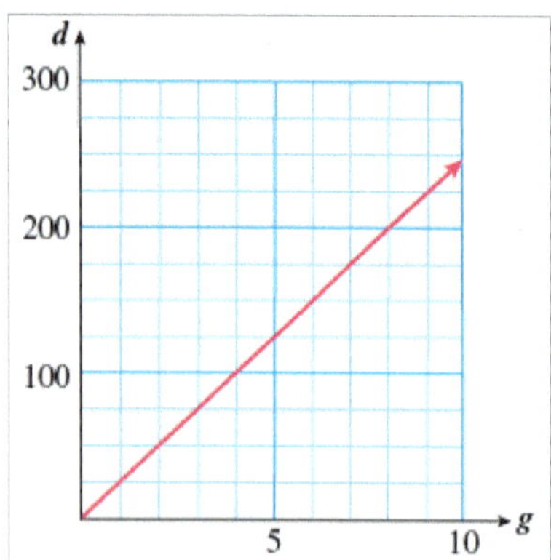

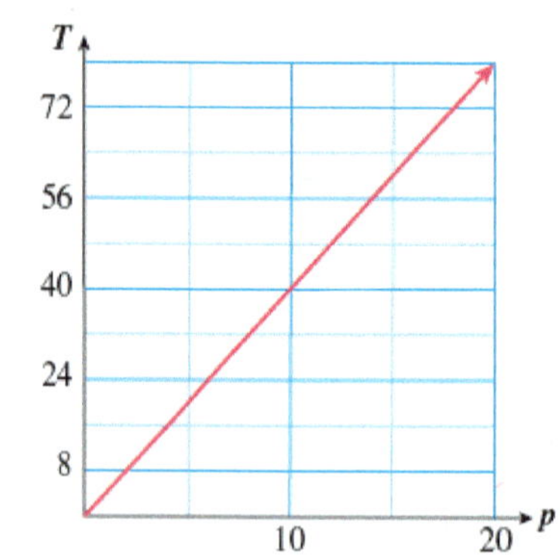

18. The sales tax on a \$15 purchase is 60 cents, and 80 cents on a \$20 purchase. Write an equation for the tax T, in cents, on a purchase of p dollars.

19. Lynette is saving money for the down payment on a new car. The figure at right shows the amount A she has saved, in dollars, w weeks after the first of the year.

a. How much does Lynette save each week?
b. Give the coordinates (w, A) of any two points on the graph. Use those coordinates to compute the slope of the graph, $\dfrac{\Delta A}{\Delta w}$.
c. What are the units of the slope? What does the slope tell you about the problem?

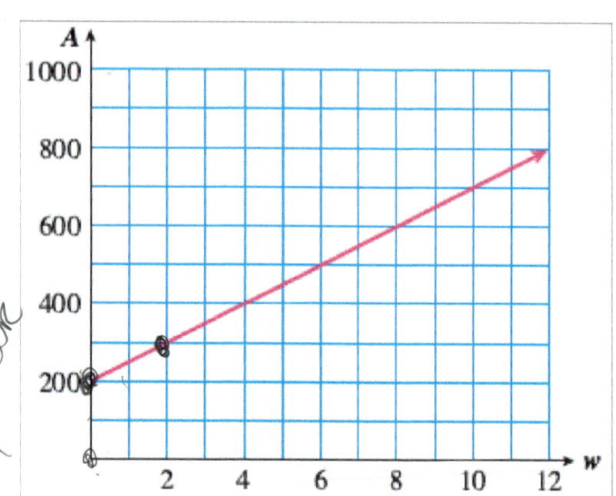

20. Jason is raising a rabbit for the county fair. The figure at right shows the rabbit's weight W when it was t weeks old.
   a. How much did the rabbit's weight increase from the fourth week to the twelfth week? From the second week to the eighth week?
   b. Compute the rabbit's rate of growth, including units.
   c. Illustrate the rate of growth, $\frac{\Delta W}{\Delta t}$, on the graph.

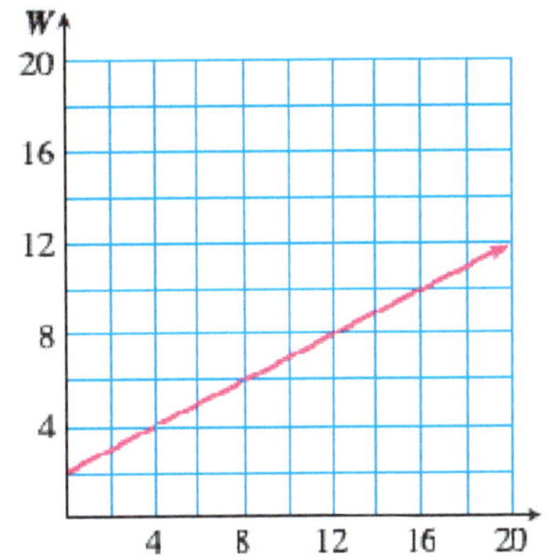

For Problems 21-22, draw and label a sketch for the situation. Use the definition of slope to answer the questions.

21. A sign on the highway says "6% grade, next 3 miles." This means that the slope of the road ahead is $\frac{6}{100}$. How much will you climb in elevation (in feet) over the next 3 miles?

22. A wheelchair ramp must have a slope of 0.125. If the ramp must reach a door whose base is 2 feet off the ground, how far from the building should the base of the ramp be placed?

23. a. Calculate the slope of the line in Figure (a) below.
   b. Explain why $\Delta y = 0$ for any two points on the line.
   c. Explain why the slope of any horizontal line is zero.

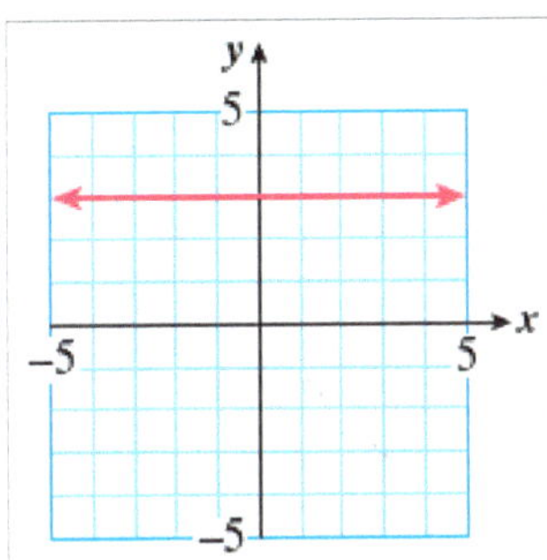

a.

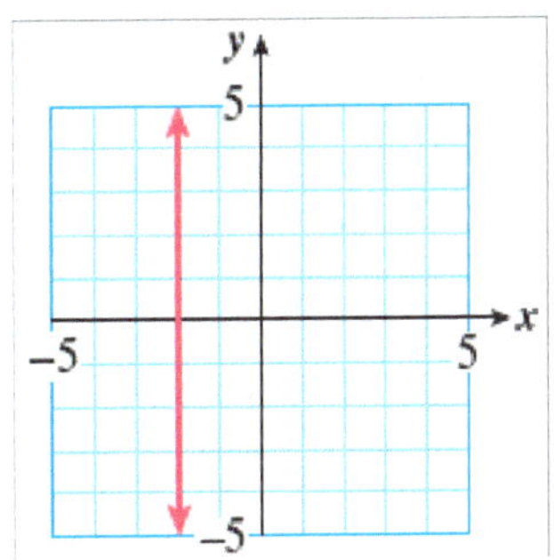

b.

24. a. Calculate the slope of the line in Figure (b) above.
   b. Explain why $\Delta x = 0$ for any two points on the line.
   c. Explain why the slope of any vertical line is undefined.

25. a. Which of the two graphs in the figure below appears steeper?

b. Compute the slopes of the two graphs. Which has the greater slope?

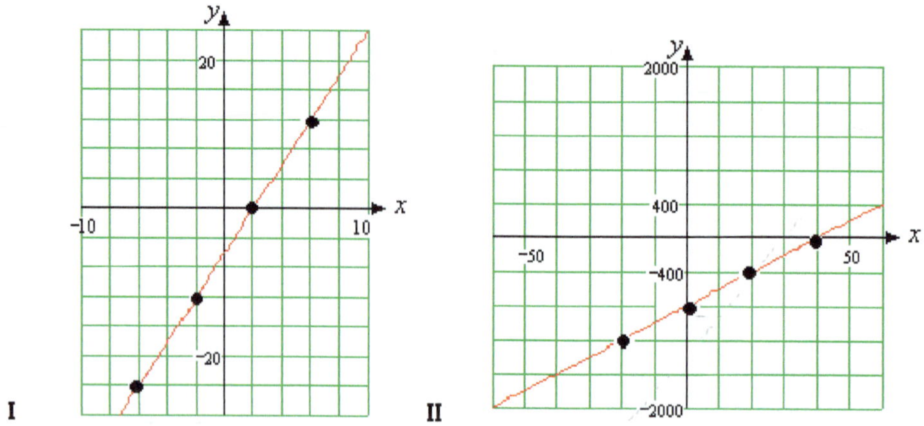

26. Kira buys granola in bulk at the health food store. There are two standard containers customers can use. The size of each container in ounces and its price in dollars are shown in the graph.

a. Read the coordinates of the two points shown on the graph.

b. Calculate the slope of the graph, including units. What does the slope tell us about the granola?

c. Extend the graph to include 25 ounces of granola. How much taller must you make the vertical axis?

d. Extend the graph to include $9.00 worth of granola. How far must you extend the horizontal axis?

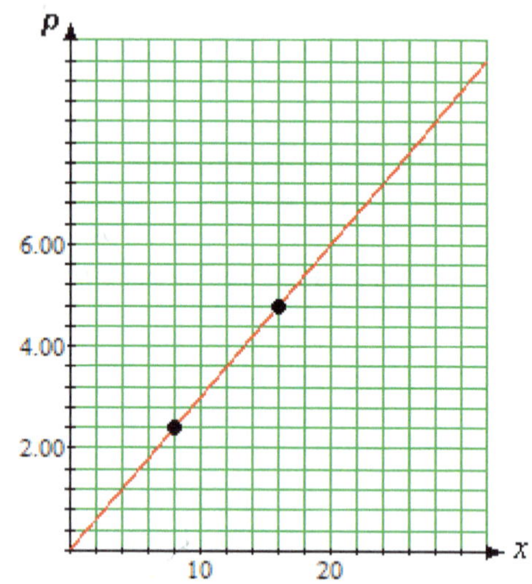

3.4 Slope-Intercept Form

Writing a Linear Equation

In Lesson 3.3, we plotted data for the temperature inside an oven used to cure pottery.

Time, x	0	10	20	30	40	50	60
Temperature, y	70	74	78	82	86	90	94

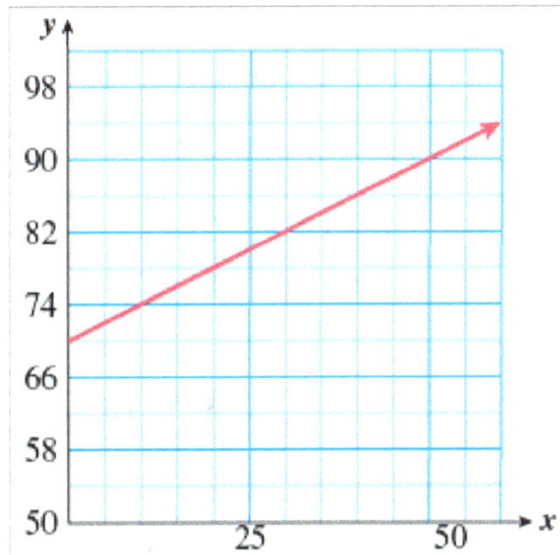

We computed the slope of the graph, shown in the figure. The slope is

$$\frac{\text{change in temperature}}{\text{change in time}} = \frac{\Delta y}{\Delta x} = \frac{4 \text{ degrees}}{10 \text{ minutes}}$$

or $m = 0.4$ degrees per minute.

You can see from either the table or the graph that the y-intercept is the point $(0, 70)$. This means that the initial temperature inside the drying oven was 70 degrees.

Look Ahead: If we know two pieces of information about a line: its slope and its initial value or y- intercept, we can write its equation.

Example 1 Write an equation for the temperature, H, inside the pottery drying oven t minutes after the oven is turned on.

Solution The initial temperature in the oven was 70 degrees, so $H = 70$ when $t = 0$. The temperature rose at a rate of 0.4 degrees per minute, so we add 0.4 degrees to H for each minute that passes. After t minutes, we have added $0.4t$ degrees, giving us a temperature of

$$H = 70 + 0.4t$$

We can also write the equation as $H = 0.4t + 70$.

RQ1. What does the constant term in an equation tell us about the graph?
RQ2. What does the coefficient of the input variable tell us?

From Example 1, we see that the coefficients of a linear equation tell us something about its graph.

1. The constant term tells us the vertical intercept of the graph.
2. The coefficient of the input variable tells us the slope of the graph.

Slope-Intercept Form

A linear equation written in the form

$$y = mx + b$$

is said to be in **slope-intercept** form. The coefficient m is the **slope** of the graph, and b is the **y-intercept**.

RQ3. What does the y-intercept tell us in Example 1?
RQ4. What is the slope-intercept form of an equation?

Slope-Intercept Method of Graphing

A linear equation has the form $Ax + By = C$, and its graph is a straight line. We have already studied two methods for graphing linear equations:

1. Make a table of values and plot points
2. Find and plot the intercepts (the intercept method)

There is a third graphing method that makes use of the slope of the line. We can use the slope-intercept form to sketch a graph quickly, without having to plot a lot of points.

Example 2 Graph the equation $y = \frac{3}{4}x - 2$

Solution The slope of the line is $\frac{3}{4}$ and its y-intercept is the point $(0, -2)$. We begin by plotting the y-intercept, as shown in the figure. Next, we use the slope to find another point on the line. The slope,

$$m = \frac{\Delta y}{\Delta x} = \frac{3}{4}$$

gives the ratio of the change in y-coordinate to the change in x-coordinate as we move from any point on the line to another. Thus, starting at the point $(0, -2)$, we move:

3 units up (the positive y-direction),
then **4 units right** (the positive x-direction)

to locate another point on the line. The coordinates of this new point are $(4, 1)$. Finally,we draw a line through the two points, as shown in the figure.

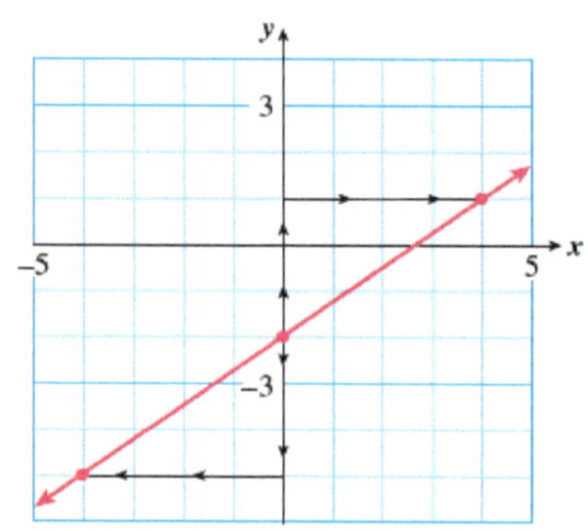

Look Closer: To improve the accuracy of the graph in Example 2, we can find a third point on the line by writing the slope in an equivalent form. We change the sign of both numerator and denominator of the slope to get

$$m = \frac{\Delta y}{\Delta x} = \frac{-3}{-4}$$

Starting again from the y-intercept $(0, -2)$, we now move 3 units down and 4 units left, and find the point $(-4, -5)$ on the graph.

RQ5. Name two methods of graphing that we have already studied.
RQ6. When using the slope-intercept method, what is the first point we plot?
RQ7. How do we use the slope to find a second point on the line?

The **slope-intercept method** can be used to graph any non-vertical line.

To Graph a Line Using the Slope-Intercept Method

1. Write the equation in the form $y = mx + b$.
2. Plot the y-intercept, $(0, b)$.
3. Write the slope as a fraction, $m = \dfrac{\Delta y}{\Delta x}$.
4. Use the slope to find a second point on the graph: Starting at the y-intercept, move Δy units in the y-direction, then Δx units in the x-direction.
5. Find a third point by moving $-\Delta y$ units in the y-direction, then $-\Delta x$ units in the x-direction, starting from the y-intercept.
6. Draw a line through the three plotted points.

Finding the Slope-Intercept Form

Not all linear equations appear in slope-intercept form. However, we can write the equation of any non-vertical line in slope-intercept form by solving the equation for y in terms of x.

Example 3 Find the slope and y-intercept of the graph of $3x - 4y = 8$

Solution To write the equation in slope-intercept form, we solve for y in terms of x.

$3x - 4y = 8$	Subtract $3x$ from both sides.
$-4y = -3x + 8$	Divide both sides by -4.
$\dfrac{-4y}{-4} = \dfrac{-3x + 8}{-4}$	Divide each term of the right side by -4.
$y = \dfrac{-3x}{-4} + \dfrac{8}{-4}$	Simplify each quotient.
$y = \dfrac{3}{4}x - 2$	

The equation is now in slope-intercept form, with $m = \frac{3}{4}$ and $b = -2$. Thus, the slope of the graph is $\frac{3}{4}$ and the y-intercept is the point $(0, -2)$.

Caution! Do not confuse solving for y with finding the y-intercept. In Example 3, we do **not** set $x = 0$ before solving for y. When we find the y-intercept, we are looking for a specific point, namely, the point with x-coordinate zero, so we replace x by 0. When we "solve for y," we are writing the equation in another form, so both variables, x and y, still appear in the equation.

RQ8. How do we put an equation into slope-intercept form?

Skills Warm-Up

Solve for the indicated variable.

1. $2q + p = 10$ for p
2. $2l + 2w = 18$ for l
3. $3a + 9 = -6b$ for a
4. $2c = 2d + 22$ for d
5. $5r - 4s = 24$ for s
6. $2m = 11 - 3n$ for n

Answers: **1.** $10 - 2q$ **2.** $9 - w$ **3.** $-3 - 2b$ **4.** $c - 11$ **5.** $\dfrac{5r - 24}{4}$ **6.** $\dfrac{11 - 2m}{3}$

Homework 3.4

Skills Practice

For Problems 1-4,

a. Write the equation in slope-intercept form.
b. State the slope and y-intercept of the graph.

1. $y = 3x + 4$

2. $6x + 3y = 5$

3. $2x - 3y = 6$

4. $5x = 4y$

For Problems 5-8,

a. Find the slope and the y-intercept of the line.
b. Write an equation for the line.

5.

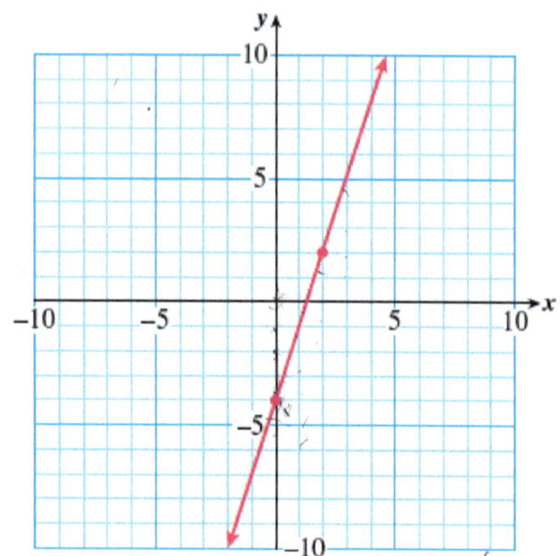

6.

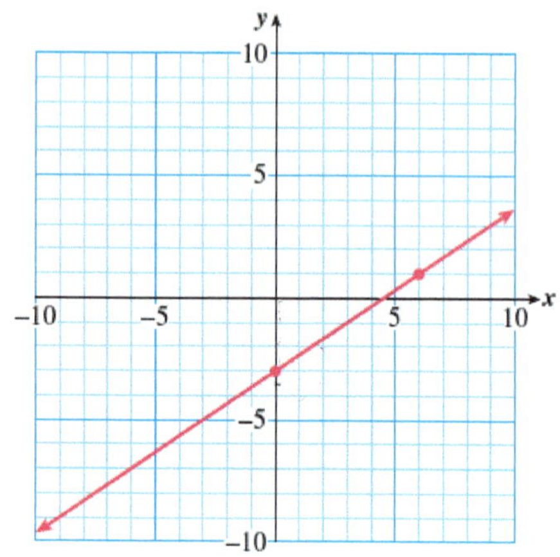

7.

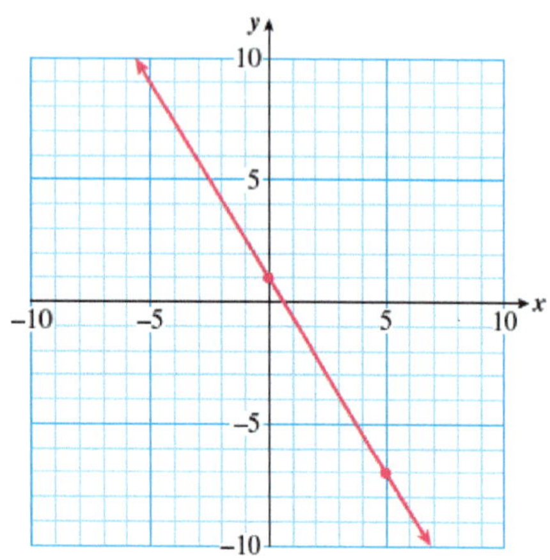

8.

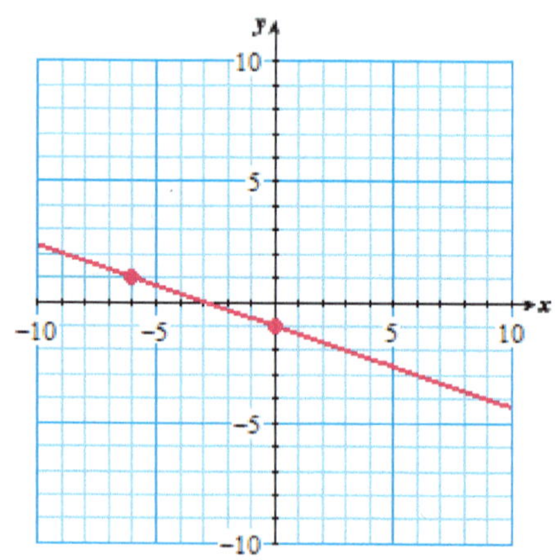

For Problems 9-12,

a. Fill in the y-values in the tables and graph the lines.

b. Choose two points on each line and compute its slope.

c. What is the y-intercept of each line?

9. I. $y = 2x - 6$

x	−1	0	1	2	3
y					

II. $y = 2x + 1$

x	−1	0	1	2	3
y					

III. $y = 2x + 3$

x	−1	0	1	2	3
y					

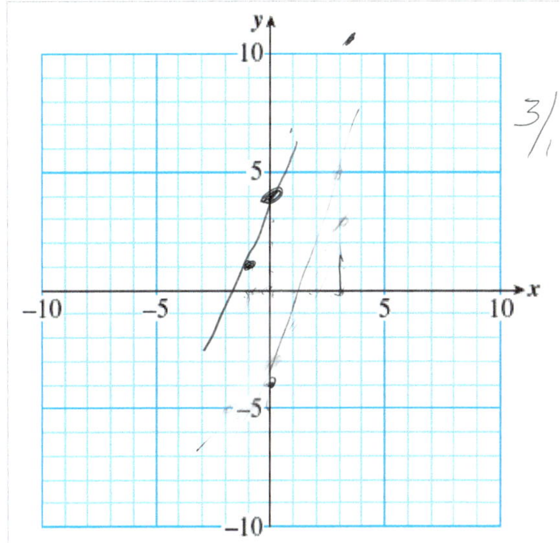

All three lines have the same ______.

10. I. $y = \frac{-3}{2}x - 4$

x	−6	−4	−2	0	2
y					

II. $y = \frac{-3}{2}x + 2$

x	−6	−4	−2	0	2
y					

III. $y = \frac{-3}{2}x + 6$

x	−6	−4	−2	0	2
y					

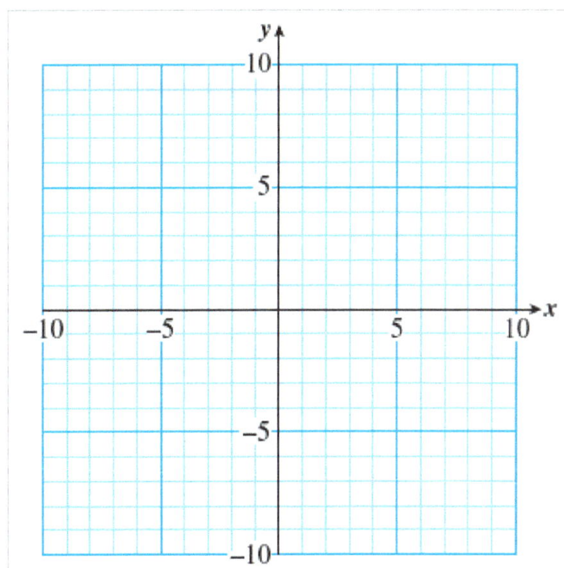

All three lines have the same ______.

11. I. $y = \frac{1}{4}x + 2$

x	−4	−2	0	2	4
y					

II. $y = \frac{1}{2}x + 2$

x	−4	−2	0	2	4
y					

III. $y = x + 2$

x	−4	−2	0	2	4
y					

All three lines have the same ______.

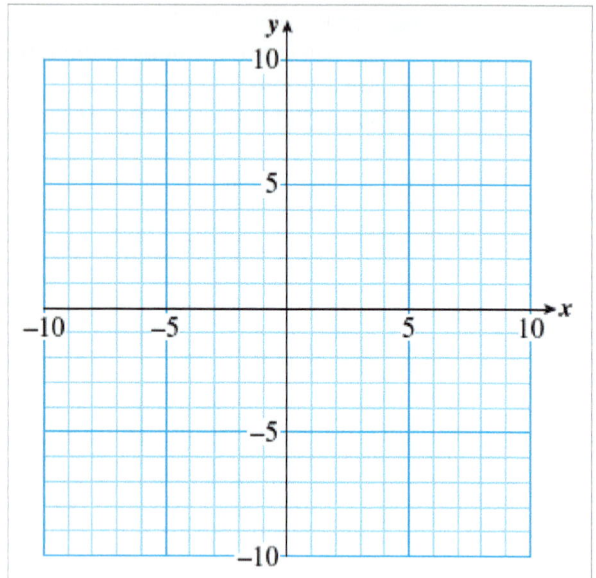

12. I. $y = -3x - 2$

x	−6	−3	0	3	6
y					

II. $y = -2x - 2$

x	−6	−3	0	3	6
y					

III. $y = \frac{-5}{3}x - 2$

x	−6	−3	0	3	6
y					

All three lines have the same ______.

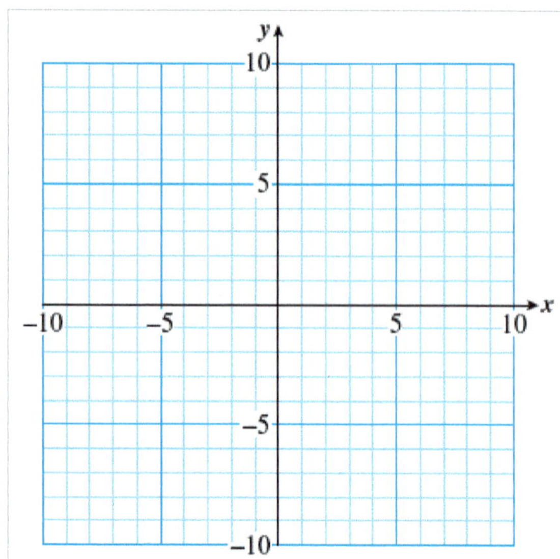

For Problems 13-14,

a. Find the intercepts of the graph and graph the line.
b. Compute the slope of the line.
c. Put the equation in slope-intercept form.

13. $3x + 4y = 12$

x	0	
y		0

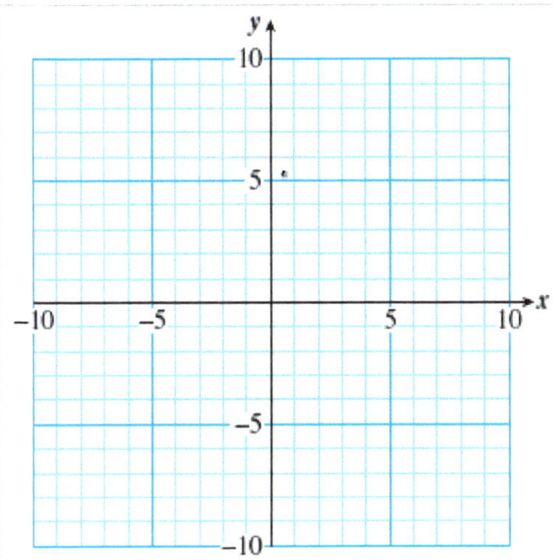

14. $y + 3x - 8 = 0$

x	0	
y		0

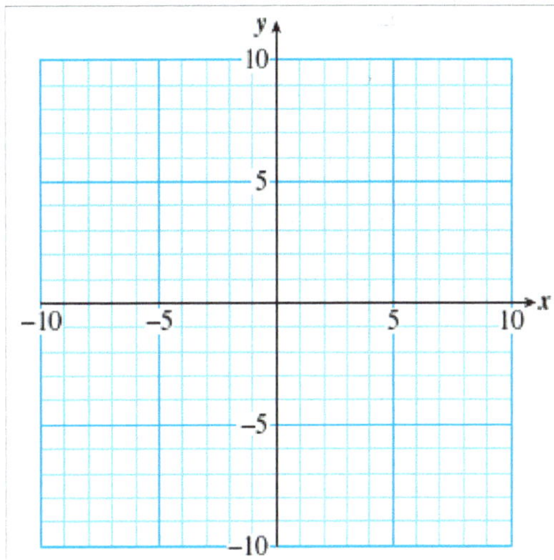

In Problems 15-16,

a. Put the equation in slope-intercept form.
b. What is the y-intercept of each line? What is its slope?
c. Use the slope to find two more points on the line.
d. Graph the line.

15. $3x - 5y = 0$

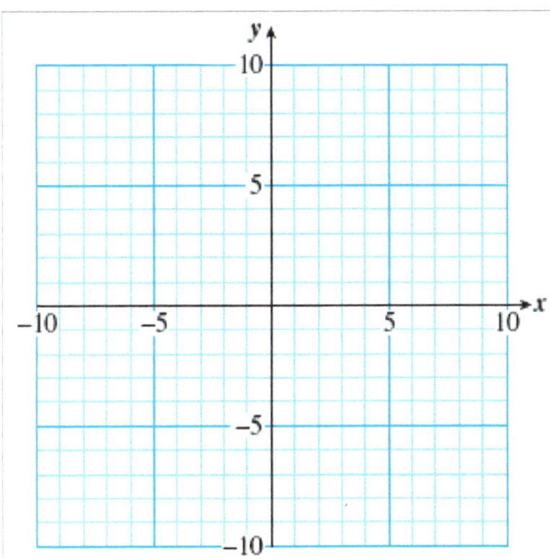

16. $5x + 4y = 0$

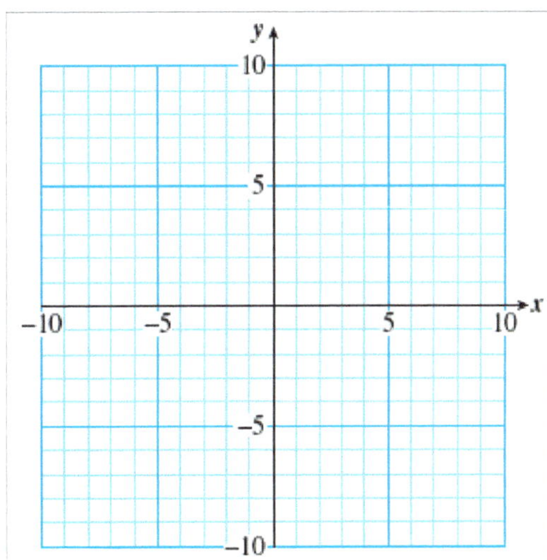

Applications

For Problems 17-20, graph the equation by using the slope-intercept method.

17. $y = 3 - x$

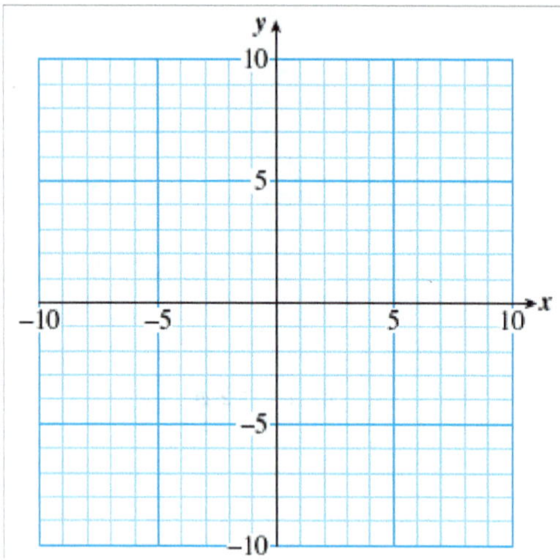

18. $y = 3x - 1$

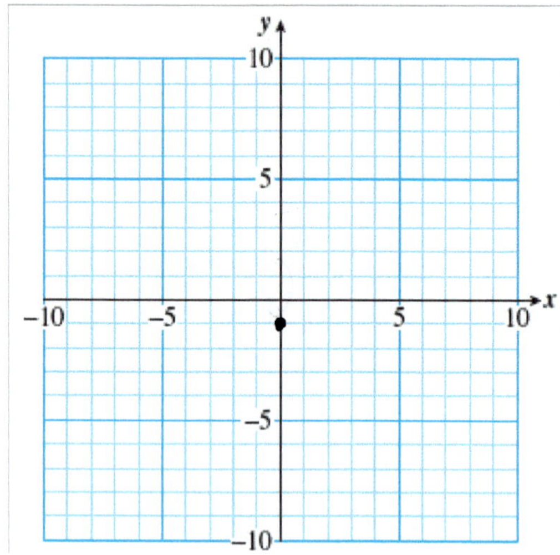

19. $y = \frac{3}{4}x + 2$

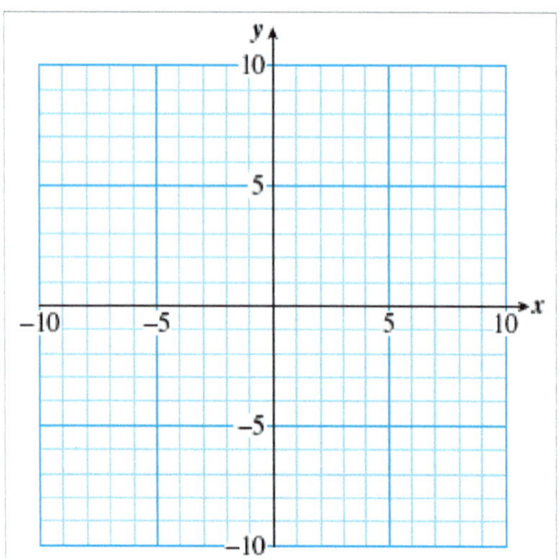

20. $y = -2 - \frac{4}{3}x$

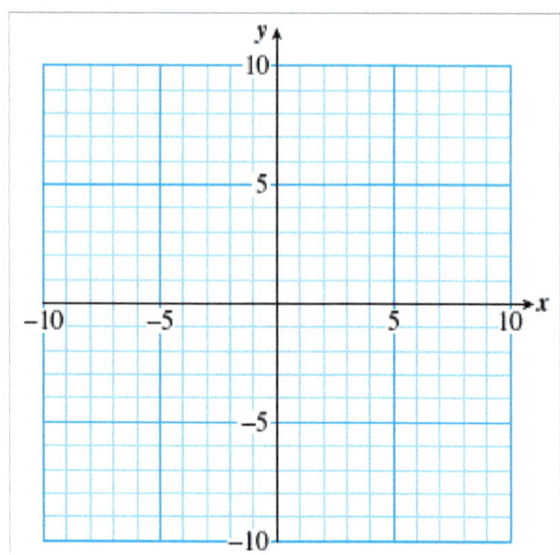

21. Robin opened a yogurt smoothie shop near campus. The graph shows Robin's profit P after selling s smoothies.

a. What is the P-intercept of the line?
b. Calculate the slope of the line.
c. Write an equation for the line in slope-intercept form.
d. What do the slope and the P-intercept tell us about the problem?

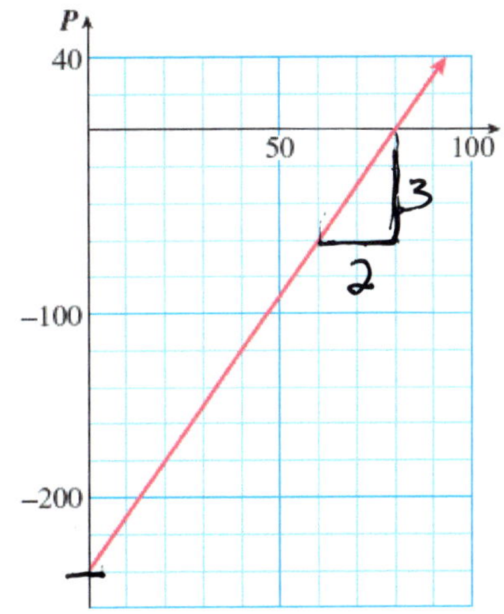

For Problems 22-24,

a. Graph the line by the slope-intercept method.
b. Explain what the slope and the vertical intercept tell us about the problem.

22. Serda's score on her driving test is computed by the equation $S = 120 - 4n$, where n is the number of wrong answers she gives.

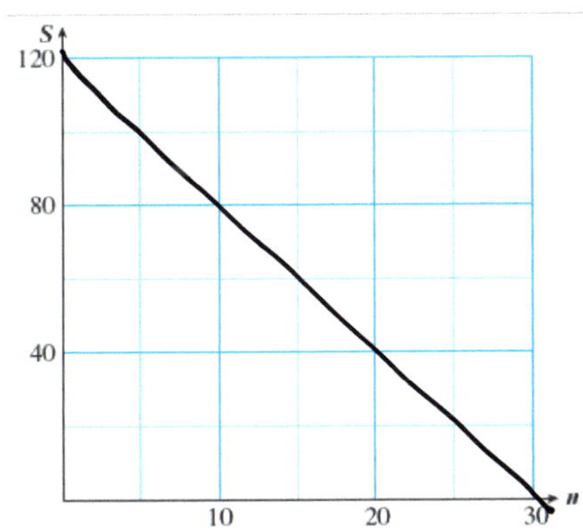

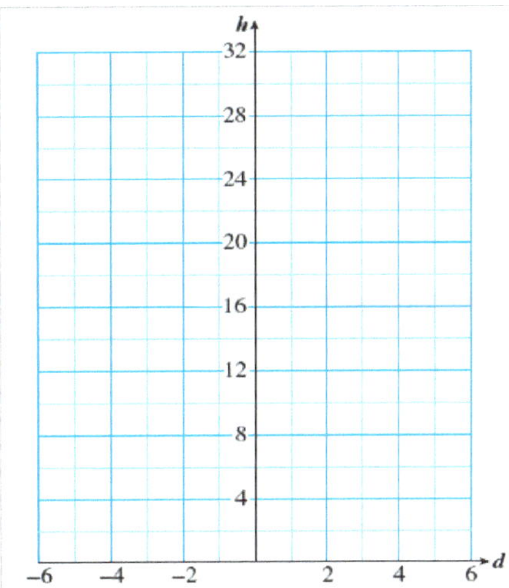

23. Greg is monitoring the growth of a new variety of string beans. The height of the vine each day is given in inches by $h = 18 + 3d$, where $d = 0$ represents today.

24. Cliff's score was negative at the end of the first round of College Quiz, but in Double Quiz his score improved according the equation $S = -400 + 20q$, where q is the number of questions he answered.

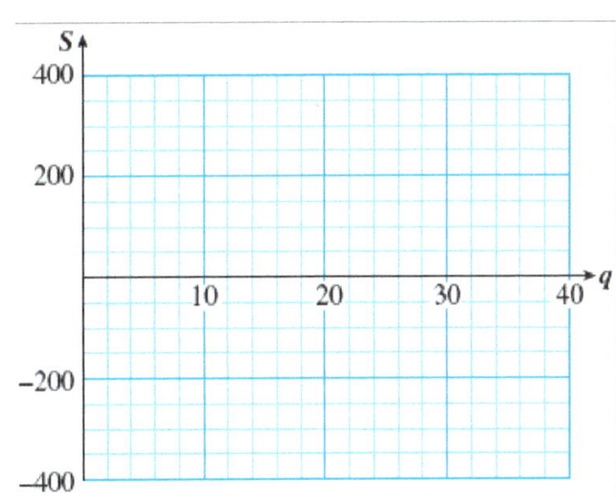

3.5 Properties of Lines

Parallel and Perpendicular Lines

Lines that lie in the same plane but never intersect are called **parallel lines**. It is easy to understand that parallel lines have the same slope.

Exercise Calculate the slopes of the two parallel lines in the figure. (We denote the slope of the first line by m_1 and the slope of the second line by m_2.)

$$m_1 = \frac{\Delta y}{\Delta x} =$$

$$m_2 = \frac{\Delta y}{\Delta x} =$$

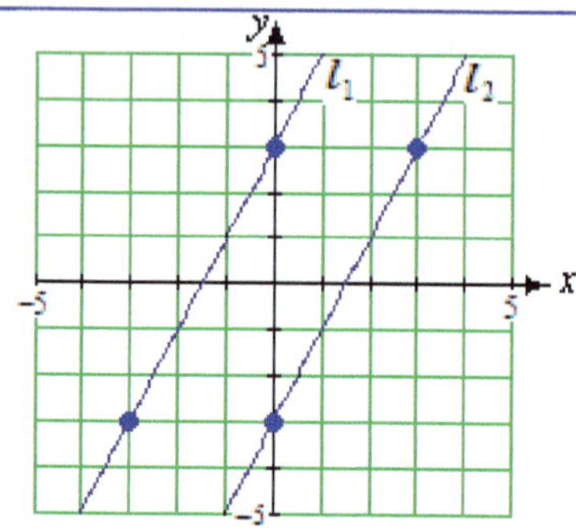

Lines that intersect at right angles are called **perpendicular lines**. It is a little harder to see the relationship between the slopes of perpendicular lines.

Exercise Calculate the slopes of the two perpendicular lines in the figure.

$$m_1 = \frac{\Delta y}{\Delta x} =$$

$$m_2 = \frac{\Delta y}{\Delta x} =$$

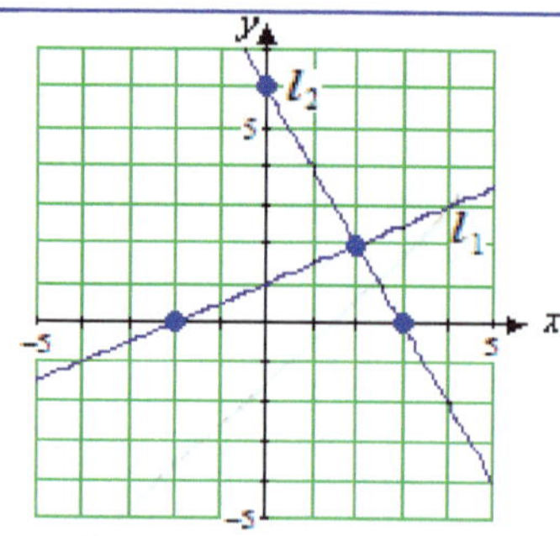

In the Exercise above, note that the product of m_1 and m_2 is -1, that is,

$$m_1 m_2 = \frac{1}{2}(-2) = -1$$

This relationship holds for any pair of perpendicular lines.

Parallel and Perpendicular Lines

1. Two lines are **parallel** if their slopes are equal, that is, if
$$m_1 = m_2$$
or if both lines are vertical.
2. Two lines are **perpendicular** if the product of their slopes is -1, that is, if
$$m_1 m_2 = -1$$
or if one of the lines is horizontal and one is vertical.

RQ1. What do we call two lines that lie in the same plane but never intersect?
RQ2. What are perpendicular lines?

Example 1 Decide whether the lines

$$2x + 3y = 6 \qquad \text{and} \qquad 3x - 2y = 6$$

are parallel, perpendicular, or neither.

Solution We could graph the lines, but we can't be sure from a graph if the lines are exactly parallel or exactly perpendicular. A more accurate way to answer the question is to find the slope of each line. To do this we write each equation in slope-intercept form, that is, we solve for y.

$$\begin{aligned} 2x + 3y &= 6 \\ 3y &= -2x + 6 \\ y &= \frac{-2}{3}x + 2 \end{aligned} \qquad\qquad \begin{aligned} 3x - 2y &= 6 \\ -2y &= -3x + 6 \\ y &= \frac{3}{2}x - 3 \end{aligned}$$

The slope of the first line is $m_1 = \frac{-2}{3}$, and the slope of the second line is $m_2 = \frac{3}{2}$. The slopes are not equal, so the lines are not parallel. However, the product of the slopes is

$$m_1 m_2 = \left(\frac{-2}{3}\right)\left(\frac{2}{3}\right) = -1$$

so the lines are perpendicular.

Look Closer: Another way to state the condition for perpendicular lines is

$$m_2 = \frac{-1}{m_1}$$

Because of this relationship, we often say that the slope of one perpendicular line is the **negative reciprocal** of the other.

RQ3. What is true about the slopes of parallel lines?
RQ4. What do we call the slopes of perpendicular lines?

Equations for Horizontal and Vertical Lines

In Lesson 3.3 we learned that the slope of a horizontal line is zero. What does this tell us about the equation of a horizontal line that passes through a particular point?

Example 2 Find the equation of the horizontal line that passes through $(5, 3)$.

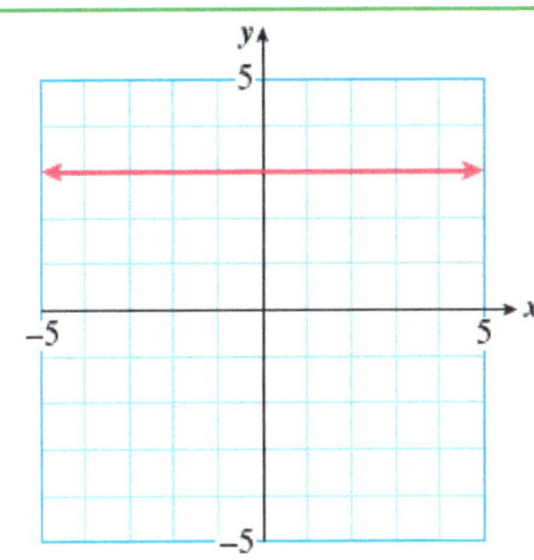

Solution Looking at the graph of the line shown at right, we see that the y-coordinate of every point on the line is 3. In particular, the y-intercept of the line is the point $(0, 3)$, so $b = 3$. As we noted above, the slope of the line is $m = 0$. Substituting these values into the slope-intercept form gives us the equation $y = 0 \cdot x + 3$, or just $y = 3$. The fact that x does not appear in the equation means that $y = 3$ for every point on the line, no matter what the value of x is.

Look Closer: What about the equation of a vertical line? The slope of a vertical line is undefined; a vertical line does not have a slope. We cannot use the slope-intercept form to write the equation of a vertical line. However, we can use what we learned about horizontal lines in Example 2.

Look at the graph of the vertical line at right. Every point on the line has x-coordinate -2, no matter what the y-coordinate is. An equation for this line is $x = -2$. The value of x does not depend upon y; it is constant, so y does not appear in the equation.

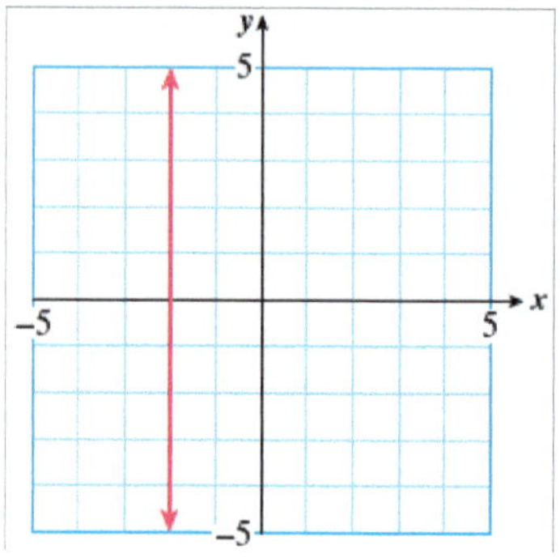

The y-intercept of the horizontal line $y = 3$ is $(0, 3)$; it has no x-intercept. The x-intercept of the vertical line $x = -2$ is $(-2, 0)$; it has no y-intercept.

Horizontal and Vertical Lines

1. The equation of the **horizontal line** passing through $(0, b)$ is

$$\boldsymbol{y = b}$$

2. The equation of the **vertical line** passing through $(a, 0)$ is

$$\boldsymbol{x = a}$$

Caution! The equation for a line is not the same thing as the slope of the line! The slope of every horizontal line is zero, but the equation of a horizontal line has the form $y = b$, where b is the y-coordinate of every point on the line. Similarly, the equation of a vertical line has the form $x = a$, but the slope of a vertical line is undefined.

RQ5. Give an example of an equation of a vertical line.
RQ6. Give an example of an equation of a horizontal line.

Distance Between Points

It is easy to compute the distance between two points that lie on the same horizontal or vertical line. We subtract the smaller coordinate from the larger one.

Example 3 The distance between the points $A(6, 5)$ and $B(6, -7)$ in the figure is

$$\Delta y = 5 - (-7) = 12$$

and the distance between points $C(-7, -3)$ and $D(-2, -3)$ is

$$\Delta x = -2 - (-7) = 5$$

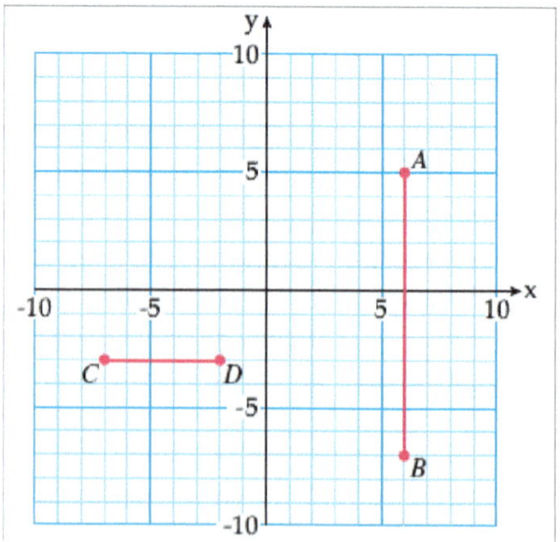

Note that the distance between two points is always a positive number.

Look Closer: When we compute slope, the direction in which we move on the line makes a difference. We will call this the **directed distance**, and it can be either positive or negative. If we move from C to D, we have moved in the positive x-direction, so the directed distance is positive:

$$\Delta x = -2 - (-7) = 5$$

If we move from D to C, we have moved in the negative x-direction, so the directed distance is negative:

$$\Delta x = -7 - (-2) = -5$$

> To find the **directed distance** between two points on a number line, we subtract the initial coordinate from the final coordinate.

RQ7. How do we find the distance between two points that lie on the same vertical line?

RQ8. A directed distance can be either ________ or __________.

Look Ahead: So far we have computed the slope of a line by finding Δy and Δx on the graph. Now we use directed distance to develop a formula for slope.

Example 4 Compute the slope of the line segment joining P and R in two ways:

a. Find Δy and Δx using the graph.

b. Find Δy and Δx using coordinates.

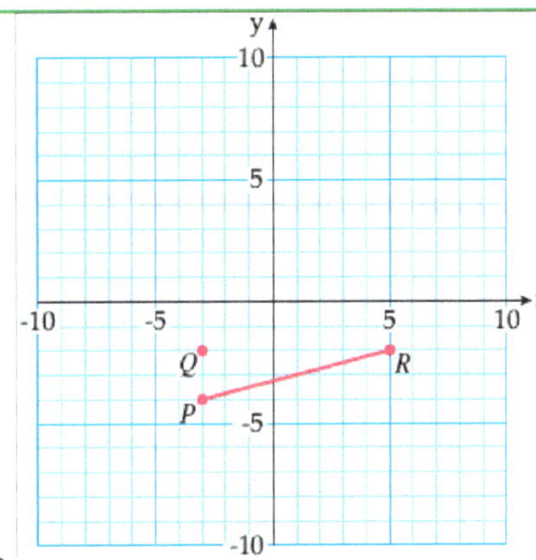

Solutions a. As we move from P to Q, we move up 2 squares on the graph, so $\Delta y = 2$. As we move from Q to R, we move 8 squares to the right, so $\Delta x = 8$. Thus, the slope of the line is

$$m = \frac{\Delta y}{\Delta x} = \frac{2}{8} = \frac{1}{4}$$

b. First we write down the coordinates of P and Q:

$$P(-3, -4) \quad \text{and} \quad Q(-3, -2)$$

and compute the directed distance from P to Q:

$$\Delta y = -2 - (-4) = 2$$

final – initial

Then we write down the coordinates of Q and R:

$$Q(-3, -2) \quad \text{and} \quad R(5, -2)$$

and compute the directed distance from Q to R:

$$\Delta x = 5 - (-3) = 8$$

final – initial

We get the same value for the slope as in part (a), $m = \dfrac{\Delta y}{\Delta x} = \dfrac{2}{8} = \dfrac{1}{4}$.

Subscript Notation

When computing the slope of a line joining two points, we must be careful to subtract their x-coordinates and their y-coordinates in the same order, final minus initial. To distinguish between the coordinates of different points, we use a new notation called **subscripts**.

> We denote the coordinates of a particular point by (x_1, y_1) and the coordinates of a second point by (x_2, y_2).

Example 5 Compute the slope between points D and F in the figure.

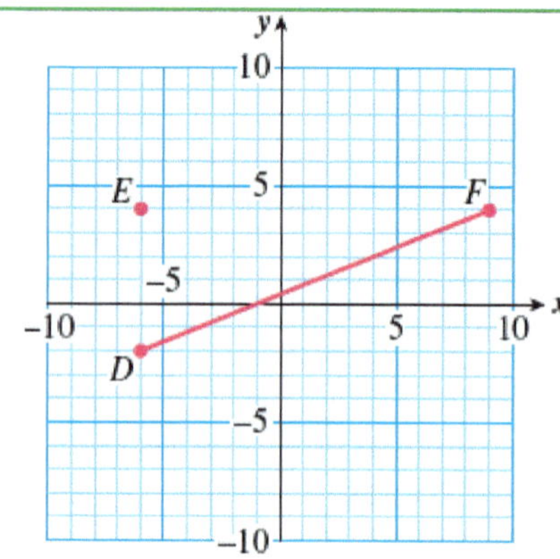

Solution We'll call D the first point and F the second point. Then their coordinates are

$$D: \quad x_1 = -6 \quad \text{and} \quad y_1 = -2$$
$$F: \quad x_2 = 9 \quad \text{and} \quad y_2 = 4$$

Now we can describe the formula for slope in an organized way. We compute Δy and Δx as directed distances. First, we observe that the coordinates of point E are $(-6, 4)$, or (x_1, y_2). From the figure, we can see that

$$\Delta y = y_2 - y_1 = 4 - (-2) = 6$$

and

$$\Delta x = x_2 - x_1 = 9 - (-6) = 15$$

Thus, the slope of the line segment joining D and F is

$$m = \frac{y_2 - y_1}{x_2 - x_1} = \frac{6}{15} = \frac{2}{5}$$

Caution! The subscript 1 on x_1, for instance, has nothing to do with the value of the coordinate; it merely identifies this coordinate as the x-coordinate of the first point.

RQ9. What do the subscripts in (x_1, y_1) mean?
RQ10. Write a formula for computing Δy.

A New Formula for Slope

The method described in Example 5 gives us a new formula for computing slope.

Two-Point Formula for Slope

The slope of the line joining points $P_1(x_1, y_1)$ and $P_2(x_2, y_2)$ is

$$m = \frac{y_2 - y_1}{x_2 - x_1} \quad \text{if} \quad x_2 \neq x_1$$

We don't need a graph in order to use this formula, just the coordinates of two points on the line.

Example 6 Compute the slope of the line joining the points $(-6, 2)$ and $(3, -1)$.

Solution It doesn't matter which point is P_1 and which is P_2, so we choose P_1 to be $(-6, 2)$. Then $(x_1, y_1) = (-6, 2)$ and $(x_2, y_2) = (3, -1)$. Thus,

$$\begin{aligned} m &= \frac{y_2 - y_1}{x_2 - x_1} \\ &= \frac{-1-2}{3-(-6)} = \frac{-3}{9} = \frac{-1}{3} \end{aligned}$$

Caution! In Example 6, we can reverse the order of **both** subtractions to find

$$\begin{aligned} m &= \frac{y_1 - y_2}{x_1 - x_2} \\ &= \frac{2-(-1)}{-6-3} = \frac{3}{-9} = \frac{-1}{3} \end{aligned}$$

the same answer as before. The order of the points does not matter, but we must be consistent and use the same order when computing Δy and Δx.

RQ11. What is wrong with this formula for slope: $m = \dfrac{y_2 - y_1}{x_1 - x_2}$

Skills Warm-Up

For Problems 1-8, find the negative reciprocal.

1. $\frac{3}{4}$ **2.** $\frac{-1}{3}$ **3.** -6 **4.** $\frac{11}{8}$

5. $1\frac{2}{3}$ **6.** $-2\frac{1}{2}$ **7.** -3.2 **8.** 0.625

For Problems 9-14, simplify.

9. $\dfrac{10-2}{2-8}$ **10.** $\dfrac{-5-(-5)}{2-9}$ **11.** $\frac{3}{2}(4-7)+\frac{1}{2}$

12. $\dfrac{-6-(-12)}{3-(-5)}$ **13.** $-3(-4-2)-6$ **14.** $\frac{5}{3}(5-8)+3$

Answers: **1.** $\frac{-4}{3}$ **2.** 3 **3.** $\frac{1}{6}$ **4.** $\frac{-8}{11}$ **5.** $\frac{-3}{5}$ **6.** $\frac{2}{5}$ **7.** 0.3125 **8.** -1.6
9. $\frac{-4}{3}$ **10.** 0 **11.** -4 **12.** $\frac{3}{4}$ **13.** 12 **14.** -2

Homework 3.5

Skills Practice

In Problems 1-3, decide whether the lines are parallel perpendicular, or neither.

1. $3x - 4y = 2$
$8y - 6x = 6$

2. $5x + 3y = 1$
$3y - 5x = 5$

3. $2x = 4 - 5y$
$2y = 4x - 5$

4. The slopes of several lines are given below. Which of the lines are parallel to the graph of $y = 0.75x + 2$, and which are perpendicular to it?

a. $m = \dfrac{3}{4}$ **b.** $m = \dfrac{8}{6}$ **c.** $m = \dfrac{-20}{15}$ **d.** $m = \dfrac{-39}{52}$

e. $m = \dfrac{4}{3}$ **f.** $m = \dfrac{-16}{12}$ **g.** $m = \dfrac{36}{48}$ **h.** $m = \dfrac{9}{12}$

In Problems 5-7,

a. Sketch a rough graph of each equation, and label its intercept.
b. State the slope of each line.

5. $y = -3$ **6.** $2x = 8$ **7.** $x = 0$

In Problems 8-11, give the equation of the line described.

8. Horizontal, passes through $(6, -5)$
9. m is undefined, passes through $(2, 1)$
10. Parallel to $x = 4$, passes through $(-8, 3)$
11. The x-axis
12. Perpendicular to the y-axis, passes through $(4, -12)$

In Problems 13-16, state the y-intercept and the slope of the line.

13. $y = -6$ **14.** $y = -6x$ **15.** $x = -6y$ **16.** $x = 6 - y$

For Problems 17-20, compute the slope of the line joining the given points.

17. $(5, 2), (8, 7)$ **18.** $(3, -2), (0, 1)$ **19.** $(6, -2), (-3, -3)$ **20.** $(3, -5), (8, -5)$

For Problems 21-23,

a. Find the x- and y-intercepts of each line.
b. Use the intercepts to compute the slope of the line.

21. $\dfrac{x}{5} + \dfrac{y}{7} = 1$ **22.** $\dfrac{-x}{2.4} + \dfrac{y}{1.6} = 1$ **23.** $3x + \dfrac{2}{7}y = 1$

Applications

For Problems 24-25, you are given the slope of a line and one point on it. Use the grids to help you find the missing coordinates of the other points on the line.

24. $m = 2, \quad (1, -1)$

a. $(0, ?)$ **b.** $(?, -7)$

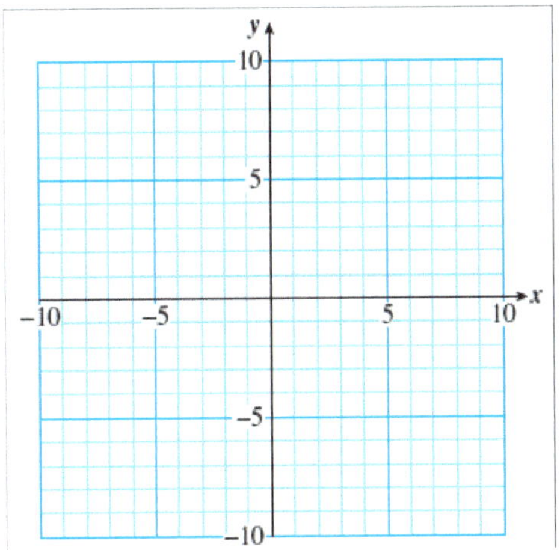

25. $m = \frac{-1}{2}, \quad (2, -6)$

a. $(6, ?)$ **b.** $(?, -3)$

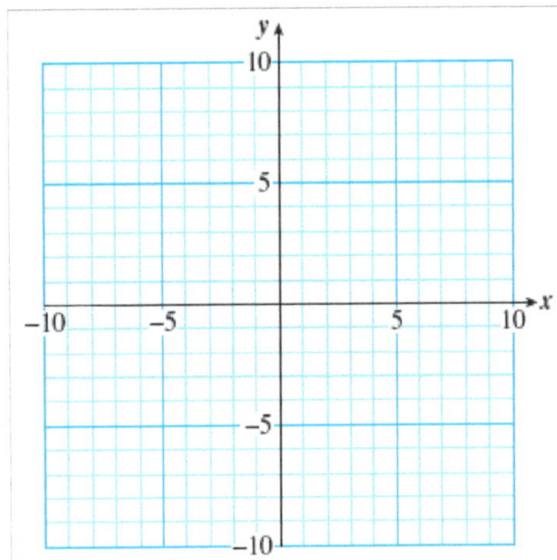

For Problems 26-29, calculate the slope of the line described, including units, and explain its meaning for the problem.

26. Rani and Larry start a college fund for their son, Colby. When Colby is five years old, the fund has \$11,000. When he is 10 years old, the fund has grown to \$17,000. The graph of the amount, A, in the fund when Colby is n years old is a straight line.

27. It cost Melanie \$6.88 to copy and bind her 26-page paper at CopyQuik. It cost Nelson \$8.80 to copy and bind his 50-page portfolio. The graph of the cost, C, of producing a document of length p pages at CopyQuik is a straight line.

28. Dean is cycling down Mt. Whitney. After five minutes, his elevation is 9200 feet, and after 21 minutes, his elevation is 1200 feet. The graph of Dean's elevation, h, after cycling for t minutes is approximately a straight line.

29. At 2 pm the temperature was 16°, and by midnight it had dropped to −14°. The graph of temperature, T, at h hours after noon is approximately a straight line.

30. The figure shows the wholesale cost, C, in dollars, of p pounds of dry-roasted peanuts.

a. What does the slope of this graph tell you about dry-roasted peanuts?

b. Compute the slope, including units.

c. If the Lone Star Barbecue increases its weekly order of peanuts from 100 to 120 pounds, how much will the cost increase?

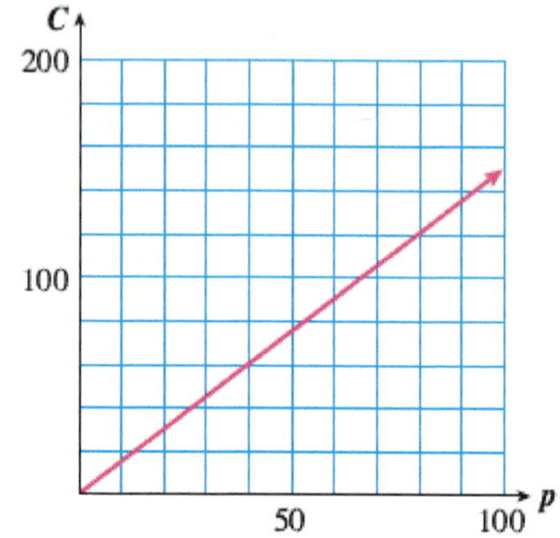

31. Roy is traveling home by train. The figure shows his distance d from home in miles h hours after the train started.

a. Compute the slope of the graph, including units.

b. What does the slope tell you about Roy's journey?

c. Find the vertical intercept of the graph. What is its meaning for the problem?

d. Compute the horizontal intercept of the graph. What is its meaning for the problem?

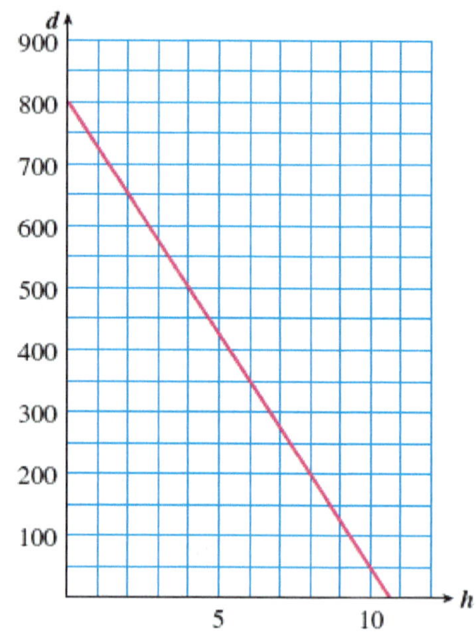

Chapter 3 Summary and Review

Section 3.1 Intercepts

- The **intercepts** of a line are the points where the graph crosses the axes.
- The intercepts of a graph often represent the starting or ending values for a particular variable.

Intercepts of a Graph
The points where a graph crosses the axes are called the **intercepts**.
1. To find the x-intercept, set $y = 0$ and solve for x.
2. To find the y-intercept, set $x = 0$ and solve for y.

To Graph a Linear Equation Using the Intercept Method:
1. Find the x- and y-intercepts of the graph.
2. Draw the line through the two intercepts.
3. Find a third point on the graph as a check.
   (Choose any convenient value for x and solve for y.)

Section 3.2 Ratio and Proportion

- A **ratio** is a type of quotient that allows us to compare two numerical quantities.

Ratio
The **ratio of** a **to** b is written $\frac{a}{b}$.

- A **rate** is a ratio that compares two quantities with different units.
- A **proportion** is a statement that two ratios are equal. In other words, a proportion is a type of equation in which both sides are ratios.
- We use cross-multiplying to solve proportions.

Property of Proportions

$$\text{If} \quad \frac{a}{b} = \frac{c}{d}, \quad \text{then} \quad ad = bc.$$

- Two variables are said to be **proportional** if their ratios are always the same.

Section 3.3 Slope

- The slope of a line measures the rate of change of y with respect to x.

Slope
The slope of a line is defined by the ratio

$$\textbf{slope} = \frac{\textbf{change in } y\textbf{-coordinate}}{\textbf{change in } x\textbf{-coordinate}}$$

as we move from one point to another on the line. In symbols,

$$m = \frac{\Delta y}{\Delta x}$$

- Lines have constant slope. No matter what two points on a line we use to calculate its slope, we always get the same result.
- The slope of a horizontal line is zero, and the slope of a vertical line is undefined.

Section 3.4 Slope-Intercept Form

Slope-Intercept Form

A linear equation written in the form

$$y = mx + b$$

is said to be in **slope-intercept** form. The coefficient m is the slope of the graph, and b is the y-intercept.

- The constant term in the slope-intercept form tells us the initial value of y, and the coefficient of x tells us the rate of change of y with respect to x.

To Graph a Line Using the Slope-Intercept Method

1. Write the equation in the form $y = mx + b$.
2. Plot the y-intercept, $(0, b)$.
3. Write the slope as a fraction, $m = \dfrac{\Delta y}{\Delta x}$.
4. Use the slope to find a second point on the graph: Starting at the y-intercept, move Δy units in the y-direction, then Δx units in the x-direction.
5. Find a third point by moving $-\Delta y$ units in the y-direction, then $-\Delta x$ units in the x-direction, starting from the y-intercept.
6. Draw a line through the three plotted points.

Section 3.5 Properties of Lines

Parallel and Perpendicular Lines

1. Two lines are **parallel** if their slopes are equal, that is, if

$$m_1 = m_2$$

or if both lines are vertical.

2. Two lines are **perpendicular** if the product of their slopes is -1, that is, if

$$m_1 m_2 = -1$$

or if one of the lines is horizontal and one is vertical.

Horizontal and Vertical Lines

1. The equation of the **horizontal line** passing through $(0, b)$ is

$$y = b$$

2. The equation of the **vertical line** passing through $(a, 0)$ is

$$x = a$$

Two-Point Formula for Slope

The slope of the line joining points $P_1(x_1, y_1)$ and $P_2(x_2, y_2)$ is

$$m = \frac{y_2 - y_1}{x_2 - x_1} \quad \text{if} \quad x_2 \neq x_1$$

Review Questions

Use complete sentences to answer the questions in Problems 1-10.

1. Explain how to find the intercepts of a graph.
2. What is the difference between a ratio and a proportion?
3. State the fundamental property of proportions.
4. How can we tell if two variables are proportional?
5. Give a formula for slope using the Δ notation.
6. State the slope-intercept formula, and explain the meaning of the coefficients.
7. Explain the slope-intercept method for graphing a line.
8. What is the easiest way to find the slope of a line from its equation?
9. How can we tell if two lines are perpendicular?
10. What is the difference between a line whose slope is zero and a line whose slope is undefined?

Review Problems

Find the x-and y-intercept of each line, then graph the line.

1. $6x - 4y = 12$

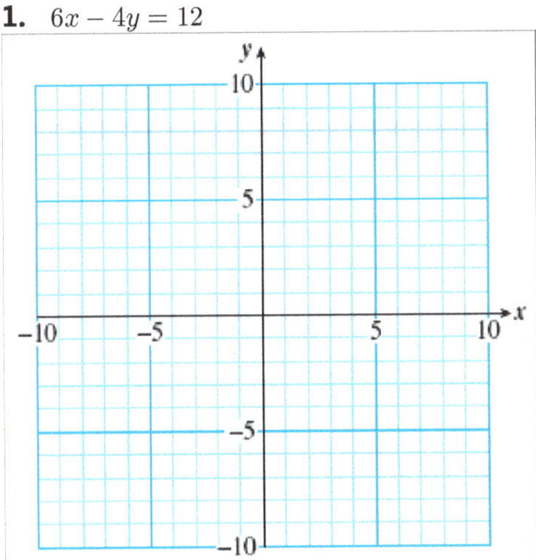

2. $y = -2x + 8$

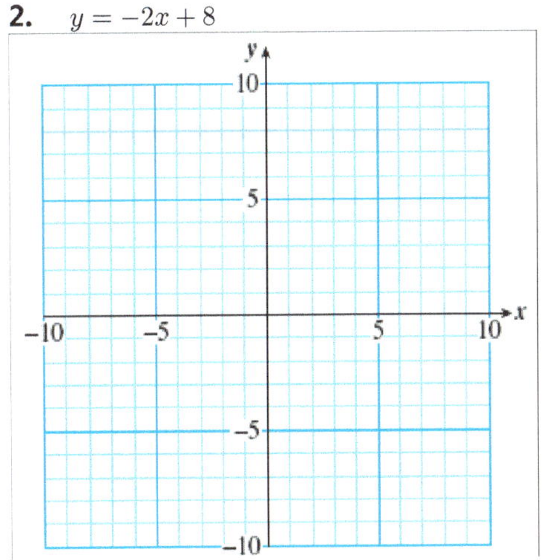

3. Monica has saved \$7800 to live on while she attends college. She spends \$600 a month.
   a. Write an equation for the amount, S, in Monica's savings account after m months.
   b. Write and solve an equation to answer the question: How long will it be before her savings are reduced to \$2400?
   c. Explain what the intercepts of your equation mean in terms of the problem.

4. Ronen is on a rock-climbing expedition. He is climbing out of a deep gorge at a rate of 4 feet per minute, and right now his elevation is -156 feet.

a. Write an equation for Ronen's elevation, h, after m minutes.

b. Write and solve an equation to answer the question: When will his elevation be -20 feet?

c. Explain what the intercepts of your equation mean in terms of the problem.

For Problems 5-6, write ratios or rates.

5. In an anthropology class of 35 students, 14 are men. What is the ratio of men to women in the class?

6. Zach's car went 210 miles on 8.4 gallons of gasoline, and Tasha's car went 204 miles on 8 gallons of gasoline. Which car had the higher rate of fuel consumption?

For Problems 7-8, solve the proportion.

7. $\frac{105}{y} = \frac{15}{17}$

8. $\frac{16}{q} = \frac{52}{86.125}$

For Problems 9-14, write and solve a proportion to answer the question.

9. At Van's Hardware, 36 metric frimbles cost \$4.86. How many metric frimbles can you buy for \$6.75?

10. Bob's weekly diet includes 70 grams of fat, 200 grams of protein, and 1800 grams of carbohydrates. Jenni's weekly diet includes 112 grams of fat. How many grams of protein and carbohydrates should she consume so that her diet has the same proportions as Bob's diet?

11. Simon joined a co-op that provides access to the Internet. His user bill, B, is proportional to the number of hours, h, that he is logged on to the net. Last month he was logged on for 28 hours, and his bill was \$21. This month Simon's bill was \$39. How many hours did he log on the Internet?

12. A small orchard that is 150 yards on each side contains 100 apple trees. The number of trees in an orchard is proportional to the area of the orchard. How many trees are in a larger orchard that is 600 yards on each side?

13. On a map of Arenac County, 3 centimeters represents 5 miles.

a. What are the true dimensions of a rectangular township whose dimensions on the map are 6 centimeters by 9 centimeters?

b. What is the perimeter of the township? What is the perimeter of the corresponding region on the map?

c. What is the ratio of the perimeter of the actual township to the perimeter of the corresponding region on the map?

d. What is the area of the township? What is the area of the corresponding region on the map?

e. What is the ratio of the area of the actual township to the area of the corresponding region on the map?

14. On a map of Euclid County, $\frac{1}{3}$ inch represents 2 miles. Lake Pythagoras is represented on the map by a right triangle with sides 1 inch, $\frac{4}{3}$ inch, and $\frac{5}{3}$ inch.

- **a.** What are the true dimensions of Lake Pythagoras?
- **b.** What is the perimeter of Lake Pythagoras? What is the perimeter of the corresponding region on the map?
- **c.** What is the ratio of the perimeter of the actual lake to the perimeter of the corresponding region on the map?
- **d.** What is the area of Lake Pythagoras? What is the area of the corresponding region on the map?
- **e.** What is the ratio of the area of the actual lake to the are of the corresponding region on the map?

For Problems 15-16, decide whether the two variables are proportional.

15. a.

D	0.5	1	2	4
V	0.25	2	16	64

b.

s	0.2	0.8	1.6	2.5
M	0.08	0.32	0.64	1

16. a.

Time	5	10	15	20
Cost	1.50	2.50	3.50	4.50

b.

Speed	30	40	50	60
Distance	42	56	70	84

For Problems 17-18, use the graph to find the slope of each line, and illustrate Δx and Δy on the graph.

17. $3x + 2y = -7$

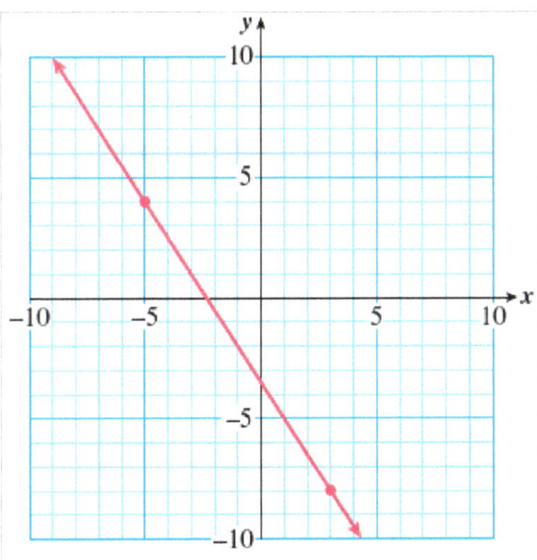

18. $3y = 5x$

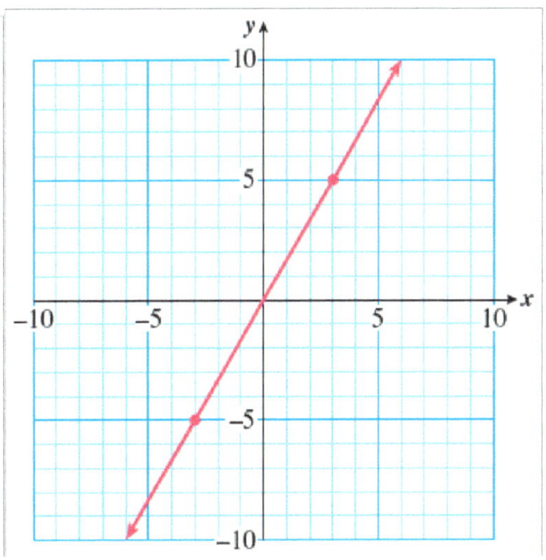

19. The Palm Springs aerial tramway ascends Mt. San Jacinto at a slope of approximately 0.516. It traverses a horizontal distance of about 11,380 feet. The elevation at the bottom of the tramway is 2643 feet. What is the elevation at the top?

20. The ruins of the Pyramid of Cholula in Guatemala have a square base 1132 feet on each side. The sides of the pyramid rise at a slope of about 0.32. How tall was the pyramid originally?

For Problems 21-24,

a. Find the intercepts of each line.
b. Use the intercepts to graph the line.
c. Use the intercepts to find the slope of the line.

21. $6x - 3y = 18$

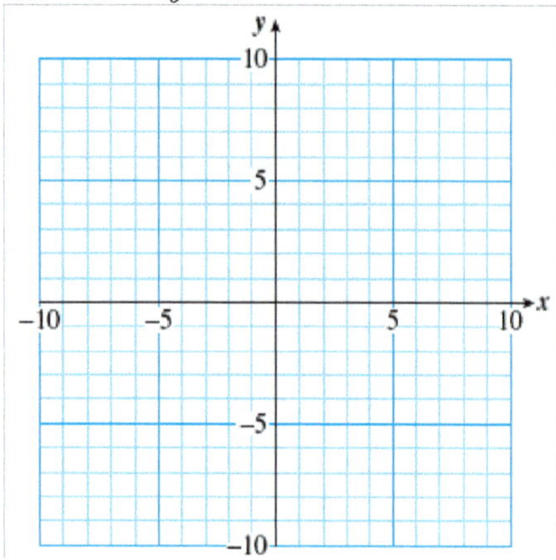

22. $4y + 9x = 36$

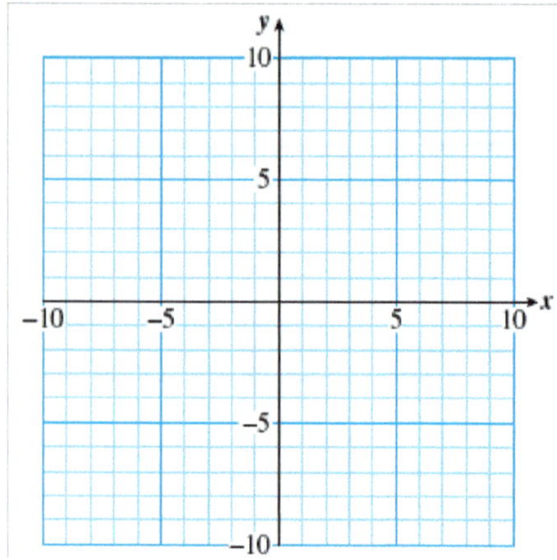

23. $\frac{x}{2} - \frac{y}{3} = 1$

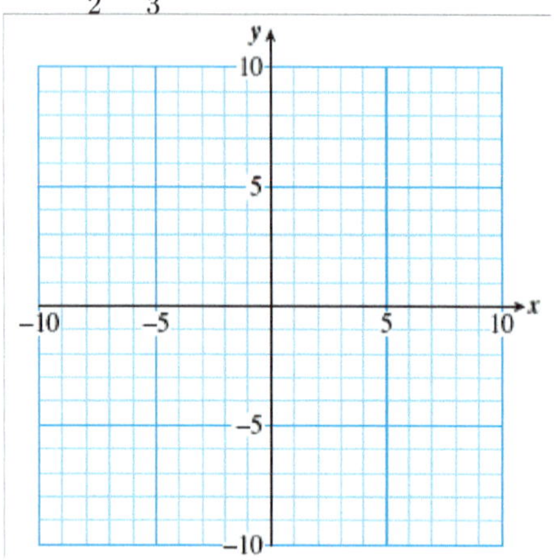

24. $y = 3x - 8$

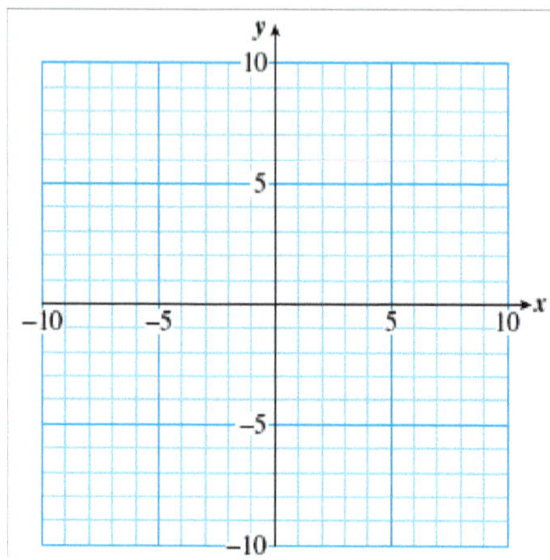

For Problems 25-26,

a. Find the slope and y-intercept of the line.
b. Write an equation for the line.

25.

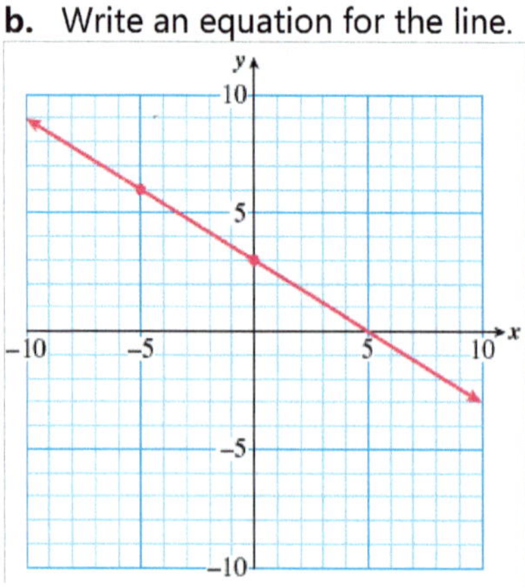

26.

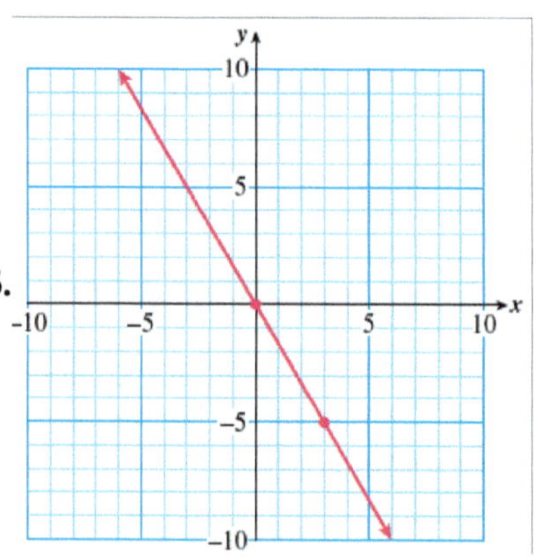

For Problems 27-30,

a. Find the slope and y-intercept of the line.

b. Graph the line using the slope-intercept method.

27. $2y + 5x = -10$

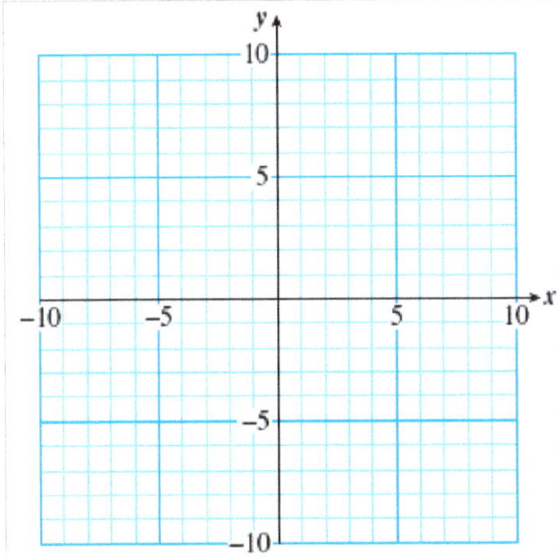

28. $y + 3x = 0$

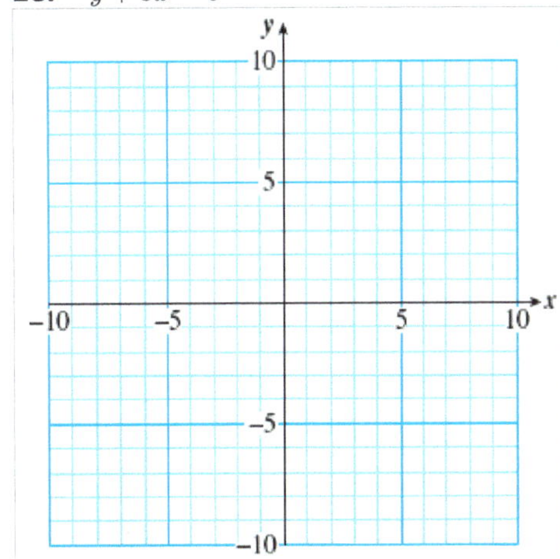

29. $3x = 4y - 8$

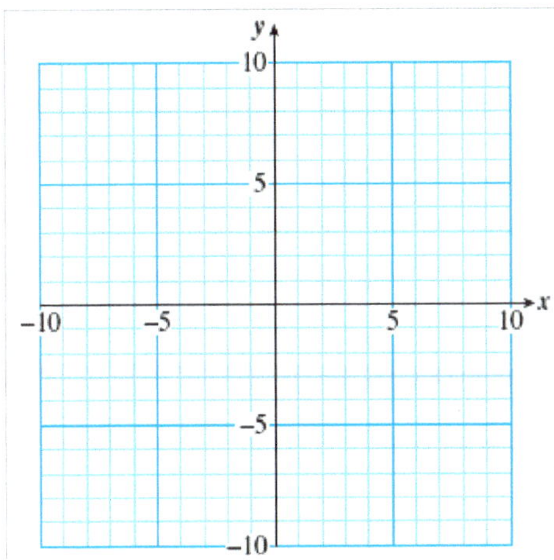

30. $\frac{x}{4} - \frac{y}{5} = 1$

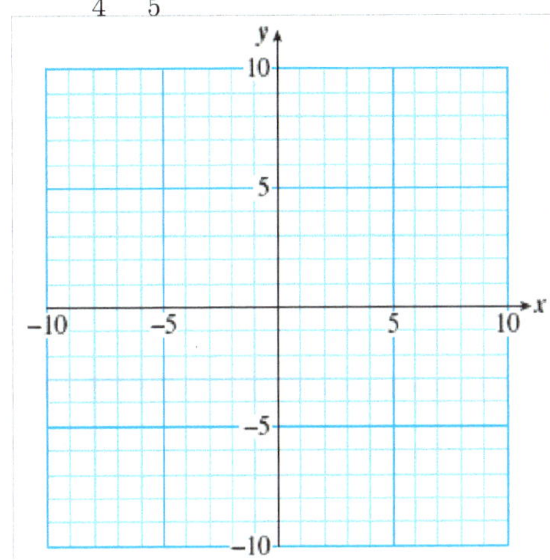

31. This year Francine bought a 3-foot tall blue spruce sapling for her front yard, and it is supposed to grow about 6 inches per year. Write an equation for the height, h, of the tree t years from now.

32. The price of a medium bowl of frozen yogurt is given by the equation $y = 1.35 + 0.85t$, where t is the number of toppings you select. Find the slope and the y-intercept of the graph, and explain what each means in terms of the problem.

33. Beryl is sailing in a hot air balloon. Her altitude at t minutes after 2 pm is given in feet by $h = 500 - 15t$. Find the slope and the y-intercept of the graph, and explain what each means in terms of the problem.

34. The amount of water in the municipal swimming pool is given in gallons by

$$y = 500,000 - 5000h$$

where h is the number of hours since they started draining the pool for the winter. State the slope and y-intercept of the equation, and explain their meaning in terms of the problem.

For Problems 35-36, graph the line with the given slope and passing through the given point.

35. $m = -\frac{3}{4}$, $(2, -1)$

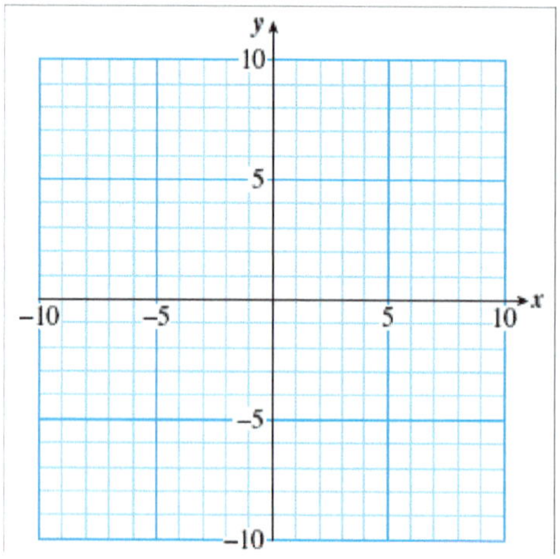

36. $m = \frac{1}{3}$, $(0, -3)$

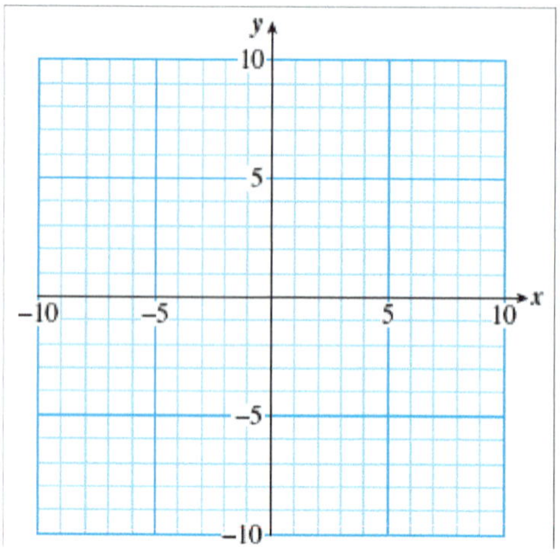

For Problems 37-40, find the slope of the line segment joining the points.

37. $(-1, 4)$, $(3, -2)$

38. $(5, 0)$, $(2, -6)$

39. $(6.2, 1.4)$, $(-2.1, 4.8)$

40. $(0, -6.4)$, $(-5.6, 3.2)$

For Problems 41-42, find the slope of the line described.

41. Vertical with x-intercept $(-5, 0)$

42. Perpendicular to the line $y = \frac{1}{3}x - 2$

43. It costs Delbert \$50.40 to fill up the 12-gallon gas tank in his sports car, and \$105 to fill the 25-gallon tank in his recreational vehicle.

a. If you plot price, p, on the vertical axis, and gallons, g, on the horizontal axis, compute the slope of the line segment joining the two points.

b. What does the slope tell you about the problem?

44. Find the slope and y-intercept of the line $y = 4$.

For Problems 45-48, find an equation for the line described. Write your answers in slope-intercept form if possible.

45. With x-intercept $(5, 0)$ and y-intercept $(0, -1)$.

46. Passing through the points $(0, -2)$ and $(3, -5)$.

47. Parallel to the y-axis and passing through the point $(-4, 2)$.

48. Horizontal and passing through the point $(-3, 8)$.

For Problems 49-50, decide whether the lines are parallel, perpendicular, or neither.

49. **a.** $y = \frac{1}{2}x + 3;\quad x - 2y = 8$ **b.** $4x - y = 6;\quad x + 4y = -2$

50. **a.** $5x + 3y = 1;\quad 5y - 3x = 6$ **b.** $4y - 5x = 1;\quad 2y - \frac{5}{2}x = 2$

For Problems 51-58, use the most convenient method to graph the equation.

51. $4x - 3y = 12$

52. $\frac{x}{6} - \frac{y}{12} = 1$

53. $50x = 40y - 20,000$

54. $1.4x + 2.1y = 8.4$

55. $3x - 4y = 0$

56. $x = -4y$

57. $4x = -12$

58. $2y - x = 0$

Chapter 4 Applications of Linear Equations

Lesson 4.1 The Distributive Law

- Apply the distributive law
- Write and simplify expressions using the distributive law
- Write an equation to model a situation
- Solve equations that involve the distributive law

Lesson 4.2 Systems of Linear Equations

- Decide whether an ordered pair is a solution of a system
- Solve a system of equations by graphing
- Identify a system as consistent, inconsistent, or dependent
- Write a system of equations to solve an applied problem

Lesson 4.3 Algebraic Solution of Systems

- Solve systems by the substitution method
- Solve systems by the elimination method
- Identify a system as consistent, inconsistent, or dependent
- Write a system of equations to solve an applied problem

Lesson 4.4 Applied Problems

- Use a system to solve problems about interest
- Use a system to solve problems about mixtures
- Use a system to solve problems about motion

Lesson 4.5 Point-Slope Form

- Use the point-slope form to graph a line
- Use the point-slope form to find the equation of a line
- Find the equation of a line through two points
- Use the point-slope form to find a linear model

4.1 The Distributive Law

Products

So far we have considered sums and differences of algebraic expressions. Now we take a look at simple products. To simplify a product such as $3(2a)$, we multiply the numerical factors to find

$$3(2a) = (3 \cdot 2)a = 6a$$

This calculation is an application of the associative law for multiplication. We can see why the product is reasonable by recalling that multiplication is repeated addition, so that

$$3(2a) = 2a + 2a + 2a = 6a$$

Three terms

Example 1a. $5(3x) = 15x$ **b.** $-2(-4h) = 8h$

RQ1. Explain the difference between the expressions $2b + 3b$ and $2(3b)$.

The Distributive Law

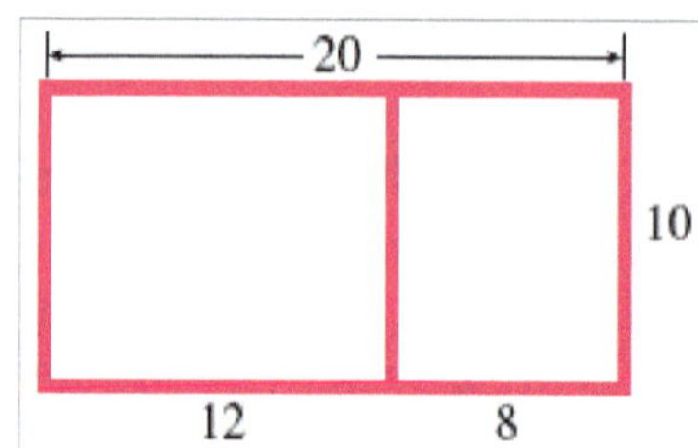

Another useful property for simplifying expressions is the distributive law. First consider a numerical example. Suppose we would like to find the area of the two rooms shown in the figure.

We can do this in two different ways:

Method 1: Think of the rooms as one large rectangle, with width 10 feet and length $12 + 8$ feet. The area is then given by

$$\begin{aligned} \text{Area} &= \text{width} \times \text{length} \\ &= \mathbf{10\,(12 + 8)} = 10\,(20) = 200 \text{ square feet} \end{aligned}$$

Method 2: Find the area of each room separately and add the two areas together. This gives us

$$\mathbf{10(12) + 10(8)} = 120 + 80 = 200 \text{ square feet}$$

Of course, we get the same answer with either method. However, looking at the first step in each calculation, we see that

$$10(12 + 8) = 10(12) + 10(8)$$

These are two ways to compute the expression $10(12 + 8)$. The first way follows the order of operations, adding $12 + 8$ first. The second way distributes the multiplication to each term inside parentheses; that is, we multiply each term inside the parentheses by 10.

$$\mathbf{10}(12 + 8) = \mathbf{10}(12) + \mathbf{10}(8)$$

This equation is an example of the distributive law.

Distributive Law

If a, b, and c are any numbers, then

$$a(b + c) = ab + ac$$

If the terms inside parentheses are not like terms, we have no choice but to use the distributive law to simplify the expression.

Example 2 Simplify $-2(3x-1)$

Solution We multiply each term inside parentheses by -2:

$$\begin{aligned} -2(3x-1) &= -2(3x)-2(-1) \\ &= -6x+2 \end{aligned}$$

Caution! Compare the three similar expressions:

$$-3x-5x, \qquad -3(-5x), \qquad -3(-5-x)$$

To simplify these expressions, we must first recognize the operations involved. The first is a sum of terms, the second is a product, and the third is an application of the distributive law. Thus,

$$-3x-5x=-8x, \qquad -3(-5x)=15x, \qquad -3(-5-x)=15+3x$$

RQ2. When do we need the distributive law to simplify an expression?

Solving Equations and Inequalities

If an equation or inequality contains parentheses, we use the distributive law to simplify each side before we begin to solve.

Example 3 Solve $25-6x=3x-2(4-x)$

Solution We apply the distributive law to the right side of the equation.

$25-6x=3x-2(4-x)$	Apply the distributive law.
$25-6x=\mathbf{3x}-8+\mathbf{2x}$	Combine like terms.
$25-6x=5x-8$	Subtract $5x$ from both sides.
$25-11x=-8$	Subtract 25 from both sides.
$-11x=-33$	Divide both sides by -11.
$x=3$	The solution is 3.

You should verify each step of the solution, and check the answer.

Caution! The expression on the right side of the equation in Example 3 is made up of two terms:

$$3x \quad \text{and} \quad -2(4-x)$$

Remember that we think of all terms as *added* together, so the minus sign in front of the 2 tells us that 2 is negative. When we apply the distributive law, we multiply each term inside parentheses by -2 to get $-8+2x$.

We know that a fraction bar often serves as a grouping device, like parentheses.

Example 4 Solve the proportion $\dfrac{7.6}{1.2} = \dfrac{x+3}{2x-1}$

Solution We apply the property of proportions and cross-multiply to get

$$7.6\,(2x-1) = 1.2\,(x+3)$$ Apply the Distributive Law.

$$15.2\,x - 7.6 = 1.2\,x + 3.6$$ Subtract $1.2x$ from both sides and add 7.6 to both sides.

$$14x = 11.2$$ Divide both sides by 14.

$$x = \frac{11.2}{14} = 0.8$$

We can check the solution by substituting $x = 0.8$ into the original proportion.

RQ3. In Example 4, why did we put parentheses around $2x-1$ and $x+3$?

Algebraic Expressions

We can often simplify algebraic expressions by combining like terms.

Example 5 Delbert and Francine are collecting aluminum cans to recycle. They will be paid x dollars for every pound of cans they collect. At the end of three weeks, Delbert collected 23 pounds of aluminum cans, and Francine collected 47 pounds.

a. Write algebraic expressions for the amount of money Delbert made, and the amount Francine made.

b. Write and simplify an expression for the total amount of money Delbert and Francine made from aluminum cans.

Solutions a. We multiply the number of pounds collected by the price per pound. Delbert made $23x$ dollars, and Francine made $47x$ dollars.

b. We add the amount of money Delbert made to the amount Francine made.

$$23x + 47x = 70x$$

RQ4. In Example 5, what does x stand for? What does $23x$ stand for?

The distributive law is also helpful for writing algebraic expressions.

Example 6 The length of a rectangle is 3 feet less than twice its width, w.

a. Write an expression for the length of the rectangle in terms of w.

b. Write and simplify an expression for the perimeter of the rectangle in terms of w.

c. Suppose the perimeter of the rectangle is 36 feet. Write and solve an equation to find the dimensions of the rectangle.

Solutions a. "Three feet less than twice the width" tells us to subtract 3 from twice the width. The length of the rectangle is $2w-3$.

b. The perimeter of a rectangle is given by the formula $P = 2l + 2w$. We substitute our expression for the length to get

$$
\begin{aligned}
P &= 2(2w - 3) + 2w && \text{Apply the distributive law.} \\
&= 4w - 6 + 2w && \text{Combine like terms.} \\
&= 6w - 6
\end{aligned}
$$

c. We set our expression for the perimeter equal to 36.

$$
\begin{aligned}
6w - 6 &= 36 && \text{Add 6 to both sides.} \\
6w &= 42 && \text{Divide both sides by 7.} \\
w &= 7
\end{aligned}
$$

The width of the rectangle is 7 feet, and its length is $2(7) - 3 = 11$ feet.

RQ5. In Example 6, why did we multiply $2w - 3$ by 2?

Skills Warm-Up

Use a formula to write an equation. Then solve the equation.

1. 32.5% of the class are engineering majors. If there are 91 engineering majors, how many students are in the class?

2. A rectangular cookie sheet is 40 cm long and has a perimeter of 116 cm. How wide is the cookie sheet?

Write an equation you could solve to answer the question.

3. Erika bought a 50-pound bag of dog food and after 28 days she had 29 pounds left. How much dog food did she use per day?

4. A hamburger contains 60 calories less than two bags of fries. A hamburger and one bag of fries contains 780 calories. How many calories are in a bag of fries?

Answers: **1.** $0.325x = 91;$ 280 **2.** $2w + 2(40) = 116;$ 18 cm
3. $50 - 28d = 29;$ $\dfrac{3}{4}$ lb **4.** $x + (2x - 60) = 780;$ 280

Homework 4.1

Skills Practice

In Problems 1-4, use the distributive law to remove parentheses. For some of these exercises you will use the distributive law in the form

$$(b+c)a = ba + ca$$

1. $5(2y-3)$

2. $-2(4x+8)$

3. $-(5b-3)$

4. $(-6+2t)(-6)$

In problems 5-6, simplify each product. Which product in each pair requires the distributive law?

5. **a.** $8(4c)$ **b.** $8(4+c)$

6. **a.** $2(-8-t)$ **b.** $2(-8t)$

7. Which of the following is a correct application of the distributive law?

a. $5(3a) = 15a$ **b.** $5(3+a) = 15 + a$

c. $5(3a) = 15(5a)$ **d.** $5(3+a) = 15 + 5a$

8. Simplify each expression if possible. Then evaluate for $x = 3,\ y = 9$.

a. $2(xy)$ **b.** $2(x+y)$

c. $2 - xy$ **d.** $-2xy$

In Problems 9-12, simplify and combine like terms.

9. $-6(x+1) + 2x$

10. $5 - 2(4x-9) + 9x$

11. $-4(3+2z) + 2z - 3(2z+1)$

12. $3(-3a-4) - 3a - 4(3a-4)$

In Problems 13-20, solve each equation or inequality.

13. $6(3y-4) = -60$

14. $5w - 64 = -2(3w-1)$

15. $-22c + 5(3c+4) = 20 + 8c$

16. $6 - 4(2a+3) \geq -6a + 2$

17. $4 - 3(2t-4) > -2(4-3t)$

18. $0.25(x+3) - 0.45(x-3) = 0.30$

19. $\dfrac{a}{a+2} = \dfrac{2}{3}$

20. $\dfrac{0.3}{0.5} = \dfrac{b+2}{12-b}$

21. Choose a value for the variable and show that the following pairs of expressions are **not** equivalent.

a. $5(x+3)$ and $5x+3$ **b.** $-4(c-3)$ and $-4c-12$

22. a. Evaluate the expression $(2x+7) - 4(4x-2) - (-2x+3)$ for $x = -3$.

b. Simplify the expression in part (a) by combining like terms.

c. Evaluate your answer to part (b) for $x = -3$. Check that you got the same answer for part (a).

d. Are the expressions in parts (a) and (b) equivalent? Why or why not?

Applications

23. The length of a rectangle is 8 feet shorter than twice its width. Write expressions in terms of w, the width of the rectangle.
- **a.** The length of the rectangle
- **b.** The perimeter of the rectangle
- **c.** The area of the rectangle

24. Risa bought a skirt and a sweater that together cost \$108. Then she bought a pair of shoes that cost twice as much as the sweater. Write and simplify expressions in terms of c, the cost of the skirt:
- **a.** The cost of the sweater
- **b.** The cost of the shoes
- **c.** The total amount Risa spent

25. Albert and Isaac left the same hotel at the same time, and each drove for 3 hours, but Albert drove 15 miles per hour faster than Isaac. Write and simplify expressions in terms of Isaac's speed, s.
- **a.** How far Isaac drove
- **b.** How far Albert drove
- **c.** If they drove in the same direction, how far apart they are now
- **d.** If they drove in opposite directions, how far apart they are now

26. A box of AlmondOats contains 15 ounces of cereal made of oat flakes and sliced almonds. Oat flakes cost 15 cents per ounce, and almonds cost 35 cents per ounce. Write and simplify expressions in terms of a, the number of ounces of almonds in the box.
- **a.** The number of ounces of oat flakes in the box
- **b.** The cost of the almonds
- **c.** The cost of the oat flakes
- **d.** The total cost of a box of AlmondOats

27. Trader Jim's Pomegranate Punch is 25% pomegranate juice, and Pomajoy is 60% pomegranate juice. Vera mixes some of each to make 8 quarts of juice drink. Write and simplify expressions in terms of x, the number of quarts of Pomajoy she uses.
- **a.** The number of quarts of Pomegranate Punch
- **b.** The number of quarts of pomegranate juice in the Pomajoy
- **c.** The number of quarts of pomegranate juice in the Pomegranate Punch
- **d.** The number of quarts of pomegranate juice in Vera's mixture

28. Premium ice cream is 12% butterfat, and chocolate syrup is 55% butterfat. A chocolate sundae, without whipped cream and a cherry, weighs 10 ounces. Write and simplify expressions in terms of x, the number of ounces of ice cream in the sundae.
- **a.** The number of ounces of chocolate syrup in the sundae
- **b.** The amount of fat in the ice cream
- **c.** The amount of fat in the chocolate syrup
- **d.** The total amount of fat in the sundae

In problems 29-30, find an algebraic expression for the area of the rectangle. Use the distributive law to write the expression in two ways.

29. 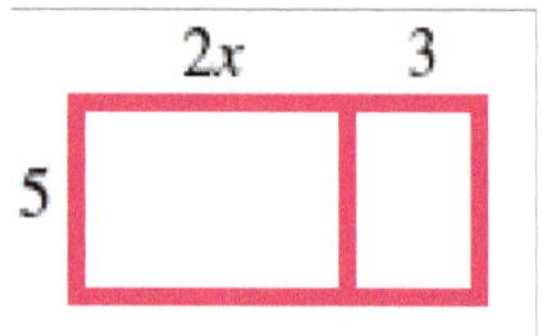

30. 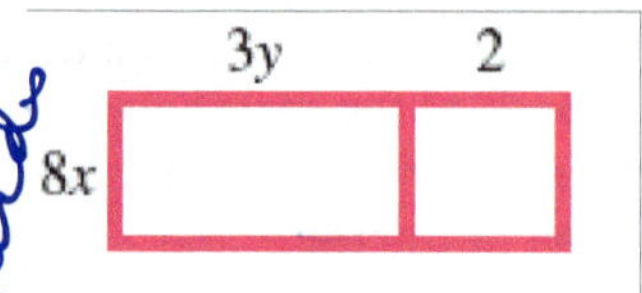

31. The length of a rectangular vegetable garden is 6 yards more than twice its width.

a. Write an expression for the perimeter of the garden in terms of its width.

b. Ann bought 42 yards of fence to enclose the garden. What are its dimensions?

32. Revenue from the state lottery is divided between education and administrative costs in the ratio of 4 to 3. If the lottery revenue this year is $24,000,000, how much money will go to education?

33. An apple and a glass of milk together contain 260 calories.

a. If an apple contains a calories, how many calories are in a glass of milk?

b. Write an expression for the number of calories in two apples and three glasses of milk.

c. If two apples and three glasses of milk contain 660 calories, find the number of calories in an apple.

34. Melody sold 47 tickets to a charity concert. Reserved seats cost $10 and open seating was $6 a ticket. Let x represent the number of reserved seats she sold. Write expressions in terms of x for:

a. The number of open seating tickets Melody sold

b. The amount of money Melody collected from reserved seating tickets

c. The amount of money Melody collected from open seating tickets

d. The total amount of money Melody collected

35. Melody from Problem 34 collected $330 from the sale of concert tickets. Write and solve an equation to answer the question: How many reserved seats did she sell?

4.2 Systems of Linear Equations

What is a System of Equations?

In many applications it is useful to write two or more linear equations.

Example 1 Delbert and Francine are buying appliances for their new home. They have a choice of two different refrigerators: a standard model that sells for \$1000, or an energy-efficient model at a price of \$1200. The standard model costs \$6 per month to run, and the energy-efficient model costs \$2 per month. Write linear equations for the total cost of each refrigerator after t months.

Solution Let S stand for the total cost of running the standard refrigerator for t months, so $S = mt + b$. The initial cost of the standard model is \$1000, so $b = 1000$. The cost increases at a rate of \$6 per month, so $m = 6$. Thus,

$$S = 6t + 1000$$

Let E stand for the total cost of running the energy-efficient refrigerator. For this model, the initial cost is $b = 1200$, and the total cost increases at a rate of $m = 2$ dollars per month. Thus,

$$E = 2t + 1200$$

Look Closer: When we consider two equations together, as in the example above, we often use the same variables for both equations, like this:

$$y = 6x + 1000$$
$$y = 2x + 1200$$

A pair of linear equations with the same variables is called a **system of linear equations**.

In Example 1, the standard refrigerator costs less initially, but costs more to run each month. Delbert and Francine want to know when the energy-efficient model will begin to pay for itself, or in other words, when the total cost of running the standard model will exceed the total cost of the energy-efficient model.

Example 2 When is the total cost of running the two refrigerators in Example 1 equal?

Solution We would like to know the value of t when $S = E$. We could use trial-and-error by evaluating both S and E for various values of t. The results of such a search are given in the table below.

t	10	20	30	40	50
S	1060	1120	1180	1240	1300
E	1220	1240	1260	1280	1300

We see that when $t = 50$, the total cost of both models is \$1300. Thus, at the end of 50 months, or 4 years 2 months, the cost of the standard model equals the cost of the energy-efficient model.

In Example 2 we found an ordered pair, $(50, 1300)$, that makes both equations true.

A **solution** to a system of equations is an ordered pair (x, y) that satisfies each equation in the system.

To check whether $(50, 1300)$ is really a solution of the system, we substitute the coordinates into each equation.

$S = 6t + 1000$ Does $1300 = 6(50) + 1000$? **True.**
$E = 2t + 1200$ Does $1300 = 2(50) + 1200$? **True.**

We have verified that the ordered pair $(50, 1300)$ is a solution to the system in Example 1.

RQ1. What is a solution to a system of equations?
RQ2. Decide whether the ordered pair $(2, 3)$ is a solution to the system

$$3x - 4y = -6$$
$$x + 2y = -4$$

Solving a System of Equations by Graphing

The solution of a system satisfies both equations in the system, so the point that represents the solution must lie on both graphs. It is the intersection point of the two lines described by the system. Thus, we can solve a system of equations by graphing the equations and looking for the point (or points) where the graphs intersect.

Example 3 Solve the system in Example 1 by graphing.

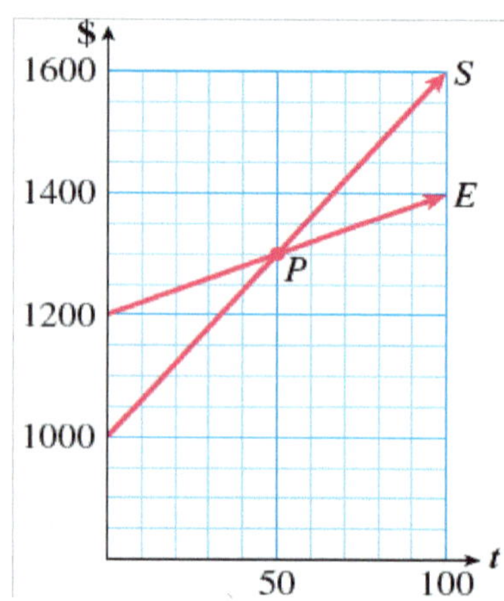

Solution The graph of the two equations is shown at right. The graph of S has a smaller initial value than E, but it increases more rapidly because its slope is greater. The two graphs intersect at the point labeled P. At this point the values of S and E are the same, approximately 1300. The t-coordinate of point P is 50, so the solution of the system is $(50, 1300)$. The total cost of each model is about \$1300 over the first 50 months of operation.

After 50 months, the total cost of the energy-efficient model will be less than the cost of the standard model.

RQ3. How do we find the solution to a system of equations by graphing?

Inconsistent and Dependent Systems

Not every system of equations has a solution.

Example 4 Delbert and Francine are also shopping for homeowners insurance. HomeLife offers a policy for semiannual premiums of \$408, or they can join a co-op for

$50 and pay just $68 per month for comparable coverage. Which option is less expensive in the long run?

Solution We'll write a system of equations and graph them, as we did in Example 3. If t stands for the number of months, then the total cost of the co-op insurance is

$$C = 50 + 68t$$

and the HomeLife policy costs

$$H = \frac{408}{6}t = 68t$$

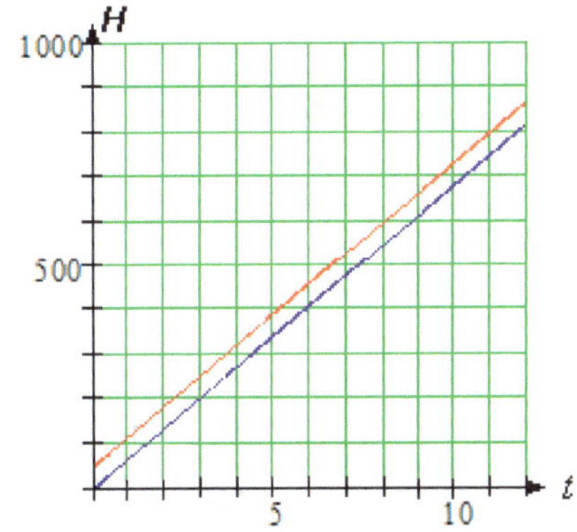

(because a semiannual premium covers 6 months). The solution of the system represents the time when the costs of the two policies are equal. As you can see, the two lines are parallel, so the graphs do not intersect. This system has no solution. There is no point where the two policies cost the same amount; the HomeLife policy is always less expensive.

In Example 4, the two graphs never intersect, so the system has no solution. Such a system is called **inconsistent**. There is a third possibility: Both equations may have the same graph.

Example 5 Solve the system

$$y = 2x + 2$$
$$6x - 3y = -6$$

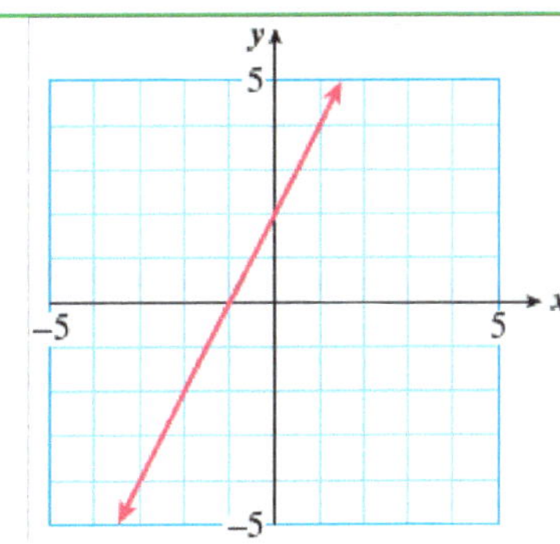

Solution We graph the first equation by the slope-intercept method:

$$b = 2 \quad \text{and} \quad m = \frac{\Delta y}{\Delta x} = \frac{2}{1}$$

Its graph is shown at right. The second equation, which has the form $Ax + By = C$, is easier to graph by the intercept method. You can verify that its intercepts are $(0, 2)$ and $(-1, 0)$. The graph of the second line is identical to the first graph. Every solution to the first equation is also a solution to the second equation, so every point on the line is a solution of the system as well. The system has infinitely many solutions.

We can verify that the two equations in Example 5 are actually equivalent by solving the second equation for y in terms of x.

$6x - 3y = -6$	Add $3y$ and 6 to both sides.
$6x + 6 = 3y$	Divide both sides by 3.
$2x + 2 = y$	

The second equation is the same as the first equation. A system in which the two equations are equivalent is called **dependent**. A dependent system has infinitely many solutions: Every point on the graph is a solution to the system.

We have now seen three types of systems. These are illustrated in the figure on the next page. Most of the systems we will study have exactly one solution. Such systems are called **consistent and independent**.

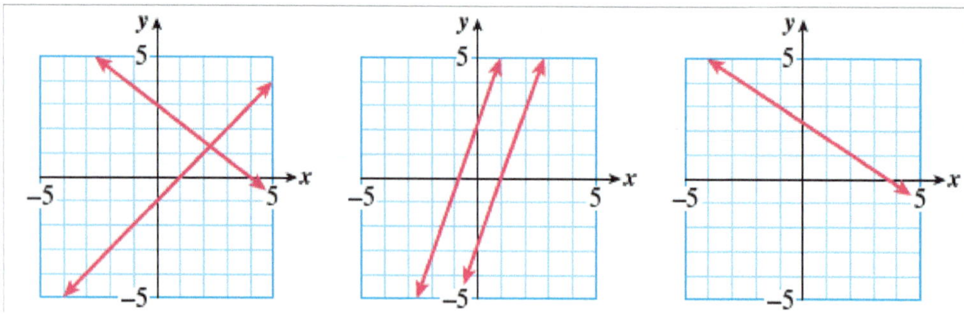

Consistent and independent; one solution

Inconsistent; no solution

Dependent; infinitely many solutions

1. **Consistent and independent system.** The graphs of the two lines intersect in exactly one point. The system has exactly one solution.
2. **Inconsistent system**. The graphs of the equations are parallel lines and hence do not intersect. An inconsistent system has no solutions.
3. **Dependent system**. All the solutions of one equation are also solutions to the second equation, and hence are solutions of the system. The graphs of the two equations are the same line. A dependent system has infinitely many solutions.

RQ4. What does the graph of an inconsistent system look like? How many solutions does it have?

RQ5. What does the graph of a dependent system look like? How many solutions does it have?

Problem Solving with Systems

Many practical problems involve two or more unknown quantities. Often it is easier to solve these problems by using two variables and writing a system of equations.

Example 6 Allen has been asked to design a rectangular Plexiglas plate whose perimeter is 28 inches and whose length is three times its width. What should the dimensions of the plate be?

Solution **Step 1** The dimensions of a rectangle are its length and width, so we let x represent the width of the plate and y represent its length.

Step 2 We must write two equations about the length and width of the plate. We know a formula for the perimeter of a rectangle, $P = 2l + 2w$, so the first equation is

$$2x + 2y = 28$$

We also know that the length is three times the width, or $y = 3x$. These two equations make a system.

$$\begin{aligned} 2x + 2y &= 28 \\ y &= 3x \end{aligned}$$

Step 3 We use the intercept method to graph the first equation, $2x + 2y = 28$:

x	y
0	14
14	0

When $x = 0$, $y = 14$, and when $y = 0$, $x = 14$.

We use the slope-intercept method to graph $y = 3x$:

$$b = 0 \qquad m = \frac{\Delta y}{\Delta x} = \frac{3}{1}$$

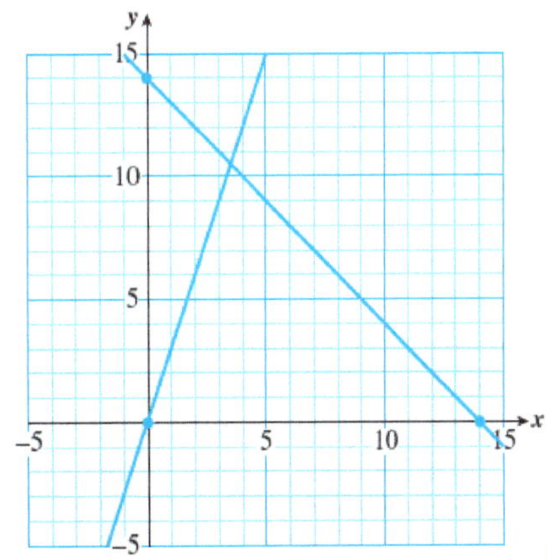

Their graphs are shown at right. The graphs intersect at approximately $(3.5, 10.5)$. You can verify that $x = 3.5$ and $y = 10.5$ are a solution to the system by checking that these values make *both* equations true.

Step 4 The width of the plate should be 3.5 inches and its length should be 10.5 inches.

RQ6. When might you use a system of equations to solve an applied problem?

Skills Warm-Up

Write algebraic expressions to answer the questions.

1. How much interest is earned after 2 years on $d + 50$ dollars at 6% annual interest rate?
2. How much manganese is there in $8 - z$ grams of an alloy that is 35% manganese?
3. How far will a boat travel in 2 hours at a speed of $r + 3$ miles per hour?
4. How far will a train moving at 40 miles per hour travel in $3 + x$ hours?
5. A small plane has a top airspeed of v miles per hour. How far can the plane travel in 5 hours against a headwind of 15 miles per hour?
6. A fishing boat has a top speed in still water of 26 miles per hour. How far can the boat travel downstream in 3 hours if the speed of the current is w miles per hour?

Answers: **1.** $0.12(d + 50)$ **2.** $0.35(8 - z)$ **3.** $2(r + 3)$ **4.** $40(3 + x)$ **5.** $5(v - 15)$ **6.** $3(26 + w)$

Homework 4.2

Skills Practice

In Problems 1-2, decide whether the given ordered pair is a solution of the system.

1. $x + 2y = -8$ $(4, -2)$
$2x - y = 4$

2. $3x + 2y = -5$ $(-3, -2)$
$x + 3y = -9$

In Problems 3-5, decide which graphing technique, the intercept method or the slope-intercept method, would be easier to use. Explain why you made the choice you did.

3. $3x - 2y = 18$

4. $y = \dfrac{5x}{3} - 8$

5. $y = \dfrac{-5}{6}x$

In Problems 6-10, solve the system by graphing.

6. $y = -3x$
$y = 3x - 6$

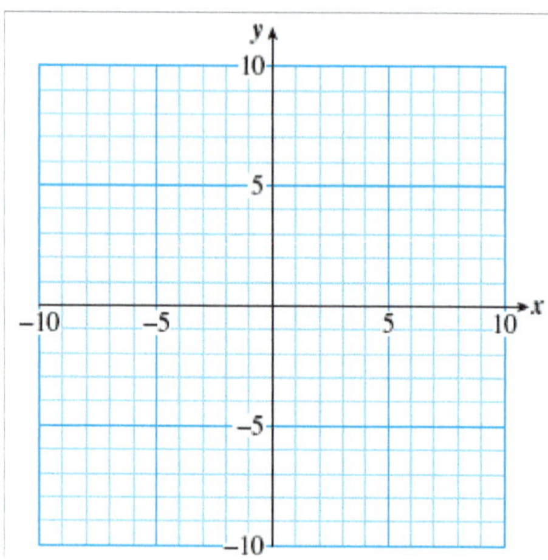

7. $y = \dfrac{3}{4}x - 2$
$2y - 5x = 10$

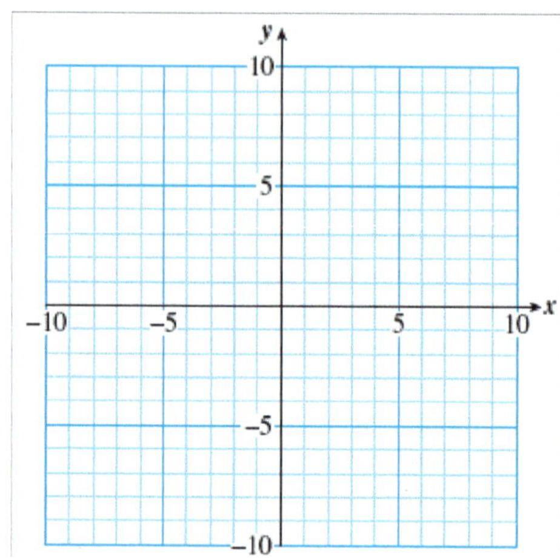

8. $x + y = 4$
$3x - 3y = 12$

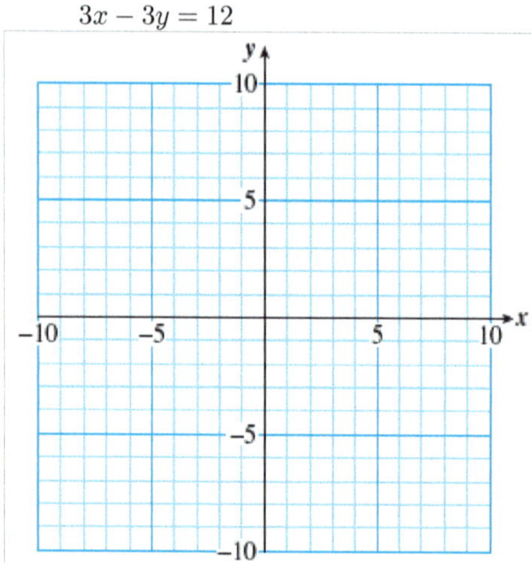

9. $2y = -3x$
$y = -2x - 1$

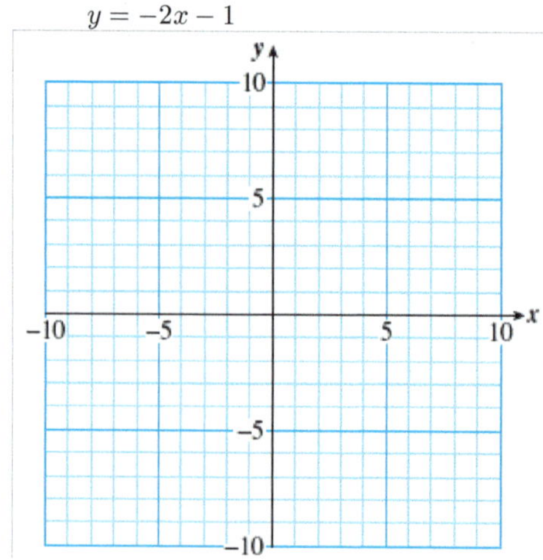

In Problems 10-11, decide whether the system is dependent or inconsistent.

10. $x + y = 4$
$2x + 2y = 12$

11. $y = -2x + 4$
$4x + 2y = 2$

12. Without graphing, explain how you can tell that the system has no solution.

$$3x = y + 3$$
$$6x - 2y = 12$$

Applications

13. Delbert has accepted a sales job and is offered a choice of two salary plans. Under Plan A he receives $20,000 a year plus a 3% commission on his sales. Plan B offers a $15,000 annual salary plus a 5% commission.

a. Let x stand for the amount of Delbert's sales in one year. Write equations for his total annual earnings, E, under each plan.

Plan A: $E =$

Plan B: $E =$

b. Fill in the tables, where x is given in thousands of dollars.

x	Earnings Under Plan A
0	
50	
100	
150	
200	
250	
300	
350	
400	

x	Earnings Under Plan B
0	
50	
100	
150	
200	
250	
300	
350	
400	

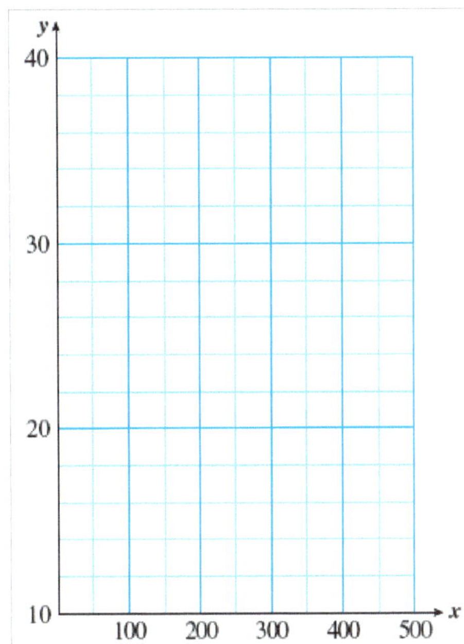

c. Graph both of your equations on the grid. (Both axes are scaled in thousands of dollars.)

d. For what sales amount do the two plans result in equal earnings for Delbert?

14. Orpheus Music plans to manufacture clarinets for schools. Their startup costs are $6000, and each clarinet costs $60 to make. They plan to sell the clarinets for $80 each.

a. Let x stand for the number of clarinets Orpheus manufactures. Write equations for the total cost, C, of producing x clarinets, and the revenue, R, earned from selling x clarinets.

Cost: $C =$ 80X +7600

Revenue: $R =$ 100X

b. Fill in the tables.

x	**Cost**
0	
50	
100	
150	
200	
250	
300	
350	
400	

x	**Revenue**
0	
50	
100	
150	
200	
250	
300	
350	
400	

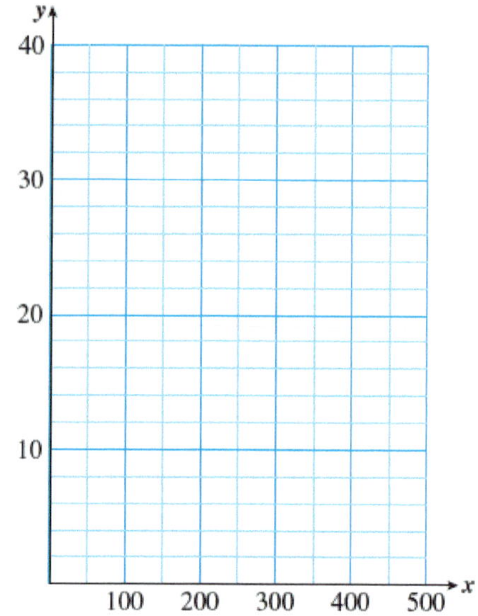

c. Graph both your equations on the grid. (The vertical axis is scaled in thousands of dollars.)

d. How many clarinets must Orpheus sell in order to break even?

e. What is their profit if they sell 500 clarinets? 200 clarinets?

f. Illustrate your answers to part (e) on your graph.

In Problems 15-16,

a. Choose a variable for each of the unknown quantities.

b. Write a system of equations in two variables to model the problem.

c. Solve the system graphically.

d. Verify your solution algebraically.

15. A bouquet of 4 roses and 8 carnations costs \$14. A bouquet of 6 roses and 9 carnations costs \$18. How much do one rose and one carnation cost?

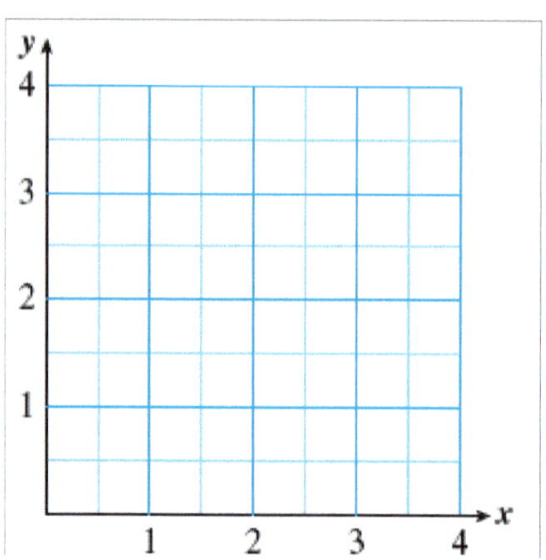

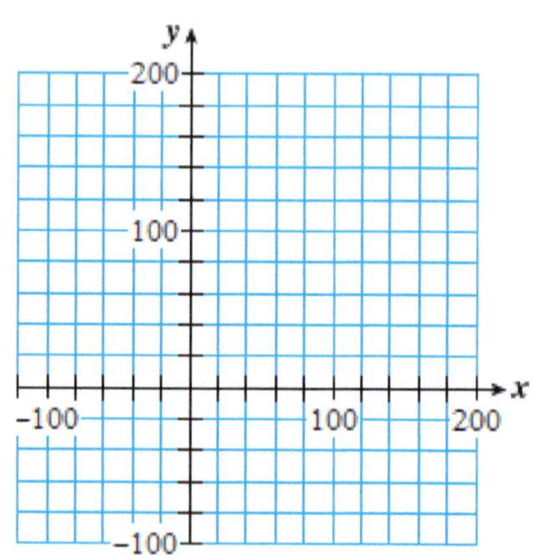

16. The vertex angle of an isosceles triangle is 15° less than each base angle. (In an isosceles triangle, the base angles are equal.) Find the measure of each angle of the triangle. (**Hint:** For your second equation, recall that the sum of the angles in any triangle is 180°.)

4.3 Algebraic Solution of Systems

In the previous Lesson, we compared the costs of operating two different refrigerators, a standard model and an energy-efficient model. We wrote a system of equations for this problem,

$$y = 6x + 1000$$
$$y = 2x + 1200$$

where x is the number of months the refrigerator has been running. We looked for a point on the graphs where the two y-values, which represent the costs, are equal. This point is the intersection point of the graphs, as shown in the figure. The x-coordinate of the intersection point shows that the costs are equal after 50 months.

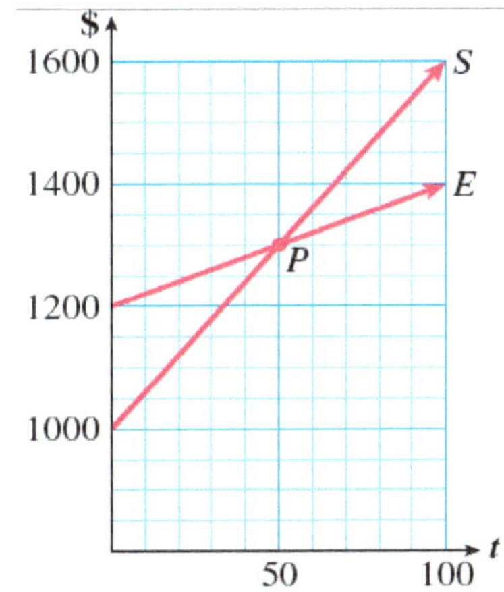

We can also solve the system using algebra.

Example 1 Solve the system algebraically.

$$y = 6x + 1000$$
$$y = 2x + 1200$$

Solution We are looking for the point where the two y-values are equal. Therefore, we set the two expressions for y equal, which gives us an equation in x to solve:

$$6x + 1000 = 2x + 1200 \qquad \text{Subtract } 2x \text{ from both sides.}$$
$$4x + 1000 = 1200 \qquad \text{Subtract 1000 from both sides.}$$
$$4x = 200 \qquad \text{Divide both sides by 4.}$$
$$x = 50$$

We find that $x = 50$, the same answer we got using graphing. Substituting $x = 50$ into either equation gives $y = 1300$. The solution is $(50, 1300)$.

Substitution Method

In Example 1, we solved the system by equating the two expressions for y. We can think of this procedure as substituting the expression for y from one equation into the other equation, like this:

$$y = 6x + 1000$$

Substitute into second equation.

$$y = 2x + 1200$$

to get:

$$6x + 1000 = 2x + 1200$$

Look Closer: The method works well in this example because both equations in the system were already in the form $y = mx + b$. Sometimes we must solve one of the equations for y in terms of x before we can substitute.

Example 2 Use substitution to solve the system

$$y + 3x = 1$$
$$2y + 5x = 5$$

Solution We first solve one of the equations for y in terms of x. It is easier to solve the first equation for y, by subtracting $3x$ from both sides to get:

$$y = 1 - 3x$$

We now substitute the expression $1 - 3x$ for y in the *other* equation:

$$y = \mathbf{1 - 3x}$$

$$2y + 5x = 5$$

$$2(\mathbf{1 - 3x}) + 5x = 5$$

This gives us an equation in one variable, which we can solve as usual. We begin by simplifying the left side.

$2(1 - 3x) + 5x = 5$	Apply the distributive law.
$2 - 6x + 5x = 5$	Combine like terms.
$2 - x = 5$	Subtract 2 from both sides.
$-x = 3$	Divide both sides by -1.
$x = -3$	

This is the x-coordinate of the solution point. To find the y-coordinate, we substitute $x = -3$ into either of the original equations. It is easiest to use the equation we solved for y, namely $y = 1 - 3x$:

$$y = 1 - 3(-3) = 10 \qquad \text{Substitute } -3 \text{ for } x.$$

The solution to the system is $x = -3$, $y = 10$, or the point $(-3, 10)$. You can check that these values satisfy both of the original equations in the system.

RQ1. What is the first step in the substitution method?
RQ2. How can you check your answer for the solution of a system?

Here is a summary of the substitution method.

To Solve a System by Substitution:

1. Choose one of the variables in one of the equations. (It is best to choose a variable whose coefficient is 1 or -1.) Solve the equation for that variable.
2. Substitute the result of Step 1 into the *other* equation. This gives an equation in one variable.
3. Solve the equation obtained in Step 2. This gives the solution value for one of the variables.
4. Substitute this value into the result of Step 1 to find the solution value of the other variable.

Elimination Method

A second algebraic method for solving systems is called **elimination**. As with the substitution method, we try to obtain an equation in a single variable, but we do it by eliminating one of the variables in the system. We first put both equations into the general linear form, $Ax + By = C$.

Example 3 Solve the system

$$\begin{aligned} 5x &= 2y + 21 \\ 2y &= 19 - 3x \end{aligned}$$

Solution First, we rewrite each equation in the form $Ax + By = C$.

$$\begin{array}{rcl} 5x = & 2y + 21 \\ -2y = & -2y \\ \hline 5x - 2y = & 21 \end{array} \quad \textbf{Subtract } 2y. \qquad \begin{array}{rcl} 2y = & 19 - 3x \\ +3x = & +3x \\ \hline 3x + 2y = & 19 \end{array} \quad \textbf{Add } 3x.$$

We add the equations together by adding the left side of the first equation to the left side of the second equation, and then adding the two right sides together, as follows:

$$\begin{array}{l} 5x - 2y = 21 \\ 3x + 2y = 19 \\ \hline 8x \qquad = 40 \end{array}$$

Note that the y-terms canceled, or were eliminated. We are left with an equation in x that is easy to solve:

$$\begin{aligned} 8x &= 40 \qquad \textbf{Divide both sides by 8.} \\ x &= 5 \end{aligned}$$

We are not finished yet, because we must still find the value of y. We can substitute our value for x into either of the original equations, and solve for y. We'll use the second equation, $3x + 2y = 19$:

$$\begin{aligned} 3(5) + 2y &= 19 \qquad \textbf{Subtract 15 from both sides.} \\ 2y &= 4 \qquad \textbf{Divide by 2.} \\ y &= 2 \end{aligned}$$

Thus, the solution is the point $(5, 2)$.

Look Ahead: You may have noticed that this method worked because the coefficients of y in the two equations were opposites, 2 and -2. This caused the y-terms to cancel out when we added the two equations together. What if the coefficients of neither x nor y are opposites? Then we must multiply one or both of the equations in the system by a suitable constant.

Example 4 Solve the system

$$\begin{aligned} 4x + 3y &= 7 \\ 3x + y &= -1 \end{aligned}$$

Solution We can choose to eliminate either the x-terms or the y-terms in a system. For this example, it will be faster to eliminate the y-terms. If we multiply each term of the second equation by -3, then the coefficients of y will be opposites:

$$\mathbf{-3}(3x + y = -1) \quad \rightarrow \quad -9x - 3y = 3$$

Because we are applying the Multiplication Property of Equality, we must multiply *each* term by -3, not just the y-term. We can then replace the second equation by its new version to obtain a new system, and we add the equations together:

$$\begin{array}{r} 4x + 3y = 7 \\ \underline{-9x - 3y = 3} \\ -5x \qquad = 10 \end{array}$$

The y-terms were eliminated, and we solve the resulting equation for x to get $x = -2$. Finally, we substitute $x = -2$ into either of the equations to find $y = 5$. The solution is $(-2, 5)$.

Caution! In Example 4, we multiplied both sides of the equation $3x + y = -1$ by -3. (This is an application of the Multiplication Property of Equality.) We must be careful to multiply *every term on both sides of the equation* by the same constant. Otherwise, we won't have an equivalent equation -- its solutions will not be the same.

When we add a multiple of one equation to the other we are making a **linear combination** of the equations.

The method of elimination is also called the method of linear combinations. Sometimes it is necessary to multiply *both* equations by suitable constants in order to eliminate one of the variables.

Example 5 Use linear combinations to solve the system

$$\begin{aligned} 5x - 2y &= 22 \\ 2x - 5y &= 13 \end{aligned}$$

Solution This time we choose to eliminate the x-terms. We must arrange things so that the coefficients of the x-terms are opposites, so we look for the smallest integer that both 2 and 5 divide into evenly. (This number is called the *lowest common multiple*, or LCM, of 2 and 5.) The LCM of 2 and 5 is 10. We want one of the coefficients of x to be 10, and the other to be -10. To achieve this, we multiply the first equation by 2 and the second equation by -5.

$$\begin{aligned} \mathbf{2}\,(5x - 2y = 22) \quad &\rightarrow \quad 10x - 4y = 44 \\ \mathbf{-5}\,(2x - 5y = 13) \quad &\rightarrow \quad -10x + 25y = -65 \end{aligned}$$

Adding these new equations eliminates the x-term and yields an equation in y.

$$\begin{array}{r} 10x - \ \ 4y = 44 \\ \underline{-10x + 25y = -65} \\ 21y = -21 \end{array}$$

We solve for y to find $y = -1$. Finally, we substitute $y = -1$ into the first equation and solve for x.

$$\begin{aligned} 5x - 2(-1) &= 22 \\ 5x + 2 &= 22 \\ x &= 4 \end{aligned}$$

The solution to the system is $(4, -1)$.

RQ3. What is a linear combination of expressions?
RQ4. What is the first step in the elimination method?

Here are the steps for solving a system by elimination.

To Solve a System by Elimination:

1. Write each equation in the form $Ax + By = C$.
2. Decide which variable to eliminate. Multiply each equation by an appropriate constant so that the coefficients of that variable are opposites.
3. Add the equation from Step 2 and solve for the remaining variable.
4. Substitute the value found in Step 3 into one of the original equations and solve for the other variable.

Look Ahead: How do you know which method to use to solve a system, substitution or elimination? Both methods will work on any linear system. However, substitution will be easier if one of the variables in one of the equations has a coefficient of 1 or -1. Otherwise, the elimination method is usually more efficient.

RQ5. How do we eliminate a variable from a system?
RQ6. When is the substitution method easier than elimination?

Inconsistent and Dependent Systems

Recall that a system of two parallel lines has no solution and is called inconsistent. If the two equations in a system have the same graph, then every point on the graph is a solution and the system is called dependent. The elimination method will reveal whether the system falls into one of these two cases.

Example 6 Solve each system by elimination.

a. $3x - y = 2$
$-6x + 2y = 3$

b. $x - 2y = 3$
$2x - 4y = 6$

Solutions a. To eliminate the y-terms, we multiply the first equation by 2 and add:

$$\begin{array}{r} 6x - 2y = 4 \\ -6x + 2y = 3 \\ \hline 0x + 0y = 7 \end{array}$$

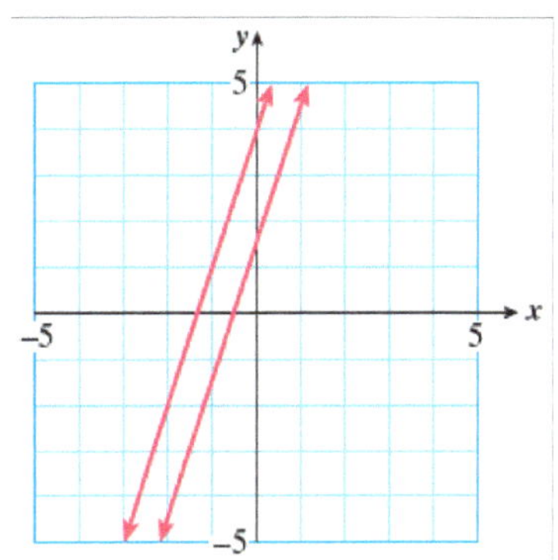

Both variables are eliminated, and we are left with the false statement $0 = 7$. There are no values of x or y that will make this equation true, so the system has no solutions. The graph shows that the system is inconsistent.

b. To eliminate the x-terms we multiply the first equation by -2 and add:

$$\begin{aligned} -2x + 4y &= -6 \\ 2x - 4y &= \ \ 6 \\ \hline 0x + 0y &= \ \ 0 \end{aligned}$$

We are left with the true but unhelpful equation $0 = 0$. The two equations are in fact equivalent (one is a constant multiple of the other), so the system is dependent. The graph of both equations is shown in the figure.

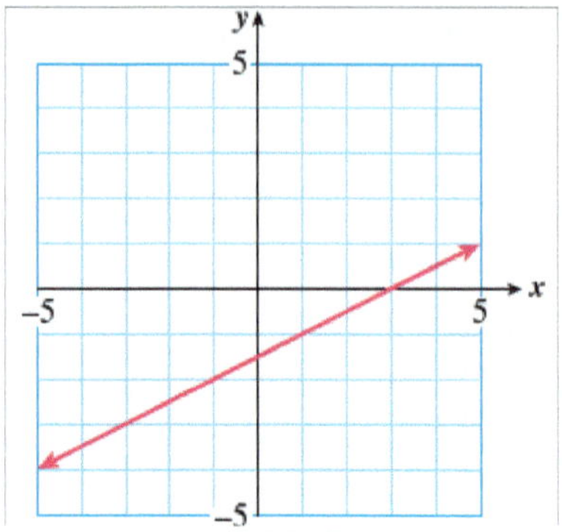

Example 5 illustrates a rule for identifying inconsistent and dependent systems.

When Using Elimination to Solve a System:

1. If combining the two equations results in an equation of the form

$$0x + 0y = k \quad (k \neq 0)$$

then the system is inconsistent.

2. If combining the two equations results in an equation of the form

$$0x + 0y = 0$$

then the system is dependent.

RQ7. How can you tell if a system is inconsistent?
RQ8. How can you tell if a system is dependent?

Skills Warm-Up

Simplify each expression. Write your answers in the form $ax + b$.

1. $2x + 3(4x - 2) + 1$ **2.** $7x - 5(2x + 3) - 2$ **3.** $\frac{3}{2}(4x - 6) - 3x$

4. $\frac{2}{3}(6x + 3) + x$ **5.** $\frac{2x + 5}{3} - 1$ **6.** $\frac{3x - 1}{4} + 2$

7. $x + 4 - \frac{3 - 5x}{2}$ **8.** $8 - x - \frac{4 + 2x}{3}$

Answers: **1.** $14x - 5$ **2.** $-3x - 17$ **3.** $3x - 9$ **4.** $5x + 2$ **5.** $\frac{2}{3}x + \frac{2}{3}$ **6.** $\frac{3}{4}x + \frac{7}{4}$ **7.** $\frac{7}{2}x + \frac{5}{2}$ **8.** $\frac{-5}{3}x + \frac{20}{3}$

Homework 4.3

Skills Practice

In Problems 1-6, solve the system by the substitution method.

1. $y = 2x$
   $3x + y = 10$
2. $2a + 3b = 4$
   $b = 2a + 4$
3. $2x - 3y = 4$
   $x + 1 = 3y$
4. $2a + b = 16$
   $4a - 4 = -3b$
5. $8r + s = 4$
   $2r + 7s = -8$
6. $36a = 6 - 3b$
   $2b = 3a - 5$

In Problems 7-12, solve the system by elimination.

7. $x + y = 5$
   $x - y = 1$
8. $2a + b = 3$
   $a + b = 2$
9. $3x + 2y = 7$
   $x + y = 3$
10. $3a = 1 + 5b$
    $2b = 14 + 6a$
11. $2x + 3y = -1$
    $3x + 5y = -2$
12. $5z = 1 - 3w$
    $5w = 2 - 7z$

In Problems 13-15, decide whether the system is inconsistent, dependent, or consistent.

13. $3x - 6y = 6$
    $x - 2y = 3$
14. $8a = 6 + 12b$
    $4 = 6a - 9b$
15. $3p = 1 + 7q$
    $21q = 9p - 3$

In Problems 16-17, which method, elimination or substitution, would be easier to apply to each system?

16. $2u + 3v = -8$
    $2u - 3v = -6$
17. $17a - b = 0$
    $11a + 3b = 4$

Applications

For Problems 18-21,

a. Choose variables for the two unknown quantities in the problem.
b. Write a system of equations to model the problem.
c. Solve the system and answer the question.

18. A hamburger and a chocolate shake together contain 1030 calories. Two shakes and three hamburgers contain 2710 calories. How many calories are there in one hamburger and in one chocolate shake?

19. Darryl is in charge of buying furniture for a new restaurant. Chairs cost \$175 and tables cost \$250. Darryl's budget allows \$11,400 for tables and chairs, and he plans to buy four chairs for each table. How many of each can he buy?

20. The perimeter of a rectangle is 42 meters, and its width is 13 meters less than its length. Find the dimensions of the rectangle.

21. Three pounds of bacon and 2 pounds of coffee costs \$17.80. Two pounds of bacon and 5 pounds of coffee costs \$32.40. How much do 1 pound of bacon and 1 pound of coffee cost?

4.4 Problem Solving with Systems

In this section we explore some applications of systems of equations. We begin by reviewing some familiar formulas.

Interest

Recall the formula for calculating interest,

$$I = Prt$$

I is the interest you will earn after t years if you invest a principal of P dollars in an account that earns simple interest at an annual rate r. If we make two or more investments at the same time, then the total interest we earn is the sum of the interests earned on each investment separately.

Caution! Notice the difference between interest, I, and interest rate, r. The **interest rate** is a *percentage*, such as 5%. The **interest** is the *amount* of money you earn, usually in dollars.

Warm-Up 1 Harvey deposited $1200 in two accounts. He put $700 in a savings account that pays 6% annual interest rate, and the rest in his credit union, which pays 7% annual interest rate. How much will Harvey's investments earn in two years?

Step 1 Calculate the interest earned by each account.

Savings: $I = Prt =$

Credit union: $I = Prt =$

Step 2 Add the earnings from the two investments.

$$\begin{pmatrix}\text{Total}\\\text{interest}\end{pmatrix} = \begin{pmatrix}\text{Interest from}\\\text{stocks}\end{pmatrix} + \begin{pmatrix}\text{Interest from}\\\text{bonds}\end{pmatrix}$$

Here is a similar example with variables.

Warm-Up 2 You have $5000 to invest for one year. You want to put part of the money into bonds that pay 7% interest rate and the rest of the money into stocks, which involve some risk but will pay 12% if the investment is successful.

a. Fill in the table.

Amount invested in stocks	Amount invested in bonds	Interest from stocks	Interest from bonds	Total interest
$500				
$1000				
$3200				
$4000				

Now suppose you invest x dollars in the stocks and y dollars in the bonds.

b. Sum of amounts invested: $x + y =$

c. Interest earned on the stocks: $I = Prt =$

Interest earned on the bonds: $I = Prt =$

d. Finally, suppose that you earned a total of \$345 in interest from your two investments. Write an equation about this, using your expressions from part (c).

$$\begin{pmatrix}\text{Total}\\\text{interest}\end{pmatrix} = \begin{pmatrix}\text{Interest from}\\\text{stocks}\end{pmatrix} + \begin{pmatrix}\text{Interest from}\\\text{bonds}\end{pmatrix}$$

RQ1. What formula do we need for interest problems?
RQ2. What is the difference between principal and interest?

Example 1 Mort invested money in two accounts, a savings plan that pays 8% interest and a mutual fund that pays 7% interest. He put twice as much money in the savings plan as in the mutual fund. At the end of the year Mort's total interest income was \$345. How much did he invest in each account?

Solution The unknown quantities are the amounts Mort invested in his two accounts.

Amount invested in savings: x
Amount invested in mutual fund: y

Do not confuse the amount Mort invested in each account (the principal) with the amount he earned in interest! A table is a good way to keep these amounts straight.

	Principal	Interest Rate	Interest
Savings plan	x	0.08	$0.08x$
Mutual fund	y	0.07	$0.07y$

We must write two equations for the problem, one about the principal and one about the interest. From the statement "He put twice as much money in the savings plan as in the mutual fund," we write the equation

$$\begin{pmatrix}\text{Amount invested}\\\text{in savings plan}\end{pmatrix} = 2\cdot\begin{pmatrix}\text{Amount invested}\\\text{in mutual fund}\end{pmatrix}$$

Equation about principal: $x = 2y$

From the statement "Mort's total interest income was \$345," we write the equation

$$\begin{pmatrix}\text{Interest from}\\\text{savings plan}\end{pmatrix} + \begin{pmatrix}\text{Interest from}\\\text{mutual fund}\end{pmatrix} = \text{Total interest}$$

Equation about interest: $0.08x + 0.07y = 345$

This gives us a system of equations.

$$\begin{aligned} x &= 2y \\ 0.08x + 0.07y &= 345 \end{aligned}$$

Because the first equation is already solved for x in terms of y, we'll use the substitution method to solve the system. We substitute $2y$ for x in the second equation and solve for y.

$$\begin{aligned} 0.08(2y) + 0.07y &= 345 \\ 0.16y + 0.07y &= 345 && \text{Combine like terms.} \\ 0.23y &= 345 && \text{Divide both sides by 0.23.} \\ y &= 1500 \end{aligned}$$

Finally, we substitute $y = 1500$ into the first equation to find

$$x = 2y = 2(1500) = 3000$$

Thus, Mort invested \$3000 in the savings plan and \$1500 in the mutual fund.

RQ3. What were the two equation in Example 1 about?
RQ4. Which method did we use to solve the system in Example 1?

Mixtures

A pharmacist has on hand 20 ounces of a certain drug at 40% strength, but she needs a small quantity of the drug at 75% strength for a prescription. She decides to add a pure form of the drug to the 40% solution. How much should she add to make a mixture of 75% strength?

To solve such problems, we review some properties of percent. We will need the percent formula,

$$P = rW$$

P stands for the part obtained when we take r percent of a whole amount, W.

Warm-Up 3 You have two jars of marbles. The first contains 40 marbles, of which 10 are red, and the second contains 60 marbles, of which 30 are red.

a. What percent of the marbles in the first jar are red? $r = \frac{P}{W} =$

What percent of the marbles in the second jar are red? $r = \frac{P}{W} =$

You pour both jars of marbles into a larger jar and mix them together.

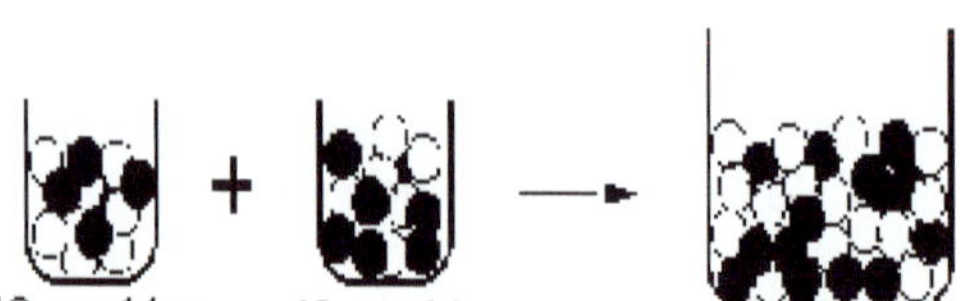

40 marbles, 10 red   60 marbles, 30 red

b. How many marbles total are in the larger jar?

How many red marbles are in the larger jar?

What percent of marbles in the larger jar are red? $r = \frac{P}{W} =$

c. Can we add the percents for the first two jars to get the percent red marbles in the mixture?

Caution! In Warm-Up 3, you should find that adding the percent of red marbles in the first two jars does **not** give the percent of red marbles in the mixture. That is,

$$25\% + 50\% \neq 40\%$$

In general, **we cannot add percents** unless they are percents of the same whole amount.

RQ5. What formula do we need for mixture problems?
RQ6. When can we add percents?

Warm-Up 4 In a local city council election, your favored candidate, Justine Honest, ran in a small district with two precincts. Ms. Honest won 30% of the 500 votes cast in Precinct 1 and 70% of the 300 votes cast in Precinct 2. Did Candidate Honest win a majority (more than 50%) of the votes in her district?

Step 1 Fill in the first two rows of the table, using information from the problem and the formula $P = rW$.

	Total votes (W)	Percent for Honest (r)	Votes for Honest (P)
Precinct 1			
Precinct 2			
Entire district			

Step 2 Add down to complete the first and third columns of the table.

Step 3 Fill in the last entry in the table to answer the question in the problem. Use the formula $P = rW$ again.

So far we have considered mixture problems involving discrete objects, such as marbles or votes. The same methods apply to mixtures of liquids.

Example 2 A chemist wants to produce 45 milliliters of a 40% solution of carbolic acid by mixing a 20% solution with a 50% solution. How many milliliters of each should he use?

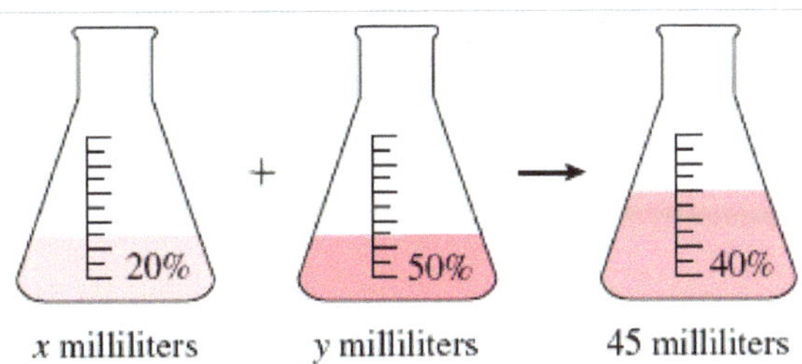

Solution We let x represent the number of milliliters of the 20% solution he needs and y the number of milliliters of the 50% solution. We use a table to organize the information. The first two columns contain the variables and information given in the problem: the number of milliliters of each solution and its strength as a percent.

	Number of Milliliters (W)	Percent Acid (r)	Amount of Acid (P)
20% Solution	x	0.20	
50% Solution	y	0.50	
Mixture	45	0.40	

We fill in the last column of the table by using the formula $P = rW$. The entries in this last column give the amount of the important ingredient (in this case, milliliters of acid) in each component solution and in the mixture.

	Number of Milliliters (W)	Percent Acid (r)	Amount of Acid (P)
20% Solution	x	0.20	$\mathbf{0.20x}$
50% Solution	y	0.50	$\mathbf{0.50x}$
Mixture	45	0.40	$\mathbf{0.40(45)}$

Now we can write two equations about the mixture problem. The first equation is about the total number of milliliters mixed together. The chemist must mix x milliliters of one solution with y milliliters of the other solution and end up with 45 milliliters of the mixture, so

Total amount of mixture: $x + y = 45$

The second equation uses the fact that the acid in the mixture can only come from the acid in each of the two original solutions. We used the last column of the table to calculate how much acid was in each component, and we add these quantities to get the amount of acid in the mixture.

Amount of acid: $0.20x + 0.50y = 0.40(45)$

These two equation make up a system:

$$\begin{aligned} x + y &= 45 \\ 0.20x + 0.50y &= 18 \end{aligned}$$

To simplify the system we first multiply the second equation by 100 to clear the decimals.

$$\begin{aligned} x + y &= 45 \\ 20x + 50y &= 1800 \end{aligned}$$

We solve the system by elimination. Multiply the first equation by -20, and add the equations together.

$$\begin{aligned} -20x - 20y &= -900 \\ 20x + 50y &= 1800 \\ \hline 30y &= 900 \end{aligned}$$

Solving for y, we find $y = 30$. Substitute $y = \mathbf{30}$ into the first equation to find

$$x + \mathbf{30} = 45$$

or $x = 15$. The chemist needs 15 milliliters of the 20% solution and 30 milliliters of the 50% solution for the mixture.

Look Closer: Notice that, in Example 2, once we have completed the table, it is easy to write a system of equations; we simply add down the first and third columns of the table. (Remember that we cannot add down the middle column, because percents don't add!)

RQ7. What were the two equations in Example 2 about?
RQ8. Which method did we use to solve the system in Example 2?

Mixture problems are easy to solve if you complete a table first. The rows of the table represent the two components and the final mixture, and the columns are used to calculate the amount of the important ingredient in each. Here is a sample table that you can customize for the specifics of a particular problem.

	Total Amount (W)	Percent of Important Ingredient (r)	Amount of Important Ingredient (P)
First Component			
Second Component			
Mixture			

Motion

We can also use systems of equations to solve problems involving motion at a constant speed. We'll need the formula $D = RT$.

Example 3 Geologists can calculate the distance from their seismograph to the epicenter of an earthquake by timing the arrival of the P and S waves. They know that P waves travel at about 5.4 miles per second and S waves travel at 3.0 miles per second. If the P waves arrived 3 minutes before the S waves, how far away is the epicenter of the quake?

Solution Let x represent the distance from the seismograph to the epicenter. The time it took for the waves to arrive is also unknown, so we'll let y be the travel time for the P waves, in seconds. The travel time for the S waves is then $y + 180$ seconds.

We organize all this information in a table.

	Rate	Time	Distance
P waves	5.4	y	x
S waves	3.0	$y+180$	x

We can now write two equations about the problem, one for the P waves and one for the S waves, using the formula $RT = D$.

$$5.4y = x$$
$$3.0(y+180) = x$$

We solve the system by substitution. Substitute $5.4y$ for x in the second equation, and then solve for y.

$$\begin{aligned} 3.0(y+180) &= 5.4y && \text{Apply the distributive law.}\\ 3y + 540 &= 5.4y && \text{Subtract } 3y \text{ from both sides.}\\ 540 &= 2.4y && \text{Divide both sides by 2.4.}\\ 225 &= y \end{aligned}$$

Thus, it took the P waves 225 seconds to arrive at the seismograph. To solve for the distance x, substitute $y = 225$ into the first equation to find

$$x = 5.4(225) = 1215$$

The epicenter is located 1215 miles from the seismograph.

RQ9. What formula do we need for motion problems?
RQ10. What did x and y represent in Example 3?

Warm-up Answers: **1.** $\$154$ **2b.** $\$5000$ **c.** $0.07x,\ 0.12y$ **d.** $345 = 0.07x + 0.12y$
3a. $25\%,\ 50\%$ **b.** $100,\ 40,\ 40\%$ **c.** No **4.** No, 45%

Skills Warm-Up

For part (b) of each problem, write an algebraic expression in two variables.

1. a. How much interest will you earn in 1 year if you invest \$2400 in a T-bill that pays $7\frac{1}{2}$% interest and \$800 in a savings account that pays 4.8% interest?
   b. How much will you earn if you invest x dollars in the T-bill and y dollars in the savings account?
2. a. How far will you travel if you jog for 40 minutes at 9 miles per hour, and then jog for 30 minutes at 5 miles per hour?
   b. How far will you travel if you jog for 40 minutes at x miles per hour, and the jog for 30 minutes at y miles per hour?
3. a. How long will it take you to drive 180 miles on the highway at an average speed of 60 miles per hour, and then 30 miles on a gravel road at an average speed of 40 miles per hour?
   b. How long will it take you if you drive on the highway at x miles per hour and on the gravel road at y miles per hour?
4. a. How much nitrogen is in a mixture of 10 pounds of fertilizer that is 6% nitrogen and 4 pounds of fertilizer that is 60% nitrogen?
   b. How much nitrogen is in a mixture of x pounds of the first fertilizer and y pounds of the second fertilizer?

Answers: **1a.** \$218.40 **b.** $0.075x + 0.048y$ **2a.** 8.5 miles **b.** $\frac{2}{3}x + \frac{1}{2}y$
3a. 3.75 hours **b.** $\frac{180}{x} + \frac{30}{y}$ **4a.** 3 lbs. **b.** $0.06x + 0.60y$

Homework 4.4

Applications

Use systems to solve Problems 1-3 about interest.

1. Goodlife Insurance Company has \$150,000 in fees from its clients to invest. They deposited part of the money into bonds that pay 6.5% annual interest and the rest into a mutual fund that pays 11.8% annual interest.
   - **a.** Assign variables to the amount of money Goodlife deposited into each account, and write an equation about the sum of the deposits.
   - **b.** Write expressions for the interest earned on each account after 1 year.
   - **c.** Goodlife earned \$12,930 interest in 1 year. Write an equation about that.
   - **d.** Solve your system of equations to find out how much Goodlife invested in each account.

2. Mario borrowed \$30,000 from two banks to open a print shop. The first loan charges 12% annual interest, and the second charges 15% interest. Mario's annual interest payment on both loans together is \$3750. How much did he borrow at each rate?
   - **a.** Assign variables to the unknown quantities and make a table like the one in Example 1.
   - **b.** Write two equations for the problem, one about the principals and one about the interests.
   - **c.** Solve your system and answer the question in the problem.

3. Stefan borrowed twice as much on his 7% car loan as on his 4% student loan. The annual interest on the car loan is \$500 more than the interest on the student loan. How much did Stefan borrow on each loan?

Solve Problems 4-5 about mixtures. (No variables are needed!)

4. The chemistry department has 80 students, of whom 35% are women. The physics department has 60 students, of whom 15% are women.
   - **a.** How many chemistry students are women? How many physics students are women?
   - **b.** How many students are there in both departments? How many of them are women?
   - **c.** What percent of the students in chemistry and physics are women?

	Number of Students (W)	Percent Women (r)	Number of Women (P)
Chemistry			
Physics			
Total			

5. Pipette, a French chemistry student, has 30 milliliters of a 50% solution of acid. She wants to reduce the strength by adding 12 milliliters of a 15% solution of the same acid.
   a. How much acid is in the 30 milliliters of 50% solution? How much acid is in the 12 milliliters of 15% solution?
   b. How much acid is in the mixture? How many milliliters of the mixture are there?
   c. What percent of the mixture is acid?
   d. Fill in the table with your answers to parts (a)-(c).

	Number of Milliliters (W)	Percent Acid (r)	Amount of Acid (P)
50% Solution			
15% Solution			
Mixture			

Use a system of equations to solve Problems 6-7 about mixtures.

6. A pet store owner wants to mix a 12% saltwater solution and a 30% saltwater solution to obtain 45 liters of a 24% solution. How many liters of each ingredient does he need?
   a. Choose variables and make a table for the problem.
   b. Use your table to write two equations about the mixture.
   c. Solve your system and answer the question in the problem.

7. A newspaper poll of 400 people stated that 58% were in favor of a recycling program. It also said that 50% of the men and 70% of the women polled favored the program. How many women were polled?

Solve Problems 8-9 about motion.

8. Delbert and Francine leave Cedar Rapids at the same time and drive in opposite directions for 6 hours.
   a. Choose variables for Delbert and Francine's speeds and fill in the table.

	Rate	Time	Distance
Delbert			
Francine			

   b. Make a sketch showing Cedar Rapids, Delbert, and Francine. Label your sketch with the distance that each traveled.
   c. Francine drove 5 miles per hour slower than Delbert. After 6 hours, they are 570 miles apart. Write two equations about the problem.
   d. Solve your system to find Delbert's and Francine's speeds.

9. Bonnie left Dallas and drove north at 40 miles per hour. Three hours later Clyde headed north form Dallas on the same road at 70 miles per hour until he caught up with Bonnie.

a. Make a sketch showing Dallas, Bonnie, and Clyde.

b. Complete the table.

	Rate	Time	Distance
Bonnie		t	d
Clyde			

c. Use your table to write two equations about the problem.

d. Solve your system. How long did Bonnie drive before Clyde caught up? How far had she driven?

Use a system of equations to solve Problems 10-11.

10. A yacht leaves San Diego and heads south, traveling at 25 miles per hour. Six hours later a Coast Guard cutter leaves San Diego traveling at 40 miles per hour and pursues the yacht. How long will it take the cutter to catch the yacht? How far will they have traveled?

11. Byron and Ada conduct sight-seeing tours of New England by bicycle. Byron leads the tour group at an average speed of 10 miles per hour, while Ada goes ahead at a speed of 12 miles per hour to prepare lunch. Ada arrives at the lunch stop 40 minutes ($\frac{2}{3}$ of an hour) before the tour group. How far did the tour bicycle that morning?

Solve by writing a single equation in two variables.

12. In Julio's history class, the final grade is computed by adding 60% of the test average to 40% of the term paper grade. Julio's test average is 72.

a. Write an equation for Julio's final grade, g, in terms of his term paper grade, t.

b. What will Julio's final grade be if he earns a grade of 65 on the term paper? What if he earns a grade of 80?

c. Use the graph at right to estimate what grade Julio must make on the term paper in order to earn afinal grade of 80 in the class.

d. Use your equation from part (a) to verify your answer algebraically.

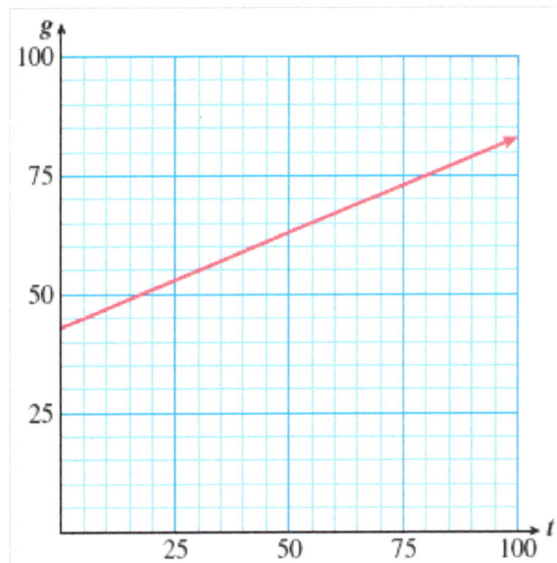

Chapter 4 Summary and Review

Section 4.1 The Distributive Law

Distributive Law

If a, b, and c are any numbers, then

$$a(b+c) = ab + ac$$

Steps for Solving Linear Equations

1. Use the distributive law to remove any parentheses.
2. Combine like terms on each side of the equation.
3. By adding or subtracting the same quantity on both sides of the equation, get all the variable terms on one side and all the constant terms on the other.
4. Divide both sides by the coefficient of the variable to obtain an equation of the form $x = a$.

Section 4.2 Systems of Linear Equations

- A pair of linear equations in two variables

$$a_1x + b_1y = c_1$$
$$a_2x + b_2y = c_2$$

considered together is called a **system of linear equations**, or a **linear system**.

- A **solution** to a system of linear equations is an ordered pair (x, y) that satisfies each equation in the system. It is the intersection point of the two lines described by the system.

Linear Systems

There are three types of linear systems:

1. **Consistent and independent system.** The graphs of the two lines intersect in exactly one point. The system has exactly one solution.
2. **Inconsistent system**. The graphs of the equations are parallel lines and hence do not intersect. An inconsistent system has no solutions.
3. **Dependent system**. All the solutions of one equation are also solutions to the second equation, and hence are solutions of the system. The graphs of the two equations are the same line. A dependent system has infinitely many solutions.

Section 4.3 Algebraic Solution of Systems

To Solve a System by Substitution:

1. Choose one of the variables in one of the equations. (It is best to choose a variable whose coefficient is 1 or -1.) Solve the equation for that variable.
2. Substitute the result of Step 1 into the *other* equation. This gives an equation in one variable.
3. Solve the equation obtained in Step 2. This gives the solution value for one of the variables.

4. Substitute this value into the result of Step 1 to find the solution value of the other variable.

To Solve a System by Elimination:

1. Write each equation in the form $Ax + By = C$.
2. Decide which variable to eliminate. Multiply each equation by an appropriate constant so that the coefficients of that variable are opposites.
3. Add the equation from Step 2 and solve for the remaining variable.
4. Substitute the value found in Step 3 into one of the original equations and solve for the other variable.

When Using Elimination to Solve a System:

1. If combining the two equations results in an equation of the form
$$0x + 0y = k \quad (k \neq 0)$$
then the system is inconsistent.
2. If combining the two equations results in an equation of the form
$$0x + 0y = 0$$
then the system is dependent.

Section 4.4 Applications of Systems

- We can use tables to organize the information in applications involving interest, mixtures, or motion.
- The formula for calculating interest is

$$I = Prt$$

I is the interest you will earn after t years if you invest a principal of P dollars in an account that earns simple interest at an annual rate r.

- The formula for calculating percents is

$$P = rW$$

P stands for the part obtained when we take r percent of a whole amount, W.

- In general, **we cannot add percents** unless they are percents of the same whole amount.
- To solve problems involving motion at a constant speed we use the formula

$$D = RT$$

D stands for the distance you travel at speed R in time T.

Section 4.5 Point-Slope Form

Point-slope Formula

The equation for the line that passes through the point (x_1, y_1) and has slope m is

$$y = y_1 + m(x - x_1)$$

or

$$y - y_1 = m(x - x_1) \quad \text{or} \quad \frac{y - y_1}{x - x_1} = m$$

To Fit a Line through Two Points :
1. Compute the slope between the two points.
2. Substitute the slope and either point into the point-slope formula.

Review Questions

Use complete sentences to answer the questions in Problems 1-10.

1. State the two-point formula for slope.
2. How can you find the equation of a line when you know the slope of the line and one point on the line?
3. How can you find the solution to a linear system by graphing?
4. Name two algebraic methods for solving linear systems.
5. Suppose you are using the elimination method to solve a system. How can you tell if the system is dependent or inconsistent?
6. How would you label the columns when making a table for a problem about interest?
7. How would you label the columns when making a table for a problem about a mixture?
8. How would you label the columns when making a table for a problem about motion?
9. Under what conditions can we add percents?
10. Explain why the distributive law does not apply to the expression $-3(2ab)$.

Review Problems

For Problems 1-2, simplify. Which product requires the distributive law?

1. **a.** $-5(-6m)$
   b. $-5(6-m)$
2. **a.** $9(-3-w)$
   b. $9(-3w)$

For Problems3-6, simplify.

3. $(4m+2n)-(2m-5n)$
4. $(-5c-6)+(-11c+15)$
5. $-7w-2(4w-13)$
6. $4(3z-10)+5(-z-6)$

For Problems 7-10, solve.

7. $5p+10(17-p)=2p-5$
8. $-3(k-2)-4(2k+5)=10+3k$
9. $4(3a-7)<-18+2a$
10. $\dfrac{2x-1}{x+3}=\dfrac{3}{2}$

11. The length of a rectangle is 3 times its width.
    a. If the width of a rectangle is x, what is its length?
    b. Express the perimeter of the rectangle in terms of x.
    c. Suppose the perimeter of the rectangle is 48 centimeters. Find the dimensions of the rectangle.

12. Last Saturday, a total of 620 people attended the Gaslamp Theater at its two performances.

a. If p people attended the matinee last Saturday, how many attended the evening performance?

b. This Saturday, attendance at the matinee increased by 5%, and attendance at the evening performance increased by 8%. Write expressions in terms of p for the attendance at each performance.

c. This Saturday, the total attendance was 663 people. How many people attended the matinee last Saturday?

13. Wheaton Elementary school plans to buy 30 computers. The computers with speakers cost \$1200 each, and those without speakers cost \$800 each. Let x represent the number of computers with speakers. Write expressions in terms of x for:

a. The number of computers without speakers.

b. The total cost of the computers with speakers.

c. The total cost of the computers without speakers.

d. The total cost of all 30 computers.

e. If Wheaton Elementary has \$28,800 to spend on the computers, how many of each kind can they buy?

14. The current in the Lazy River flows at 4 miles per hour.

a. If your motorboat is traveling at speed v miles per hour, write expressions for your speed traveling upstream and your speed traveling downstream.

b. You travel upstream for $\frac{3}{2}$ hours and stop for lunch at an island. Write an expression in terms of v for the distance you traveled upstream.

c. You return to your starting point downstream in $\frac{1}{2}$ hour. Write an expression in terms of v for the distance you traveled downstream.

d. Write an equation and solve it to find the speed, v, of your motorboat.

For Problems 15-16, decide whether the given point is a solution of the system.

15. $x + y = 8$, $x - y = 2$ $\quad (-2, 10)$

16. $8x + 3y = 21$, $5x = y - 16$ $\quad (-3, 1)$

For Problems 17-18, solve the system by graphing.

17. $x + y = 5$, $2x - y = 4$

18. $x - y = 7$, $y = \frac{-2}{3}x - 2$

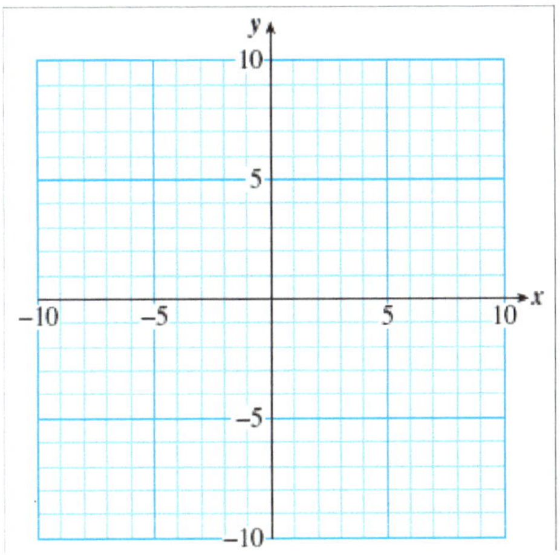

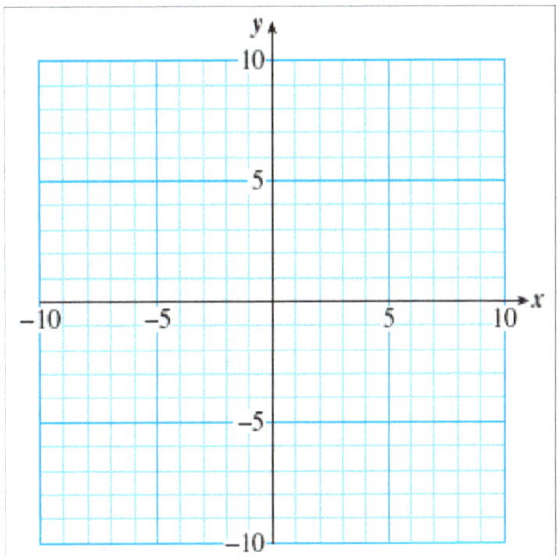

For Problems 19-20, solve by substitution.

19. $y = 2x + 1$
$2x + 3y = -21$

20. $x + 4y = 1$
$2x + 3y = -3$

For Problems 21-22, solve by elimination.

21. $2x + 7y = -19$
$5x - 3y = 14$

22. $4x + 3y = -19$
$5x + 15 = -2y$

For Problems 23-26, solve using substitution or elimination.

23. $x + 5y = 18$
$x - y = -3$

24. $x + 5y = 11$
$2x + 3y = 8$

25. $\frac{2}{3}x - 3y = 8$
$x + \frac{3}{4}y = 12$

26. $3x = 5y - 6$
$3y = 10 - 11x$

For Problems 27-30, decide whether each system is inconsistent, dependent, or consistent and independent.

27. $2x - 3y = 4$
$x + 2y = 7$

28. $2x - 3y = 4$
$6x - 9y = 4$

29. $2x - 3y = 4$
$6x - 9y = 12$

30. $x - y = 6$
$x + y = 6$

For Problems 30-42, solve by writing and solving a system of equations.

31. A health food store wants to produce 30 pounds of granola worth 80 cents per pound. They plan to mix cereal worth 65 cents per pound with dried fruit worth 90 cents per pound. How much of each should they use?

32. The perimeter of a rectangle is 50 yards and its length is 9 yards greater than its width. Find the dimensions of the rectangle.

33. Last year Veronica made \$93 in interest from her savings accounts, one of which paid 6% interest, and the other paid 9%. This year her interest rates dropped to 4% and 8%, respectively, and she made \$76 interest. How much does Veronica have invested in each account?

34. Marvin invested \$300 more at 6% than he invested at 8%. His total annual income from his two investments is \$242. How much did he invest at each rate?

35. How many pounds of an alloy containing 60% copper must be melted with an alloy containing 20% copper to obtain 8 pounds of an alloy containing 30% copper?

36. Jerry Glove came to bat only 20 times in the first half of the season and got hits 15% of the time. During the second half of the season, Jerry came to bat 140 times and got hits 35% of the time.

a. How many hits did Jerry get in the first half of the season? How many hits did he get in the second half?

b. How many times did Jerry come to bat all season? How many hits did he get?
c. What percent of Jerry's at-bats resulted in hits?
d. Who had a better batting average in the first half of the season, Joe Cleat from Problem 8 in Homework 4.3, or Jerry? Who had a better batting average in the second half? Who had the better batting average overall?

37. Alida and Steve are moving to San Diego. Alida is driving their car, and Steve is driving a rental truck. They start together, but Alida drives twice as fast as Steve. After 3 hours they are 93 miles apart. How fast is each traveling?

38. Jake rides for the Pony Express, covering his route in 6 hours and returning home, 8 miles per hour slower, in 9 hours. How far does Jake ride?

39. A math contest exam has 40 questions. A contestant scores 5 points for each correct answer, but loses 2 points for each wrong answer. Lupe answered all the questions and her score was 102. How many questions did she answer correctly?

40. A game show contestant wins \$25 for each correct answer he gives but loses \$10 for each incorrect response. Roger answered 24 questions and won \$355. How many answers did he get right?

41. Barbara wants to earn \$500 a year by investing \$5000 in two accounts, a savings plan that pays 8% annual interest and a high-risk option that pays 13.5% interest. How much should she invest in each account?

42. An investment broker promises his client a 12% return on her funds. If the broker invests \$3000 in bonds paying 8% interest, how much must he invest in stocks paying 15% interest to keep his promise?

43. a. Graph the line with slope $\frac{-5}{3}$ that passes through the point $(-2, 1)$.
b. Find an equation in point-slope form for the line in part (a)

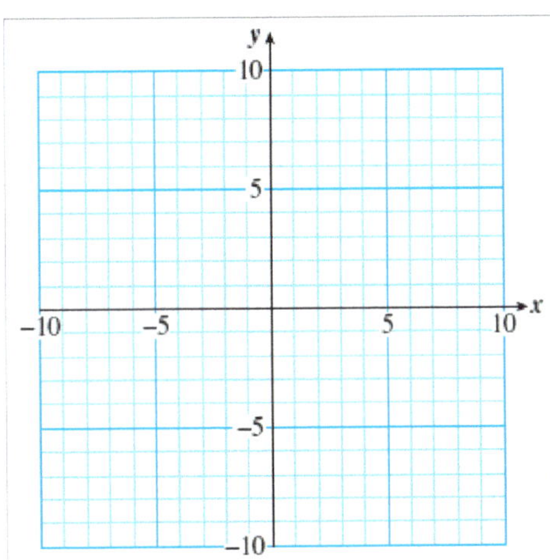

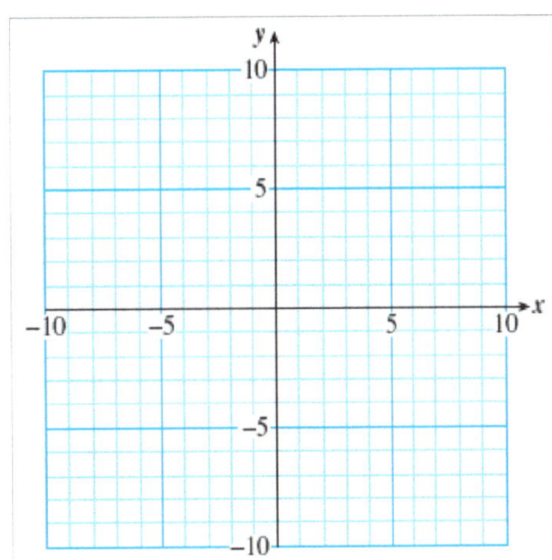

44. a. Graph the line of slope $\frac{6}{5}$ that passes through the point $(-3, -4)$.
b. Find an equation in point-slope form for the line in part (a).

For Problems 45-46, find an equation for the line passing through the two points.

45. $(3, -5), (-2, 4)$ **46.** $(0, 8), (4, -2)$

47. An interior decorator bases her fee on the cost of a remodeling job. The table below shows her fee, F, for jobs of various costs, C, both given in dollars.

C	5000	10,000	20,000	50,000
F	1000	1500	2500	5500

a. Write a linear equation for F in terms of C.
b. Give the slope of the graph, and explain the meaning of the slope in terms of the decorator's fee.

48. Auto registration fees in Connie's home state depend on the value of the automobile. The table below shows the registration fee R for a car whose value is V, both given in dollars.

V	5000	10,000	15,000	20,000
R	135	235	335	435

a. Write a linear equation for R in terms of V.
b. Give the slope of the graph, and explain the meaning of the slope in terms of the registration fee.

For Problems 49-50,
a. Make a table of values showing two data points.
b. Find a linear equation relating the variables.
c. State the slope of the line, including units, and explain its meaning in the context of the problem.

49. The population of Maple Rapids was 4800 in 1982 and had grown to 6780 by 1997. Assume that the population increases at a constant rate. Express the population P of Maple Rapids in terms of the number of years t since 1982.

50. Cicely's odometer read 112 miles when she filled up her 14-gallon gas tank and 308 when the gas gauge read half full. Express her odometer reading m in terms of the amount of gas g she used.

51. a. Write an equation for any line that is parallel to $2y = 5x - 3$.
b. Write an equation for any line that is perpendicular to $2y = 5x - 3$.

52. a. Write an equation for the vertical line that passes through $(3, -6)$
b. Write an equation for the horizontal line that passes through $(3, -6)$

53. Write an equation for the line that is parallel to the graph of $2x + 3y = 6$ and passes through the point $(1, 4)$.

54. Write an equation for the line that is perpendicular to the graph of $2x + 3y = 6$ and passes through the point $(1, 4)$.

5 Measurement

Chapter Outline

The speedometer on your car may give your speed in both miles per hour and kilometers per hour. If you are moving at 60 miles per hour, the speedometer will also read 97 kilometers per hour. The table and chart below show the relationship between these two quantities. The numbers in the second column of the table have been rounded to the nearest whole number.

Comparing Speeds

Miles Per Hour	Kilometers Per Hour
0	0
20	32
40	64
60	97
80	129
100	161

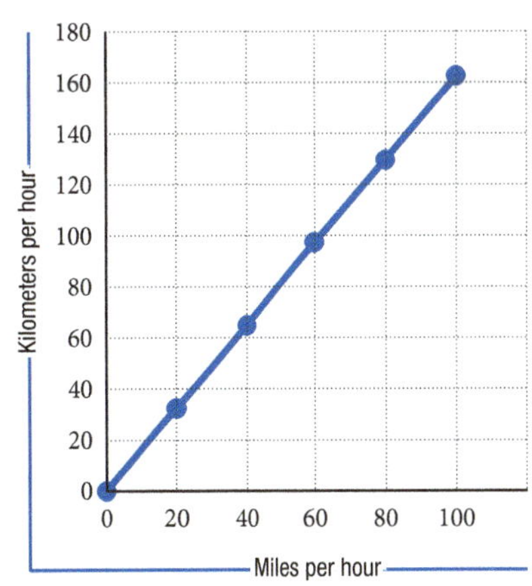

In this chapter we look at the process we use to convert from one set of units, such as miles per hour, to another set of units, such as kilometers per hour. You will be interested to know that regardless of the units in question, the method we use is the same in all cases. The method is called *unit analysis* and it is the foundation of this chapter.

Success Skills

If you have made it this far, then you have the study skills necessary to be successful in this course. Success skills are more general in nature and will help you with all your classes and ensure your success in college as well.

Let's start with a question:

Question: What quality is most important for success in any college course?

Answer: Independence. You want to become an independent learner.

We all know people like this. They are generally happy. They don't worry about getting the right instructor, or whether or not things work out every time. They have a confidence that comes from knowing that they are responsible for their success or failure in the goals they set for themselves.

Here are some of the qualities of an independent learner:

- Intends to succeed.
- Doesn't let setbacks deter them.
- Knows their resources.
  - Instructor's office hours
  - Math lab
  - Student Solutions Manual
  - Group study
  - Internet
- Doesn't mistake activity for achievement.
- Has a positive attitude.

There are other traits as well. The first step in becoming an independent learner is doing a little self-evaluation and then making of list of traits that you would like to acquire. What skills do you have that align with those of an independent learner? What attributes do you have that keep you from being an independent learner? What qualities would you like to obtain that you don't have now?

Unit Analysis I: Length

5.1

Introduction . . .

In this section we will become more familiar with the units used to measure length. We will look at the U.S. system of measurement and the metric system of measurement.

Measuring the length of an object is done by assigning a number to its length. To let other people know what that number represents, we include with it a unit of measure. The most common units used to represent length in the U.S. system are inches, feet, yards, and miles. The basic unit of length is the foot. The other units are defined in terms of feet, as Table 1 shows.

TABLE 1

12 inches (in.)	=	1 foot (ft)
1 yard (yd)	=	3 feet
1 mile (mi)	=	5,280 feet

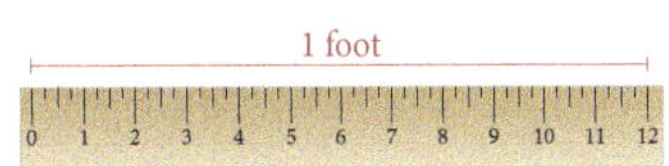

As you can see from the table, the abbreviations for inches, feet, yards, and miles are in., ft, yd, and mi, respectively. What we haven't indicated, even though you may not have realized it, is what 1 foot represents. We have defined all our units associated with length in terms of feet, but we haven't said what a foot is.

There is a long history of the evolution of what is now called a foot. At different times in the past, a foot has represented different arbitrary lengths. Currently, a foot is defined to be exactly 0.3048 meter (the basic measure of length in the metric system), where a meter is 1,650,763.73 wavelengths of the orange-red line in the spectrum of krypton-86 in a vacuum (this doesn't mean much to me either). The reason a foot and a meter are defined this way is that we always want them to measure the same length. Because the wavelength of the orange-red line in the spectrum of krypton-86 will always remain the same, so will the length that a foot represents.

Now that we have said what we mean by 1 foot (even though we may not understand the technical definition), we can go on and look at some examples that involve converting from one kind of unit to another.

EXAMPLE 1 Convert 5 feet to inches.

SOLUTION Because 1 foot = 12 inches, we can multiply 5 by 12 inches to get

$$\begin{aligned} 5 \text{ feet} &= 5 \times 12 \text{ inches} \\ &= 60 \text{ inches} \end{aligned}$$

This method of converting from feet to inches probably seems fairly simple. But as we go further in this chapter, the conversions from one kind of unit to another will become more complicated. For these more complicated problems, we need another way to show conversions so that we can be certain to end them with the correct unit of measure. For example, since 1 ft = 12 in., we can say that there are 12 in. per 1 ft or 1 ft per 12 in. That is:

$$\frac{12 \text{ in.}}{1 \text{ ft}} \longleftarrow \text{Per} \qquad \text{or} \qquad \frac{1 \text{ ft}}{12 \text{ in.}} \longleftarrow \text{Per}$$

We call the expressions $\frac{12 \text{ in.}}{1 \text{ ft}}$ and $\frac{1 \text{ ft}}{12 \text{ in.}}$ **conversion factors**. The fraction bar is

read as "per." Both these conversion factors are really just the number 1. That is:

$$\frac{12 \text{ in.}}{1 \text{ ft}} = \frac{12 \text{ in.}}{12 \text{ in.}} = 1$$

We already know that multiplying a number by 1 leaves the number unchanged. So, to convert from one unit to the other, we can multiply by one of the conversion factors without changing value. Both the conversion factors above say the same thing about the units feet and inches. They both indicate that there are 12 inches in every foot. The one we choose to multiply by depends on what units we are starting with and what units we want to end up with. If we start with feet and we want to end up with inches, we multiply by the conversion factor

$$\frac{12 \text{ in.}}{1 \text{ ft}}$$

The units of feet will divide out and leave us with inches.

$$\begin{aligned} 5 \text{ feet} &= 5 \text{ ft} \times \frac{12 \text{ in.}}{1 \text{ ft}} \\ &= 5 \times 12 \text{ in.} \\ &= 60 \text{ in.} \end{aligned}$$

The key to this method of conversion lies in setting the problem up so that the correct units divide out to simplify the expression. We are treating units such as feet in the same way we treated factors when reducing fractions. If a factor is common to the numerator and the denominator, we can divide it out and simplify the fraction. The same idea holds for units such as feet.

We can rewrite Table 1 so that it shows the conversion factors associated with units of length, as shown in Table 2.

Note

We will use this method of converting from one kind of unit to another throughout the rest of this chapter. You should practice using it until you are comfortable with it and can use it correctly. However, it is not the only method of converting units. You may see shortcuts that will allow you to get results more quickly. Use shortcuts if you wish, so long as you can consistently get correct answers and are not using your shortcuts because you don't understand our method of conversion. Use the method of conversion as given here until you are good at it; then use shortcuts if you want to.

TABLE 2

UNITS OF LENGTH IN THE U.S. SYSTEM

The Relationship Between	Is	To Convert From One To The Other, Multiply By
feet and inches	12 in. = 1 ft	$\frac{12 \text{ in.}}{1 \text{ ft}}$ or $\frac{1 \text{ ft}}{12 \text{ in.}}$
feet and yards	1 yd = 3 ft	$\frac{3 \text{ ft}}{1 \text{ yd}}$ or $\frac{1 \text{ yd}}{3 \text{ ft}}$
feet and miles	1 mi = 5,280 ft	$\frac{5{,}280 \text{ ft}}{1 \text{ mi}}$ or $\frac{1 \text{ mi}}{5{,}280 \text{ ft}}$

EXAMPLE 2 The most common ceiling height in houses is 8 feet. How many yards is this?

SOLUTION To convert 8 feet to yards, we multiply by the conversion factor $\frac{1\text{ yd}}{3\text{ ft}}$ so that feet will divide out and we will be left with yards.

$$8\text{ ft} = 8\text{ ft} \times \frac{1\text{ yd}}{3\text{ ft}} \quad \text{Multiply by correct conversion factor}$$

$$= \frac{8}{3}\text{ yd} \quad 8 \times \frac{1}{3} = \frac{8}{3}$$

$$= 2\tfrac{2}{3}\text{ yd} \quad \text{Or 2.67 yd to the nearest hundredth}$$

EXAMPLE 3 A football field is 100 yards long. How many inches long is a football field?

100 yd

SOLUTION In this example we must convert yards to feet and then feet to inches. (To make this example more interesting, we are pretending we don't know that there are 36 inches in a yard.) We choose the conversion factors that will allow all the units except inches to divide out.

$$100\text{ yd} = 100\text{ yd} \times \frac{3\text{ ft}}{1\text{ yd}} \times \frac{12\text{ in.}}{1\text{ ft}}$$

$$= 100 \times 3 \times 12\text{ in.}$$

$$= 3{,}600\text{ in.}$$

Metric Units of Length

In the metric system the standard unit of length is a meter. A meter is a little longer than a yard (about 3.4 inches longer). The other units of length in the metric system are written in terms of a meter. The metric system uses prefixes to indicate what part of the basic unit of measure is being used. For example, in *millimeter* the prefix *milli* means "one thousandth" of a meter. Table 3 gives the meanings of the most common metric prefixes.

TABLE 3

THE MEANING OF METRIC PREFIXES

Prefix	Meaning
milli	0.001
centi	0.01
deci	0.1
deka	10
hecto	100
kilo	1,000

TABLE 4

METRIC UNITS OF LENGTH

The Relationship Between	Is	To Convert From One To The Other, Multiply By
millimeters (mm) and meters (m)	1,000 mm = 1 m	$\frac{1{,}000\text{ mm}}{1\text{ m}}$ or $\frac{1\text{ m}}{1{,}000\text{ mm}}$
centimeters (cm) and meters	100 cm = 1 m	$\frac{100\text{ cm}}{1\text{ m}}$ or $\frac{1\text{ m}}{100\text{ cm}}$
decimeters (dm) and meters	10 dm = 1 m	$\frac{10\text{ dm}}{1\text{ m}}$ or $\frac{1\text{ m}}{10\text{ dm}}$
dekameters (dam) and meters	1 dam = 10 m	$\frac{10\text{ m}}{1\text{ dam}}$ or $\frac{1\text{ dam}}{10\text{ m}}$
hectometers (hm) and meters	1 hm = 100 m	$\frac{100\text{ m}}{1\text{ hm}}$ or $\frac{1\text{ hm}}{100\text{ m}}$
kilometers (km) and meters	1 km = 1,000 m	$\frac{1{,}000\text{ m}}{1\text{ km}}$ or $\frac{1\text{ km}}{1{,}000\text{ m}}$

We can use these prefixes to write the other units of length and conversion factors for the metric system, as given in Table 4.

We use the same method to convert between units in the metric system as we did with the U.S. system. We choose the conversion factor that will allow the units we start with to divide out, leaving the units we want to end up with.

EXAMPLE 4 Convert 25 millimeters to meters.

SOLUTION To convert from millimeters to meters, we multiply by the conversion factor $\frac{1\text{ m}}{1{,}000\text{ mm}}$:

$$\begin{aligned} 25\text{ mm} &= 25\text{ mm} \times \frac{1\text{ m}}{1{,}000\text{ mm}} \\ &= \frac{25\text{ m}}{1{,}000} \\ &= 0.025\text{ m} \end{aligned}$$

EXAMPLE 5 Convert 36.5 centimeters to decimeters.

SOLUTION We convert centimeters to meters and then meters to decimeters:

$$\begin{aligned} 36.5\text{ cm} &= 36.5\text{ cm} \times \frac{1\text{ m}}{100\text{ cm}} \times \frac{10\text{ dm}}{1\text{ m}} \\ &= \frac{36.5 \times 10}{100}\text{ dm} \\ &= 3.65\text{ dm} \end{aligned}$$

The most common units of length in the metric system are millimeters, centimeters, meters, and kilometers. The other units of length we have listed in our table of metric lengths are not as widely used. The method we have used to convert from one unit of length to another in Examples 1–5 is called *unit analysis*. If you take a chemistry class, you will see it used many times. The same is true of many other science classes as well.

We can summarize the procedure used in unit analysis with the following steps:

Steps Used in Unit Analysis

1. Identify the units you are starting with.
2. Identify the units you want to end with.
3. Find conversion factors that will bridge the starting units and the ending units.
4. Set up the multiplication problem so that all units except the units you want to end with will divide out.

EXAMPLE 6 A sheep rancher is making new lambing pens for the upcoming lambing season. Each pen is a rectangle 6 feet wide and 8 feet long. The fencing material he wants to use sells for $1.36 per foot. If he is planning to build five separate lambing pens (they are separate because he wants a walkway between them), how much will he have to spend for fencing material?

SOLUTION To find the amount of fencing material he needs for one pen, we find the perimeter of a pen.

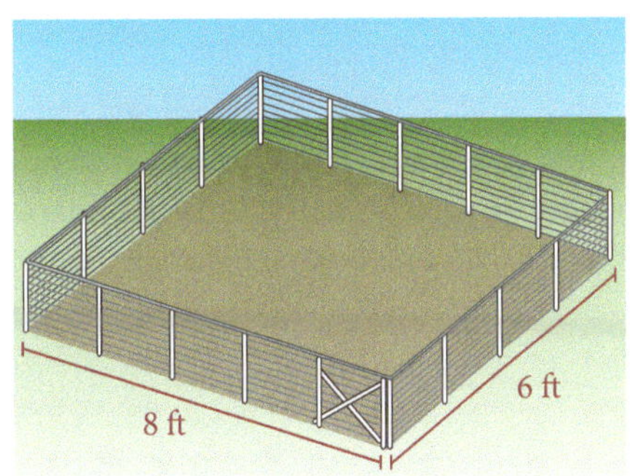

$$\text{Perimeter} = 6 + 6 + 8 + 8 = 28 \text{ feet}$$

We set up the solution to the problem using unit analysis. Our starting unit is *pens* and our ending unit is *dollars*. Here are the conversion factors that will form a bridge between pens and dollars:

$$1 \text{ pen} = 28 \text{ feet of fencing}$$
$$1 \text{ foot of fencing} = 1.36 \text{ dollars}$$

Next we write the multiplication problem, using the conversion factors, that will allow all the units except dollars to divide out:

$$5 \text{ pens} = 5 \text{ pens} \times \frac{28 \text{ feet of fencing}}{1 \text{ pen}} \times \frac{1.36 \text{ dollars}}{1 \text{ foot of fencing}}$$

$$= 5 \times 28 \times 1.36 \text{ dollars}$$
$$= \$190.40$$

EXAMPLE 7 In 1993, a ski resort in Vermont advertised their new high-speed chair lift as "the world's fastest chair lift, with a speed of 1,100 feet per second." Show why the speed cannot be correct.

SOLUTION To solve this problem, we can convert feet per second into miles per hour, a unit of measure we are more familiar with on an intuitive level. Here are the conversion factors we will use:

$$1 \text{ mile} = 5{,}280 \text{ feet}$$

$$1 \text{ hour} = 60 \text{ minutes}$$

$$1 \text{ minute} = 60 \text{ seconds}$$

$$1{,}100 \text{ ft/second} = \frac{1{,}100 \text{ }\cancel{\text{feet}}}{1 \text{ }\cancel{\text{second}}} \times \frac{1 \text{ mile}}{5{,}280 \text{ }\cancel{\text{feet}}} \times \frac{60 \text{ }\cancel{\text{seconds}}}{1 \text{ }\cancel{\text{minute}}} \times \frac{60 \text{ }\cancel{\text{minutes}}}{1 \text{ hour}}$$

$$= \frac{1{,}100 \times 60 \times 60 \text{ miles}}{5{,}280 \text{ hours}}$$

$$= 750 \text{ miles/hour}$$

GETTING READY FOR CLASS

After reading through the preceding section, respond in your own words and in complete sentences.

1. Write the relationship between feet and miles. That is, write an equality that shows how many feet are in every mile.
2. Give the metric prefix that means "one hundredth."
3. Give the metric prefix that is equivalent to 1,000.
4. As you know from reading the section in the text, conversion factors are ratios. Write the conversion factor that will allow you to convert from inches to feet. That is, if we wanted to convert 27 inches to feet, what conversion factor would we use?

Problem Set 5.1

Make the following conversions in the U.S. system by multiplying by the appropriate conversion factor. Write your answers as whole numbers or mixed numbers.

1. 5 ft to inches
2. 9 ft to inches
3. 10 ft to inches
4. 20 ft to inches
5. 2 yd to feet
6. 8 yd to feet
7. 4.5 yd to inches
8. 9.5 yd to inches
9. 27 in. to feet
10. 36 in. to feet
11. 2.5 mi to feet
12. 6.75 mi to feet
13. 48 in. to yards
14. 56 in. to yards

Make the following conversions in the metric system by multiplying by the appropriate conversion factor. Write your answers as whole numbers or decimals.

15. 18 m to centimeters
16. 18 m to millimeters
17. 4.8 km to meters
18. 8.9 km to meters
19. 5 dm to centimeters
20. 12 dm to millimeters
21. 248 m to kilometers
22. 969 m to kilometers
23. 67 cm to millimeters
24. 67 mm to centimeters
25. 3,498 cm to meters
26. 4,388 dm to meters
27. 63.4 cm to decimeters
28. 89.5 cm to decimeters

Applying the Concepts

29. **Softball** If the distance between first and second base in softball is 60 feet, how many yards is it from first to second base?

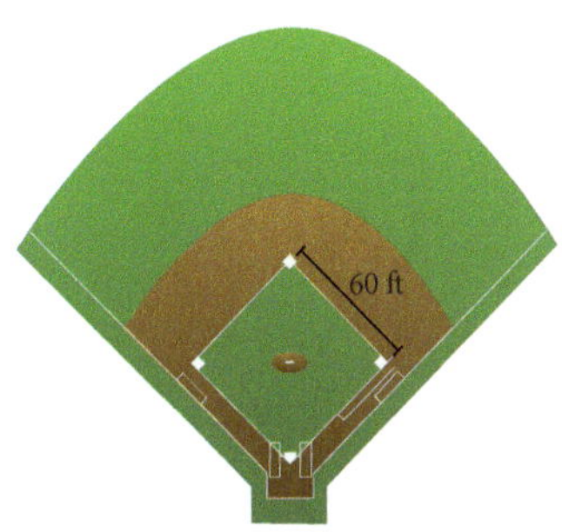

30. **Tower Height** A transmitting tower is 100 feet tall. How many inches is that?
31. **High Jump** If a person high jumps 6 feet 8 inches, how many inches is the jump?
32. **Desk Width** A desk is 48 inches wide. What is the width in yards?
33. **Ceiling Height** Suppose the ceiling of a home is 2.44 meters above the floor. Express the height of the ceiling in centimeters.

34. **Notebook Width** Standard-sized notebook paper is 21.6 centimeters wide. Express this width in millimeters.

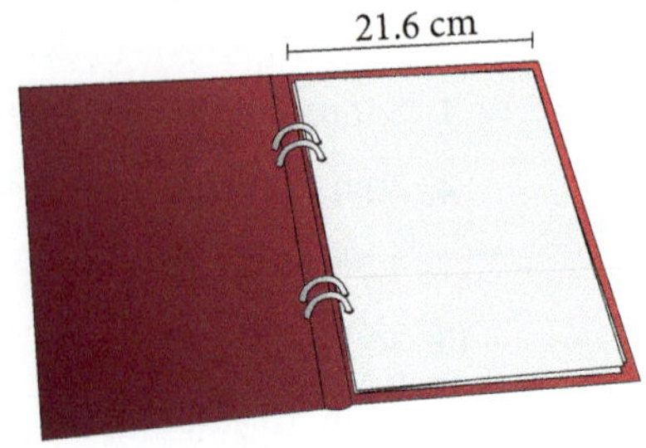

35. **Dollar Width** A dollar bill is about 6.5 centimeters wide. Express this width in millimeters.

36. **Pencil Length** Most new pencils are 19 centimeters long. Express this length in meters.

37. **Surveying** A unit of measure sometimes used in surveying is the *chain*. There are 80 chains in 1 mile. How many chains are in 37 miles?

38. **Surveying** Another unit of measure used in surveying is a *link;* 1 link is about 8 inches. About how many links are there in 5 feet?

39. **Metric System** A very small unit of measure in the metric system is the *micron* (abbreviated μm). There are 1,000 μm in 1 millimeter. How many microns are in 12 centimeters?

40. **Metric System** Another very small unit of measure in the metric system is the *angstrom* (abbreviated Å). There are 10,000,000 Å in 1 millimeter. How many angstroms are in 15 decimeters?

41. **Horse Racing** In horse racing, 1 *furlong* is 220 yards. How many feet are in 12 furlongs?

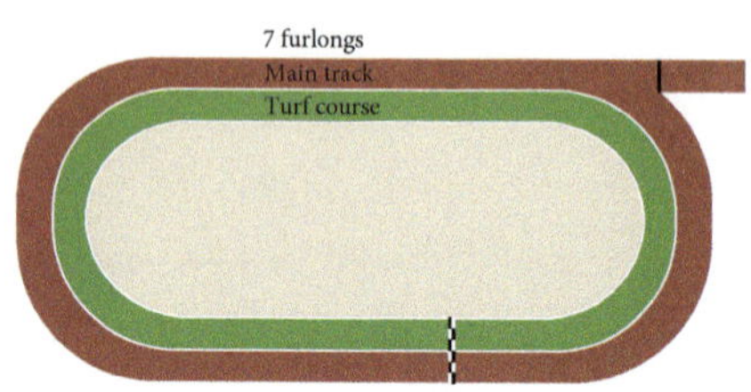

42. **Sailing** A *fathom* is 6 feet. How many yards are in 19 fathoms?

43. **Speed Limit** The maximum speed limit on part of Highway 101 in California is 55 miles/hour. Convert 55 miles/hour to feet/second. (Round to the nearest tenth.)

44. **Speed Limit** The maximum speed limit on part of Highway 5 in California is 65 miles/hour. Convert 65 miles/hour to feet/second. (Round to the nearest tenth.)

45. **Track and Field** A person who runs the 100-yard dash in 10.5 seconds has an average speed of 9.52 yards/second. Convert 9.52 yards/second to miles/hour. (Round to the nearest tenth.)

46. **Track and Field** A person who runs a mile in 8 minutes has an average speed of 0.125 miles/minute. Convert 0.125 miles/minute to miles/hour.

47. **Speed of a Bullet** The bullet from a rifle leaves the barrel traveling 1,500 feet/second. Convert 1,500 feet/second to miles/hour. (Round to the nearest whole number.)

iStockPhoto/©Glenn Frank

48. **Speed of a Bullet** A bullet from a machine gun on a B-17 Flying Fortress in World War II had a muzzle speed of 1,750 feet/second. Convert 1,750 feet/second to miles/hour. (Round to the nearest whole number.)

49. **Farming** A farmer is fencing a pasture that is $\frac{1}{2}$ mile wide and 1 mile long. If the fencing material sells for \$1.15 per foot, how much will it cost him to fence all four sides of the pasture?

50. **Cost of Fencing** A family with a swimming pool puts up a chain-link fence around the pool. The fence forms a rectangle 12 yards wide and 24 yards long. If the chain-link fence sells for \$2.50 per foot, how much will it cost to fence all four sides of the pool?

51. **Farming** A 4-H Club group is raising lambs to show at the County Fair. Each lamb eats $\frac{1}{8}$ of a bale of alfalfa a day. If the alfalfa costs \$10.50 per bale, how much will it cost to feed one lamb for 120 days?

52. **Farming** A 4-H Club group is raising pigs to show at the County Fair. Each pig eats 2.4 pounds of grain a day. If the grain costs \$5.25 per pound, how much will it cost to feed one pig for 60 days?

Calculator Problems

Set up the following conversions as you have been doing. Then perform the calculations on a calculator.

53. Change 751 miles to feet.

54. Change 639.87 centimeters to meters.

55. Change 4,982 yards to inches.

56. Change 379 millimeters to kilometers.

57. Mount Whitney is the highest point in California. It is 14,494 feet above sea level. Give its height in miles to the nearest tenth.

58. The tallest mountain in the United States is Mount McKinley in Alaska. It is 20,320 feet tall. Give its height in miles to the nearest tenth.

59. California has 3,427 miles of shoreline. How many feet is this?

60. The tip of the television tower at the top of the Empire State Building in New York City is 1,472 feet above the ground. Express this height in miles to the nearest hundredth.

Getting Ready for the Next Section

Perform the indicated operations.

61. 12×12

62. 36×24

63. $1 \times 4 \times 2$

64. $5 \times 4 \times 2$

65. $10 \times 10 \times 10$

66. $100 \times 100 \times 100$

67. $75 \times 43{,}560$

68. $55 \times 43{,}560$

69. $864 \div 144$

70. $1{,}728 \div 144$

71. $256 \div 640$

72. $960 \div 240$

73. $45 \times \frac{9}{1}$

74. $36 \times \frac{9}{1}$

75. $1{,}800 \times \frac{1}{4}$

76. $2{,}000 \times \frac{1}{4} \times \frac{1}{10}$

77. 1.5×30

78. 1.5×45

79. $2.2 \times 1{,}000$

80. $3.5 \times 1{,}000$

81. 67.5×9

82. 43.5×9

Unit Analysis II: Area and Volume 5.2

Figure 1 below gives a summary of the geometric objects we have worked with in previous chapters, along with the formulas for finding the area of each object.

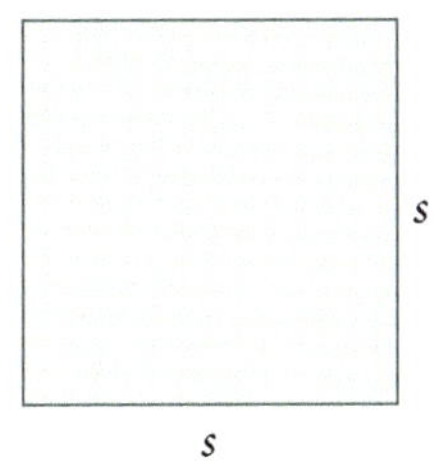

Area = (side)(side)
= (side)^2
$A = s^2$

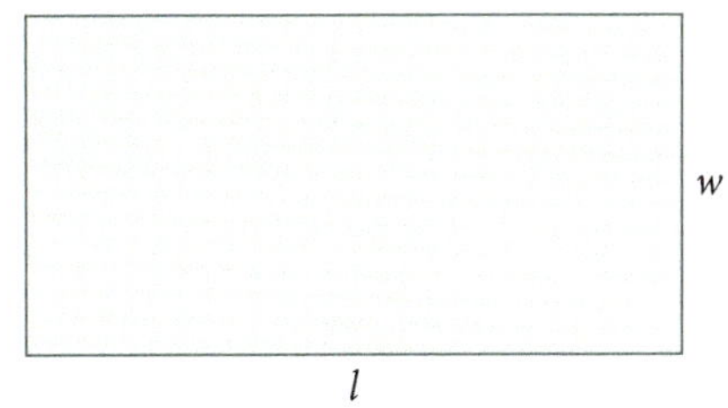

Area = (length)(width)
$A = lw$

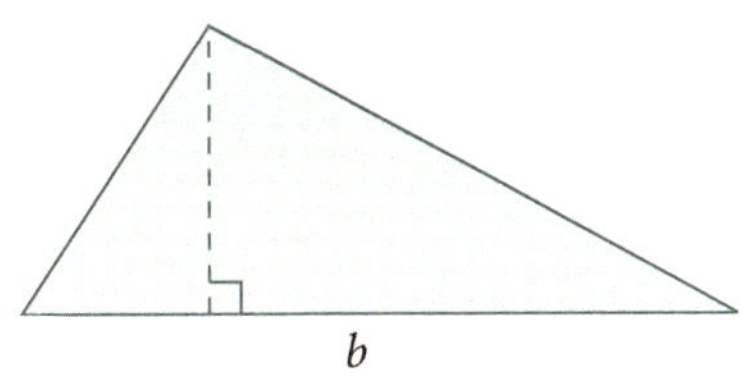

Area = $\frac{1}{2}$(base)(height)
$A = \frac{1}{2}bh$

FIGURE 1 Areas of common geometric shapes

EXAMPLE 1 Find the number of square inches in 1 square foot.

SOLUTION We can think of 1 square foot as 1 ft^2 = 1 ft × ft. To convert from feet to inches, we use the conversion factor 1 foot = 12 inches. Because the unit foot appears twice in 1 ft^2, we multiply by our conversion factor twice.

$$1 \text{ ft}^2 = 1 \cancel{\text{ft}} \times \cancel{\text{ft}} \times \frac{12 \text{ in.}}{1 \cancel{\text{ft}}} \times \frac{12 \text{ in.}}{1 \cancel{\text{ft}}} = 12 \times 12 \text{ in.} \times \text{in.} = 144 \text{ in}^2$$

Now that we know that 1 ft^2 is the same as 144 in^2, we can use this fact as a conversion factor to convert between square feet and square inches. Depending on which units we are converting from, we would use either

$$\frac{144 \text{ in}^2}{1 \text{ ft}^2} \quad \text{or} \quad \frac{1 \text{ ft}^2}{144 \text{ in}^2}$$

EXAMPLE 2 A rectangular poster measures 36 inches by 24 inches. How many square feet of wall space will the poster cover?

SOLUTION One way to work this problem is to find the number of square inches the poster covers, and then convert square inches to square feet.

$$\text{Area of poster} = \text{length} \times \text{width} = 36 \text{ in.} \times 24 \text{ in.} = 864 \text{ in}^2$$

To finish the problem, we convert square inches to square feet:

$$\begin{aligned} 864 \text{ in}^2 &= 864 \text{ in}^2 \times \frac{1 \text{ ft}^2}{144 \text{ in}^2} \\ &= \frac{864}{144} \text{ ft}^2 \\ &= 6 \text{ ft}^2 \end{aligned}$$

iStockPhoto/©Duncan Walker

Table 1 gives the most common units of area in the U.S. system of measurement, along with the corresponding conversion factors.

TABLE 1

U.S. UNITS OF AREA

The Relationship Between	Is	To Convert From One To The Other, Multiply By
square inches and square feet	$144 \text{ in}^2 = 1 \text{ ft}^2$	$\frac{144 \text{ in}^2}{1 \text{ ft}^2}$ or $\frac{1 \text{ ft}^2}{144 \text{ in}^2}$
square yards and square feet	$9 \text{ ft}^2 = 1 \text{ yd}^2$	$\frac{9 \text{ ft}^2}{1 \text{ yd}^2}$ or $\frac{1 \text{ yd}^2}{9 \text{ ft}^2}$
acres and square feet	$1 \text{ acre} = 43{,}560 \text{ ft}^2$	$\frac{43{,}560 \text{ ft}^2}{1 \text{ acre}}$ or $\frac{1 \text{ acre}}{43{,}560 \text{ ft}^2}$
acres and square miles	$640 \text{ acres} = 1 \text{ mi}^2$	$\frac{640 \text{ acres}}{1 \text{ mi}^2}$ or $\frac{1 \text{ mi}^2}{640 \text{ acres}}$

EXAMPLE 3 A dressmaker orders a bolt of material that is 1.5 yards wide and 30 yards long. How many square feet of material were ordered?

SOLUTION The area of the material in square yards is

$$\begin{aligned} A &= 1.5 \times 30 \\ &= 45 \text{ yd}^2 \end{aligned}$$

Converting this to square feet, we have

$$\begin{aligned} 45 \text{ yd}^2 &= 45 \text{ yd}^2 \times \frac{9 \text{ ft}^2}{1 \text{ yd}^2} \\ &= 405 \text{ ft}^2 \end{aligned}$$

EXAMPLE 4 A farmer has 75 acres of land. How many square feet of land does the farmer have?

SOLUTION Changing acres to square feet, we have

$$75 \text{ acres} = 75 \text{ acres} \times \frac{43{,}560 \text{ ft}^2}{1 \text{ acre}}$$

$$= 75 \times 43{,}560 \text{ ft}^2$$

$$= 3{,}267{,}000 \text{ ft}^2$$

EXAMPLE 5 A new shopping center is to be constructed on 256 acres of land. How many square miles is this?

SOLUTION Multiplying by the conversion factor that will allow acres to divide out, we have

$$256 \text{ acres} = 256 \text{ acres} \times \frac{1 \text{ mi}^2}{640 \text{ acres}}$$

$$= \frac{256}{640} \text{ mi}^2$$

$$= 0.4 \text{ mi}^2$$

Units of area in the metric system are considerably simpler than those in the U.S. system because metric units are given in terms of powers of 10. Table 2 lists the conversion factors that are most commonly used.

TABLE 2

METRIC UNITS OF AREA

The Relationship Between	Is	To Convert From One To The Other, Multiply By
square millimeters and square centimeters	$1 \text{ cm}^2 = 100 \text{ mm}^2$	$\frac{100 \text{ mm}^2}{1 \text{ cm}^2}$ or $\frac{1 \text{ cm}^2}{100 \text{ mm}^2}$
square centimeters and square decimeters	$1 \text{ dm}^2 = 100 \text{ cm}^2$	$\frac{100 \text{ cm}^2}{1 \text{ dm}^2}$ or $\frac{1 \text{ dm}^2}{100 \text{ cm}^2}$
square decimeters and square meters	$1 \text{ m}^2 = 100 \text{ dm}^2$	$\frac{100 \text{ dm}^2}{1 \text{ m}^2}$ or $\frac{1 \text{ m}^2}{100 \text{ dm}^2}$
square meters and ares (a)	$1 \text{ a} = 100 \text{ m}^2$	$\frac{100 \text{ m}^2}{1 \text{ a}}$ or $\frac{1 \text{ a}}{100 \text{ m}^2}$
ares and hectares (ha)	$1 \text{ ha} = 100 \text{ a}$	$\frac{100 \text{ a}}{1 \text{ ha}}$ or $\frac{1 \text{ ha}}{100 \text{ a}}$

EXAMPLE 6 How many square millimeters are in 1 square meter?

SOLUTION We start with 1 m² and end up with square millimeters:

$$1\text{ m}^2 = 1\text{ m}^2 \times \frac{100\text{ dm}^2}{1\text{ m}^2} \times \frac{100\text{ cm}^2}{1\text{ dm}^2} \times \frac{100\text{ mm}^2}{1\text{ cm}^2}$$

$$= 100 \times 100 \times 100\text{ mm}^2$$

$$= 1{,}000{,}000\text{ mm}^2$$

Units of Measure for Volume

Table 3 lists the units of volume in the U.S. system and their conversion factors.

TABLE 3

UNITS OF VOLUME IN THE U.S. SYSTEM

The Relationship Between	Is	To Convert From One To The Other, Multiply By
cubic inches (in^3) and cubic feet (ft^3)	$1\text{ ft}^3 = 1{,}728\text{ in}^3$	$\frac{1{,}728\text{ in}^3}{1\text{ ft}^3}$ or $\frac{1\text{ ft}^3}{1{,}728\text{ in}^3}$
cubic feet and cubic yards (yd^3)	$1\text{ yd}^3 = 27\text{ ft}^3$	$\frac{27\text{ ft}^3}{1\text{ yd}^3}$ or $\frac{1\text{ yd}^3}{27\text{ ft}^3}$
fluid ounces (fl oz) and pints (pt)	1 pt = 16 fl oz	$\frac{16\text{ fl oz}}{1\text{ pt}}$ or $\frac{1\text{ pt}}{16\text{ fl oz}}$
pints and quarts (qt)	1 qt = 2 pt	$\frac{2\text{ pt}}{1\text{ qt}}$ or $\frac{1\text{ qt}}{2\text{ pt}}$
quarts and gallons (gal)	1 gal = 4 qt	$\frac{4\text{ qt}}{1\text{ gal}}$ or $\frac{1\text{ gal}}{4\text{ qt}}$

EXAMPLE 7 What is the capacity (volume) in pints of a 1-gallon container of milk?

SOLUTION We change from gallons to quarts and then quarts to pints by multiplying by the appropriate conversion factors as given in Table 3.

$$1\text{ gal} = 1\text{ gal} \times \frac{4\text{ qt}}{1\text{ gal}} \times \frac{2\text{ pt}}{1\text{ qt}}$$

$$= 1 \times 4 \times 2\text{ pt}$$

$$= 8\text{ pt}$$

A 1-gallon container has the same capacity as 8 1-pint containers.

EXAMPLE 8 A dairy herd produces 1,800 quarts of milk each day. How many gallons is this equivalent to?

SOLUTION Converting 1,800 quarts to gallons, we have

$$1{,}800 \text{ qt} = 1{,}800 \text{ qt} \times \frac{1 \text{ gal}}{4 \text{ qt}}$$

$$= \frac{1{,}800}{4} \text{ gal}$$

$$= 450 \text{ gal}$$

We see that 1,800 quarts is equivalent to 450 gallons.

In the metric system the basic unit of measure for volume is the liter. A liter is the volume enclosed by a cube that is 10 cm on each edge, as shown in Figure 2. We can see that a liter is equivalent to 1,000 cm³.

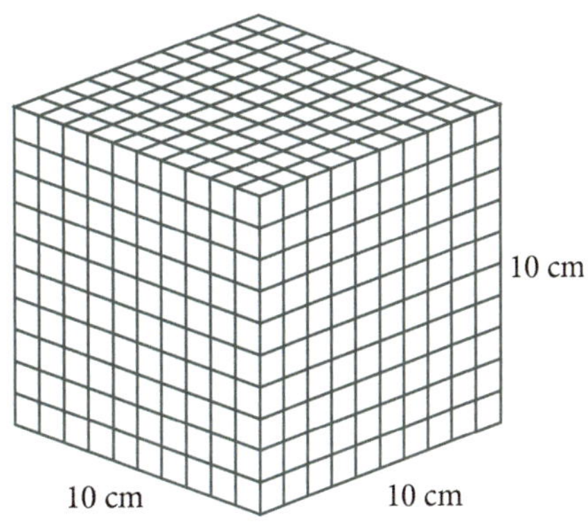

$$1 \text{ liter} = 10 \text{ cm} \times 10 \text{ cm} \times 10 \text{ cm}$$
$$= 1{,}000 \text{ cm}^3$$

FIGURE 2

The other units of volume in the metric system use the same prefixes we encountered previously. The units with prefixes centi, deci, and deka are not as common as the others, so in Table 4 we include only liters, milliliters, hectoliters, and kiloliters.

Note

As you can see from the table and the discussion above, a cubic centimeter (cm³) and a milliliter (mL) are equal. Both are one thousandth of a liter. It is also common in some fields (like medicine) to abbreviate the term cubic centimeter as cc. Although we will use the notation mL when discussing volume in the metric system, you should be aware that 1 mL = 1 cm³ = 1 cc.

TABLE 4

METRIC UNITS OF VOLUME

The Relationship Between	Is	To Convert From One To The Other, Multiply By
milliliters (mL) and liters	1 liter (L) = 1,000 mL	$\frac{1{,}000 \text{ mL}}{1 \text{ liter}}$ or $\frac{1 \text{ liter}}{1{,}000 \text{ mL}}$
hectoliters (hL) and liters	100 liters = 1 hL	$\frac{100 \text{ liters}}{1 \text{ hL}}$ or $\frac{1 \text{ hL}}{100 \text{ liters}}$
kiloliters (kL) and liters	1,000 liters (L) = 1 kL	$\frac{1{,}000 \text{ liters}}{1 \text{ kL}}$ or $\frac{1 \text{ kL}}{1{,}000 \text{ liters}}$

Here is an example of conversion from one unit of volume to another in the metric system.

EXAMPLE 9 A sports car has a 2.2-liter engine. What is the displacement (volume) of the engine in milliliters?

SOLUTION Using the appropriate conversion factor from Table 4, we have

$$2.2 \text{ liters} = 2.2 \text{ liters} \times \frac{1{,}000 \text{ mL}}{1 \text{ liter}}$$

$$= 2.2 \times 1{,}000 \text{ mL}$$

$$= 2{,}200 \text{ mL}$$

GETTING READY FOR CLASS

After reading through the preceding section, respond in your own words and in complete sentences.

1. Write the formula for the area of each of the following:
   a. a square of side s.
   b. a rectangle with length l and width w.
2. What is the relationship between square feet and square inches?
3. Fill in the numerators below so that each conversion factor is equal to 1.

   a. $\frac{\quad \text{qt}}{1 \text{ gal}}$ b. $\frac{\quad \text{mL}}{1 \text{ liter}}$ c. $\frac{\quad \text{acres}}{1 \text{ mi}^2}$

4. Write the conversion factor that will allow us to convert from square yards to square feet.

Problem Set 5.2

Use the tables given in this section to make the following conversions. Be sure to show the conversion factor used in each case.

1. 3 ft² to square inches
2. 5 ft² to square inches
3. 288 in² to square feet
4. 720 in² to square feet
5. 30 acres to square feet
6. 92 acres to square feet
7. 2 mi² to acres
8. 7 mi² to acres
9. 1,920 acres to square miles
10. 3,200 acres to square miles
11. 12 yd² to square feet
12. 20 yd² to square feet
13. 17 cm² to square millimeters
14. 150 mm² to square centimeters
15. 2.8 m² to square centimeters
16. 10 dm² to square millimeters
17. 1,200 mm² to square meters
18. 19.79 cm² to square meters
19. 5 a to square meters
20. 12 a to square centimeters
21. 7 ha to ares
22. 3.6 ha to ares
23. 342 a to hectares
24. 986 a to hectares

Applying the Concepts

25. **Sports** The diagrams below show the dimensions of playing fields for the National Football League (NFL), the Canadian Football League, and Arena Football. Find the area of each field and then convert each area to acres. Round answers to the nearest hundredth.

Football Fields

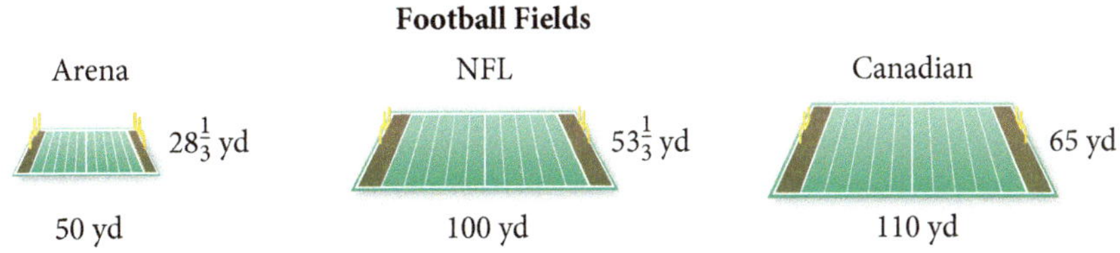

26. **Soccer** The rules for soccer state that the playing field must be from 100 to 120 yards long and 55 to 75 yards wide. The 1999 Women's World Cup was played at the Rose Bowl on a playing field 116 yards long and 72 yards wide. The diagram below shows the smallest possible soccer field, the largest possible soccer field, and the soccer field at the Rose Bowl. Find the area of each one and then convert the area of each to acres. Round to the nearest tenth, if necessary.

Soccer Fields

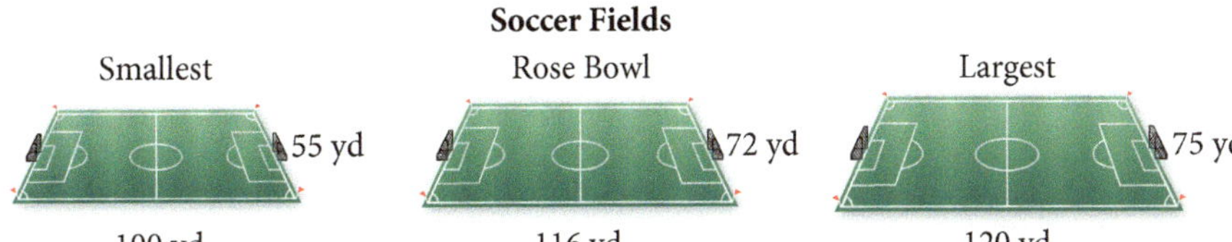

27. Swimming Pool A public swimming pool measures 100 meters by 30 meters and is rectangular. What is the area of the pool in ares?

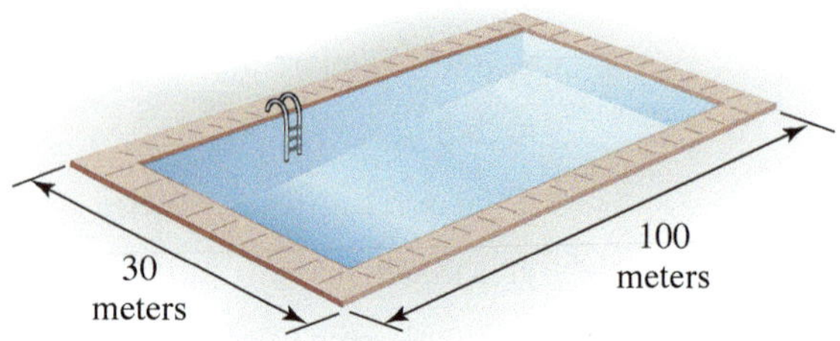

28. Construction A family decides to put tiles in the entryway of their home. The entryway has an area of 6 square meters. If each tile is 5 centimeters by 5 centimeters, how many tiles will it take to cover the entryway?

29. Landscaping A landscaper is putting in a brick patio. The area of the patio is 110 square meters. If the bricks measure 10 centimeters by 20 centimeters, how many bricks will it take to make the patio? Assume no space between bricks.

30. Sewing A dressmaker is using a pattern that requires 2 square yards of material. If the material is on a bolt that is 54 inches wide, how long must a piece of material be cut from the bolt to be sure there is enough material for the pattern?

Make the following conversions using the conversion factors given in Tables 3 and 4.

31. 5 yd^3 to cubic feet
32. 3.8 yd^3 to cubic feet
33. 3 pt to fluid ounces
34. 8 pt to fluid ounces
35. 2 gal to quarts
36. 12 gal to quarts
37. 2.5 gal to pints
38. 7 gal to pints
39. 15 qt to fluid ounces
40. 5.9 qt to fluid ounces
41. 64 pt to gallons
42. 256 pt to gallons
43. 12 pt to quarts
44. 18 pt to quarts
45. 243 ft^3 to cubic yards
46. 864 ft^3 to cubic yards
47. 5 L to milliliters
48. 9.6 L to milliliters
49. 127 mL to liters
50. 93.8 mL to liters
51. 4 kL to milliliters
52. 3 kL to milliliters
53. 14.92 kL to liters
54. 4.71 kL to liters

Applying the Concepts

55. Filling Coffee Cups If a regular-size coffee cup holds about $\frac{1}{2}$ pint, about how many cups can be filled from a 1-gallon coffee maker?

56. Filling Glasses If a regular-size drinking glass holds about 0.25 liter of liquid, how many glasses can be filled from a 750-milliliter container?

57. Capacity of a Refrigerator A refrigerator has a capacity of 20 cubic feet. What is the capacity of the refrigerator in cubic inches?

58. Volume of a Tank The gasoline tank on a car holds 18 gallons of gas. What is the volume of the tank in quarts?

59. Filling Glasses How many 8-fluid-ounce glasses of water will it take to fill a 3-gallon aquarium?

60. Filling a Container How many 5-milliliter test tubes filled with water will it take to fill a 1-liter container?

Calculator Problems

Set up the following problems as you have been doing. Then use a calculator to perform the actual calculations. Round all answers to two decimal places where appropriate.

61. Geography Lake Superior is the largest of the Great Lakes. It covers 31,700 square miles of area. What is the area of Lake Superior in acres?

62. Geography The state of California consists of 156,360 square miles of land and 2,330 square miles of water. Write the total area (both land and water) in acres.

63. Geography Death Valley National Monument contains 2,067,795 acres of land. How many square miles is this?

64. Geography The Badlands National Monument in South Dakota was established in 1929. It covers 243,302 acres of land. What is the area in square miles?

65. Convert 93.4 qt to gallons.

66. Convert 7,362 fl oz to gallons.

67. How many cubic feet are contained in 796 cubic yards?

68. The engine of a car has a displacement of 440 cubic inches. What is the displacement in cubic feet?

69. Volume of Water The Grand Coulee Dam holds 10,585,000 cubic yards of water. What is the volume of water in cubic feet?

70. Volume of Water Hoover Dam was built in 1936 on the Colorado River in Nevada. It holds a volume of 4,400,000 cubic yards of water. What is this volume in cubic feet?

Getting Ready for the Next Section

Perform the indicated operations.

71. 12×16

72. 15×16

73. $3 \times 2{,}000$

74. $5 \times 2{,}000$

75. $3 \times 1{,}000 \times 100$

76. $5 \times 1{,}000 \times 100$

77. 50×250

78. 75×200

79. $12{,}500 \times \dfrac{1}{1{,}000}$

80. $15{,}000 \times \dfrac{1}{1{,}000}$

Unit Analysis III: Weight

5.3

The most common units of weight in the U.S. system are ounces, pounds, and tons. The relationships among these units are given in Table 1.

TABLE 1

UNITS OF WEIGHT IN THE U.S. SYSTEM

The Relationship Between	Is	To Convert From One To The Other, Multiply By
ounces (oz) and pounds (lb)	1 lb = 16 oz	$\frac{16\text{ oz}}{1\text{ lb}}$ or $\frac{1\text{ lb}}{16\text{ oz}}$
pounds and tons (T)	1 T = 2,000 lb	$\frac{2{,}000\text{ lb}}{1\text{ T}}$ or $\frac{1\text{ T}}{2{,}000\text{ lb}}$

EXAMPLE 1 Convert 12 pounds to ounces.

SOLUTION Using the conversion factor from the table, and applying the method we have been using, we have

$$\begin{aligned} 12\text{ lb} &= 12\text{ lb} \times \frac{16\text{ oz}}{1\text{ lb}} \\ &= 12 \times 16\text{ oz} \\ &= 192\text{ oz} \end{aligned}$$

12 pounds is equivalent to 192 ounces.

EXAMPLE 2 Convert 3 tons to pounds.

SOLUTION We use the conversion factor from the table. We have

$$\begin{aligned} 3\text{ T} &= 3\text{ T} \times \frac{2{,}000\text{ lb}}{1\text{ T}} \\ &= 6{,}000\text{ lb} \end{aligned}$$

6,000 pounds is the equivalent of 3 tons.

In the metric system the basic unit of weight is a gram. We use the same prefixes we have already used to write the other units of weight in terms of grams. Table 2 lists the most common metric units of weight and their conversion factors.

TABLE 2

METRIC UNITS OF WEIGHT

The Relationship Between	Is	To Convert From One To The Other, Multiply By
milligrams (mg) and grams (g)	1 g = 1,000 mg	$\frac{1{,}000\text{ mg}}{1\text{ g}}$ or $\frac{1\text{ g}}{1{,}000\text{ mg}}$
centigrams (cg) and grams	1 g = 100 cg	$\frac{100\text{ cg}}{1\text{ g}}$ or $\frac{1\text{ g}}{100\text{ cg}}$
kilograms (kg) and grams	1,000 g = 1 kg	$\frac{1{,}000\text{ g}}{1\text{ kg}}$ or $\frac{1\text{ kg}}{1{,}000\text{ g}}$
metric tons (t) and kilograms	1,000 kg = 1 t	$\frac{1{,}000\text{ kg}}{1\text{ t}}$ or $\frac{1\text{ t}}{1{,}000\text{ kg}}$

EXAMPLE 3 Convert 3 kilograms to centigrams.

SOLUTION We convert kilograms to grams and then grams to centigrams:

$$3 \text{ kg} = 3 \text{ kg} \times \frac{1{,}000 \text{ g}}{1 \text{ kg}} \times \frac{100 \text{ cg}}{1 \text{ g}}$$
$$= 3 \times 1{,}000 \times 100 \text{ cg}$$
$$= 300{,}000 \text{ cg}$$

EXAMPLE 4 A bottle of vitamin C contains 50 tablets. Each tablet contains 250 milligrams of vitamin C. What is the total number of grams of vitamin C in the bottle?

SOLUTION We begin by finding the total number of milligrams of vitamin C in the bottle. Since there are 50 tablets, and each contains 250 mg of vitamin C, we can multiply 50 by 250 to get the total number of milligrams of vitamin C:

$$\text{Milligrams of vitamin C} = 50 \times 250 \text{ mg}$$
$$= 12{,}500 \text{ mg}$$

Next we convert 12,500 mg to grams:

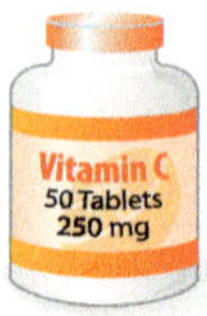

$$12{,}500 \text{ mg} = 12{,}500 \text{ mg} \times \frac{1 \text{ g}}{1{,}000 \text{ mg}}$$
$$= \frac{12{,}500}{1{,}000} \text{ g}$$
$$= 12.5 \text{ g}$$

The bottle contains 12.5 g of vitamin C.

GETTING READY FOR CLASS

After reading through the preceding section, respond in your own words and in complete sentences.

1. What is the relationship between pounds and ounces?
2. Write the conversion factor used to convert from pounds to ounces.
3. Write the conversion factor used to convert from milligrams to grams.
4. What is the relationship between grams and kilograms?

Problem Set 5.3

Use the conversion factors in Tables 1 and 2 to make the following conversions.

1. 8 lb to ounces

2. 5 lb to ounces

3. 2 T to pounds

4. 5 T to pounds

5. 192 oz to pounds

6. 176 oz to pounds

7. 1,800 lb to tons

8. 10,200 lb to tons

9. 1 T to ounces

10. 3 T to ounces

11. $3\frac{1}{2}$ lb to ounces

12. $5\frac{1}{4}$ lb to ounces

13. $6\frac{1}{2}$ T to pounds

14. $4\frac{1}{5}$ T to pounds

15. 2 kg to grams

16. 5 kg to grams

17. 4 cg to milligrams

18. 3 cg to milligrams

19. 2 kg to centigrams

20. 5 kg to centigrams

21. 5.08 g to centigrams

22. 7.14 g to centigrams

23. 450 cg to grams

24. 979 cg to grams

25. 478.95 mg to centigrams

26. 659.43 mg to centigrams

27. 1,578 mg to grams

28. 1,979 mg to grams

29. 42,000 cg to kilograms

30. 97,000 cg to kilograms

Applying the Concepts

31. Fish Oil A bottle of fish oil contains 60 soft gels, each containing 800 mg of the omega-3 fatty acid. How many total grams of the omega-3 fatty acid are in this bottle?

32. Fish Oil A bottle of fish oil contains 50 soft gels, each containing 300 mg of the omega-6 fatty acid. How many total grams of the omega-6 fatty acid are in this bottle?

33. B-Complex A certain B-complex vitamin supplement contains 50 mg of riboflavin, or vitamin B_2. A bottle contains 80 vitamins. How many total grams of riboflavin are in this bottle?

34. B-Complex A certain B-complex vitamin supplement contains 30 mg of thiamine, or vitamin B_1. A bottle contains 80 vitamins. How many total grams of thiamine are in this bottle?

35. Aspirin A bottle of low-strength aspirin contains 120 tablets. Each tablet contains 81 mg of aspirin. How many total grams of aspirin are in this bottle?

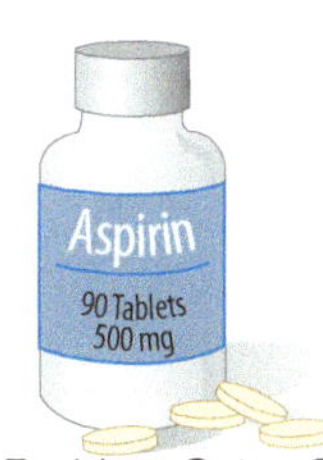

36. Aspirin A bottle of maximum-strength aspirin contains 90 tablets. Each tablet contains 500 mg of aspirin. How many total grams of aspirin are in this bottle?

37. Vitamin C A certain brand of vitamin C contains 500 mg per tablet. A bottle contains 240 vitamins. How many total grams of vitamin C are in this bottle?

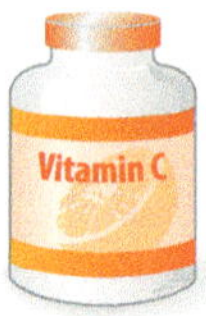

38. Vitamin C A certain brand of vitamin C contains 600 mg per tablet. A bottle contains 150 vitamins. How many total grams of vitamin C are in this bottle?

Coca Cola Bottles The soft drink Coke is sold throughout the world. Although the size of the bottle varies between different countries, a "six-pack" is sold everywhere. For each of the problems below, find the number of liters in a "6-pack" from the given bottle size.

Write each fraction or mixed number as a decimal.

Country	Bottle size	Liters in a 6-pack
39. Estonia	500 mL	
40. Israel	350 mL	
41. Jordan	250 mL	
42. Kenya	300 mL	

Getting Ready for the Next Section

Perform the indicated operations.

43. 8×2.54 **44.** 9×3.28 **45.** $3 \times 1.06 \times 2$ **46.** $3 \times 5 \times 3.79$

47. $80.5 \div 1.61$ **48.** $96.6 \div 1.61$ **49.** $125 \div 2.20$ **50.** $165 \div 2.20$

51. $2{,}000 \div 16.39$ (Round your answer to the nearest whole number.)

52. $2{,}200 \div 16.39$ (Round your answer to the nearest whole number.)

53. $\frac{9}{5}(120) + 32$ **54.** $\frac{9}{5}(40) + 32$ **55.** $\frac{5(102 - 32)}{9}$ **56.** $\frac{5(101.6 - 32)}{9}$

Converting Between the Two Systems and Temperature 5.4

Because most of us have always used the U.S. system of measurement in our everyday lives, we are much more familiar with it on an intuitive level than we are with the metric system. We have an intuitive idea of how long feet and inches are, how much a pound weighs, and what a square yard of material looks like. The metric system is actually much easier to use than the U.S. system. The reason some of us have such a hard time with the metric system is that we don't have the feel for it that we do for the U.S. system. We have trouble visualizing how long a meter is or how much a gram weighs. The following list is intended to give you something to associate with each basic unit of measurement in the metric system:

1. A meter is just a little longer than a yard.
2. The length of the edge of a sugar cube is about 1 centimeter.
3. A liter is just a little larger than a quart.
4. A sugar cube has a volume of approximately 1 milliliter.
5. A paper clip weighs about 1 gram.
6. A 2-pound can of coffee weighs about 1 kilogram.

TABLE 1

ACTUAL CONVERSION FACTORS BETWEEN THE METRIC AND U.S. SYSTEMS OF MEASUREMENT

The Relationship Between	Is	To Convert From One To The Other, Multiply By
Length		
inches and centimeters	2.54 cm = 1 in.	$\frac{2.54\text{ cm}}{1\text{ in.}}$ or $\frac{1\text{ in.}}{2.54\text{ cm}}$
feet and meters	1 m = 3.28 ft	$\frac{3.28\text{ ft}}{1\text{ m}}$ or $\frac{1\text{ m}}{3.28\text{ ft}}$
miles and kilometers	1.61 km = 1 mi	$\frac{1.61\text{ km}}{1\text{ mi}}$ or $\frac{1\text{ mi}}{1.61\text{ km}}$
Area		
square inches and square centimeters	$6.45\text{ cm}^2 = 1\text{ in}^2$	$\frac{6.45\text{ cm}^2}{1\text{ in}^2}$ or $\frac{1\text{ in}^2}{6.45\text{ cm}^2}$
square meters and square yards	$1.196\text{ yd}^2 = 1\text{ m}^2$	$\frac{1.196\text{ yd}^2}{1\text{ m}^2}$ or $\frac{1\text{ m}^2}{1.196\text{ yd}^2}$
acres and hectares	1 ha = 2.47 acres	$\frac{2.47\text{ acres}}{1\text{ ha}}$ or $\frac{1\text{ ha}}{2.47\text{ acres}}$
Volume		
cubic inches and milliliters	$16.39\text{ mL} = 1\text{ in}^3$	$\frac{16.39\text{ mL}}{1\text{ in}^3}$ or $\frac{1\text{ in}^3}{16.39\text{ mL}}$
liters and quarts	1.06 qt = 1 liter	$\frac{1.06\text{ qt}}{1\text{ liter}}$ or $\frac{1\text{ liter}}{1.06\text{ qt}}$
gallons and liters	3.79 liters = 1 gal	$\frac{3.79\text{ liters}}{1\text{ gal}}$ or $\frac{1\text{ gal}}{3.79\text{ liters}}$
Weight		
ounces and grams	28.3 g = 1 oz	$\frac{28.3\text{ g}}{1\text{ oz}}$ or $\frac{1\text{ oz}}{28.3\text{ g}}$
kilograms and pounds	2.20 lb = 1 kg	$\frac{2.20\text{ lb}}{1\text{ kg}}$ or $\frac{1\text{ kg}}{2.20\text{ lb}}$

There are many other conversion factors that we could have included in Table 1. We have listed only the most common ones. Almost all of them are approximations. That is, most of the conversion factors are decimals that have been rounded to the nearest hundredth. If we want more accuracy, we obtain a table that has more digits in the conversion factors.

EXAMPLE 1 Convert 8 inches to centimeters.

SOLUTION Choosing the appropriate conversion factor from Table 1, we have

$$8 \text{ in.} = 8 \text{ in.} \times \frac{2.54 \text{ cm}}{1 \text{ in.}}$$

$$= 8 \times 2.54 \text{ cm} \quad \text{This calculation should be done on a calculator}$$

$$= 20.32 \text{ cm}$$

EXAMPLE 2 Convert 80.5 kilometers to miles.

SOLUTION Using the conversion factor that takes us from kilometers to miles, we have

$$80.5 \text{ km} = 80.5 \text{ km} \times \frac{1 \text{ mi}}{1.61 \text{ km}}$$

$$= \frac{80.5}{1.61} \text{ mi} \quad \text{This calculation should be done on a calculator}$$

$$= 50 \text{ mi}$$

So 50 miles is equivalent to 80.5 kilometers. If we travel at 50 miles per hour in a car, we are moving at the rate of 80.5 kilometers per hour.

EXAMPLE 3 Convert 3 liters to pints.

SOLUTION Because Table 1 doesn't list a conversion factor that will take us directly from liters to pints, we first convert liters to quarts, and then convert quarts to pints.

$$3 \text{ liters} = 3 \text{ liters} \times \frac{1.06 \text{ qt}}{1 \text{ liter}} \times \frac{2 \text{ pt}}{1 \text{ qt}}$$

$$= 3 \times 1.06 \times 2 \text{ pt} \quad \text{This calculation should be done on a calculator}$$

$$= 6.36 \text{ pt}$$

EXAMPLE 4 The engine in a car has a 2-liter displacement. What is the displacement in cubic inches?

SOLUTION We convert liters to milliliters and then milliliters to cubic inches:

$$2 \text{ liters} = 2 \text{ liters} \times \frac{1{,}000 \text{ mL}}{1 \text{ liter}} \times \frac{1 \text{ in}^3}{16.39 \text{ mL}}$$

$$= \frac{2 \times 1{,}000}{16.39} \text{ in}^3 \quad \text{This calculation should be done on a calculator}$$

$$= 122 \text{ in}^3 \quad \text{To the nearest cubic inch}$$

EXAMPLE 5 If a person weighs 125 pounds, what is her weight in kilograms?

SOLUTION Converting from pounds to kilograms, we have

$$125 \text{ lb} = 125 \text{ lb} \times \frac{1 \text{ kg}}{2.20 \text{ lb}}$$

$$= \frac{125}{2.20} \text{kg}$$

$$= 56.8 \text{ kg} \quad \text{To the nearest tenth}$$

Temperature

We end this section with a discussion of temperature in both systems of measurement. In the U.S. system we measure temperature on the Fahrenheit scale. On this scale, water boils at 212 degrees and freezes at 32 degrees. When we write 32 degrees measured on the Fahrenheit scale, we use the notation

32°F (read, "32 degrees Fahrenheit")

In the metric system the scale we use to measure temperature is the Celsius scale (formerly called the centigrade scale). On this scale, water boils at 100 degrees and freezes at 0 degrees. When we write 100 degrees measured on the Celsius scale, we use the notation

100°C (read, "100 degrees Celsius")

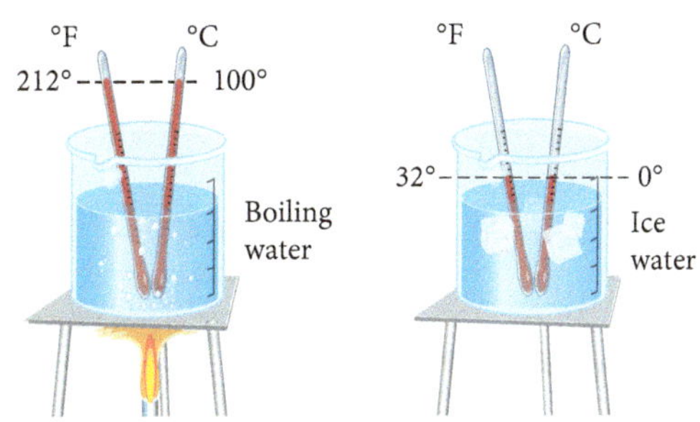

Table 2 is intended to give you a feel for the relationship between the two temperature scales. Table 3 gives the formulas, in both symbols and words, that are used to convert between the two scales.

TABLE 2

Situation	Temperature Fahrenheit	Temperature Celsius
Water freezes	32°F	0°C
Room temperature	68°F	20°C
Normal body temperature	98.6°F	37°C
Water boils	212°F	100°C
Bake cookies	350°F	176.7°C
Broil meat	554°F	290°C

TABLE 3

To Convert From	Formula In Symbols	Formula In Words
Fahrenheit to Celsius	$C = \frac{5(F - 32)}{9}$	Subtract 32, multiply by 5, and then divide by 9.
Celsius to Fahrenheit	$F = \frac{9}{5}C + 32$	Multiply by $\frac{9}{5}$, and then add 32.

The following examples show how we use the formulas given in Table 3.

EXAMPLE 6 Convert 120°C to degrees Fahrenheit.

SOLUTION We use the formula

$$F = \frac{9}{5}C + 32$$

and replace C with 120:

When	$C = 120$
the formula	$F = \frac{9}{5}C + 32$
becomes	$F = \frac{9}{5}(120) + 32$
	$F = 216 + 32$
	$F = 248$

We see that 120°C is equivalent to 248°F; they both mean the same temperature.

EXAMPLE 7 A man with the flu has a temperature of 102°F. What is his temperature on the Celsius scale?

SOLUTION

When	$F = 102$	
the formula	$C = \frac{5(F - 32)}{9}$	
becomes	$C = \frac{5(102 - 32)}{9}$	
	$C = \frac{5(70)}{9}$	
	$C = 38.9$	Rounded to the nearest tenth

The man's temperature, rounded to the nearest tenth, is 38.9°C on the Celsius scale.

GETTING READY FOR CLASS

After reading through the preceding section, respond in your own words and in complete sentences.

1. Write the equality that gives the relationship between centimeters and inches.
2. Write the equality that gives the relationship between grams and ounces.
3. Fill in the numerators below so that each conversion factor is equal to 1.

   a. $\frac{\text{ft}}{1\text{ meter}}$ b. $\frac{\text{qt}}{1\text{ liter}}$ c. $\frac{\text{lb}}{1\text{ kg}}$
4. Is it a hot day if the temperature outside is 37°C?

Problem Set 5.4

Use Tables 1 and 3 to make the following conversions.

1. 6 in. to centimeters
2. 1 ft to centimeters
3. 4 m to feet
4. 2 km to feet
5. 6 m to yards
6. 15 mi to kilometers
7. 20 mi to meters (round to the nearest hundred meters)
8. 600 m to yards
9. 5 m^2 to square yards
10. 2 in^2 to square centimeters
11. 10 ha to acres
12. 50 a to acres
13. 500 in^3 to milliliters
14. 400 in^3 to liters
15. 2 L to quarts
16. 15 L to quarts
17. 20 gal to liters
18. 15 gal to liters
19. 12 oz to grams
20. 1 lb to grams (round to the nearest 10 grams)
21. 15 kg to pounds
22. 10 kg to ounces
23. 185°C to degrees Fahrenheit
24. 20°C to degrees Fahrenheit
25. 86°F to degrees Celsius
26. 122°F to degrees Celsius

Calculator Problems

Set up the following problems as we have set up the examples in this section. Then use a calculator for the calculations and round your answers to the nearest hundredth.

27. 10 cm to inches
28. 100 mi to kilometers
29. 25 ft to meters
30. 400 mL to cubic inches
31. 49 qt to liters
32. 65 L to gallons
33. 500 g to ounces
34. 100 lb to kilograms
35. **Weight** Give your weight in kilograms.
36. **Height** Give your height in meters and centimeters.
37. **Sports** The 100-yard dash is a popular race in track. How far is 100 yards in meters?
38. **Engine Displacement** A 351-cubic-inch engine has a displacement of how many liters?
39. **Sewing** 25 square yards of material is how many square meters?
40. **Weight** How many grams does a 5 lb 4 oz roast weigh?
41. **Speed** 55 miles per hour is equivalent to how many kilometers per hour?
42. **Capacity** A 1-quart container holds how many liters?

43. **Sports** A high jumper jumps 6 ft 8 in. How many meters is this?

44. **Farming** A farmer owns 57 acres of land. How many hectares is that?

45. **Body Temperature** A person has a temperature of 101°F. What is the person's temperature, to the nearest tenth, on the Celsius scale?

46. **Air Temperature** If the temperature outside is 30°C, is it a better day for water skiing or for snow skiing?

Getting Ready for the Next Section

Perform the indicated operations.

47. $15 + 60$

48. $25 + 60$

49. $37 + 45$

50. $27 + 46$

51. $3 + 0.25$

52. $2 + 0.75$

53. $82 - 60$

54. $73 - 60$

55. $75 - 34$

56. $85 - 42$

57. 12×4

58. 8×4

59. $3 \times 60 + 15$

60. $2 \times 65 + 45$

61. $3 + 16 \times \frac{1}{60}$

62. $2 + 45 \times \frac{1}{60}$

63. If fish costs $6.00 per pound, find the cost of 15 pounds.

64. If fish costs $5.00 per pound, find the cost of 14 pounds.

Operations with Time and Mixed Units 5.5

Many occupations require the use of a time card. A time card records the number of hours and minutes at work. At the end of a work week the hours and minutes are totaled separately, and then the minutes are converted to hours.

In this section we will perform operations with mixed units of measure. Mixed units are used when we use 2 hours 30 minutes, rather than 2 and a half hours, or 5 feet 9 inches, rather than five and three-quarter feet. As you will see, many of these types of problems arise in everyday life.

The relationship between	is	To convert from one to the other, multiply by
minutes and seconds	1 min = 60 sec	$\frac{1\text{ min}}{60\text{ sec}}$ or $\frac{60\text{ sec}}{1\text{ min}}$
hours and minutes	1 hr = 60 min	$\frac{1\text{ hr}}{60\text{ min}}$ or $\frac{60\text{ min}}{1\text{ hr}}$

EXAMPLE 1 Convert 3 hours 15 minutes to

a. Minutes **b.** Hours

SOLUTION **a.** To convert to minutes, we multiply the hours by the conversion factor then add minutes:

$$3\text{ hr }15\text{ min} = 3\,\cancel{\text{hr}} \times \frac{60\text{ min}}{1\,\cancel{\text{hr}}} + 15\text{ min}$$
$$= 180\text{ min} + 15\text{ min}$$
$$= 195\text{ min}$$

b. To convert to hours, we multiply the minutes by the conversion factor then add hours:

$$3\text{ hr }15\text{ min} = 3\text{ hr} + 15\,\cancel{\text{min}} \times \frac{1\text{ hr}}{60\,\cancel{\text{min}}}$$
$$= 3\text{ hr} + 0.25\text{ hr}$$
$$= 3.25\text{ hr}$$

EXAMPLE 2 Add 5 minutes 37 seconds and 7 minutes 45 seconds.

SOLUTION First, we align the units properly

$$\begin{array}{rr} 5\text{ min} & 37\text{ sec} \\ +\ 7\text{ min} & 45\text{ sec} \\ \hline 12\text{ min} & 82\text{ sec} \end{array}$$

Since there are 60 seconds in every minute, we write 82 seconds as 1 minute 22 seconds. We have

$$12\text{ min }82\text{ sec} = 12\text{ min} + 1\text{ min }22\text{ sec}$$
$$= 13\text{ min }22\text{ sec}$$

The idea of adding the units separately is similar to adding mixed fractions. That is, we align the whole numbers with the whole numbers and the fractions with the fractions.

Similarly, when we subtract units of time, we "borrow" 60 seconds from the minutes column, or 60 minutes from the hours column.

EXAMPLE 3 Subtract 34 minutes from 8 hour 15 minutes.

SOLUTION Again, we first line up the numbers in the hours column, and then the numbers in the minutes column:

$$\begin{array}{rr} 8\text{ hr} & 15\text{ min} \\ - & 34\text{ min} \\ \hline \end{array} \qquad \begin{array}{rr} 7\text{ hr} & 75\text{ min} \\ - & 34\text{ min} \\ \hline 7\text{ hr} & 41\text{ min} \end{array}$$

Next we see how to multiply and divide using units of measure.

EXAMPLE 4 Jake purchases 4 halibut. The fish cost \$6.00 per pound, and each weighs 3 lb 12 oz. What is the cost of the fish?

SOLUTION First, we multiply each unit by 4:

$$\begin{array}{rr} 3\text{ lb} & 12\text{ oz} \\ \times & 4 \\ \hline 12\text{ lb} & 48\text{ oz} \end{array}$$

To convert the 48 ounces to pounds, we multiply the ounces by the conversion factor

$$\begin{aligned} 12\text{ lb } 48\text{ oz} &= 12\text{ lb} + 48\text{ oz} \times \frac{1\text{ lb}}{16\text{ oz}} \\ &= 12\text{ lb} + 3\text{ lb} \\ &= 15\text{ lb} \end{aligned}$$

Finally, we multiply the 15 lb and \$6.00/lb for a total price of \$90.00.

GETTING READY FOR CLASS

After reading through the preceding section, respond in your own words and in complete sentences.

1. Explain the difference between saying *2 and a half hours* and saying *2 hours and 50 minutes.*
2. How are operations with mixed units of measure similar to operations with mixed numbers?
3. Why do we borrow a 60 from the minutes column for the seconds column when subtracting in Example 3?
4. Give an example of when you may have to use multiplication with mixed units of measure.

Problem Set 5.5

Use the tables of conversion factors given in this section and other sections in this chapter to make the following conversions. (Round your answers to the nearest hundredth.)

1. 4 hours 30 minutes to
a. Minutes
b. Hours

2. 2 hours 45 minutes to
a. Minutes
b. Hours

3. 5 hours 20 minutes to
a. Minutes
b. Hours

4. 4 hours 40 minutes to
a. Minutes
b. Hours

5. 6 minutes 30 seconds to
a. Seconds
b. Minutes

6. 8 minutes 45 seconds to
a. Seconds
b. Minutes

7. 5 minutes 20 seconds to
a. Seconds
b. Minutes

8. 4 minutes 40 seconds to
a. Seconds
b. Minutes

9. 2 pounds 8 ounces to
a. Ounces
b. Pounds

10. 3 pounds 4 ounces to
a. Ounces
b. Pounds

11. 4 pounds 12 ounces to
a. Ounces
b. Pounds

12. 5 pounds 16 ounces to
a. Ounces
b. Pounds

13. 4 feet 6 inches to
a. Inches
b. Feet

14. 3 feet 3 inches to
a. Inches
b. Feet

15. 5 feet 9 inches to
a. Inches
b. Feet

16. 3 feet 4 inches to
a. Inches
b. Feet

17. 2 gallons 1 quart
a. Quarts
b. Gallons

18. 3 gallons 2 quarts
a. Quarts
b. Gallons

Perform the indicated operation. Again, remember to use the appropriate conversion factor.

19. Add 4 hours 47 minutes and 6 hours 13 minutes.

20. Add 5 hours 39 minutes and 2 hours 21 minutes.

21. Add 8 feet 10 inches and 13 feet 6 inches.

22. Add 16 feet 7 inches and 7 feet 9 inches.

23. Add 4 pounds 12 ounces and 6 pounds 4 ounces.

24. Add 11 pounds 9 ounces and 3 pounds 7 ounces.

25. Subtract 2 hours 35 minutes from 8 hours 15 minutes.

26. Subtract 3 hours 47 minutes from 5 hours 33 minutes.

27. Subtract 3 hours 43 minutes from 7 hours 30 minutes.

28. Subtract 1 hour 44 minutes from 6 hours 22 minutes.

29. Subtract 4 hours 17 minutes from 5 hours 9 minutes.

30. Subtract 2 hours 54 minutes from 3 hours 7 minutes.

Applying the Concepts

iStockPhoto/©Paul Gsell

Triathlon The Ironman Triathlon World Championship, held each October in Kona on the island of Hawaii, consists of three parts: a 2.4-mile ocean swim, a 112-mile bike race, and a 26.2-mile marathon. The table shows the results from the 2003 event.

Triathlete	Swim Time (Hr:Min:Sec)	Bike Time (Hr:Min:Sec)	Run Time (Hr:Min:Sec)	Total Time (Hr:Min:Sec)
Peter Reid	0:50:36	4:40:04	2:47:38	
Lori Bowden	0:56:51	5:09:00	3:02:10	

31. Fill in the total time column.

32. How much faster was Peter's total time than Lori's?

33. How much faster was Peter than Lori in the swim?

34. How much faster was Peter than Lori in the run?

35. **Cost of Fish** Fredrick is purchasing four whole salmon. The fish cost $4.00 per pound, and each weighs 6 lb 8 oz. What is the cost of the fish?

36. **Cost of Steak** Mike is purchasing eight top sirloin steaks. The meat costs $4.00 per pound, and each steak weighs 1 lb 4 oz. What is the total cost of the steaks?

37. **Stationary Bike** Maggie rides a stationary bike for 1 hour and 15 minutes, 4 days a week. After 2 weeks, how many hours has she spent riding the stationary bike?

38. **Gardening** Scott works in his garden for 1 hour and 5 minutes, 3 days a week. After 4 weeks, how many hours has Scott spent gardening?

39. **Cost of Fabric** Allison is making a quilt. She buys 3 yards and 1 foot each of six different fabrics. The fabrics cost $7.50 a yard. How much will Allison spend?

40. **Cost of Lumber** Trish is building a fence. She buys six fence posts at the lumberyard, each measuring 5 ft 4 in. The lumber costs $3 per foot. How much will Trish spend?

41. **Cost of Avocados** Jacqueline is buying six avocados. Each avocado weighs 8 oz. How much will they cost her if avocados cost $2.00 a pound?

42. **Cost of Apples** Mary is purchasing 12 apples. Each apple weighs 4 oz. If the cost of the apples is $1.50 a pound, how much will Mary pay?

Chapter 5 Summary

Conversion Factors [5.1, 5.2, 5.3, 5.4, 5.5]

EXAMPLES

1. Convert 5 feet to inches.

$5 \text{ ft} = 5 \text{ ft} \times \frac{12 \text{ in.}}{1 \text{ ft}}$
$= 5 \times 12 \text{ in.}$
$= 60 \text{ in.}$

To convert from one kind of unit to another, we choose an appropriate conversion factor from one of the tables given in this chapter. For example, if we want to convert 5 feet to inches, we look for conversion factors that give the relationship between feet and inches. There are two conversion factors for feet and inches:

$$\frac{12 \text{ in.}}{1 \text{ ft}} \quad \text{and} \quad \frac{1 \text{ ft}}{12 \text{ in.}}$$

Length [5.1]

2. Convert 8 feet to yards.

$8 \text{ ft} = 8 \text{ ft} \times \frac{1 \text{ yd}}{3 \text{ ft}}$
$= \frac{8}{3} \text{ yd}$
$= 2\frac{2}{3} \text{ yd}$

U.S. SYSTEM

The Relationship Between	Is	To Convert From One To The Other, Multiply By
feet and inches	12 in. = 1 ft	$\frac{12 \text{ in.}}{1 \text{ ft}}$ or $\frac{1 \text{ ft}}{12 \text{ in.}}$
feet and yards	1 yd = 3 ft	$\frac{3 \text{ ft}}{1 \text{ yd}}$ or $\frac{1 \text{ yd}}{3 \text{ ft}}$
feet and miles	1 mi = 5,280 ft	$\frac{5{,}280 \text{ ft}}{1 \text{ mi}}$ or $\frac{1 \text{ mi}}{5{,}280 \text{ ft}}$

3. Convert 25 millimeters to meters.

25 mm
$= 25 \text{ mm} \times \frac{1 \text{ m}}{1{,}000 \text{ mm}}$
$= \frac{25 \text{ m}}{1{,}000}$
$= 0.025 \text{ m}$

METRIC SYSTEM

The Relationship Between	Is	To Convert From One To The Other, Multiply By
millimeters (mm) and meters (m)	1,000 mm = 1 m	$\frac{1{,}000 \text{ mm}}{1 \text{ m}}$ or $\frac{1 \text{ m}}{1{,}000 \text{ mm}}$
centimeters (cm) and meters	100 cm = 1 m	$\frac{100 \text{ cm}}{1 \text{ m}}$ or $\frac{1 \text{ m}}{100 \text{ cm}}$
decimeters (dm) and meters	10 dm = 1 m	$\frac{10 \text{ dm}}{1 \text{ m}}$ or $\frac{1 \text{ m}}{10 \text{ dm}}$
dekameters (dam) and meters	1 dam = 10 m	$\frac{10 \text{ m}}{1 \text{ dam}}$ or $\frac{1 \text{ dam}}{10 \text{ m}}$
hectometers (hm) and meters	1 hm = 100 m	$\frac{100 \text{ m}}{1 \text{ hm}}$ or $\frac{1 \text{ hm}}{100 \text{ m}}$
kilometers (km) and meters	1 km = 1,000 m	$\frac{1{,}000 \text{ m}}{1 \text{ km}}$ or $\frac{1 \text{ km}}{1{,}000 \text{ m}}$

Area [5.2]

4. Convert 256 acres to square miles.

256 acres

$= 256 \text{ acres} \times \frac{1 \text{ mi}^2}{640 \text{ acres}}$

$= \frac{256}{640} \text{ mi}^2$

$= 0.4 \text{ mi}^2$

U.S. SYSTEM

The Relationship Between	Is	To Convert From One To The Other, Multiply By
square inches and square feet	$144 \text{ in}^2 = 1 \text{ ft}^2$	$\frac{144 \text{ in}^2}{1 \text{ ft}^2}$ or $\frac{1 \text{ ft}^2}{144 \text{ in}^2}$
square yards and square feet	$9 \text{ ft}^2 = 1 \text{ yd}^2$	$\frac{9 \text{ ft}^2}{1 \text{ yd}^2}$ or $\frac{1 \text{ yd}^2}{9 \text{ ft}^2}$
acres and square feet	$1 \text{ acre} = 43{,}560 \text{ ft}^2$	$\frac{43{,}560 \text{ ft}^2}{1 \text{ acre}}$ or $\frac{1 \text{ acre}}{43{,}560 \text{ ft}^2}$
acres and square miles	$640 \text{ acres} = 1 \text{ mi}^2$	$\frac{640 \text{ acres}}{1 \text{ mi}^2}$ or $\frac{1 \text{ mi}^2}{640 \text{ acres}}$

METRIC SYSTEM

The Relationship Between	Is	To Convert From One To The Other, Multiply By
square millimeters and square centimeters	$1 \text{ cm}^2 = 100 \text{ mm}^2$	$\frac{100 \text{ mm}^2}{1 \text{ cm}^2}$ or $\frac{1 \text{ cm}^2}{100 \text{ mm}^2}$
square centimeters and square decimeters	$1 \text{ dm}^2 = 100 \text{ cm}^2$	$\frac{100 \text{ cm}^2}{1 \text{ dm}^2}$ or $\frac{1 \text{ dm}^2}{100 \text{ cm}^2}$
square decimeters and square meters	$1 \text{ m}^2 = 100 \text{ dm}^2$	$\frac{100 \text{ dm}^2}{1 \text{ m}^2}$ or $\frac{1 \text{ m}^2}{100 \text{ dm}^2}$
square meters and ares (a)	$1 \text{ a} = 100 \text{ m}^2$	$\frac{100 \text{ m}^2}{1 \text{ a}}$ or $\frac{1 \text{ a}}{100 \text{ m}^2}$
ares and hectares (ha)	$1 \text{ ha} = 100 \text{ a}$	$\frac{100 \text{ a}}{1 \text{ ha}}$ or $\frac{1 \text{ ha}}{100 \text{ a}}$

Volume [5.2]

U.S. SYSTEM

The Relationship Between	Is	To Convert From One To The Other, Multiply By
cubic inches (in^3) and cubic feet (ft^3)	$1 \text{ ft}^3 = 1{,}728 \text{ in}^3$	$\frac{1{,}728 \text{ in}^3}{1 \text{ ft}^3}$ or $\frac{1 \text{ ft}^3}{1{,}728 \text{ in}^3}$
cubic feet and cubic yards (yd^3)	$1 \text{ yd}^3 = 27 \text{ ft}^3$	$\frac{27 \text{ ft}^3}{1 \text{ yd}^3}$ or $\frac{1 \text{ yd}^3}{27 \text{ ft}^3}$
fluid ounces (fl oz) and pints (pt)	$1 \text{ pt} = 16 \text{ fl oz}$	$\frac{16 \text{ fl oz}}{1 \text{ pt}}$ or $\frac{1 \text{ pt}}{16 \text{ fl oz}}$
pints and quarts (qt)	$1 \text{ qt} = 2 \text{ pt}$	$\frac{2 \text{ pt}}{1 \text{ qt}}$ or $\frac{1 \text{ qt}}{2 \text{ pt}}$
quarts and gallons (gal)	$1 \text{ gal} = 4 \text{ qt}$	$\frac{4 \text{ qt}}{1 \text{ gal}}$ or $\frac{1 \text{ gal}}{4 \text{ qt}}$

5. Convert 2.2 liters to milliliters.

2.2 liters

$= 2.2 \text{ liters} \times \frac{1{,}000 \text{ mL}}{1 \text{ liter}}$

$= 2.2 \times 1{,}000 \text{ mL}$

$= 2{,}200 \text{ mL}$

METRIC SYSTEM

The Relationship Between	Is	To Convert From One To The Other, Multiply By
milliliters (mL) and liters	1 liter (L) = 1,000 mL	$\frac{1{,}000 \text{ mL}}{1 \text{ liter}}$ or $\frac{1 \text{ liter}}{1{,}000 \text{ mL}}$
hectoliters (hL) and liters	100 liters = 1 hL	$\frac{100 \text{ liters}}{1 \text{ hL}}$ or $\frac{1 \text{ hL}}{100 \text{ liters}}$
kiloliters (kL) and liters	1,000 liters (L) = 1 kL	$\frac{1{,}000 \text{ liters}}{1 \text{ kL}}$ or $\frac{1 \text{ kL}}{1{,}000 \text{ liters}}$

Weight [5.3]

6. Convert 12 pounds to ounces.

$12 \text{ lb} = 12 \text{ lb} \times \frac{16 \text{ oz}}{1 \text{ lb}}$

$= 12 \times 16 \text{ oz}$

$= 192 \text{ oz}$

U.S. SYSTEM

The Relationship Between	Is	To Convert From One To The Other, Multiply By
ounces (oz) and pounds (lb)	1 lb = 16 oz	$\frac{16 \text{ oz}}{1 \text{ lb}}$ or $\frac{1 \text{ lb}}{16 \text{ oz}}$
pounds and tons (T)	1 T = 2,000 lb	$\frac{2{,}000 \text{ lb}}{1 \text{ T}}$ or $\frac{1 \text{ T}}{2{,}000 \text{ lb}}$

7. Convert 3 kilograms to centigrams.

3 kg

$= 3 \text{ kg} \times \frac{1{,}000 \text{ g}}{1 \text{ kg}} \times \frac{100 \text{ cg}}{1 \text{ g}}$

$= 3 \times 1{,}000 \times 100 \text{ cg}$

$= 300{,}000 \text{ cg}$

METRIC SYSTEM

The Relationship Between	Is	To Convert From One To The Other, Multiply By
milligrams (mg) and grams (g)	1 g = 1,000 mg	$\frac{1{,}000 \text{ mg}}{1 \text{ g}}$ or $\frac{1 \text{ g}}{1{,}000 \text{ mg}}$
centigrams (cg) and grams	1 g = 100 cg	$\frac{100 \text{ cg}}{1 \text{ g}}$ or $\frac{1 \text{ g}}{100 \text{ cg}}$
kilograms (kg) and grams	1,000 g = 1 kg	$\frac{1{,}000 \text{ g}}{1 \text{ kg}}$ or $\frac{1 \text{ kg}}{1{,}000 \text{ g}}$
metric tons (t) and kilograms	1,000 kg = 1 t	$\frac{1{,}000 \text{ kg}}{1 \text{ t}}$ or $\frac{1 \text{ t}}{1{,}000 \text{ kg}}$

Converting Between the Systems [5.4]

8. Convert 8 inches to centimeters.

$$8 \text{ in.} = 8 \text{ in.} \times \frac{2.54 \text{ cm}}{1 \text{ in.}}$$
$$= 8 \times 2.54 \text{ cm}$$
$$= 20.32 \text{ cm}$$

CONVERSION FACTORS

The Relationship Between	Is	To Convert From One To The Other, Multiply By
Length		
inches and centimeters	2.54 cm = 1 in.	$\frac{2.54 \text{ cm}}{1 \text{ in.}}$ or $\frac{1 \text{ in.}}{2.54 \text{ cm}}$
feet and meters	1 m = 3.28 ft	$\frac{3.28 \text{ ft}}{1 \text{ m}}$ or $\frac{1 \text{ m}}{3.28 \text{ ft}}$
miles and kilometers	1.61 km = 1 mi	$\frac{1.61 \text{ km}}{1 \text{ mi}}$ or $\frac{1 \text{ mi}}{1.61 \text{ km}}$
Area		
square inches and square centimeters	6.45 cm² = 1 in²	$\frac{6.45 \text{ cm}^2}{1 \text{ in}^2}$ or $\frac{1 \text{ in}^2}{6.45 \text{ cm}^2}$
square meters and square yards	1.196 yd² = 1 m²	$\frac{1.196 \text{ yd}^2}{1 \text{ m}^2}$ or $\frac{1 \text{ m}^2}{1.196 \text{ yd}^2}$
acres and hectares	1 ha = 2.47 acres	$\frac{2.47 \text{ acres}}{1 \text{ ha}}$ or $\frac{1 \text{ ha}}{2.47 \text{ acres}}$
Volume		
cubic inches and milliliters	16.39 mL = 1 in³	$\frac{16.39 \text{ mL}}{1 \text{ in}^3}$ or $\frac{1 \text{ in}^3}{16.39 \text{ mL}}$
liters and quarts	1.06 qt = 1 liter	$\frac{1.06 \text{ qt}}{1 \text{ liter}}$ or $\frac{1 \text{ liter}}{1.06 \text{ qt}}$
gallons and liters	3.79 liters = 1 gal	$\frac{3.79 \text{ liters}}{1 \text{ gal}}$ or $\frac{1 \text{ gal}}{3.79 \text{ liters}}$
Weight		
ounces and grams	28.3 g = 1 oz	$\frac{28.3 \text{ g}}{1 \text{ oz}}$ or $\frac{1 \text{ oz}}{28.3 \text{ g}}$
kilograms and pounds	2.20 lb = 1 kg	$\frac{2.20 \text{ lb}}{1 \text{ kg}}$ or $\frac{1 \text{ kg}}{2.20 \text{ lb}}$

Temperature [7.4]

9. Convert 120°C to degrees Fahrenheit.

$$F = \frac{9}{5}C + 32$$
$$F = \frac{9}{5}(120) + 32$$
$$F = 216 + 32$$
$$F = 248$$

To Convert From	Formula In Symbols	Formula In Words
Fahrenheit to Celsius	$C = \frac{5(F-32)}{9}$	Subtract 32, multiply by 5, and then divide by 9.
Celsius to Fahrenheit	$F = \frac{9}{5}C + 32$	Multiply by $\frac{9}{5}$, and then add 32.

Time [7.5]

10. Convert 3 hours 45 minutes to minutes.

$$= 3 \text{ hr} \times \frac{60 \text{ min}}{1 \text{ hr}} + 45 \text{ min}$$
$$= 180 \text{ min} + 45 \text{ min}$$
$$= 225 \text{ min}$$

The relationship between	is	To convert from one to the other, multiply by
minutes and seconds	1 min = 60 sec	$\frac{1 \text{ min}}{60 \text{ sec}}$ or $\frac{60 \text{ sec}}{1 \text{ min}}$
hours and minutes	1 hr = 60 min	$\frac{1 \text{ hr}}{60 \text{ min}}$ or $\frac{60 \text{ min}}{1 \text{ hr}}$

Chapter 5 Test

These problems are all taken from examples in your text. Work each problem and check your answers. If you have made a mistake, work the problem again. If you cannot get the correct answer after two tries, look up the correct solution in your text.

1. A football field is 100 yards long. Convert 100 yards to inches.
2. Convert 25 millimeters to meters.
3. Convert 36.5 centimeters to decimeters.
4. Find the number of square inches in 1 square foot.
5. How many square millimeters are in 1 square meter?
6. A dairy herd produces 1,800 quarts of milk each day. How many gallons is this?
7. Convert 12 pounds to ounces.
8. Convert 3 kilograms to centigrams.
9. Convert 80.5 kilometers to miles.
10. The engine in a car has a 2-liter displacement. What is the displacement in cubic inches?
11. A man with the flu has a temperature of 102 (degrees) F. What is his temperature in degrees Celsius?
12. Convert 3 hours 15 minutes to:

    a. Minutes **b.** Hours
13. Subtract 34 minutes from 8 hours 15 minutes

6 Descriptive Statistics

Chapter Outline

Baton Rouge, Louisiana

The Google Earth photo shows the Atchafalaya Basin Bridge in Louisiana. The bridge measures 96,100 feet long and is currently the third longest bridge in the United States. The following table lists the top five longest bridges in the United States.

Bridge	Length
Lake Pontchartrain Causeway, Louisiana	126,122 feet
Manchac Swamp Bridge, Louisiana	120,400 feet
Atchafalaya Basin Bridge, Louisiana	96,100 feet
Chesapeake Bay Bridge-Tunnel, Virginia	79,200 feet
Bonnet Carré Spillway Bridge of I-10, Louisiana	58,077 feet

Based on the above table, we can see the longest and shortest bridges of the given data. The difference between these two measurements gives us the **range** of the data. In this chapter, we will define range in greater detail, as well as other terms that will help us evaluate a given set of numbers. Furthermore, we will continue the work we have done previously with averages, bar charts, line graphs, and pie charts. You will also create some charts of your own.

© Silvia Boratti/iStockphoto

Success Skills

Don't complain about anything, ever.

Do you complain to your classmates about your teacher? If you do, it could be getting in the way of your success in the class.

Some of my students tell me that they like the way I teach. Others in the same class complain to each other about me. They say I don't explain things well enough. Are the complaining students giving themselves a reason for not doing well in the class? I think so. They are shifting the responsibility for their success from themselves to me. When these students are alone, trying to do homework, they start thinking about how unfair everything is, and they lose their motivation to study. Without intending to, they have set themselves up to fail.

What happens when you stop complaining? You put yourself back in charge of your success. When there is no one to blame if things don't go well, you are more likely to do well. Students have told me that, once they stopped complaining about a class, the teacher became a better teacher, and they started to actually enjoy going to class.

If you find yourself complaining to your friends about a class or a teacher, decide to stop. Don't participate in complaining sessions. Try it for a day, or a week, or for the rest of the term. It may be difficult at first, but I'm sure you will like the results.

Watch the Video

Watch the Video

6.1

Mean, Median, and Mode

Objectives

A. Find the mean given a set of numbers.

B. Find the median given a set of numbers.

C. Find the mode given a set of numbers.

D. Find the range given a set of numbers.

In the 2014 Winter Olympic Games, the United States four-man bobsleigh team of Steve Holcomb, Steve Langton, Curtis Tomasevicz, and Chris Fogt took the bronze medal for their event. The team competed on a 1365-meter-long track at the Sanki Olympic Sliding Center near Sochi, Russia. During four runs, the team logged a total time of 3:40.99 (read "3 minutes, 40 seconds, and 99 one hundredths of a second"). The following table shows the team's individual run times.

Run	Time
Run 1	54.89 seconds
Run 2	55.47 seconds
Run 3	55.30 seconds
Run 4	55.33 seconds

In this section, we will be discussing averages, which will give us the tools to further evaluate the bobsleigh team's times. For instance, we can calculate their mean run time as 55.25 seconds.

Average

In everyday language, the word *average* can represent the mean, median, or mode for a set of data. If we go online to the Merriam-Webster dictionary at www.m-w.com, we find the following definition for the word average when it is used as a noun:

> **Average**
>
> **av · er · age** *noun*: a single value (as a mean, mode, or median) that summarizes or represents the general significance of a set of unequal values …

Mean

The mean is probably the most common measure of central tendency and it refers to the arithmetic average for a set of numbers. It is typically used to describe data such as test scores, salary, and various averages having to do with sports.

> **Mean**
>
> To find the **mean** for a set of numbers, we add all the numbers and then divide the sum by the number of numbers in the set. The mean is sometimes called the **arithmetic mean.**

VIDEO EXAMPLES

SECTION 9.1

Example 1 Enrollment in college over a five-year period is shown by the following numbers. Find the mean enrollment over the five-year period.

16,911,000 17,272,000 17,487,000 17,759,000 18,248,000

Solution We add the five enrollments and then divide by 5, the number of values in the set.

$$\text{Mean} = \frac{16{,}911{,}000 + 17{,}272{,}000 + 17{,}487{,}000 + 17{,}759{,}000 + 18{,}248{,}000}{5}$$

$$= \frac{87{,}677{,}000}{5} = 17{,}535{,}400$$

The mean enrollment in college over the last five years is 17,535,400 students.

© Steve Debenport/iStockPhoto

Example 2 A firefighter in a large southern California city earned the following salaries for his first five years on the job. Find the mean of these salaries.

$59,687 $60,880 $62,098 $63,651 $65,879

Solution We add the five numbers and then divide by 5, the number of numbers in the set.

$$\text{Mean} = \frac{59{,}687 + 60{,}880 + 62{,}098 + 63{,}651 + 65{,}879}{5} = \frac{312{,}195}{5} = 62{,}439$$

The firefighter's mean salary for the first five years was $62,439 per year.

Median

The table below shows the hourly wages for some occupations in the U.S. in May 2011.

Hourly Wages

Occupation	Wage
All Occupations	$16.57
Waiters/Waitresses	$8.93
School Bus Driver	$13.51
Clergy	$21.22
Firefighter	$21.84
Social Worker	$25.91
Veterinarians	$39.86
Lawyer	$54.48

Source: U.S. Bureau of Labor Statistics
(all wages are median figures for 2011)

If you look at the type at the bottom of the table shown above, you can see that the numbers are the *median* figures for 2011. The median for a set of numbers is the number such that half of the numbers in the set are above it and half are below it. Here is the exact definition:

The **median** for a set of numbers is the number in the middle of the data, meaning half the numbers in the set have a value less than the median, and half the numbers in the set have a value greater than the median. Housing prices are frequently described by using a median.

Median

To find the **median** for a set of numbers, we write the numbers in order from smallest to largest. If there is an odd number of numbers, the median is the middle number. If there is an even number of numbers, then the median is the mean of the two numbers in the middle.

Example 3 Find the median enrollment from the numbers in Example 1.

Solution The numbers in Example 1 are already written from smallest to largest. Because there are an odd number of numbers in the set, the median is the middle number.

16,911,000 17,272,000 17,487,000 17,759,000 18,248,000

↑ Median (17,487,000)

The median enrollment for the five years is 17,487,000.

Example 4 These selling prices of four hybrid cars were listed on hybridcars.com.

\$23,063 \$31,700 \$28,600 \$24,650

Find the mean and the median for the four prices. Round to the nearest cent if necessary.

Solution To find the mean, we add the four numbers and then divide by 4:

$$\frac{23{,}063 + 31{,}700 + 28{,}600 + 24{,}650}{4} = \frac{108{,}013}{4} = 27{,}003.25$$

To find the median, we write the numbers in order from smallest to largest. Then, because there is an even number of numbers, we average the middle two numbers to obtain the median.

\$23,063 \$24,650 \$28,600 \$31,700

Median (\$24,650 and \$28,600)

↓

$$\frac{24{,}650 + 28{,}600}{2} = 26{,}625$$

The mean is \$27,003.25, and the median is \$26,625.

Mode

The **mode** is best used when we are looking for the most common eye color in a group of people, the most popular breed of dog in the United States, or the movie that was seen the most often. When we have a set of numbers in which one number occurs more often than the rest, that number is the mode. For example, consider the following set of golf caps:

Given the set of caps, the most popular color is green. We call this the mode.

> **Mode**
> The **mode** for a set of numbers is the value that occurs most frequently. If all the numbers in the set occur the same number of times, there is no mode.

Example 5 A math class with 18 students had the grades shown below on their first test. Find the mean, the median, and the mode.

77	87	100	65	79	87
79	85	87	95	56	87
56	75	79	93	97	92

Solution To find the mean, we add all the numbers and divide by 18.

$$\text{Mean} = \frac{77+87+100+65+79+87+79+85+87+95+56+87+56+75+79+93+97+92}{18}$$

$$= \frac{1{,}476}{18} = 82$$

To find the median, we must put the test scores in order from smallest to largest; then, because there are an even number of test scores, we must find the mean of the middle two scores.

56	56	65	75	77	79	79	79	85
87	87	87	87	92	93	95	97	100

$$\text{Median} = \frac{85 + 87}{2} = 86$$

The mode is the most frequently occurring score. Because 87 occurs 4 times, and no other scores occur that many times, 87 is the mode.

The mean is 82, the median is 86, and the mode is 87.

More Vocabulary

When we used the word average in the beginning of this section, we used it as a noun. It can also be used as an adjective and a verb. The following is the definition from an online dictionary of the word average when it is used as a verb:

> *av · er · age verb* **. . . 2 : to find the arithmetic mean of (a series of unequal quantities) . . .**

If you are asked for the **average** of a set of numbers, the word average can represent the mean, the median, or the mode. When used in this way, the word average is a noun. However, if you are asked to average a set of numbers, then the word average is a verb, and you are being asked to find the mean of the numbers.

Range

The **range** of a set of data is the difference between the least and greatest. For example, the following table shows some of the highest and lowest minimum wages across the country.

State	Minimum Wage
Arkansas	\$6.25
California	\$8.00
Connecticut	\$8.25
Minnesota	\$5.25
Oklahoma	\$2.00
Washington	\$8.67

Source: U.S. Department of Labor

From the information in the table, we see that the lowest minimum wage is found in Oklahoma at \$2.00 per hour for some businesses and the highest minimum wage is found in Washington at \$8.67. The range for this set of data is the difference between these two numbers.

$$\$8.67 - \$2.00 = \$6.67$$

We say that the minimum wage in the United States has a range of \$6.67.

> **Range**
> The **range** for a set of numbers is the difference between the largest number and the smallest number in the sample.

Example 6 Find the range of the test scores given in Example 5.

Solution The highest score is 100, and the lowest score is 56. The range is $100 - 56 = 44$.

Getting Ready for Class

After reading through the preceding section, respond in your own words and in complete sentences.

A. The word average can refer to what three mathematical concepts?
B. What is the median for a set of numbers?
C. What is the mode for a set of numbers?
D. What number must we use for x, if the average of 6, 8, and x is to be 8?

Problem Set 6.1

Find the mean for each set of numbers.

1. 1, 2, 3, 4, 5
2. 2, 4, 6, 8, 10
3. 1, 3, 9, 11
4. 5, 7, 9, 12, 12
5. 29,500, 10,650, 8,900, 15,120, 16,800
6. 8,040, 5,505, 4,121, 9,910
7. 12.5, 8.2, 1.8
8. 4.1, 6.9, 2.2, 3.6

Find the median for each set of numbers.

9. 5, 9, 11, 13, 15
10. 42, 48, 50, 64
11. 10, 20, 50, 90, 100
12. 500, 800, 1200, 1300
13. 900, 700, 1100
14. 850, 100, 225, 480
15. 1.0, 6.5, 3.2, 1.7, 2.1, 4.6, 3.9
16. 2.7, 3.4, 1.8, 1.1, 2.3, 3.0

Find the mode for each set of numbers.

17. 14, 18, 27, 36, 18, 73
18. 11, 27, 18, 11, 72, 11
19. 98, 87, 65, 73, 82, 87, 65, 97, 87, 77
20. 3.0, 3.2, 2.5, 4.0, 3.1, 3.1, 2.6, 1.9, 1.8, 3.4, 3.1, 2.0
21. 1, 1, 2, 3, 1, 3, 3, 2, 1, 2, 2, 3, 1
22. 5, 8, 9, 9, 6, 6, 7, 7, 5, 8, 6, 8, 9, 5, 8, 8, 9

Determine the range of the given data.

23. 15, 34, 12, 25, 27
24. 2.6, 4.1, 5.4, 3.9, 0.6
25. 1.0, 3.9, 2.1, 3.6, 2.9, 3.8
26. 12,000, 13,500, 10,120, 14,250, 11,490
27. 52, 69, 84, 81, 79, 46, 81, 73, 68
28. 4080, 2900, 1650, 1800, 1925, 690

Applying the Concepts

29. **Test Average** A student's scores for four exams in a basic math class were 79, 62, 87, and 90. What is the student's mean and median test score?

30. **Test Scores** A first-year math student had grades of 79, 64, 78, and 95 on the first four tests. What is the student's mean and median test grade?

31. **Average Salary** Over a 3-year period a woman's annual salaries were \$28,000, \$31,000, and \$34,000. What was her mean annual salary and median annual salary for this 3-year period?

© kzenon/iStockPhoto

32. **Bowling** If a person has scores of 205, 222, 174, 236, 185, and 215 for six games of bowling, what is the median score for the six games?

33. **Average** Suppose a basketball team has scores of 64, 76, 98, 55, 76, and 102 in their first six games.
   a. Find the mean score.
   b. Find the median score.
   c. Find the mode of the scores.

34. **Home Sales** Below are listed the prices paid for 10 homes that sold during the month of February in a city in Texas.

\$210,000	\$139,000	\$122,000	\$145,000	\$120,000
\$540,000	\$167,000	\$125,000	\$125,000	\$950,000

   a. Find the mean housing price for the month.
   b. Find the median housing price for the month.
   c. Find the mode of the housing prices for the month.
   d. Which measure of "average" best describes the average housing price for the month? Explain your answer.

35. **Shoe Sales** The table below shows the number of pairs of shoes sold, in various sizes, during a one-day sale.

Size	Number of Pairs
8	5
$8\frac{1}{2}$	7
9	8
$9\frac{1}{2}$	10
10	15
$10\frac{1}{2}$	12
11	6

   a. Find the mode of the shoe size for all pairs of shoes sold.
   b. Find the median shoe size for all pairs of shoes sold. (Do this by looking at the numbers in the second column to decide which shoe size is in the middle of all the sizes of the pairs of shoes sold.)

36. **Cost of College** The net cost of tuition is the cost that students actually pay when financial aid is taken into account. If the net cost for a 4-year public college for four different years was \$2,260, \$2,210, \$2,130, and \$2,850, what was the mean and median net cost? Round to the nearest cent.

37. **Financial Aid** The following are the amounts of federal grants (in millions) that were given out from 2004 to 2008: \$19,788, \$20,304, \$19,416, \$19,472, \$20,946. What was the mean and median amount given out in federal grants? Round to the nearest dollar.

© zimmytws/iStockPhoto

38. **Financial Aid** The following are the amounts of all financial aid distributed including all grants, loans, and scholarships, from 2004 to 2008: \$132,839, \$143,694, \$149,668, \$154,044, \$162,501. What was the mean and median amount of financial aid from 2004 to 2008. Round to the nearest cent.

39. **Basketball** Find the range of the basketball game scores given in Problem 33.

40. **Bowling** Find the range of the bowling scores given in Problem 32.

41. **Shoes Sales** Find the range of shoes sizes given in Problem 35.

42. **Cost of College** Find the range of the net costs of tuition given in Problem 36.

43. **Financial Aid** Find the range of the amounts of federal grants given in Problem 37.

44. **Financial Aid** Find the range of the amounts of financial aid given in Problem 38.

Getting Ready for the Next Section

45. An internet company wants to sell ad space on their website to generate revenue, but they need to average 5,000 visitors each day for a week. Here are the data for the first six days of the week:

Day	Visitors
Monday	4,432
Tuesday	5,340
Wednesday	5,895
Thursday	6,003
Friday	5,486
Saturday	4,789

 a. How many visitors do they need on Sunday?
 b. What type of average is this; a mean, a median, or a mode?

46. Luke's test scores are 84, 65, 91, 75, and 92 points out of 100. He has one more test in the semester.
 a. What score does he need on the last test to average an 83 on all six tests?
 b. If the teacher drops the lowest score, is it possible for Luke to average 90?

47. The mean of a set of numbers is 234, and the sum of the numbers in the set is 1,638. How many numbers are in the set?

48. The mean of a set of numbers is 0.68, and the sum of the numbers in the set is 5.44. How many numbers are in the set?

6.2 Displaying Information

Objectives

A. Construct bar charts.

B. Construct scatter diagrams and line graphs.

C. Construct frequency tables.

D. Construct histograms.

The table and bar chart below were introduced in Chapter 1. They both give the same information about the caffeine content of various drinks. The table is a numeric representation and the bar chart is a visual representation.

Caffeine Content of Hot Drinks

Drink (6-Ounce Cup)	Caffeine (In Milligrams)
Brewed Coffee	100
Instant Coffee	70
Tea	50
Cocoa	5
Decaffeinated Coffee	4

Table 1

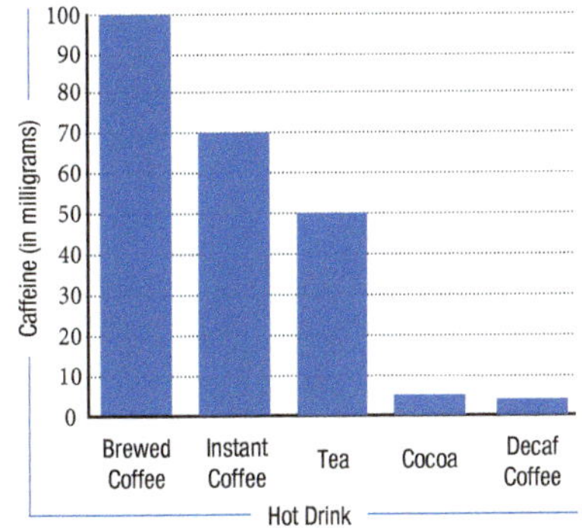

Figure 1

On the bar chart, recall that the horizontal line below which the drinks are listed is called the *horizontal axis*, while the vertical line that is labeled from 0 to 100 is called the *vertical axis*. The key to constructing a readable bar chart is in labeling the axes in a simple and straightforward manner, so that the information shown in the bar chart is easy to read.

Bar Charts

VIDEO EXAMPLES

SECTION 9.2

Example 1 The information in Table 2 was published in 2004 by the Insurance Institute for Highway Safety. It gives the total repair bill from damage done to a car from four separate crashes at 5 miles per hour. Construct a bar chart that gives a clear visual summary of the information in the table.

Damage at 5 Miles Per Hour

Car	Damage
Mitsubishi Galant	\$2,099
Nissan Maxima	\$2,451
Acura TL	\$2,923
Chevrolet Malibu	\$3,807
Chrysler Pacifica	\$5,259
Cadillac SRX	\$6,577

Table 2

Solution Although there are many ways to construct a bar chart of this information, we will construct one in which the information in the first column of the table is given along the horizontal axis and the information in the second column is associated with the vertical axis, as shown in Figure 2 on the next page.

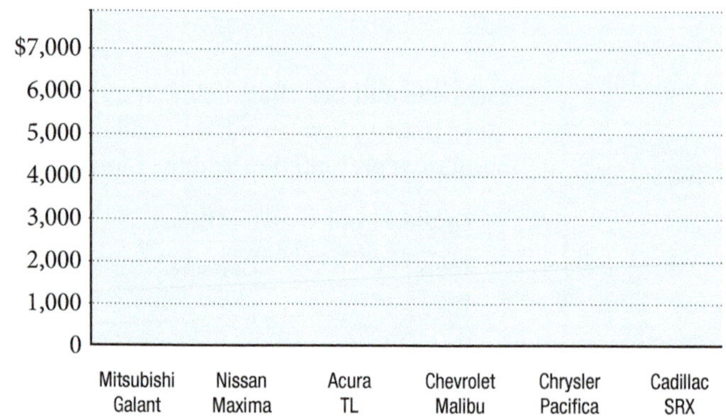

Figure 2

To complete the bar chart, we mentally round the numbers in the table to the nearest hundred to better estimate how tall we want the bars. Using the rounded numbers to draw the bars, we have the bar chart shown in Figure 3.

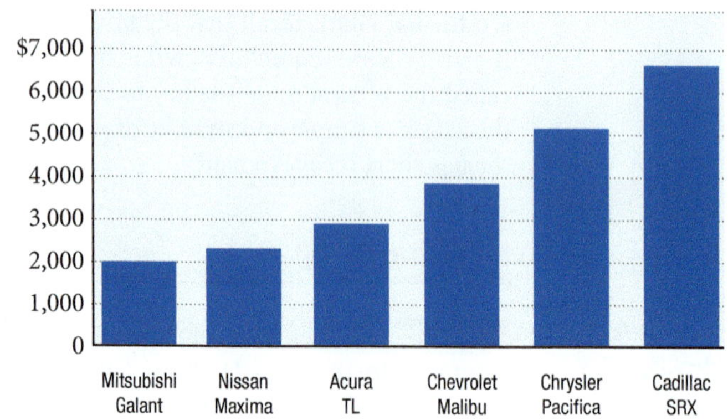

Figure 3

Example 2 Referring to Example 1, what is the mean amount of damage done to the cars listed in the table? Round to the nearest dollar.

Solution To find the mean, we add the numbers in the table and then divide by the number of cars in the table, which is 6.

$$\frac{2{,}099 + 2{,}451 + 2{,}923 + 3{,}807 + 5{,}259 + 6{,}577}{6} = \frac{23{,}116}{6} = \$3{,}853$$

The mean amount of damage for the cars in Table 2 is $3,853. If we draw a bar representing this amount on our bar chart, we have the chart in Figure 4. We can see visually that the mean is a measure of the center of our data.

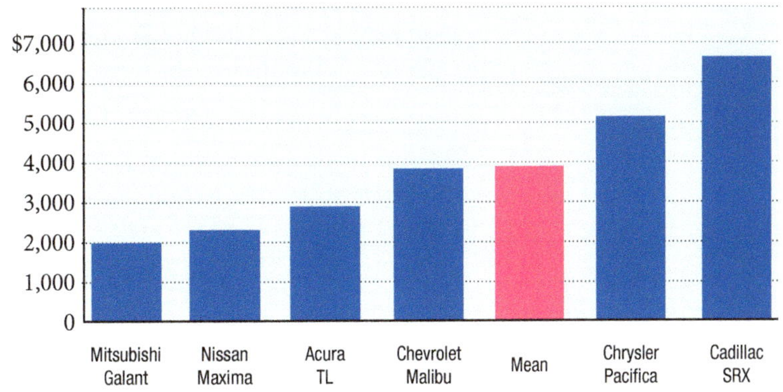

Figure 4

Scatter Diagrams and Line Graphs

This table and figure show the relationship between the table of values for the speed of a race car and the corresponding bar chart. In Figure 5, the horizontal axis shows the elapsed time in seconds, and the vertical axis shows the speed in miles per hour.

Speed of a Race Car

Time in Seconds	Speed in Miles per Hour
0	0
1	72.7
2	129.9
3	162.8
4	192.2
5	212.4
6	228.1

Table 3

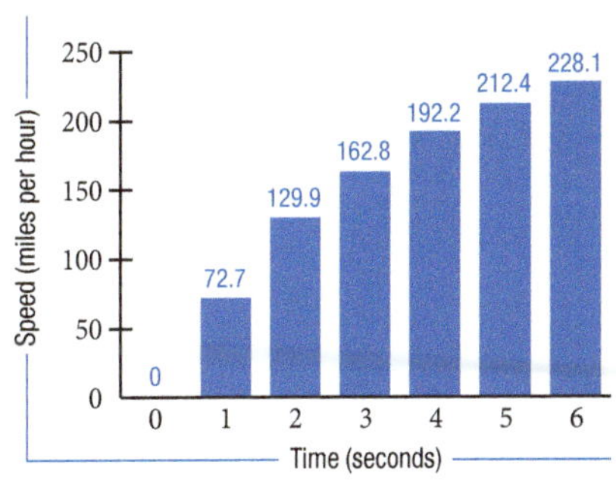

Figure 5

The information in the table can be visualized with a **scatter diagram** and **line graph** as well. A scatter diagram is created by using dots instead of bars and a line graph is constructed by connecting the dots with straight lines. Figure 6 is a scatter diagram of the information in the table above, and Figure 7 shows a line graph of the same data. Notice that we have labeled the axes in these two figures a little differently than we did with the bar chart by making the axes intersect at the number 0.

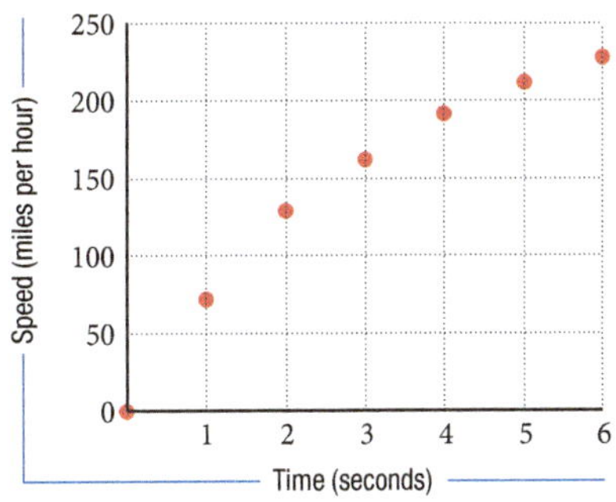

Figure 6

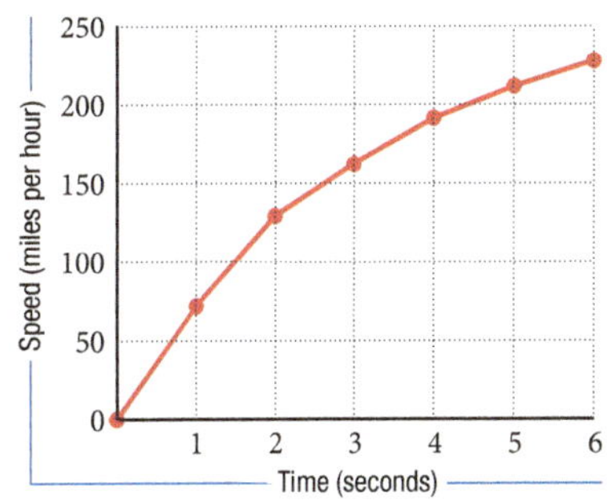

Figure 7

The number sequences we have worked with in the past can also be written as paired data by associating each number in the sequence with its position in the sequence. For instance, in the sequence of odd numbers

$$1, 3, 5, 7, 9, \ldots$$

the number 7 is the fourth number in the sequence. Its position is 4, and its value is 7. Here is the sequence of odd numbers written so that the position of each term is noted:

Position	1, 2, 3, 4, 5, …
Value	1, 3, 5, 7, 9, …

Example 3 The tables below give the first five terms of the sequence of odd numbers and the sequence of squares as paired data. In each case construct a scatter diagram.

Odd Numbers

Position	Value
1	1
2	3
3	5
4	7
5	9

Table 4

Squares

Position	Value
1	1
2	4
3	9
4	16
5	25

Table 5

Solution The two scatter diagrams are based on the data from these tables. Notice how the dots in Figure 8 seem to line up in a straight line, whereas the dots in Figure 9 give the impression of a curve. We say the points in Figure 8 suggest a linear relationship between the two sets of data, whereas the points in Figure 9 suggest a nonlinear relationship.

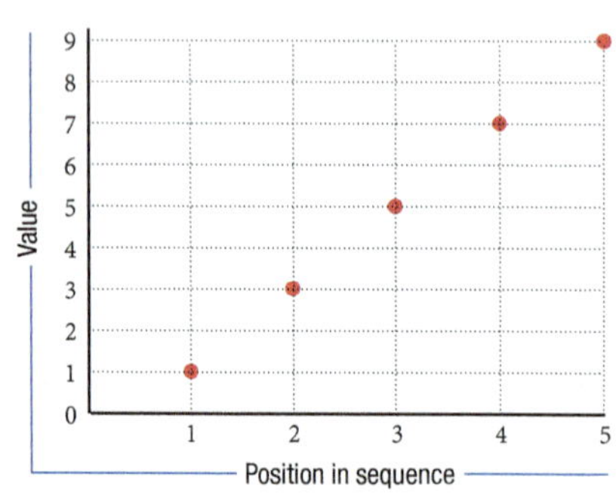

Figure 8

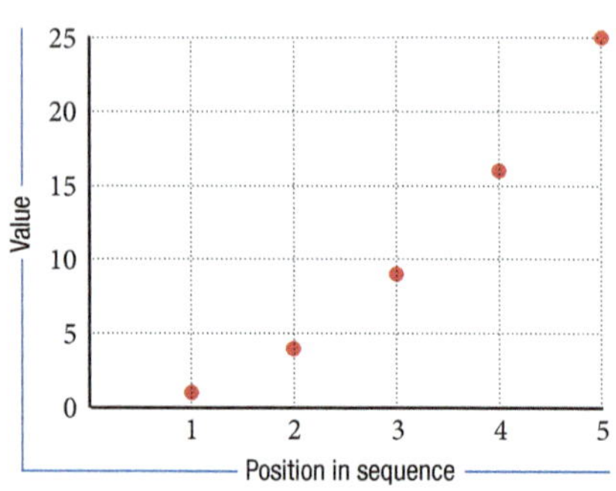

Figure 9

Frequency Tables

Frequency tables allow us to record data by how often it occurs, but without keeping track of each individual piece of data. Instead, a frequency table groups the data in intervals. Data is grouped into intervals by first deciding the number of intervals, then dividing the range of data by the number of intervals to get the range for each interval.

Example 4 A group of 20 college students was asked to indicate how many hours they spent on Facebook today. The results are listed below.

2, 6, 10, 3, 1, 0, 5, 8, 0, 2, 3, 5, 7, 9, 0, 3, 7, 2, 1, 10

Create a frequency distribution using three intervals.

Solution The solution will consist of 4 steps. In step 1, we will find the range. In step 2, we will determine the length of the intervals. In step 3, we will find the intervals, and in step 4, we tally the data and find the frequencies.

Step 1: By examining the data, we see that the lowest value is 0 and the highest value is 10. The range is 10.

Step 2: We decide to use 3 intervals. We divide the range, 10, by 3 to get $3.3\overline{3}$. We round this up to the next whole number to get 4. Each interval will be 4 units.

Step 3: We start with the lowest value, 0, and count four numbers for each interval. The intervals are 0 to 3, 4 to 7, and 8 to 11. We place these in the first column of our frequency table.

Step 4: For the next column in our frequency table, we make a tally of the number of data points in each interval. For the third column we count tallies to obtain our frequencies.

Here is our frequency table:

Interval	Tally	Frequency
0–3	卌 卌 I	11
4–7	卌	5
8–11	IIII	4

In actual practice, it is not necessary to show the tally column.

Histograms

Histograms offer a graphical view of data which, like with frequency tables, is grouped in intervals. Although a histogram looks like a bar chart, it is not the same. Histograms illustrate data that has been put into intervals or bins. These bins can be different widths, and therefore the areas of the rectangles in a histogram are used to measure the frequency of the data rather than the heights of the bars as in a bar chart.

Histogram

A **histogram** is a representation of a frequency distribution by means of rectangles whose widths represent class intervals and whose areas are proportional to the corresponding frequencies

Example 5 The data here show the critical reading SAT scores for a group of 20 incoming college freshmen. Construct a histogram from the data.

710 598 615 494 625 700 542 572 472 694
610 488 642 583 649 720 496 583 627 678

Solution To begin, we decide how large we want the interval of the bins to be. We will use bins ranging from 400 to 799, with widths of 100 points. Next, we make a frequency table showing the interval and the frequency of each interval. Then, we draw the histogram with the bins along the horizontal axis and the frequencies along the vertical axis.

Interval	Frequency
400–499	4
500–599	5
600–699	8
700–799	3

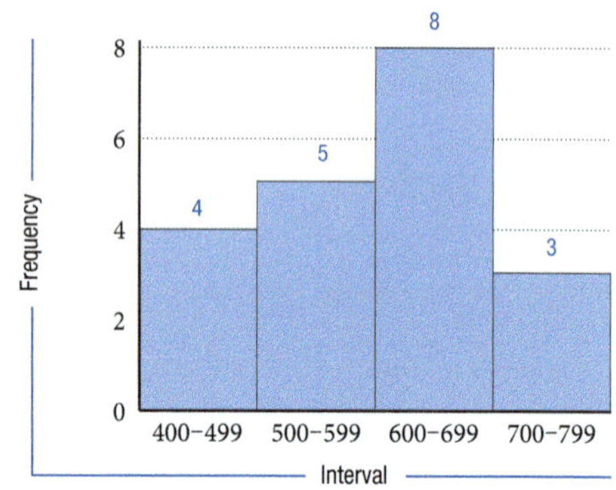

Getting Ready for Class

After reading through the preceding section, respond in your own words and in complete sentences.

A. What types of graphs can we make from a table of data?
B. When is a scatter diagram a useful tool for displaying data?
C. Explain the difference between labeling axes for a bar chart versus a line graph.
D. How are frequency tables and histograms related?

Problem Set 6.2

1. **Caffeine Content** The following table lists the amount of caffeine in five different soft drinks.

Caffeine Content of Soft Drinks

Drink	Caffeine (In Milligrams)
Mountain Dew	54
Dr Pepper	36
Coca-Cola	34
Barq's Root Beer	18
Sprite	0

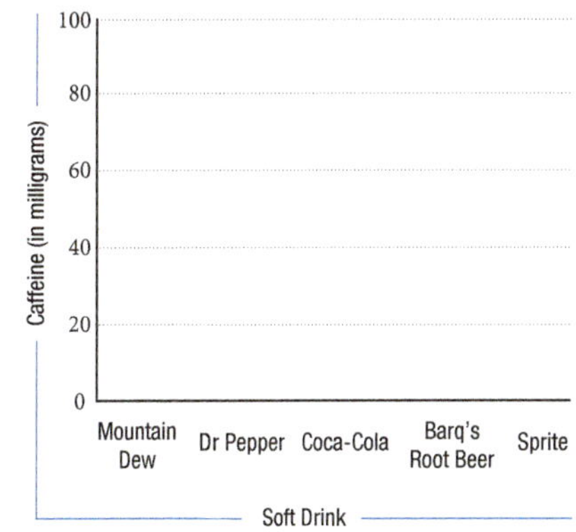

   a. Use the template to construct a bar chart from the information in the table.
   b. Find the mean caffeine content in the drinks listed in the table.
   c. Find the median caffeine content of the drinks listed in the table.
   d. What is the range of caffeine content of the drinks listed in the table?

2. **Caffeine Content** The following table lists the amount of caffeine in five different energy drinks.

Caffeine Content in Energy Drinks

Drink	Caffeine (In Milligrams)
5-Hour Energy	207
Monster	80
Rockstar	80
Amp	74
Vault	47

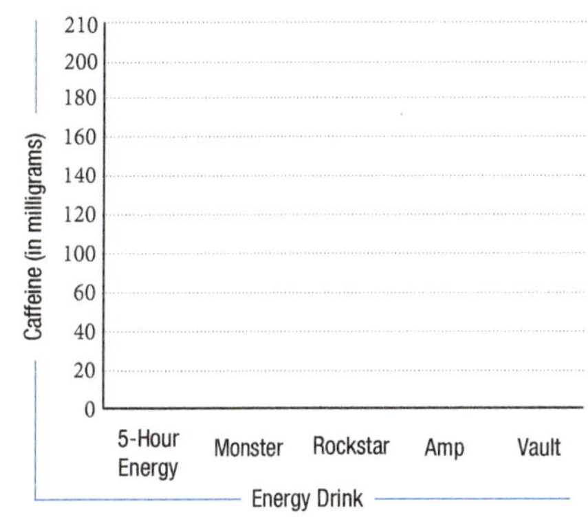

   a. Use the template to construct a bar chart from the information in the table.
   b. Find the mean caffeine content for the items listed in the table.
   c. Find the median caffeine content of the items listed in the table.
   d. What is the range of caffeine content of the items listed in the table?

3. **Exercise** The following table lists the number of calories burned in 1 hour of exercise by a person who weighs 150 pounds.

Calories Burned by a 150 Pound Person in One Hour	
Activity	**Calories**
Bicycling	374
Bowling	265
Handball	680
Jazzercize	340
Jogging	680
Skiing	544

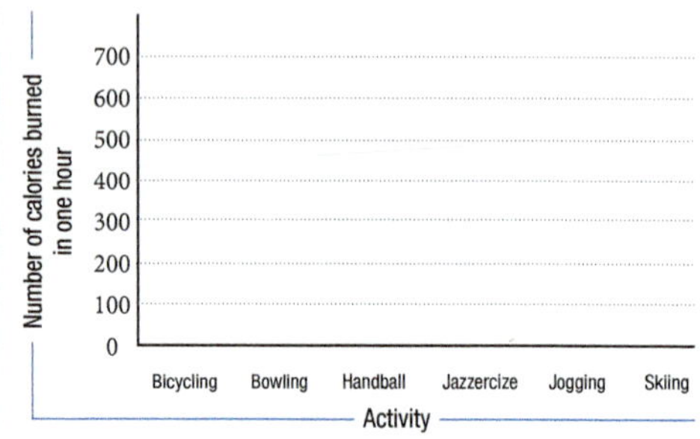

a. Use the template to construct a bar chart from the information in the table.
b. Find the mean calories burned for the activities listed in the table.
c. Find the mode for calories listed in the table.
d. What is the range of calories burned for the activities listed in the table?

4. **Fast Food** The following table lists the number of calories consumed by eating some popular fast foods.

© bpablo/iStockPhoto

Calories in Fast Food	
Food	**Calories**
McDonald's Hamburger	250
Burger King Hamburger	240
Wendy's Hamburger	260
McDonald's Big Mac	550
Burger King Whopper	630
Wendy's Baconator	940

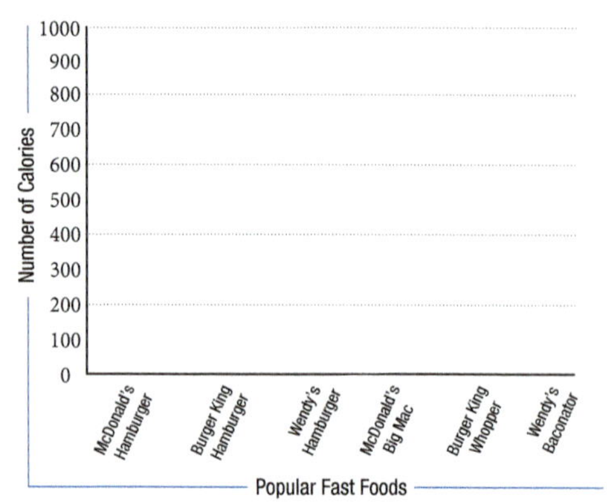

a. Use the template to construct a bar chart from the information in the table.
b. To the nearest calorie, find the mean calories contained in the foods listed in the table.
c. Find the median calories contained in the foods listed in the table.
d. What is the range of calories for the foods listed in the table?

5. **Car Speeds** The table below gives the number of seconds it takes various cars to accelerate from 0 miles per hour to 60 miles per hour. Construct a bar chart from the information in the table. Then find the difference in elapsed time between the slowest car and the fastest car.

Car	Time (In Seconds)
BMW M3	5.4
Chrysler Cirrus LXi	9.4
Ford Contour SE	9.5
Honda Civic EX	8.8
Nissan Sentra GLE	11.0
Toyota Corolla DX	10.2

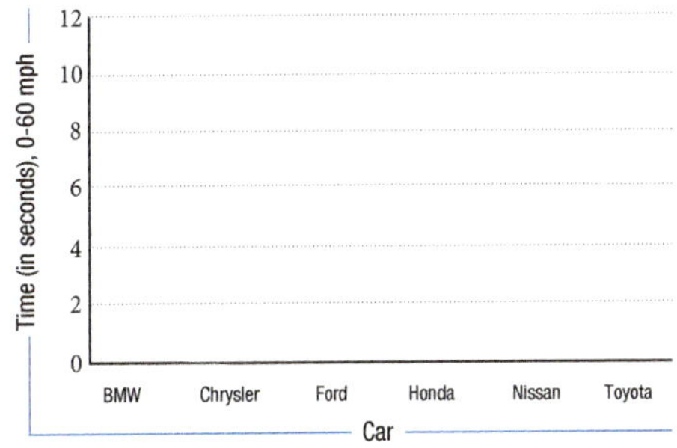

© hxdbzxy/iStockPhoto

6. **Social Media** In the summer of 2012, Twitter was buzzing with people talking about the London Olympics. The following table shows the top 10 most tweeted sports and the top 10 most tweeted athletes for the two-week-long event.

Sport	Number of Tweets (In Thousands)
Tennis	471
Swimming	397
Rowing	378
Volleyball	331
Soccer	319
Basketball	275
Badminton	216
Boxing	199
Diving	139
Track & Field	138

Athlete, Sport	Number of Tweets (In Thousands)
Michael Phelps, Swimming	493
Ryan Lochte, Swimming	133
Usain Bolt, Track	132
Tom Daley, Diving	91
Jessica Ennis, Heptathlon	65
Gabrielle Douglas, Gymnastics	48
Missy Franklin, Swimming	41
Jordyn Wieber, Gymnastics	32
Lin Dan, Badminton	21
Mark Cavendish, Cycling	16

(Source: Social Bakers)

a. Give a ratio that compares the number of diving tweets to the number of swimming tweets.

b. Give a ratio that compares the number of tweets for Mark Cavendish to the number of tweets for Michael Phelps.

c. What do you find surprising about the two tables when you compare them?

7. **Elevation** Mauna Kea is the tallest mountain on Earth, but how does it compare to other peaks in our solar system? The bar chart below shows the heights in kilometers of some of the tallest mountains in our solar system.

(Source: Popular Science Magazine)

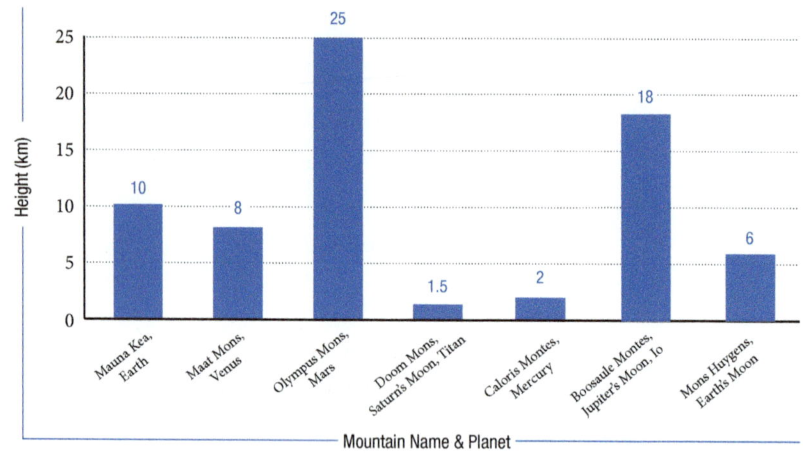

Mountain	Planet/Moon	Height (km)
Mauna Kea	Earth	
Maat Mons		8
Olympus Mons	Mars	
Doom Mons		1.5
	Mercury	2
Boosaule Montes		18
Mons Huygens	Earth's Moon	

a. Use the chart to fill in the following table.

b. Find the difference in height between Mauna Kea and Olympus Mons.

c. What percent taller is Olympus Mons than Mauna Kea?

© blackred/iStockPhoto

8. **Smartphone Sales in China** The bar chart below was created in 2013. It shows the number of smartphones sold in China for the years 2009 through 2013, and then predicts the number of smartphones sold in China for 2014 and 2015.

(Source: IDC, The Wall Street Journal)

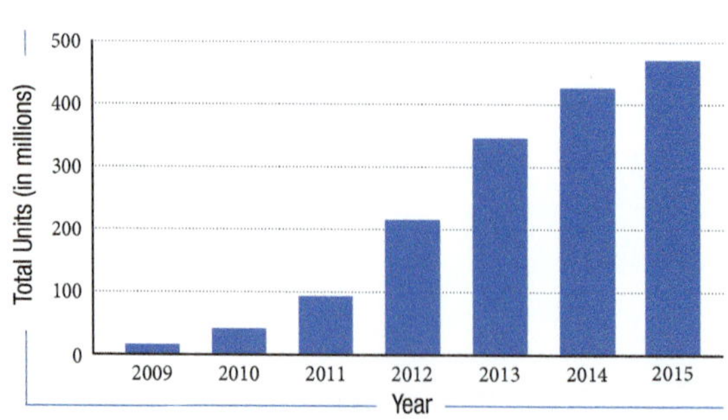

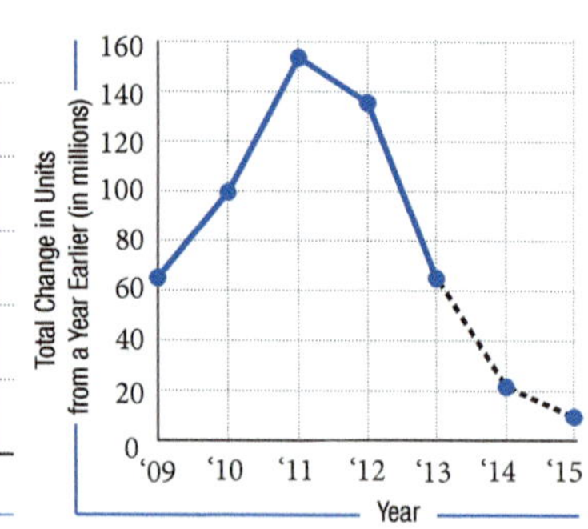

a. Use the information in the bar chart to fill in the table below. Estimate your answers and round to the nearest 10 million.

Year	Smartphone Sales in China
2009	
2010	
2011	
2012	
2013	
2014	
2015	

The line graph to the right of the bar chart shows the change in smartphone sales from the previous year for 2009 through 2013, and the predicted change in smartphone sales for the years 2014 and 2015.

b. Why do you think the line graph drops off after 2011?

c. By what percent did smartphone sales in China change from 2009 to 2010?

d. By what percent did smartphone sales in China change from 2012 to 2013?

9. Half-life of an Antidepressant The half-life of a medication tells how quickly the medication is eliminated from a person's system. The line graph below shows the fraction of an antidepressant that remains in a patient's system once the patient stops taking the antidepressant. The half-life of the antidepressant is 5 days. Use the line graph to complete the table.

Concentration of Antidepressant

Days Since Discontinuing	Fraction Remaining in Patient's System
0	1
5	
	$\frac{1}{4}$
15	
	$\frac{1}{16}$

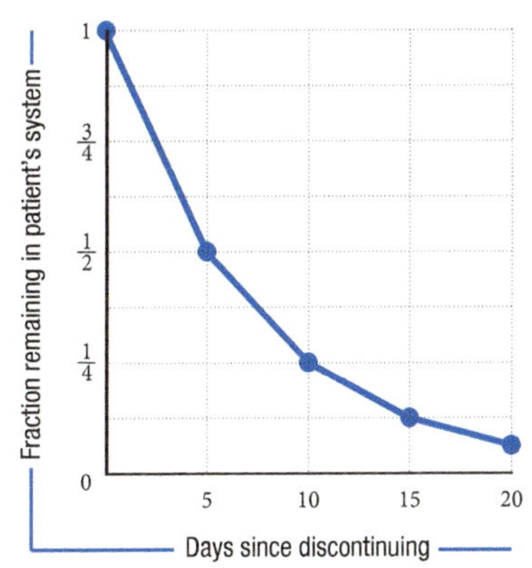

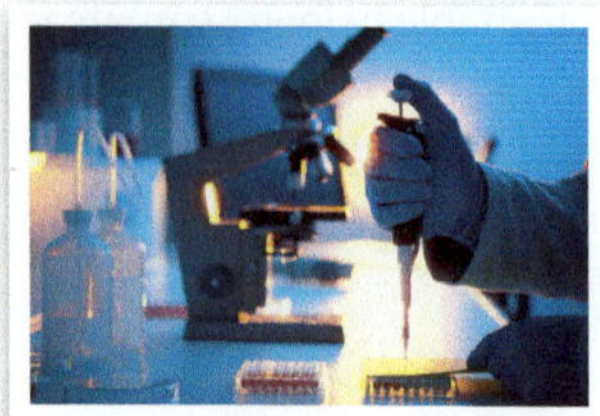

© Loran Nicolas/iStockPhoto

10. Carbon Dating All living things contain a small amount of carbon-14, which is radioactive and decays. The half-life of carbon-14 is 5,600 years. During the lifetime of an organism, the carbon-14 is replenished, but after its death the carbon-14 begins to disappear. By measuring the amount left, the age of the organism can be determined with surprising accuracy. The line graph below shows the fraction of carbon-14 remaining after the death of an organism. Use the line graph to complete the table.

Concentration of Carbon-14

Years Since Death of Organism	Fraction of Carbon-14 Remaining
0	1
5,600	
	$\frac{1}{4}$
16,800	
	$\frac{1}{16}$

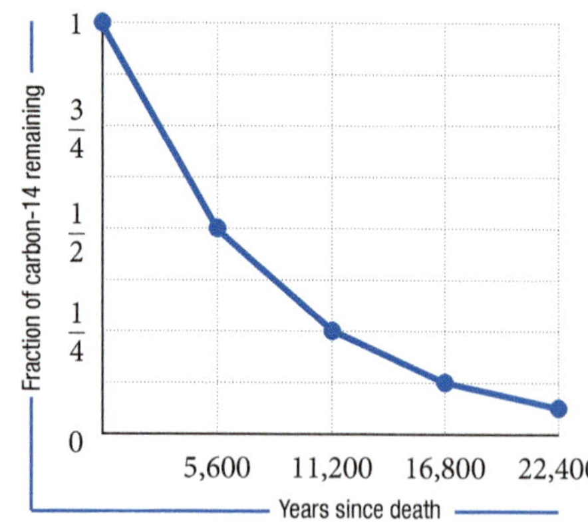

11. Running Robots In 2013, Google acquired a Massachusetts-based engineering and robotics company named Boston Dynamics, which produces running robots. The company's newest creation was the Cheetah robot, the fastest legged robot in the world. The following table compares the Cheetah robot's running speed in 2013 to other legged robots and animals.

(Source: WSJ)

Runner	Speed (mph)
Mabel (two-legged robot) University of Michigan	6.8
Usain Bolt (fastest human)	27.78
Cheetah (four-legged robot) Boston Dynamics	28.3
Ostrich (two-legged animal)	40
Cheetah (four-legged animal)	64

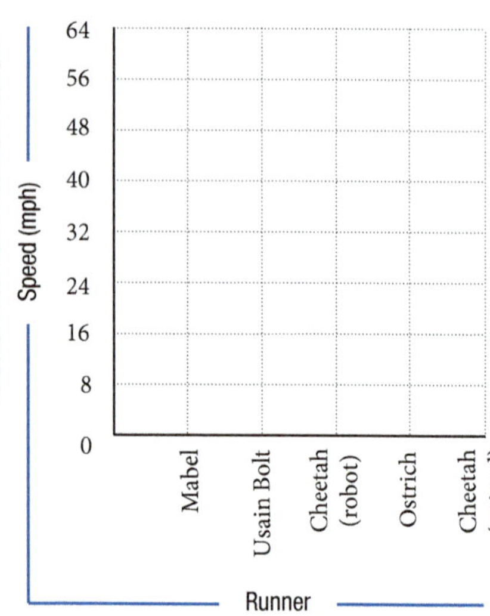

a. Use the table to construct a line graph, listing the runners from slowest to fastest

b. Find the difference in miles per hour between cheetah the animal and Cheetah the robot.

c. Find the difference in miles per hour between Cheetah the robot and Mabel the robot.

d. What percent faster is Cheetah the robot than Usain Bolt?

12. SUP Participation Stand Up Paddleboarding (SUP) is the newest trend in outdoor recreation. According to The Outdoor Foundation, the sport had the highest percentage of new participants in 2012 with 56% of participants trying it for the first time. The following table shows the percentages of first-time participants and the median ages for some of the most popular outdoor activities in 2012. (Source: The Outdoor Foundation)

Sport	First Time Percentage	Median Age
Stand Up Paddleboarding	56	28
Triathlon	38	30
Kayaking (Sea)	30	32
Climbing (Sport)	27	26
Surfing	23	27
Cross-Country Skiing	19	27

a. Use the table to construct a line graph of the new participant percentages.

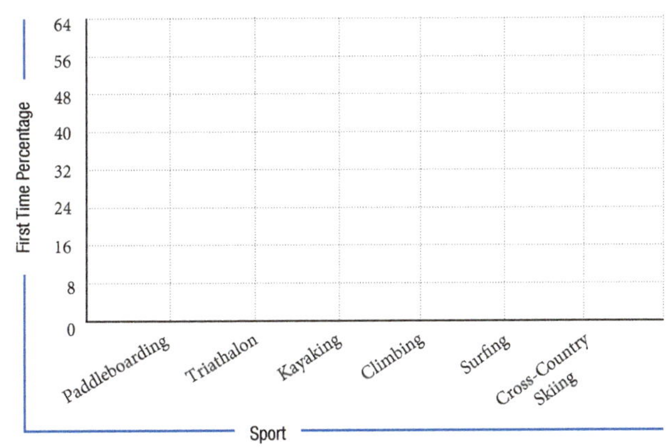

b. Give a ratio for each sport's new participation percentage compared to the percentage for new SUP participation.

c. Based on the table, what is the mean age for trying a new outdoor activity? Round to the nearest whole year of age.

d. Why do you think SUP has the most new participants?

13. Sports by Generation The following chart compares the participation rates in sports by generation in 2012. (Source: 2012 SGMA Participation Topline Report)

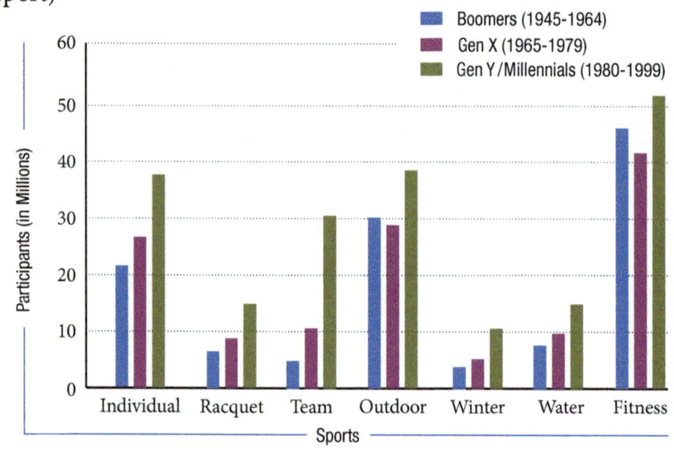

a. Use the chart to approximate values for each generation. Then construct a line graph for each generation on the same axes.

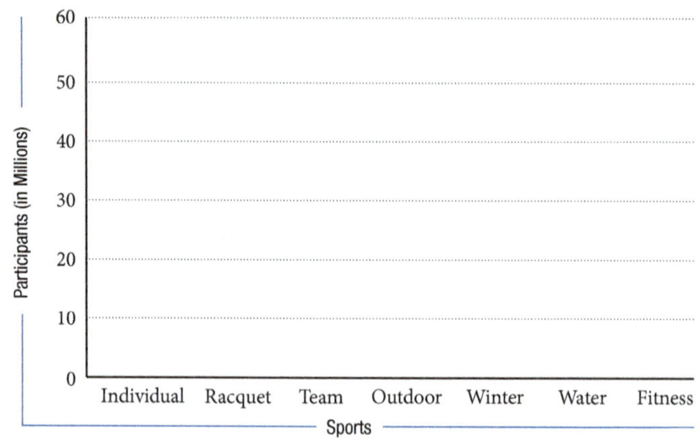

b. What do you notice about the participation rates of Gen Y / Millennials compared to the other two generations?

c. Using your approximated values, give a ratio that represents the number of Boomers who participate in team sports to the number of Gen Y / Millennials.

d. Using your approximated values, give a ratio that represents the total number of Gen X participants in team sports to Gen X participants in individual sports.

© crisserbug/iStockPhoto

14. Value of a Painting A painting is purchased in 1990 for $125. The table below shows what the painting is worth at various times, if it doubles in value every 5 years.

a. What is the range of values for the painting?

b. What is the mean value of the painting to the nearest dollar?

c. What is the median value of the painting?

d. Construct a line graph of the information in the table.

e. Use the line graph to estimate the value of the painting in 2015.

Year	Value
1990	$125
1995	$250
2000	$500
2005	$1,000
2010	$2,000

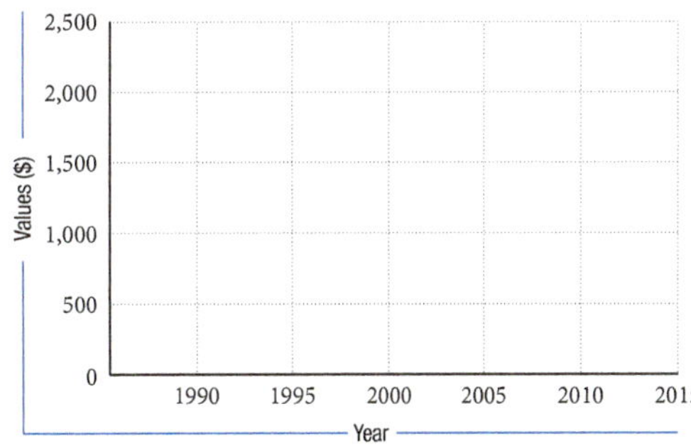

15. Social Media Facebook was the first social network to exceed 1 billion registered accounts. The following bar chart shows global use for Facebook and other social networks. (Source: Statista)

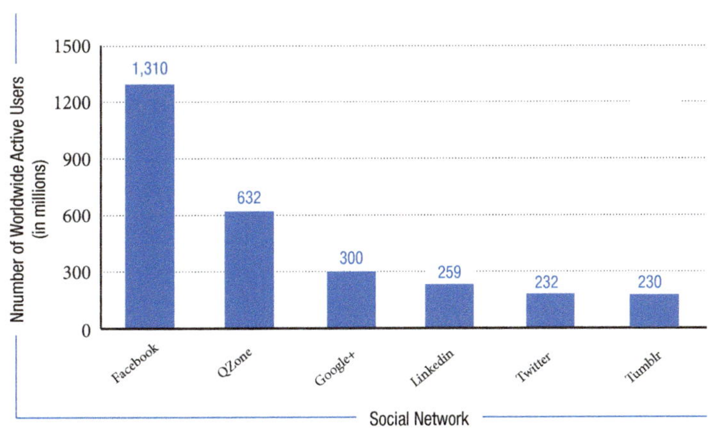

a. Use the bar chart to fill in the following table:

Social Network	Number of Worldwide Active Users (in millions; as of January 2014)
	1,310
QZone	
	300
	259
Twitter	
Tumblr	

b. The above table shows the top 6 social networks in the world. Based on the values in the chart, what percentage of the total worldwide users have Facebook accounts? What percentage have QZone accounts?

c. Now let's see how Facebook's user total has grown over the years. Create a bar chart for the following table: (Source: Statistic Brain, Facebook Inc.)

© franckreporter/iStockPhoto

Year	Number of Total Active Facebook Users (in millions)
2004	0
2006	12
2008	100
2010	608
2012	1,060
2014	1,310

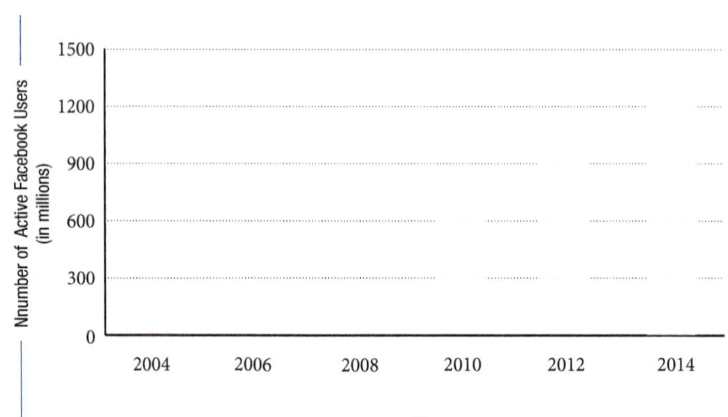

d. Using the above table, give a ratio that represents the user growth when comparing total users in 2010 with 2014. Then express in words what this ratio means.

e. Based on the table, would you expect the total number of active users to increase or decrease by 2016?

f. What is the average growth in Facebook users per year over the last 10 years?

16. **Tossing a Softball** Chaudra is tossing a softball into the air with an underhand motion. It takes exactly 2 seconds for the ball to come back to her. The table shows the distance the ball is above her hand at quarter-second intervals.

Tossing a Softball

Time (sec)	Distance (ft)
0	0
0.25	7
0.5	12
0.75	15
1	16
1.25	15
1.5	12
1.75	7
2	0

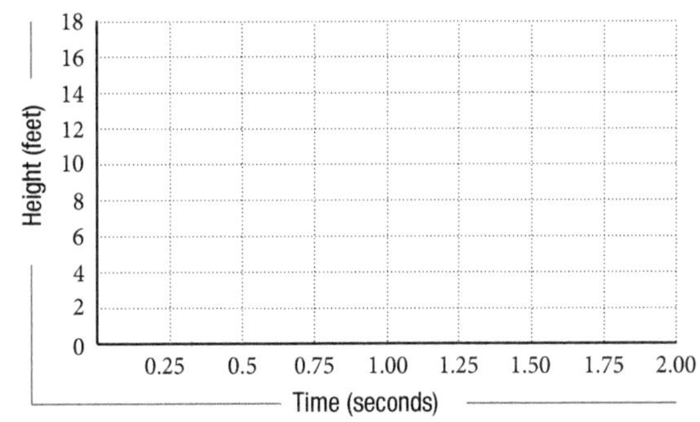

a. Use the template to construct a line graph from the information in the table.

b. To the nearest foot, how high above her hand is the ball after 0.4 seconds?

17. **Intensity of Light** The table below gives the relationship between the intensity of light that falls on a surface various distances from a 100–watt light bulb.

Light Intensity: 100–Watt Bulb

Distance Above Surface (ft)	Intensity (Lumens/sq ft)
1	120.0
2	30.0
3	13.3
4	7.5
5	4.8
6	3.3

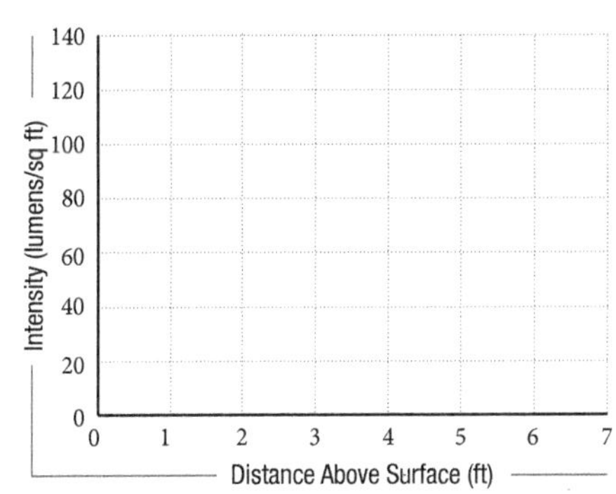

a. Use the template to construct a line graph of the information in the table.

b. Find the mean of the light intensities, to the nearest tenth.

c. Find the range of light intensities.

18. **Gravity** The effects of gravity decrease as we move further above the surface of the Earth. The table gives the acceleration due to gravity at various altitudes above the Earth.

Gravity Above the Earth

Altitude (km)	Acceleration (meters/sec^2)
0	9.80
1,000	7.33
2,000	5.68
3,000	4.53
4,000	3.70
5,000	3.08
6,000	2.60
7,000	2.23
8,000	1.93
9,000	1.69
10,000	1.49

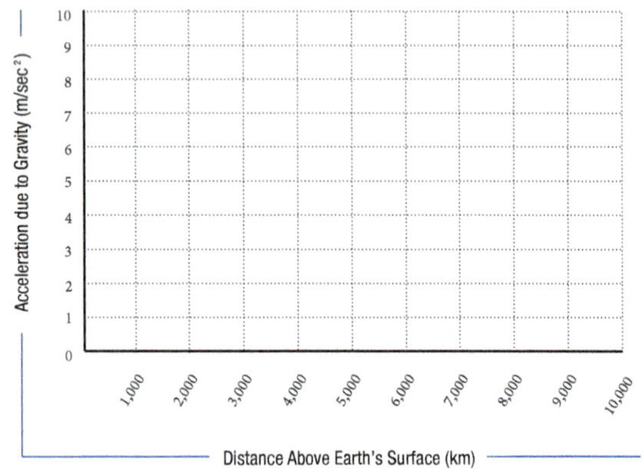

   a. Construct a line graph from the information in the table.
   b. Find the median acceleration due to gravity value.
   c. Find the range of acceleration due to gravity values.

© Matt_Brown/iStockPhoto

19. **Basketball** The college basketball team just finished their season. These were their game scores. Use the data to make a frequency table with five intervals.

76 58 82 47 73 76 81 93 68 76
72 84 64 59 67 74 80 69 84 76

Interval	Frequency

20. **Incomes** The data shows the annual incomes in thousands of a group of engineers. Use the data to create a frequency table. Use five intervals.

63 75 63 58 74 72 77 54
67 62 58 59 66 68 76

Interval	Frequency

21. Golf Scores Suppose a golf course hosted a fundraiser tournament to raise money for a local school. The data set shows the final scores of each of the participants. Use the data to create a histogram with five intervals.

69 72 76 85 67 73 78 86 81 96 103
87 96 84 100 79 84 72 79 97 84 86

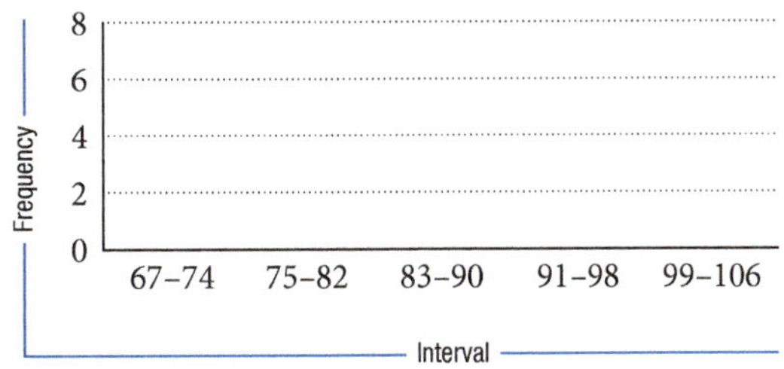

22. Oil Prices The data shows the price per barrel of oil rounded to the nearest dollar over the last 20 years. Use the data to construct a histogram with seven intervals.

18 23 20 19 17 16 17 20 19 12
17 27 23 22 28 38 50 58 64 98

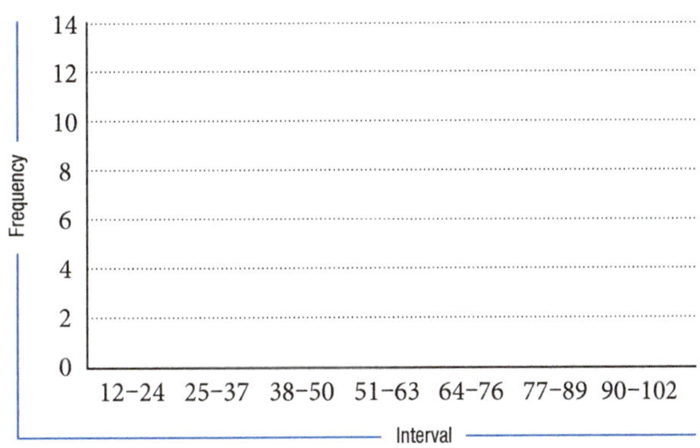

23. Candy The frequency table shows the number of M&M's in a bag of peanut M&M's. Use the table to work the following problem.

Interval	Frequency
13-16	4
17-20	6
21-24	7
25-28	5

Create a histogram from the frequency table.

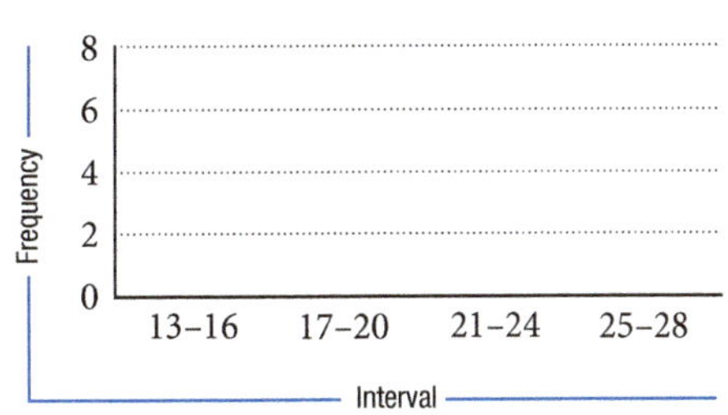

24. **Watching TV** The frequency table shows the hours per week that each person in a class watches television. Use the table to work the following problems.

Interval	Frequency
1-3	2
4-6	10
7-9	7
10-12	2

a. Create a histogram from the frequency table.

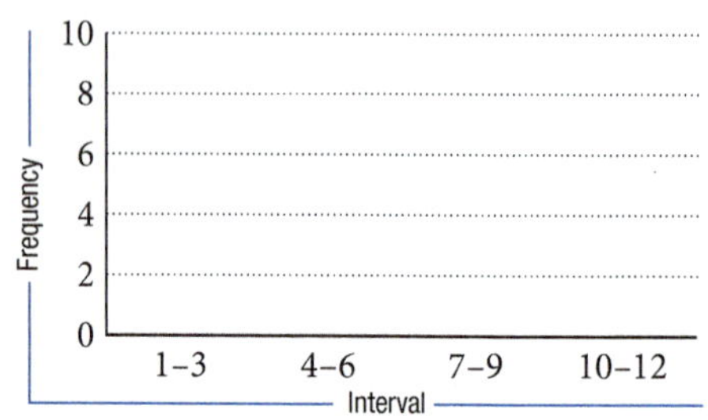

b. How many students are in the class?

c. How many students watch more than 6 hours of television?

© fotoVoyager/iStockPhoto

25. **Private College Enrollment** In 2014, one of the larger private colleges reported that their enrollment for the year was 75,000 students, which was down 14% from the previous year. The college also reported a 12 percent decline in revenue for the first three months of the year, which was a loss of $80 million.

a. How many students were enrolled in the college in 2013?

b. How much was their revenue for the first three months of 2013? Round to the nearest million.

© belterz/iStockPhoto

26. **Concert Revenue** The concert industry is looking to 2014 to bring in record-breaking revenue from ticket sales, based on the numbers comparing 2013 to previous years. For instance, the top 20 world tours in 2013 generated $2.43 billion, which was a 24% increase from 2012.

a. How much revenue did ticket sales from the top 20 tours generate in 2012?

b. The following table shows the total number of concert tickets sold per year. Suppose the average ticket price in 2012 was $85.93. Approximate the total ticket revenue for 2012 using this average price.

(Source: Pollstar, WSJ)

Year	Total Tickets Sold (in millions)
2009	52
2010	45.3
2011	35.48
2012	34.9

c. What percent of the total revenue generated in 2012 came from the top 20 world tours?

27. **Global Internet Usage** The following table shows some online statistics for the global population. (Source: US Census Bureau, InternetWorldStats, Facebook, ITU, We Are Social)

	Total Population (in millions)	Total Internet Users (in millions)	Total Active Facebook Users (in millions)
USA	316.7	254.3	178

	Total Population (in millions)	Total Internet Users (in millions)	Total Active Social Network Users (in millions)
World	7,095.5	2,484.9	1,856.7

a. What percentage of global internet users are in the United States?

b. What percentage of global active social network users are using Facebook in the United States?

c. How does the percentage of total Internet users in the United States compare to the percentage of total Internet users in the world?

Getting Ready for the Next Section

Change to percent.

28. $\frac{75}{250}$ **29.** $\frac{150}{250}$ **30.** $\frac{400}{2,400}$ **31.** $\frac{200}{2,400}$

Multiply.

32. 0.3(360) **33.** 0.4(360) **34.** 0.45(360) **35.** 0.15(360)

Divide.

36. $40 \div 5$ **37.** $45 \div 5$ **38.** $15 \div 5$ **39.** $5 \div 5$

SPOTLIGHT ON SUCCESS *Student Instructor Stephanie*

For success, attitude is equally as important as ability.
—Harry F. Banks

Math has always fascinated me. From addition to calculus, I've taken great interest in the material and great pride in my work. Whenever I struggled with concepts, I asked questions and worked problems over and over until they became second nature. I used to assume this was how everyone dealt with concepts they didn't understand. However, in high school, I noticed how easily students got discouraged with mathematics. In my senior year calculus and statistics classes, I was surrounded by bright students who simply gave up on trying to fully understand the material because it seemed confusing or difficult. Even if we shared a similar level of academic ability, the difference between these students' grades and my own reflected a difference in attitude. I noticed many students giving up without really trying to understand the concepts because they lacked confidence and didn't feel they were capable. They began coming to me for help. Though I was glad to help them with the math, I had a greater goal to help them believe they could succeed on their own. Soon the students I tutored gained more understanding and achieved success by simply paying more attention in class and working extra problems outside of class. It was amazing how much improvement I saw in both their confidence levels and their grades. It goes to show that a little extra effort and a positive attitude can truly make a difference.

6.3 Circle Graphs

Objectives

A. Read a circle graph.

B. Construct a circle graph.

In this section, we will review how to read a circle graph, or pie chart, and learn how to construct a chart using given data. Circle graphs are another way in which to visualize numerical information. They lend themselves well to information that adds up to 100% and are common in the world around us. In fact, it is hard to pick up a newspaper or magazine without seeing a circle graph.

Reading a Circle Graph

Some of this introductory material will be review. We want to begin our study of circle graphs by reading information from circle graphs.

VIDEO EXAMPLES

SECTION 9.3

Example 1 The circle graph shows where a group of freshmen students live. Use the circle graph to answer the following questions.

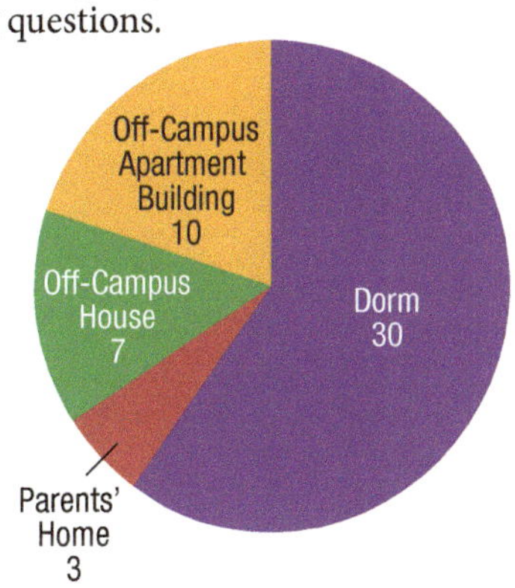

a. Find the total number of students surveyed.

b. Find the ratio of those living in the dorms to the total surveyed.

c. Find the ratio of those living in their parents' homes to those living in apartments.

Solution

a. To find the total number of student surveyed we add the numbers in all sections of the circle graph.

$$30 + 3 + 10 + 7 = 50 \text{ students surveyed}$$

b. The ratio of those living in dorms to the total surveyed is

$$\frac{\text{Number living in dorms}}{\text{Total number of students surveyed}} = \frac{30}{50} = \frac{3}{5}$$

c. The ratio of those living at home to those living in apartments is

$$\frac{\text{Number living at home}}{\text{Number living in apartments}} = \frac{3}{10}$$

Example 2 The circle graph shows the market share for different digital movie companies.

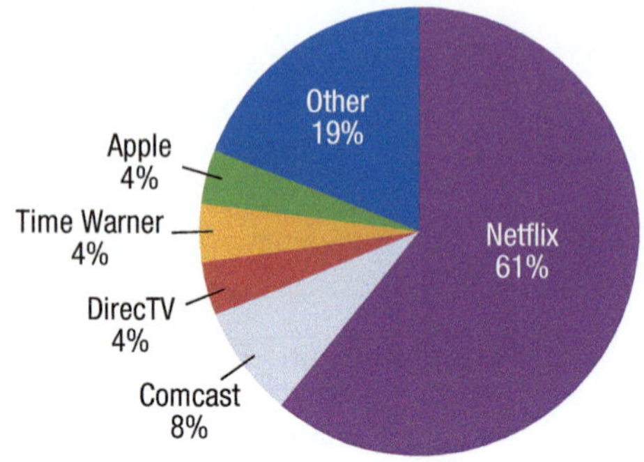

Suppose 800 people were surveyed and their answers matched the results shown above.

a. How many people in the survey stream their digital movies from Netflix?

b. How many people stream their digital movies from Apple or Comcast?

Solution

a. To find out how many people in the survey stream from Netflix, we need to find 61% of 800.

$$0.61(800) = 488 \text{ people}$$

488 of the people surveyed stream their movies from Netflix.

b. The number of people streaming from Apple or Comcast account for 4% + 8% = 12%. To find out how many of the 800 people are in this category, we must find 12% of 800.

$$0.12\ (800) = 96 \text{ people}$$

96 of the people surveyed stream their movies from Apple or Comcast.

Constructing Circle Graphs

Example 3 A recent study found that Americans spend just over six hours per day on social media sites. Construct a circle graph that shows the amount of time in a day spent on social media.

© franckreporter/iStockPhoto

Solution 1 **Using a Template** As mentioned previously, circle graphs are constructed with percents. Therefore we must first convert data to percents. To find the percent of hours from a day that are spent on social media sites, we divide the number of hours spent by the total number of hours in a day. We have

$$\frac{6}{24} = 0.25 \text{ which is } 25\%$$

The area of each section of the following template is 5% of the area of the whole circle. If we shade 5 sections of the template, we will have shaded 25% of the area of the whole circle.

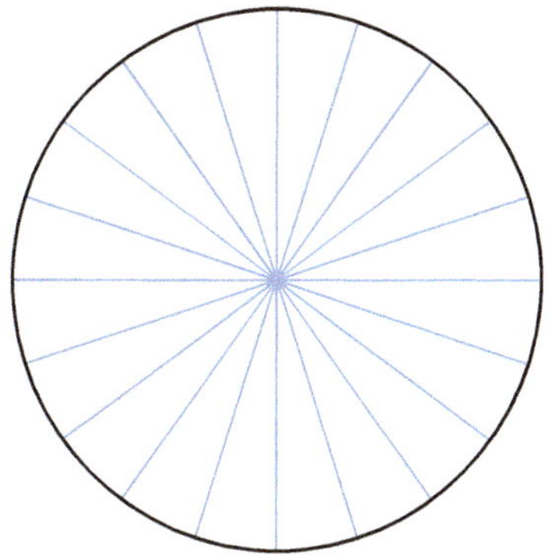

CIRCLE GRAPH TEMPLATE *Each slice is 5% of the area of the circle.*

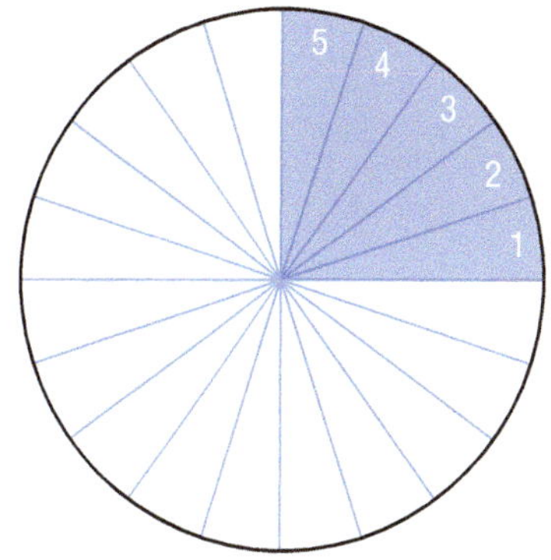

CREATING A CIRCLE GRAPH *To shade 25% of the circle, we shade 5 sections.*

The shaded area represents 25%, which is the amount of each day spent on social media sites. The rest of the circle must represent the 75% of the rest of the hours in the day. Shading each area with a different color and labeling each, we have our circle graph.

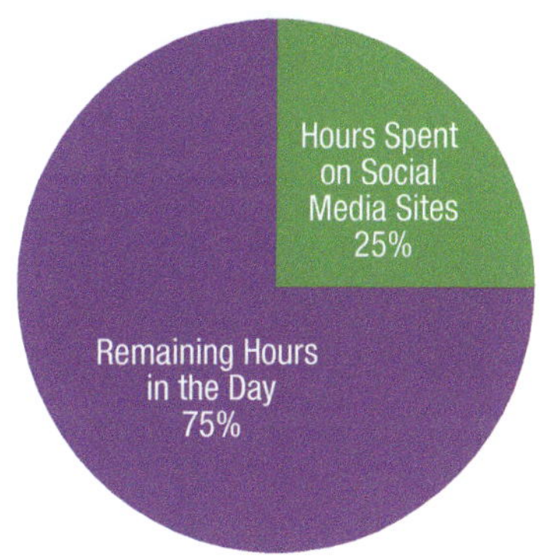

FIGURE 1

Solution 2 **Using a Protractor** Since a circle graph is a circle, and a circle contains 360°, we must now convert our data to degrees. We do this by multiplying our percents in decimal form by 360. We have

$$(0.25)360° = 90°$$

Now we place a protractor on top of a circle. First we draw a line from the center of the circle to 0° as shown in Figure 2. Now we measure and mark 90° from our starting point, as shown in Figure 3.

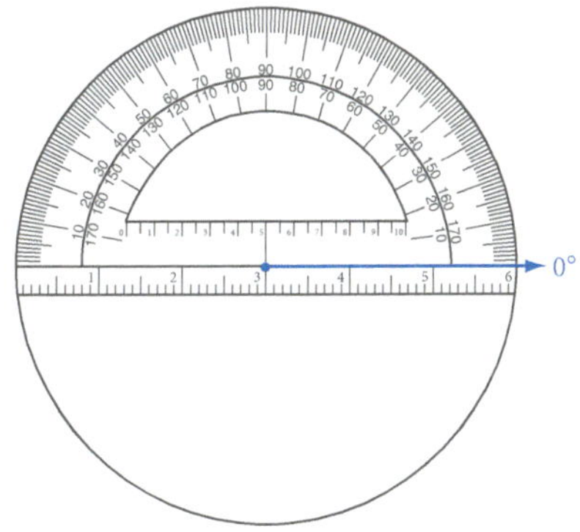

FIGURE 2

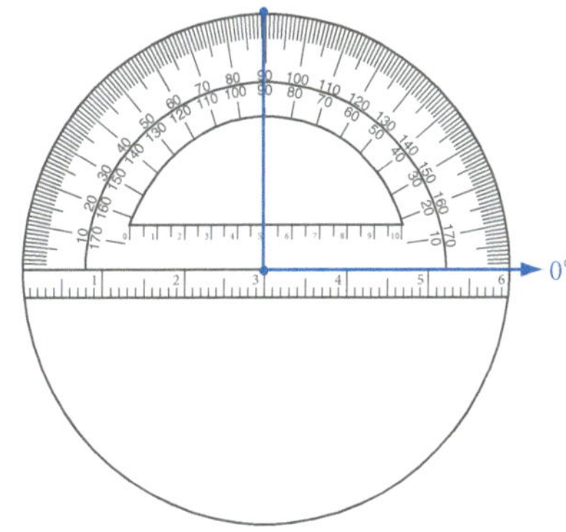

FIGURE 3

Finally we draw a line from the center of the circle to this mark, as shown in Figure 4. Then we shade and label the two regions as shown in Figure 5.

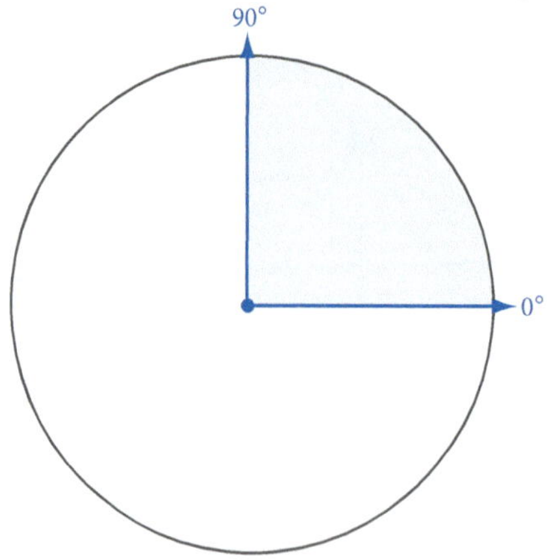

FIGURE 4

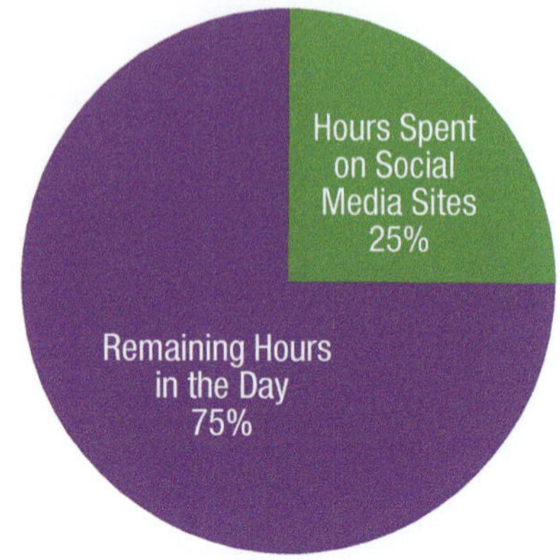

FIGURE 5

Example 4 The following table gives the ingredients for a lasagna and the breakdown of total price by percent. Construct a circle graph using this information.

Expense	Percent of Price
Ground Beef	15%
Cheese	45%
Pasta	5%
Tomato and Vegetables	30%
Other	5%

Solution Since our template uses sections that each represent 5% of the circle, we shade 3 sections, representing 15% for the price of the ground beef. We then shade 9 sections, representing 45% for the cheese. We proceed in the same manner until our entire recipe is represented.

We label each section with the appropriate information, and our circle graph is complete.

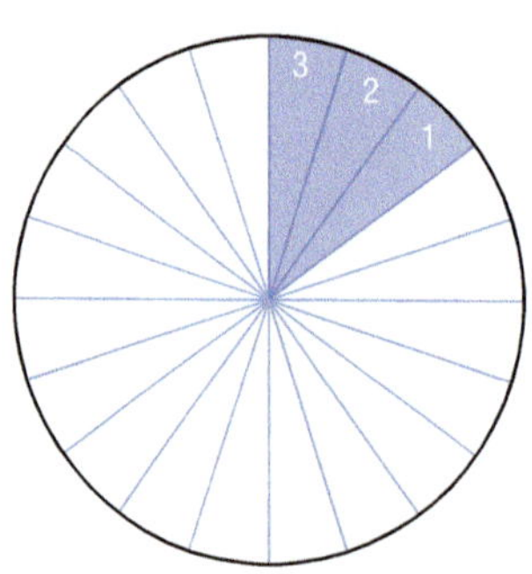

We shade 3 sections to represent the 15% for ground beef

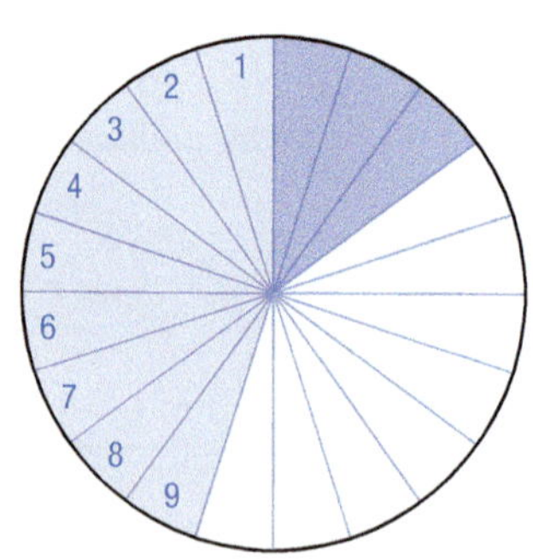

To shade 45% of the circle, we shade 9 sections of the template.

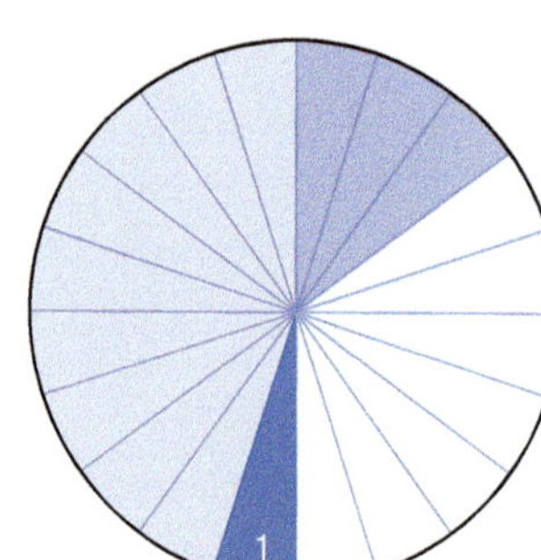

We shade 1 section to represent the 5% for the pasta.

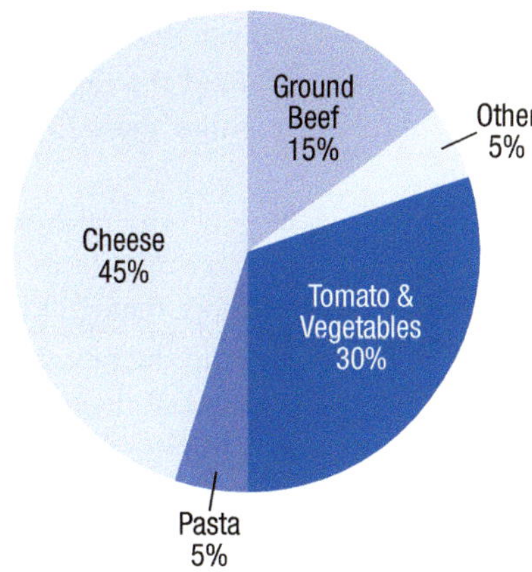

Getting Ready for Class

After reading through the preceding section, respond in your own words and in complete sentences.

A. If a circle is divided into 20 equal slices, then each of the slices is what percent of the total area enclosed by the circle?

B. If a 250 MB computer drive contains 75 MB of data, then how much of the drive is free space?

C. If a 250 MB computer drive contains 75 MB of data, then what percent of the drive contains data?

D. Explain how you would construct a pie chart of monthly expenses for a person who spends \$700 on rent, \$200 on food, and \$100 on entertainment.

Problem Set 6.3

1. **Students with Jobs** The circle graph shows the results of surveying 200 college students to find out how many hours they worked per week at a job.

    a. Find the ratio of students who work more than 15 hours a week to total students.

    b. Find the ratio of students who don't have a job to students who work more than 15 hours a week.

    c. Find the ratio of students with jobs to total students.

    d. Find the ratio of students with jobs to students without jobs.

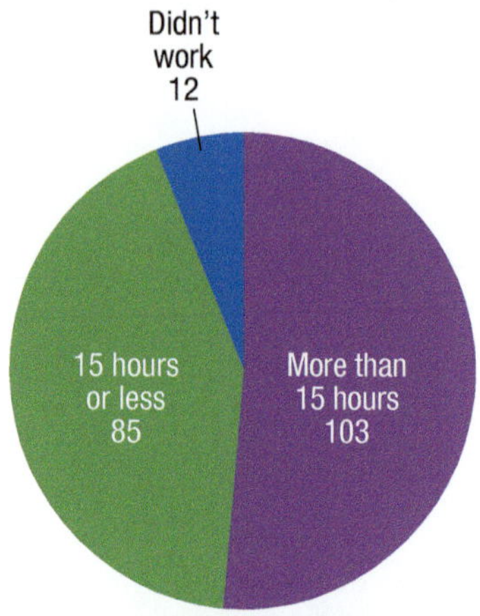

2. **Favorite Salad Dressing** The circle graph shows the results of a survey on favorite salad dressing.

    a. What is the most preferred salad dressing?

    b. Which salad dressing is preferred second most?

    c. Which salad dressing is least preferred?

    d. What percentage of people preferred ranch?

    e. What percentage of people preferred Italian or Thousand Island?

    f. If 50 people responded to the survey, how many people preferred ranch? (Round to the nearest whole number.)

    g. If 50 people responded to the survey, how many people preferred Thousand Island? (Round to the nearest whole number.)

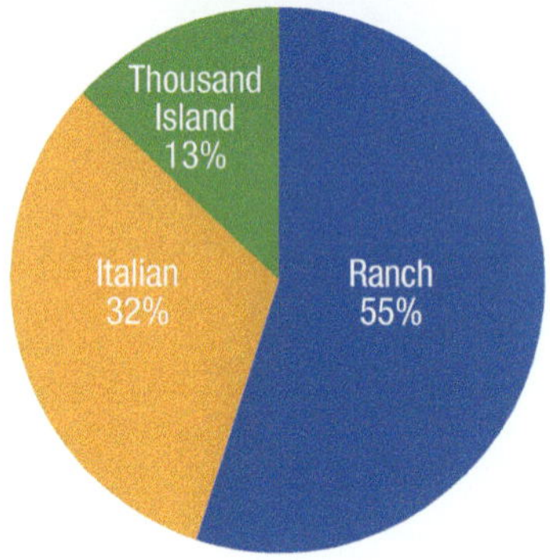

3. **Food Dropped on the Floor** The circle graph shows the results of a survey about eating food that has been dropped on the floor. Participants were asked whether they eat food that has been on the floor for 3, 5, or 10 seconds.

    a. What percentage of people say it is not safe to eat food dropped on the floor?

    b. What percentage of people believe the "three-second rule"?

    c. What percentage of people will eat food that stays on the floor for five seconds or less?

    d. What percentage of people will eat foot that stays on the floor for ten seconds or less?

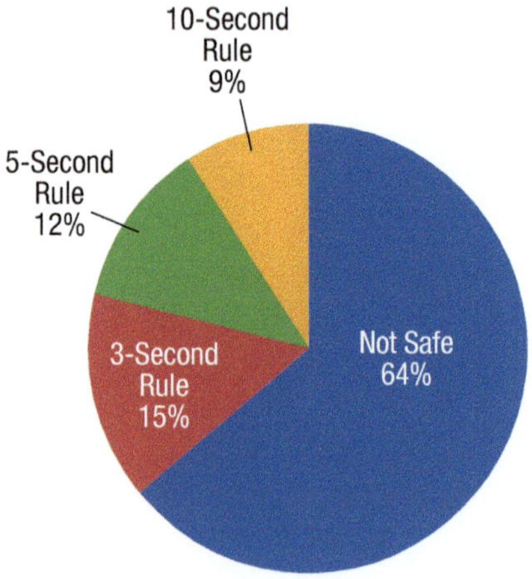

4. **Talking to Our Plants** A survey asked plant owners how often they talk to their plants.

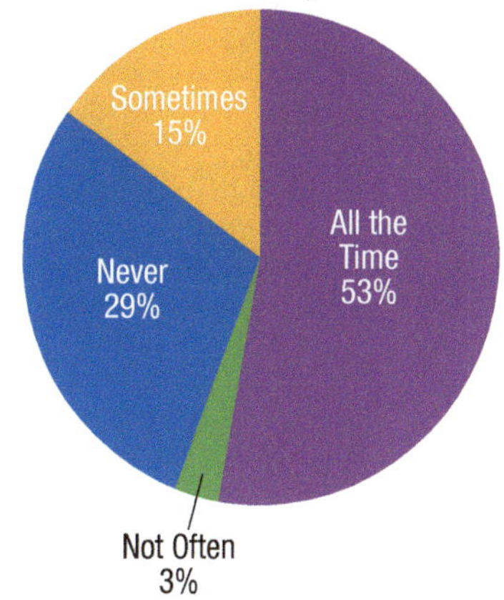

   a. What percentage of plant owners say they never talk to their plants?
   b. What percentage of plant owners say they talk to their plants all the time?
   c. What percentage of plant owners say they talk to their plants sometimes or not often?

5. **Monthly Car Payments** Suppose 3,000 people responded to a survey on car loan payments, the results of which are shown in the circle graph. Find the number of people whose monthly payments would be the following:

   a. \$700 or more
   b. Less than \$300
   c. \$500 or more
   d. \$300 to \$699

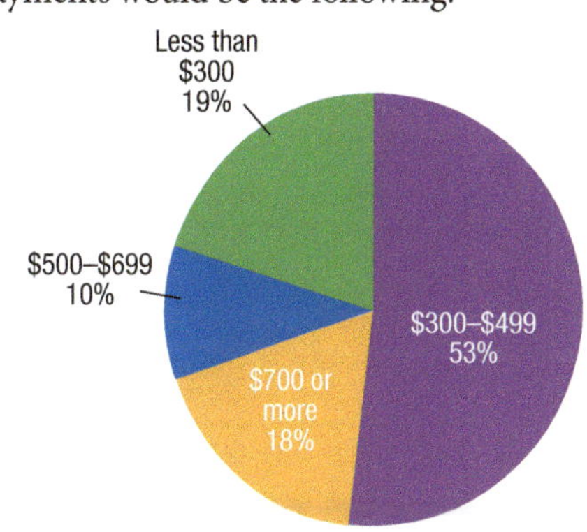

6. **Where Workers Say Germs Lurk** A survey asked workers where they thought the most germ-contaminated spot in the workplace was. Suppose the survey took place at a large company with 4,200 employees. Use the circle graph to determine the number of employees who would vote for each of the following as the most germ-contaminated areas.

   a. Keyboards
   b. Doorknobs
   c. Restrooms or other
   d. Telephones or doorknobs

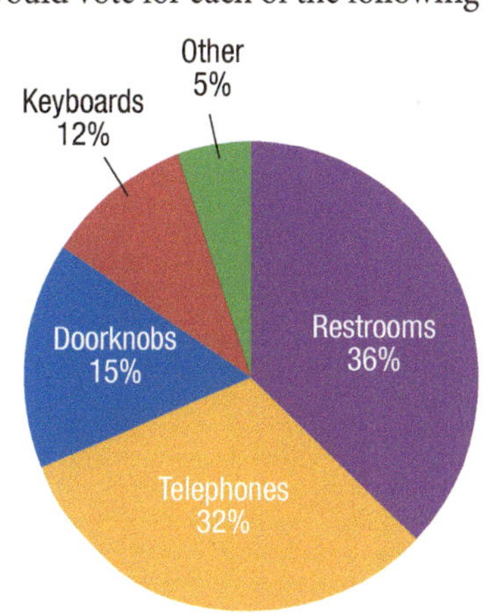

© damircudic/iStockPhoto

7. **Grade Distribution** Scores on a recent math test are shown in the table below for a class of 20 students. Construct a circle graph that shows the number of A's, B's, and C's earned on the test. Use the template provided here or use a protractor.

Grade Distribution

Grade	Number
A	5
B	8
C	7
Total	20

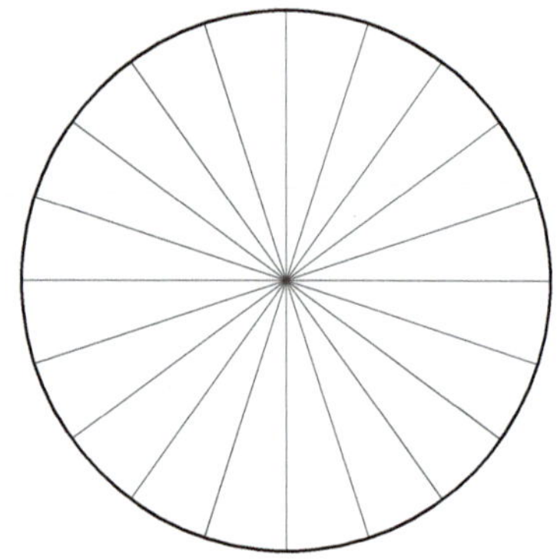

CIRCLE GRAPH TEMPLATE *Each slice is 5% of the area of the circle.*

8. **Building Sizes** The Lean and Mean Gym Company recently ran a promotion for their four locations in the county. The table shows the locations along with the amount of square feet at each location. Use the information in the table to construct a circle graph, using the template provided here or using a protractor. Round to the nearest percent if necessary.

Gym Locker and Size

Location	Square Feet
Downtown	35,000
Uptown	85,000
Lakeside	25,000
Mall	75,000
Total	220,000

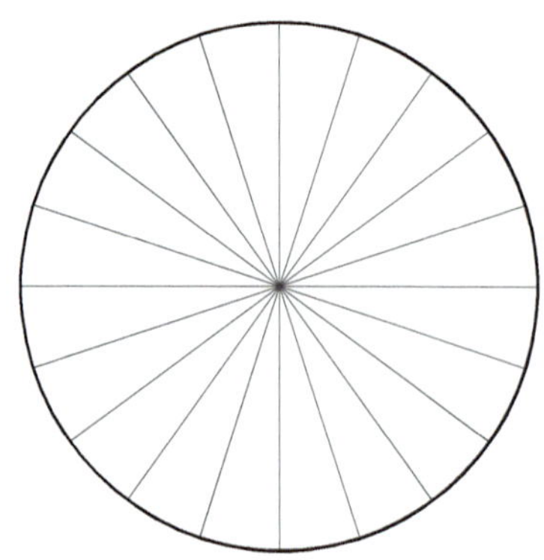

CIRCLE GRAPH TEMPLATE *Each slice is 5% of the area of the circle.*

9. Room Sizes Scott and Amy are building their dream house. The size of the house will be 2,400 square feet. The table below shows the size of each room. Use the information in the table to construct a circle graph, using the template provided here or using a protractor. Round to the nearest percent if necessary.

Room Sizes

Room	Square Feet
Kitchen	400
Dining Room	310
Bedrooms	890
Living Room	600
Bathrooms	200
Total	2,400

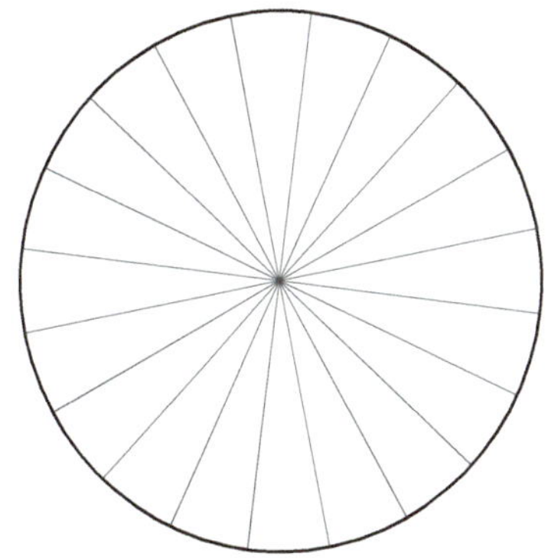

CIRCLE GRAPH TEMPLATE *Each slice is 5% of the area of the circle.*

© RASimon/iStockPhoto

10. Passenger Train Seating The table below gives the number of seats sold in the four classes of seating on a passenger train. Create a circle graph from the information in the table. Round to the nearest percent if necessary.

Passenger Train Seating

Seating Class	Number of Seats
Coach	114
Business	85
First Class	31
Sleeper Class	25
Total	255

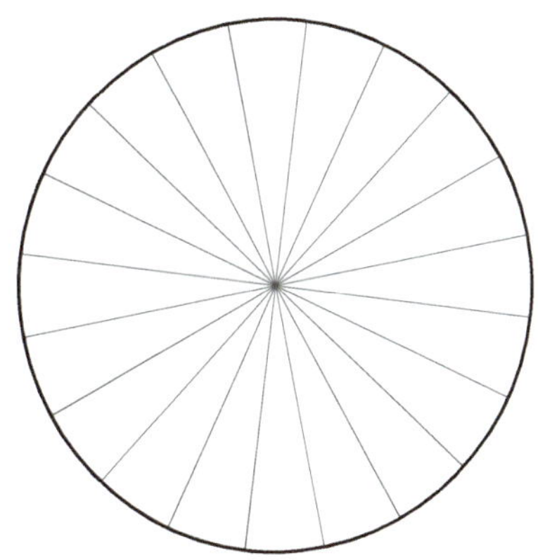

CIRCLE GRAPH TEMPLATE *Each slice is 5% of the area of the circle.*

Getting Ready For The Next Section

Simplify.

11. $\frac{0}{5}$ **12.** $\frac{3}{6}$ **13.** $\frac{4}{52}$

14. $\frac{48}{52}$ **15.** $0.06 + 0.03$ **16.** $1.00 - 0.77$

Chapter 6 Summary

Examples

1. Find the mean for test scores of

   74, 63, 74, 80, 83, 88.

   We add the numbers and divide by 6:

   $$\frac{74 + 63 + 74 + 80 + 83 + 88}{6}$$
   $$= \frac{462}{6} = 77$$

Mean [6.1]

To find the **mean** for a set of numbers, we add all the numbers and then divide the sum by the number of values in the set. The mean is sometimes called the *arithmetic mean*.

2. The median for the test scores in Example 1 will be halfway between 74 and 80.

   $$\text{Median} = \frac{74 + 80}{2} = 77$$

Median [6.1]

To find the **median** for a set of numbers, we write the numbers in order from smallest to largest. If there is an odd number of values, the median is the middle number. If there is an even number of values, then the median is the mean of the two numbers in the middle.

3. The mode for the test scores in Example 1 will be the most frequently occurring score, which is 74.

Mode [6.1]

The **mode** for a set of numbers is the value that occurs most frequently. If all the numbers in the set occur the same number of times, there is no mode.

4. The range for the test scores in Example 1 will be the difference between 88 and 63.

   $$\text{Range} = 88 - 63 = 25$$

Range [6.1]

The **range** for a set of numbers is the difference between the largest number and the smallest number in the sample.

Tables and Bar Charts [6.2]

The bar chart is a visual representation of the information in the table.

Caffeine Content of Hot Drinks

Drink (6-Ounce Cup)	Caffeine (In Milligrams)
Brewed Coffee	100
Instant Coffee	70
Tea	50
Cocoa	5
Decaffeinated Coffee	4

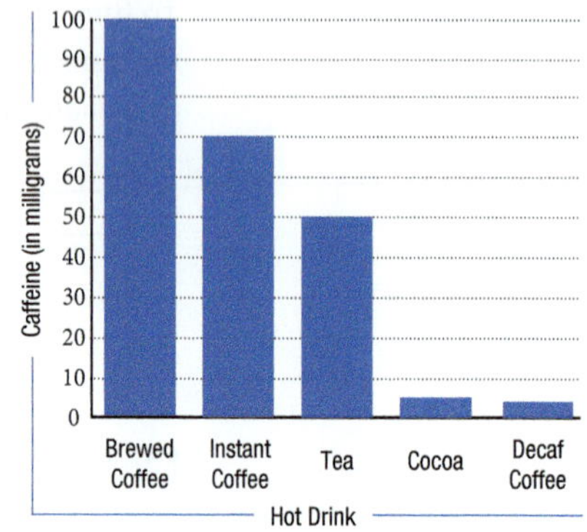

Scatter Diagrams and Line Graphs [6.2]

The diagrams below are additional ways to display information from the preceding table. Each gives the same information as the bar chart on the previous page, the format is just different.

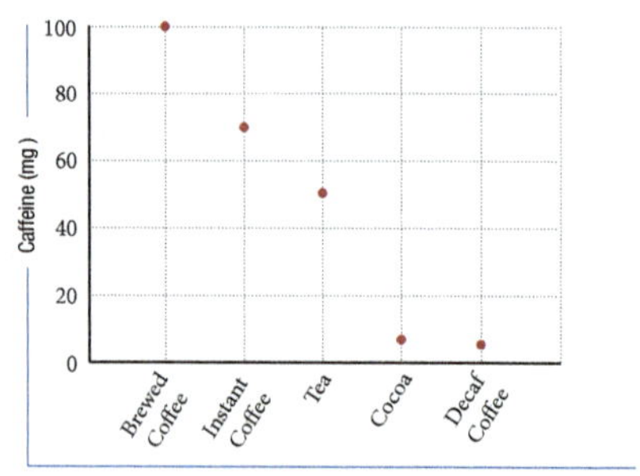

Scatter Diagram

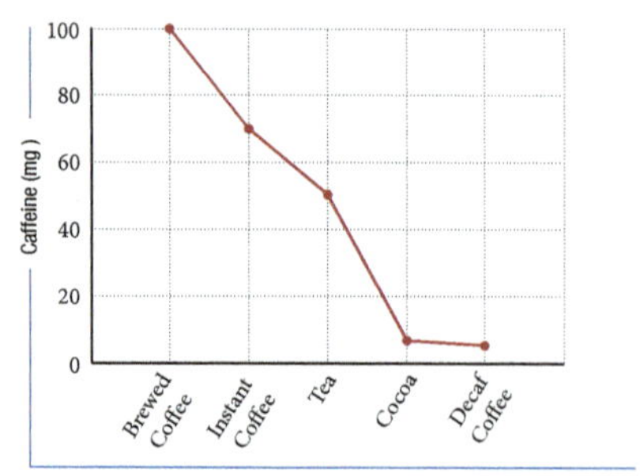

Line Graph

Frequency Tables and Histograms [6.2]

A **frequency table** shows how often data within certain intervals occurs. A **histogram** is a graph of a frequency table.

Interval	Frequency
60-69	4
70-79	5
80-89	7
90-100	4

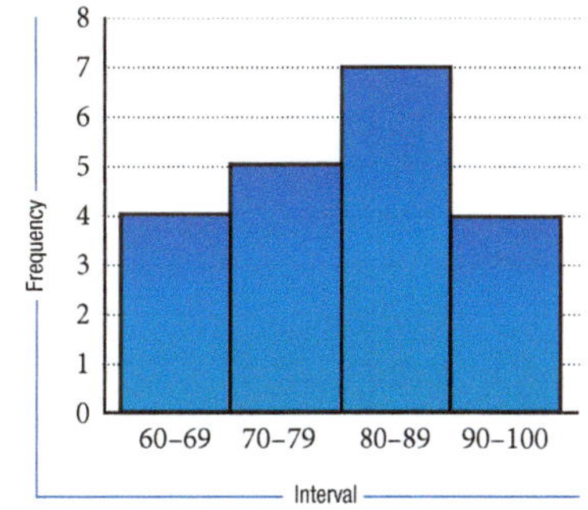

Circle Graphs [6.3]

A ***circle graph*** is another way to give a visual representation of the information in a table.

Seating Class	Number of Seats
First	18
Business	42
Coach	163

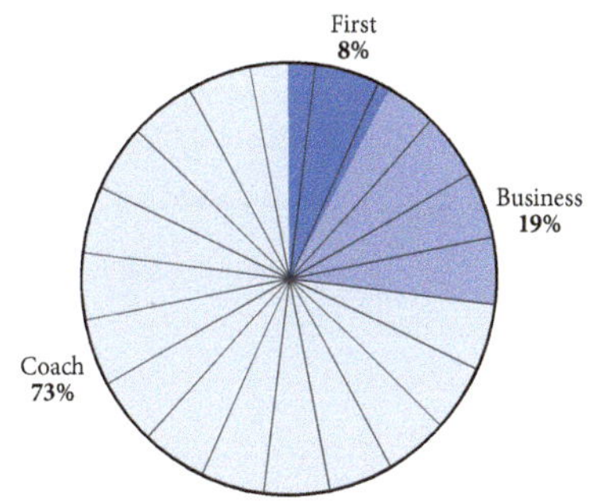

Chapter 6 Test

1. **NBA** The table below shows the active NBA players who had the most minutes played in a particular season.

Player	Times (Minutes)
Gilbert Arenas	3,384
Kobe Bryant	3,401
LeBron James	3,388
Antwon Jamison	3,394
Shawn Marion	3,373

   a. What is the range in minutes played?
   b. What is the mean number of minutes played?
   c. What is the median number of minutes played?

2. **Lighthouses** The table shows the top 10 tallest lighthouses in the United States.

Lighthouse	Height (Feet)
Cape Hatteras	207
Cape Charles	191
Ponce de Leon	175
Barnegat	172
Oak Island	169
Cape Lookout	169
Absecon	169
Fire Island	168
St. Augustine	165
Cape Henry	164

   a. Find the mean lighthouse height.
   b. Find the median lighthouse height.
   c. What is the mode of the lighthouse heights?
   d. What is the range of the lighthouse heights?

3. Coffee Prices The table below shows the unit price of some popular coffees.

Coffee	Price (cents/cup)
Green Mountain Kona Estate	32
Gevalia Kaffe Select	17
Dunkin' Donuts Original	11
Brothers Gourmet	10
Trader Joe's Colombian	7
Yuban Colombian	6
Folgers Classic	5

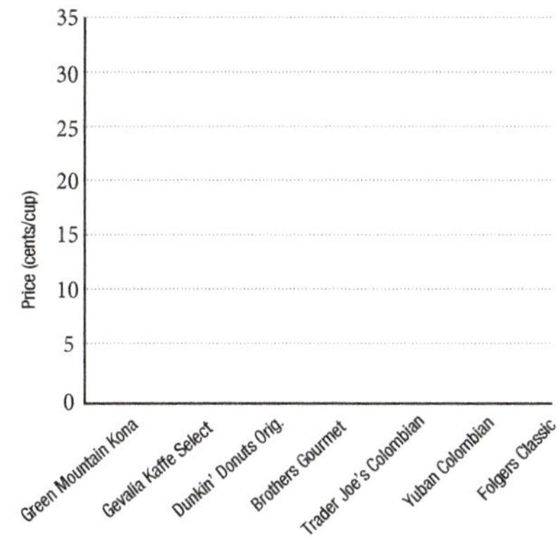

a. Use the template to construct a bar chart from the information in the table.

b. To the nearest tenth, find the mean of the unit prices listed in the table.

c. Find the median unit price of the drinks listed in the table.

d. What is the range of unit prices for the drinks listed in the table?

4. Car Speeds The table below gives the number of seconds it takes various sport-utility vehicles to accelerate from 0 miles per hour to 60 miles per hour.

Car	Time (seconds)
Chevrolet Blazer	9.1
Ford Explorer	10.7
Isuzu Trooper	10.9
Jeep Cherokee	9.7
Nissan Pathfinder	12.3
Toyota 4Runner	15.7

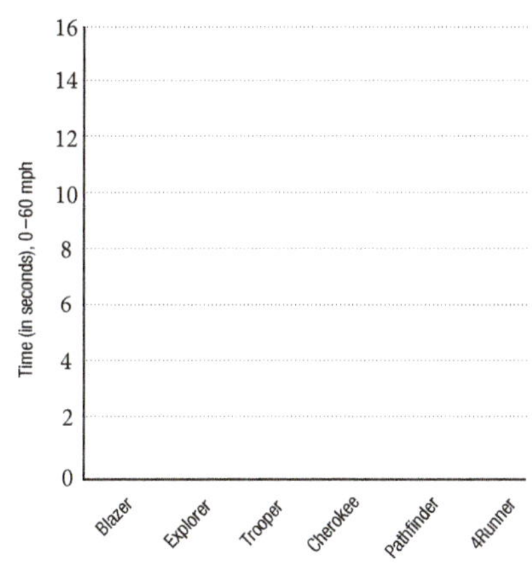

a. What is the range of acceleration times?

b. What is the mean acceleration time?

c. What is the median of acceleration times?

d. Construct a bar chart from the information in the table.

5. Used Car Prices The following prices were listed for Ford Explorers on the ebay.com car auction site.

Car Prices

Year	Price
1999	$14,500
1999	$15,150
2000	$11,000
2000	$14,500
2000	$16,900
2000	$18,950
2001	$16,900
2001	$19,500
2001	$21,450

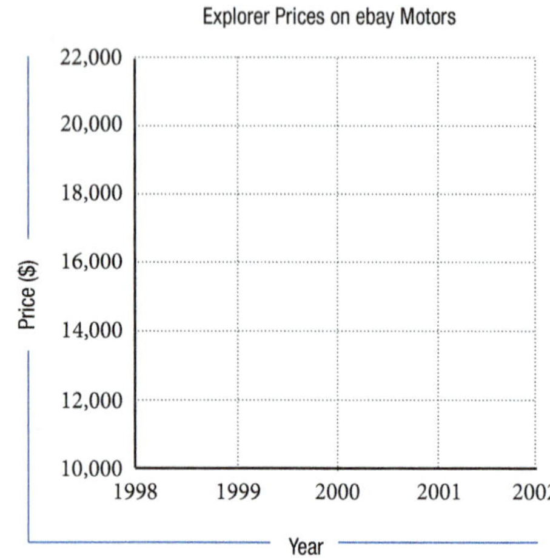

a. Prepare a scatter diagram of the information in the table.

b. What is the median car price?

c. What is the range of car prices?

6. Airline Ticket Prices The table below shows a number of different ticket prices for a round trip ticket from Seattle to Boston in July 2005.

Seating Class	Ticket Price
Coach	307
Coach	528
Coach	1,444
Coach	2,063
Business	2,251
Business	2,549
Business	2,806
First	2,292
First	3,123

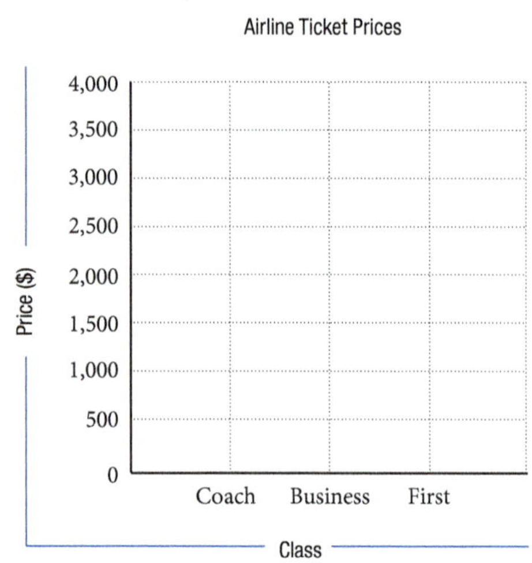

a. Create a scatter diagram from the information in the table.

b. Find the median for the ticket prices.

c. What is the range of ticket prices?

7. Sleeping Infants The table below shows the average number of hours an infant sleeps each day. Construct a line graph from the information in the table.

Age (Months)	Daily Sleep (Hours)
1	$18\frac{1}{2}$
2	17
3	16
4	15
5	$14\frac{1}{2}$
6	14

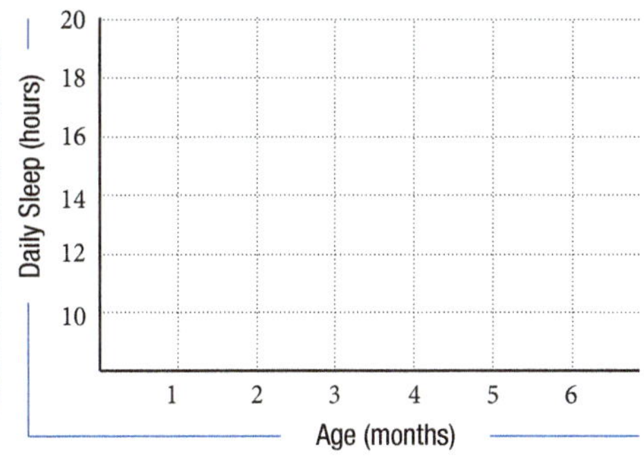

8. Used Car Prices The following prices were listed for BMW 318i cars on the ebay.com car auction site.

Car Prices

Year	Price
1992	$2,850
1993	$2,200
1995	$4,000
1995	$4,100
1995	$4,450
1995	$5,100
1996	$8,100
1997	$10,000
1997	$11,100

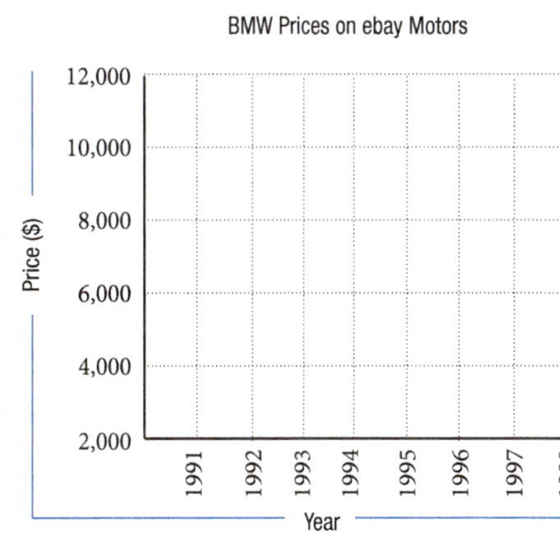

a. Prepare a scatter diagram of the information in the table.

b. To the nearest dollar, what is the mean for the car prices?

c. What is the median car price?

d. What is the range of car prices?

9. Airline Ticket Prices The table below shows a number of different ticket prices for a round trip ticket from Los Angeles to New York City in June 2002.

Seating Class	Ticket Price
Coach	261
Coach	335
Coach	1,525
Coach	2,443
Business	2,643
Business	3,637
First	1,211
First	2,443
First	3,825

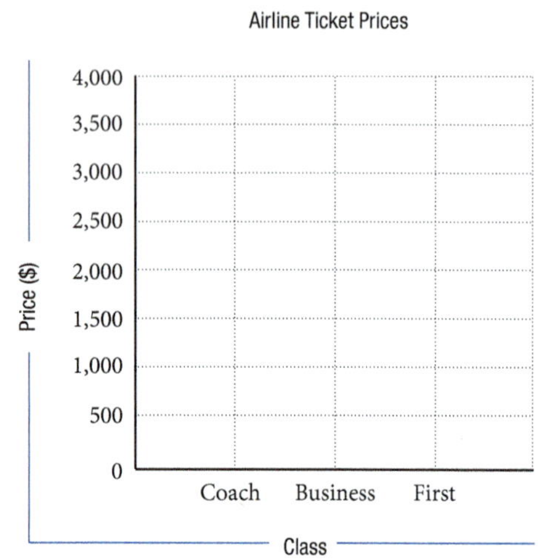

a. Find the mean ticket price to the nearest dollar.

b. Find the median for the ticket prices

c. What is the mode for the ticket prices?

d. What is the range of ticket prices?

e. Create a scatter diagram from the information in the table.

10. Computers The chart shows which operating systems people use. Use the illustration to create a circle graph.

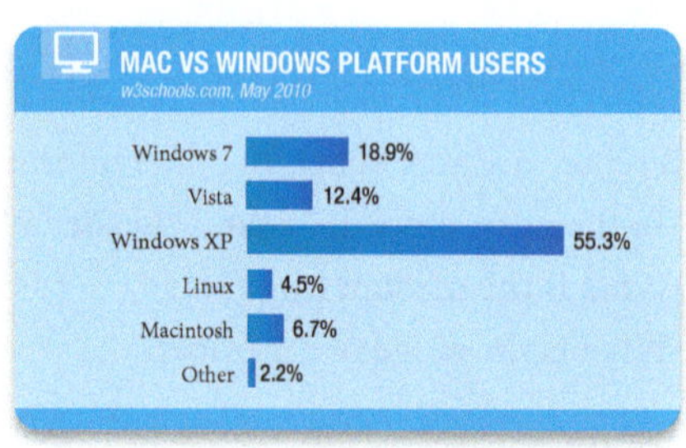

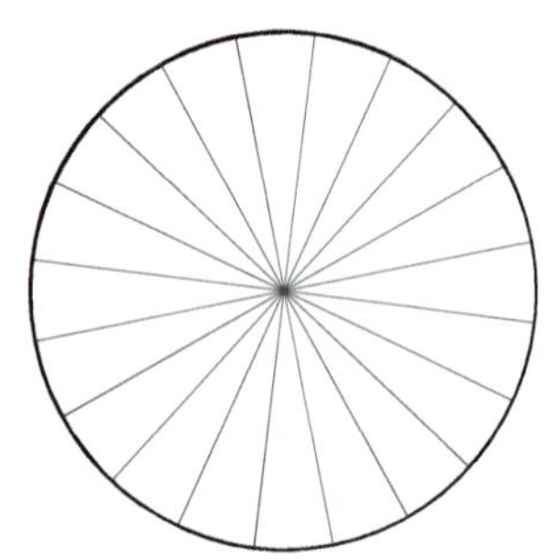

Circle Graph Template Each slice is 5% of the area of the circle.

11. Digital Storage Space A storage device can hold 250 MB of data. The amount of space available on the disk is 100 MB. Use this information to construct a circle graph.

Total Space	250 MB
Used Space	150 MB
Free Space	100 MB

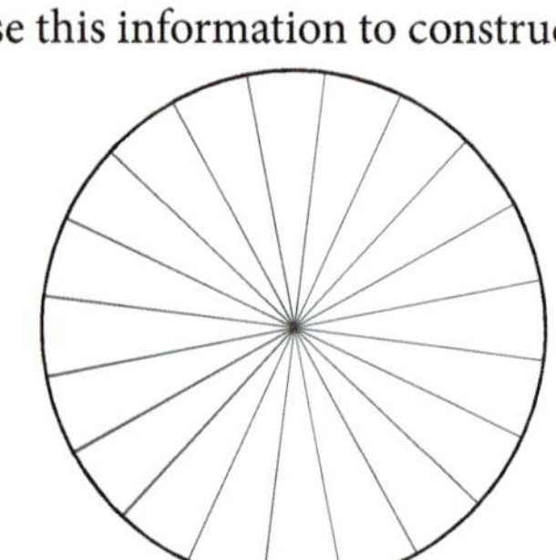

Circle Graph Template Each slice is 5% of the area of the circle.

12. Airline Seating The table below gives the number of seats in each of the three classes of seating on an American Airlines Boeing 777 airliner. Create a circle graph from the information in the table.

Seating Class	Number of Seats
First	18
Business	42
Coach	163

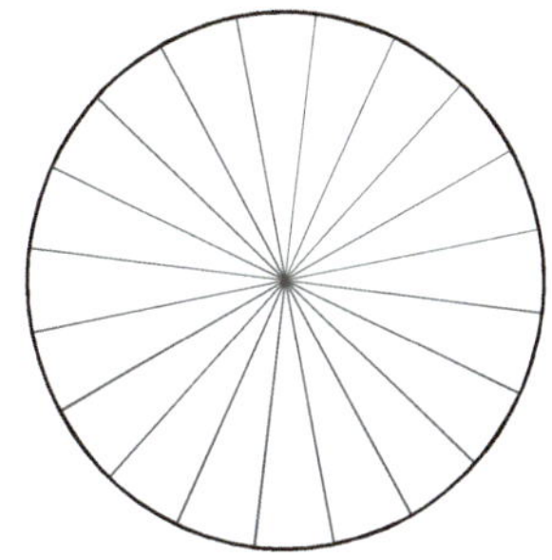

Pie Chart Template Each slice is 5% of the area of the circle.

13. What is the probability of drawing a heart from a deck of 52 cards?

14. What is the probability of drawing a three or an odd card from a deck of 52 cards?

Board Games The diagram below shows a colored 6-sided fair die. Use the illustration to answer the following questions.

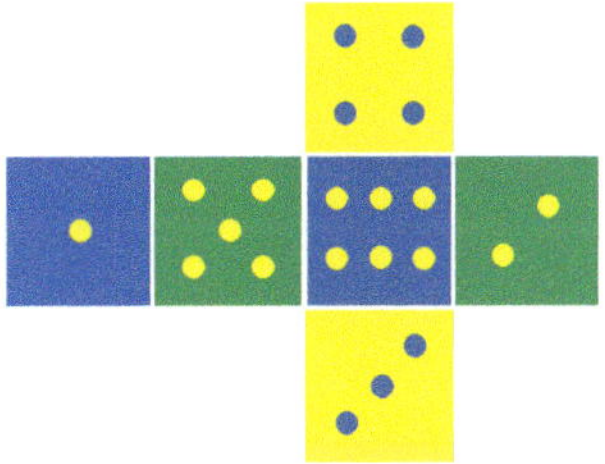

15. Find the probability of rolling a number no less than 2.

16. Find the probability of rolling a 3 or a green.

A survey was taken to determine the most popular laptop among college students. Use the table to answer the following questions.

Laptop	Number of Votes
HP	15
Dell	24
Apple	27
Sony	4
Toshiba	10

17. Find the probability someone chose an HP or Toshiba laptop.

18. Find the probability someone chose an Apple laptop.

Chapter R

Exercise Set R.1

Vocabulary Review **1.** prime number **2.** composite number **3.** numerator **4.** denominator **5.** equivalent **6.** lowest terms

Problems **1.** Prime **3.** Composite; 3, 5, and 7 are factors **5.** Composite; 3 is a factor **7.** Prime **9.** $2^2 \cdot 3$ **11.** 3^4 **13.** $5 \cdot 43$ **15.** $3 \cdot 5$ **17.** $\frac{1}{2}$ **19.** $\frac{2}{3}$ **21.** $\frac{4}{5}$ **23.** $\frac{9}{5}$ **25.** $\frac{7}{11}$ **27.** $\frac{3}{5}$ **29.** $\frac{1}{7}$ **31.** $\frac{7}{9}$ **33.** $\frac{7}{5}$ **35.** $\frac{1}{5}$ **37.** $\frac{11}{7}$ **39.** $\frac{5}{3}$ **41.** $\frac{8}{9}$ **43.** $\frac{42}{55}$ **45.** $\frac{17}{19}$ **47.** $\frac{14}{33}$ **49. a.** $\frac{2}{17}$ **b.** $\frac{3}{36}$ **c.** $\frac{1}{9}$ **d.** $\frac{3}{28}$ **e.** $\frac{2}{19}$ **51. a.** $\frac{1}{45}$ **b.** $\frac{1}{30}$ **c.** $\frac{1}{18}$ **d.** $\frac{1}{15}$ **e.** $\frac{1}{10}$ **53. a.** $\frac{1}{3}$ **b.** $\frac{5}{6}$ **c.** $\frac{1}{5}$ **55.** $\frac{9}{16}$

57. $\frac{1}{2} = \frac{2}{4} = \frac{4}{8} = \frac{8}{16}$ **59.** $\frac{5}{4} = \frac{10}{8} = \frac{20}{16}$

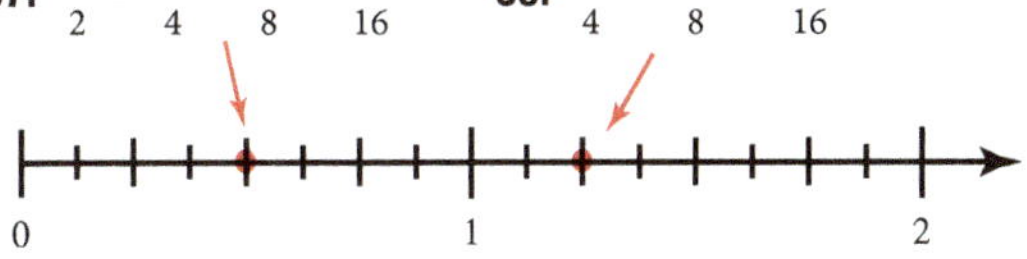

61. $\frac{1}{4}$ **63. a.** More than half **b.** Less than one-third **65.** $\frac{2}{3}$ **67.** $\frac{7}{15}$ **69.** $\frac{1}{4}$ **71.** 0

73.

Duration	Total Cost	Cost per Day
1 day	\$80	\$80
2 days	\$158	\$79
3 days	\$199	\$66.33
4 days	\$219	\$54.75
5 days	\$231	\$46.20

Source: disneyland.disney.go.com

75. a. 2.47 times **b.** 7.25 times **77.** a and b **79.** c **81.** 1,482 **83.** 39,968 **85.** 9 **87.** 7

Find the Mistake **1.** A number is a divisor of another number if it divides it without a remainder. **2.** 31 is a prime number because the only numbers that divide it without a remainder are 31 and 1.
3. Factoring 112 into the product of primes gives us $2^4 \cdot 7$.
4. To get the fraction $\frac{24}{36}$ in lowest terms, we divide the numerator and the denominator by $2 \cdot 2 \cdot 3$.

Exercise Set R.2

Vocabulary Review **1.** multiplication **2.** division **3.** reciprocal **4.** least common denominator **5.** factors

Problems **1.** $\frac{8}{15}$ **3.** $\frac{7}{8}$ **5.** 1 **7.** $\frac{27}{4}$ **9.** 1 **11.** $\frac{1}{24}$ **13.** $\frac{24}{125}$ **15.** $\frac{105}{8}$ **17.** $\frac{8}{27}$ **19.** $\frac{3}{5}$ **21.** 9 **23.** 1 **25.** 8 **27.** $\frac{1}{15}$ **29.** $\frac{15}{4}$ **31.** $\frac{4}{3}$ **33.** 9 **35.** 200 **37.** $\frac{3}{8}$ **39.** 1 **41.** $\frac{49}{64}$ **43.** $\frac{3}{4}$ **45.** $\frac{15}{16}$ **47.** $\frac{1}{6}$ **49.** 6 **51.** $\frac{5}{18}$ **53.** $\frac{9}{2}$ **55.** $\frac{2}{9}$ **57.** $\frac{2}{3}$ **59.** $\frac{1}{4}$ **61.** $\frac{1}{2}$ **63.** $\frac{1}{3}$ **65.** $\frac{3}{2}$ **67.** $\frac{x+6}{2}$ **69.** $\frac{4}{5}$ **71.** $\frac{10}{3}$ **73.** $\frac{7}{9}$ **75.** $\frac{7}{3}$ **77.** $\frac{7}{4}$ **79.** $\frac{7}{6}$ **81.** $\frac{9}{20}$ **83.** $\frac{7}{10}$ **85.** $\frac{19}{24}$ **87.** $\frac{13}{60}$ **89.** $\frac{31}{100}$ **91.** $\frac{67}{144}$

Find the Mistake **1.** To multiply the fractions $\frac{2}{8}$ and $\frac{3}{4}$, we put the product of 2 and 3 in the numerator over the product of 8 and 4 in the denominator. **2.** The reciprocal of $\frac{5}{20}$ is $\frac{20}{5}$. **3.** To divide $\frac{7}{8}$ by $\frac{2}{5}$, multiply $\frac{7}{8}$ by $\frac{5}{2}$. **4.** The least common denominator for the fractions $\frac{3}{7}$ and $\frac{5}{6}$ is 42.

Exercise Set R.3

Vocabulary Review **1.** multiply **2.** divide **3.** decimal point **4.** tenths **5.** hundredths

Problems **1.** Three tenths **3.** Fifteen thousandths **5.** Three and four tenths **7.** Fifty-two and seven tenths **9.** $405\frac{36}{100}$ **11.** $9\frac{9}{1,000}$ **13.** $1\frac{234}{1,000}$ **15.** $\frac{305}{100,000}$ **17.** Tens **19.** Tenths **21.** Hundred-thousandths **23.** Ones **25.** Hundreds **27.** Hundred-thousandths **29.** 0.55 **31.** 6.9 **33.** 11.11 **35.** 100.02 **37.** 3,000.003

	Round to the Nearest			
	Whole Number	Tenth	Hundredth	Thousandth
39.	48	47.5	47.55	47.548
41.	1	0.8	0.82	0.818
43.	0	0.2	0.16	0.156
45.	2,789	2,789.3	2,789.32	2,789.324
47.	100	100.0	100.00	100.000

49. 0.002 0.005 0.02 0.025 0.05 0.052 **51.** 7.451 and 7.54 **53.** $\frac{1}{4}$ **55.** $\frac{7}{20}$ **57.** $\frac{1}{8}$ **59.** $\frac{5}{8}$ **61.** $\frac{7}{8}$ **63.** $\frac{3}{250}$

65. Building	Height in Words
Burj Khalifa	Two thousand, seven hundred sixteen
Taipei 101	One thousand, six hundred sixty-seven
World Financial Center	One thousand, six hundred fourteen
Petronus Tower 1	One thousand, four hundred eighty-three
Petronus Tower 2	One thousand, four hundred eighty-three

67. 9.99 **69.** 10.05 **71.** 0.05 **73.** 0.01

Find the Mistake **1.** The value of each column to the left of the decimal point is 10 times the value of the column on its right. **2.** The fourth column to the right of the decimal point is called the ten thousandths column. **3.** 2,568 rounded to the nearest hundred is 2,600. **4.** 2.568 rounded to the nearest tenth is 2.6.

Exercise Set R.4

Vocabulary Review **1.** add **2.** multiplying, right **3.** divide, above

Problems **1.** 6.19 **3.** 1.13 **5.** 6.29 **7.** 9.042 **9.** 8.021 **11.** 11.7843 **13.** 24.342 **15.** 24.111 **17.** 258.8414 **19.** 666.66 **21.** 38.0916 **23.** 11.11 **25.** 3.57 **27.** 4.22 **29.** 120.41 **31.** 44.933 **33.** 7.673 **35.** 530.865 **37.** 8.058 **39.** 27.89 **41.** 35.64 **43.** 411.438 **45.** 6 **47.** 1 **49.** 3.1 **51.** 5.9 **53.** 3.272 **55.** 4.001 **57.** 0.28 **59.** 0.028 **61.** 0.0027 **63.** 0.78 **65.** 0.792 **67.** 0.0156 **69.** 24.29821 **71.** 0.03 **73.** 187.85 **75.** 0.002 **77.** 27.96 **79.** 0.43 **81.** 49,940 **83.** 9,876,540 **85.** 1 **87.** 1.89 **89.** 0.0025 **91** 5.1106 **93.** 7.3485 **95.** 4.4 **97.** 2.074 **99.** 3.58 **101.** 187.4 **103.** 116.64 **105.** 20.75 **107.** 19.7 **109.** 6.2 **111.** 5.2 **113.** 11.04 **115.** 4.8 **117.** 9.7 **119.** 2.63 **121.** 4.24 **123.** 2.55 **125.** 1.35 **127.** 6.5 **129.** 9.9 **131.** 0.05 **133.** 89 **135.** 2.2 **137.** 12 **139.** 1.35 **141.** 16.97 **143.** 0.25 **145.** 2.71 **147.** 7.5 miles **149.** $7.55/hour **151.** 22.4 miles **153.** 3 hours **155.** 7 minutes **157.** Federal budget: −$1.102 trillion; Congressional plan: −$.999 trillion. The alternative plan is better.

Find the Mistake **1.** To subtract 2.35 from 16.854, line up the decimals and subtract as usual. **2.** The product of 7.19 and 189.4 will have 3 digits to the right of the decimal. **3.** To estimate the answer to the problem 6.79×28.6, we round the numbers to 7 and 30 and multiply to get 210. **4.** When dividing with decimal numbers, the decimal point in the answer is placed directly above the decimal in the dividend.

Landmark Review **1.** Composite, 5^2 **2.** Prime **3.** Prime **4.** Composite, $3 \cdot 7$ **5.** $\frac{1}{3}$ **6.** $\frac{16}{15}$ **7.** 1 **8.** $\frac{1}{2}$ **9.** $15\frac{7}{10}$ **10.** $\frac{46}{10{,}000}$ **11.** 1.4 **12.** 11.11 **13.** 6.084 **14.** 120.12 **15.** 0.69 **16.** 6.098 **17.** 35.667 **18.** 88.667

Exercise Set R.5

Vocabulary Review **1.** percent **2.** denominator **3.** left **4.** right **5.** fraction **6.** decimal

Problems **1.** $\frac{20}{100}$ **3.** $\frac{60}{100}$ **5.** $\frac{24}{100}$ **7.** $\frac{65}{100}$ **9.** $\frac{48}{100}$ **11.** $\frac{10}{100}$ **13.** 0.23 **15.** 0.92 **17.** 0.09 **19.** 0.034 **21.** 0.0634 **23.** 0.009 **25.** 23% **27.** 92% **29.** 45% **31.** 3% **33.** 60% **35.** 80% **37.** 27% **39.** 123% **41.** 70% **43.** $\frac{3}{5}$ **45.** $\frac{3}{4}$ **47.** $\frac{1}{25}$ **49.** $\frac{53}{200}$ **51.** $\frac{7{,}187}{10{,}000}$ **53.** $\frac{3}{400}$ **55.** $\frac{1}{16}$ **57.** $\frac{1}{3}$ **59.** $\frac{1}{40}$ **61.** 50% **63.** 75% **65.** $33\frac{1}{3}\%$ **67.** 80% **69.** 87.5% **71.** 14% **73.** 325% **75.** 150% **77.** 250% **79.** 48.8% **81.** 78.4% **83.** 11.8% **85.** 72.2% **87.** 71.0% **89.** 0.50; 0.75

91. a. $\frac{49}{100}, \frac{7}{25}, \frac{7}{50}, \frac{1}{25}, \frac{3}{100}, \frac{1}{50}, \frac{1}{100}$

b. 0.49, 0.28, 0.14, 0.04, 0.03, 0.02, 0.01 **93.** 20%

95. $\frac{4}{25} = 16\%, \frac{3}{25} = 12\%, \frac{11}{50} = 22\%, \frac{1}{10} = 10\%, \frac{3}{20} = 15\%, \frac{1}{4} = 25\%$

97. $\frac{1}{17} \approx 5.9\%; \frac{1}{19} \approx 5.3\%$ **99.** 48%, 73% **101.** 38%

Find the Mistake **1.** The fraction $\frac{36}{100}$ given as a percent is 36%. **2.** Changing the percent 57.4% to a decimal gives us 0.574. **3.** Changing the decimal 0.1465 to a percent gives us 14.65%. **4.** To change a fraction to a percent, first change the fraction to a decimal and then change to a percent.

Exercise Set R.6

Vocabulary Review **1.** base, exponent **2.** factor **3.** first **4.** zero **5.** sum **6.** difference **7.** product **8.** quotient **List of steps:** 3, 2, 1, 4

Problems **1.** Base 4, Exponent 5 **3.** Base 3, Exponent 6 **5.** Base 8, Exponent 2 **7.** Base 9, Exponent 1 **9.** Base 4, Exponent 0 **11.** Base 4, Exponent 8 **13.** 36 **15.** 8 **17.** 1 **19.** 1 **21.** 81 **23.** 10 **25.** 12 **27.** 1 **29.** 27 **31.** 12 **33.** 100 **35.** 4 **37.** 43 **39.** 16 **41.** 84 **43.** 14 **45.** 74 **47.** 12,768 **49.** 104 **51.** 416 **53.** 66 **55.** 21 **57.** 7 **59.** 16 **61.** 84 **63.** 40 **65.** 41 **67.** 18 **69.** 405 **71.** 124 **73.** 11 **75.** 91 **77.** 7 **79.** $\frac{1}{2}$ **81.** $\frac{9}{100}$ **83.** 3 **85.** 40 **87.** $\frac{7}{10}$ **89.** 13 **91.** 12 **93.** 186 **95.** 646 **97.** $8(4 + 2) = 48$ **99.** $2(10 + 3) = 26$ **101.** $3(3 + 4) + 4 = 25$ **103.** $(20 \div 2) - 9 = 1$ **105.** $(8 \cdot 5) + (5 \cdot 4) = 60$

Find the Mistake **1.** The expression 8^5 is read "8 to the 5th power." **2.** In the expression 16^3, the 16 is used as a factor three times. **3.** To simplify $9^2 + 4 - 6 \cdot 7$ using the order of operations, we first work the exponent. **4.** To work the problem $5 + 6 - 2 + 18$ using the order of operations, we add and subtract left to right.

Homework 1.1

1. $g = m + 3$

m	g
2	5
3	6
5	8
10	13
12	15
16	19
18	21
m	$m+3$

2. $w = 20 - t$

t	w
0	20
2	18
4	16
5	15
6	14
10	10
12	8
t	$20-t$

3. $x = \frac{b}{2}$

b	x
0	0
2	1
4	2
5	2.5
6	3
8	4
9	4.5
b	$\frac{b}{2}$

4.

z	3	6	8	12	15	18	20	z
r	2	4	$\frac{16}{3}$	8	10	12	$\frac{40}{3}$	$\frac{2}{3}z$

$r = \frac{2}{3}z$

5. Answers will vary.

Sample Answer:

n	0	5	10	15	20
W	0	6	12	18	24

6. Answers will vary.

Sample Answer:

x	0	2	4	6	8
M	0	3	6	9	12

7. a, b, and c.

x	y
4	0.5
8	1
10	1.25
16	2

The three tables have identical values.

8. a, b, and c.

x	y
5	3
10	6
12	7.2
1	0.6

The three tables have identical values.

9.

Ridgecrest	70	75	82	86	90	R
Calculation	$70+15$	$75+15$	$82+15$	$86+15$	$90+15$	$R+15$
Sunnyvale	85	90	97	101	105	$R+15$

a. Add 15° to Ridgecrest's temperature to find Sunnyvale's temperature.

b. Temperature in Sunnyvale = Temperature in Ridgecrest + 15

c. $S = R + 15$

d.

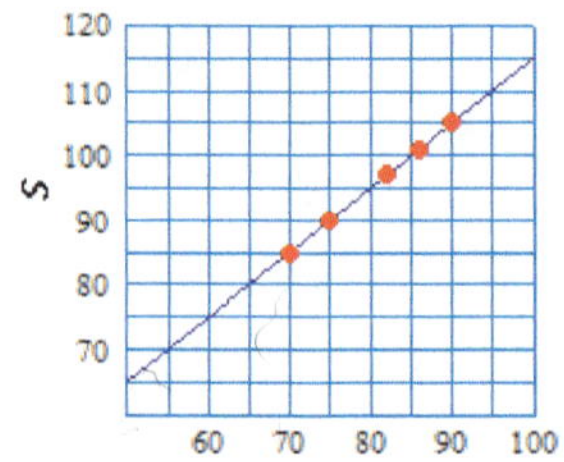

10.

driven	40	60	90	120
Calculation	$200-40$	$200-60$	$200-90$	$200-120$
remaining	160	140	110	80

driven	140	170	d
Calculation	$200-140$	$200-170$	$200-d$
remaining	60	30	$200-d$

a. Subtract the miles driven from 200 to get Jerome's remaining miles.

b. Miles remaining = 200 − miles driven

c. $r = 200 - d$

d.

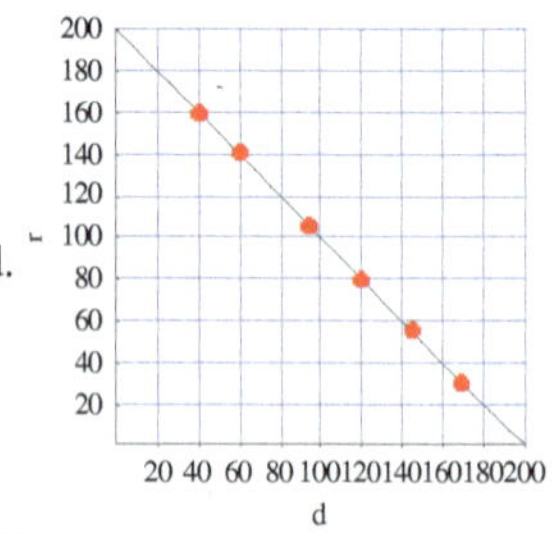

11.

Total	15	30	45	60	75	81	b
Calculation	$\frac{15}{3}$	$\frac{30}{3}$	$\frac{45}{3}$	$\frac{60}{3}$	$\frac{75}{3}$	$\frac{81}{3}$	$\frac{b}{3}$
share	5	10	15	20	25	27	$\frac{b}{3}$

a. Divide the total bill by 3 to get Milton's share.

b. Milton's share $= \frac{\text{total bill}}{3}$

c. $s = \frac{b}{3}$

d. 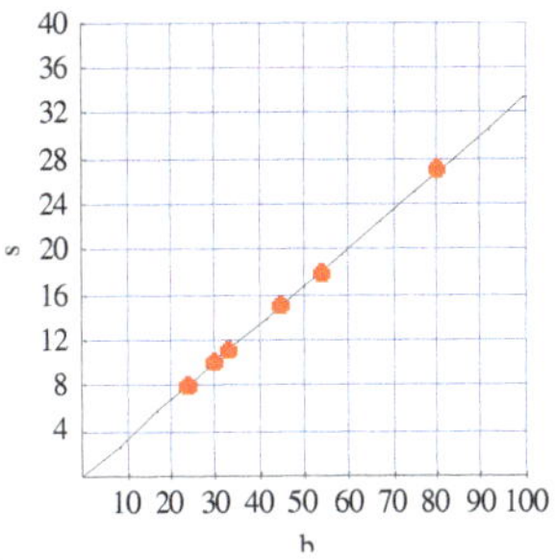

12.

Total Calories	1000	1500	2000
Calculation	0.30 × 1000	0.30 × 1500	0.30 × 2000
Fat Calories	300	450	600

Total Calories	2500	3000	C
Calculation	0.30 × 2500	0.30 × 3000	0.30 × C
remaining	750	900	0.30 × C

a. Multiply the Total Calories by 0.30
b. Fat Calories = 0.30 × Total Calories
c. $F = 0.30 \times C$

d. 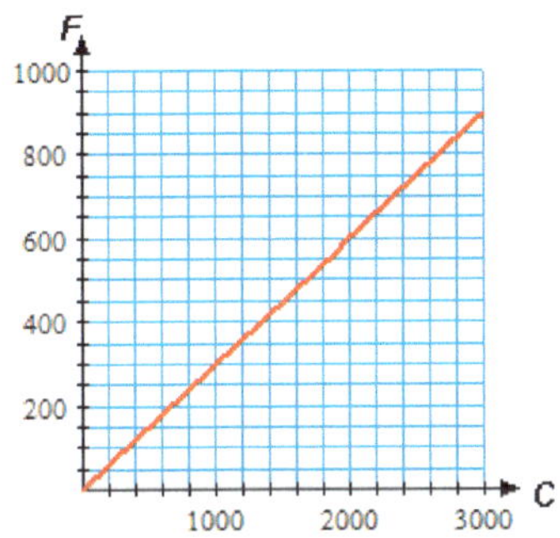

13. a. Horizontal: Number of years, Vertical: Amount in account
b. About \$3000
c. Approximately after 13.75 years
d. About \$200 e. About \$400

14. a. Horizontal: Number of minutes, Vertical: Temperature of soup
b. After about 3.5 minutes c. 70 degrees
d. About 55 degrees e. 60 degrees

15. Graph C

16. a. I. Graph C II. Graph D III. Gaph A
b. Answers will vary. Sample Answer: Delbert walks toward school, about halfway there he realizes that he left a book at home. He finds and retrieves the book, then walks back to school.

17. a. 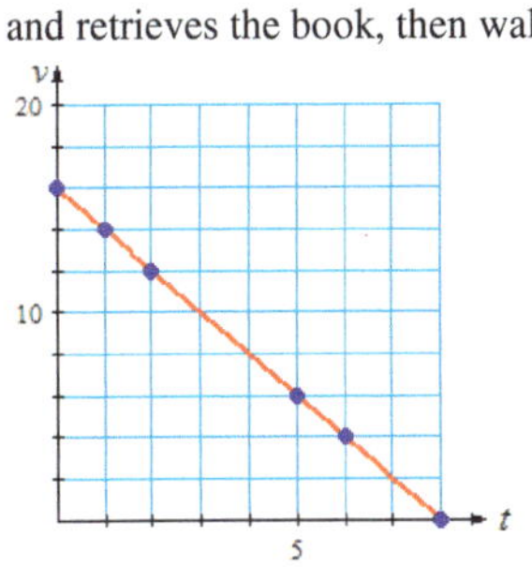

b. 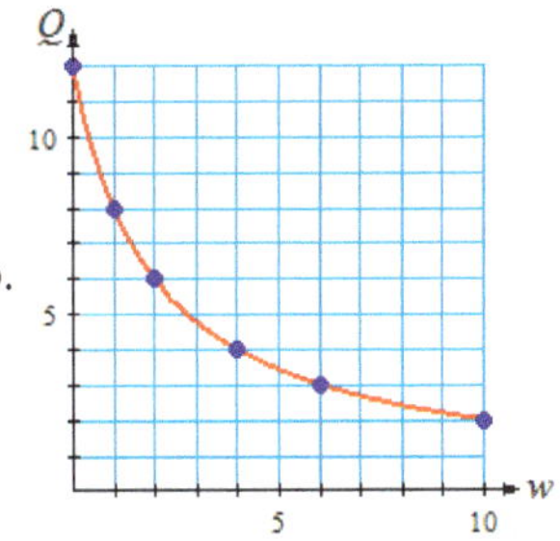

18. a.

x	0	10	30	40	60	70
y	70	60	40	30	10	0

b. $y = 70 - x$

19. a. $y = 2 - x$

b.

x	0	0.2	0.3	0.5	0.6	0.7
y	2	1.8	1.7	1.5	1.4	1.3

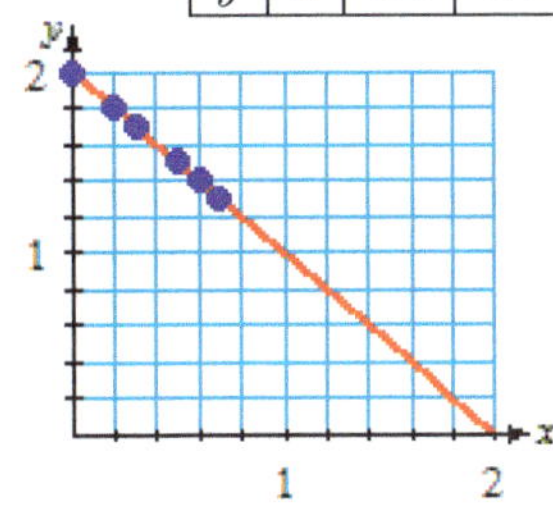

20. a. $y = \frac{120}{x}$

b.

x	1	2	3	4	6	8
y	120	60	40	30	20	15

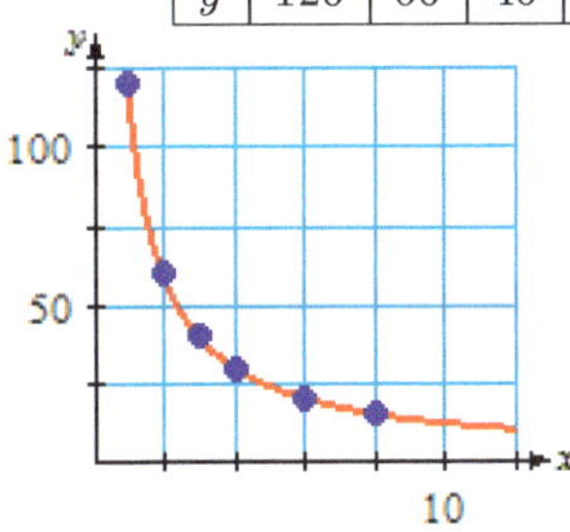

Homework 1.2

1. $4y$ 2. $2b$ 3. $1.15g$
4. $t - 5$ 5. $\frac{7}{w}$ 6. $\frac{B}{4}$
7. $T + 20$ 8. $16 - p$ 9. $\frac{15}{M}$
10. $R - 3.5$ 11. $n - 6$ 12. $6n$
13. $6 - n$ 14. $\frac{6}{n}$ 15. $2x$
16. $4b$ 17. $2v$
18. a. $\frac{x}{6}, \frac{1}{6}x$ b. $0.06, \frac{6}{100}x$
19. a. $\frac{1}{5}m$

b.

Spending Money	m	20	25	60	80
Amount Saved	$\frac{1}{5}m$	4	5	12	16

20. a. $\frac{1}{4}W$

b.

Width of photo	W	6	10	24	30
Height of photo	$\frac{1}{4}W$	1.5	2.5	6	7.5

21. a. $\frac{2}{3}p$

b.

Members present	p	90	96	120	129
Votes Needed	$\frac{2}{3}p$	60	64	80	86

22. a. $\frac{3}{5}D$

b.

Stock Dividend	D	40	85	115	170
Maria's Share	$\frac{3}{5}D$	24	51	69	102

23. Ticket Price: p; $15p$
24. Cost of a light bulb: b; $3b$
25. Savings account balance: s; $\frac{3}{5}s$
26. Price of the pizza : p; $\frac{p}{6}$
27. Weight of copper : w; $\frac{w}{16}$
28. School buses : b, $0.09b$
29. Cost of vaccine : v, $v - 16$
30. Cost of each computer : c, $32c$
31. a. $480 - x$ b. $500 - x$ c. $30 - x$
32. a. $x - 15$ b. $x - 12$ c. $x - 18$
33. Amount of milk : m, $0.70m$
34. Number of days : d, $\frac{1200}{d}$
35. Votes opponent received : v, $v - 432$
36. Budget : B, $B + 2000$
37. a. $d = 180t$
    b. 360 miles; 630 miles; 2160 miles
38. a. $C = 0.0002g$
    b. 0.08 gal, 1 gal, 10 gal
39. a. $B = 0.085R$
    b. \$8500, \$42500, \$170,000
40. a.. $P = R - 4000$
    b. \$6000, \$2500, $-$\$1500

Homework 1.3

1. a. no b. yes
   c. no d. no
2. a. yes b. yes
   c. no d. no
3. a. yes b. no
   c. no d. yes
4. a. no b. yes
   c. yes d. no
5. a. no b. no
   c. yes d. yes
6. a. yes b. yes
   c. no d. no
7. horizontal: 0.25; vertical: 4
8. horizontal: 50; vertical: 0.5

9. a.

x	0	2	5	6	10	12	16
y	16	14	11	10	6	4	0

b. $y = 16 - x$

10. a.

x	0	1	4	5	7	8	10
y	0	4	16	20	28	32	40

b. $y = 4x$

11. f 12. g 13. c, e 14. b

15. a. \$35

b.

w	12	16	20	30	36	40
p	15	20	25	37.50	45	50

c.

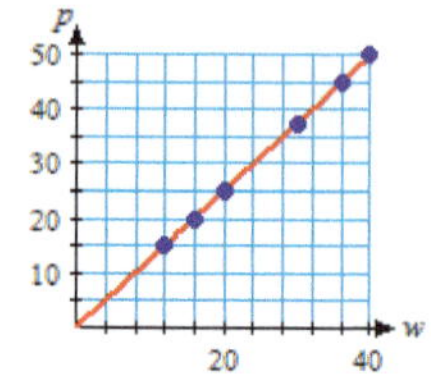

16. a. $D = 1000 - L$

b.

L	250	300	450	600
D	750	700	550	400

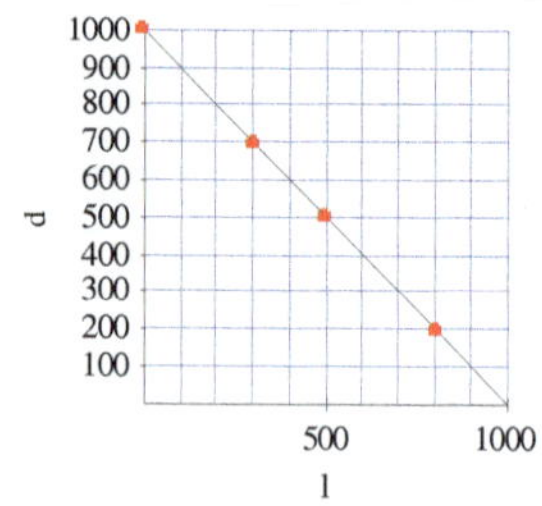

17. a. $S = \dfrac{3600}{m}$

b.

m	20	40	60	100
S	180	90	60	36

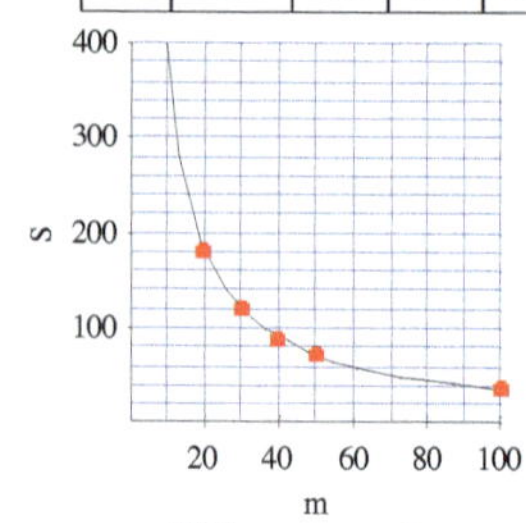

18. a. $p = 110$ b. $p = 60$
19. a. $g \approx 13.9$ b. $g \approx 11.7$
20 a. $(50, 4)$
    b. The tax on a \$50 item is \$4.
21 a. $(4, 16)$
    b. A 4-inch tall bunny weighs 16 ounces.
22. a $I = 80t$

b.

t	2	4	7
I	\$160	\$320	\$560

c.

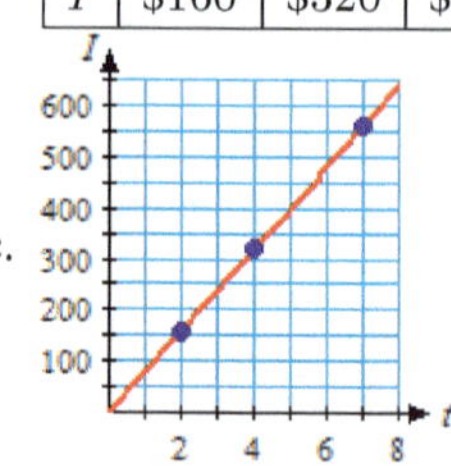

23. a $P = 6000 - C$

b.

C	\$800	\$1000	\$2500
P	\$5200	\$5000	\$3500

c.

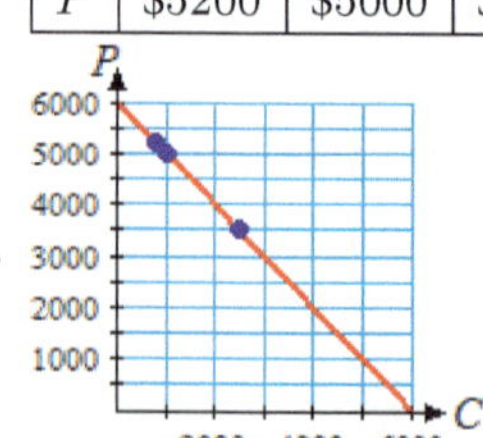

24. a $I = 15000r$

b.

r	8%	$10\frac{1}{4}$%	12%
I	\$1200	\$1537.50	\$1800

c. 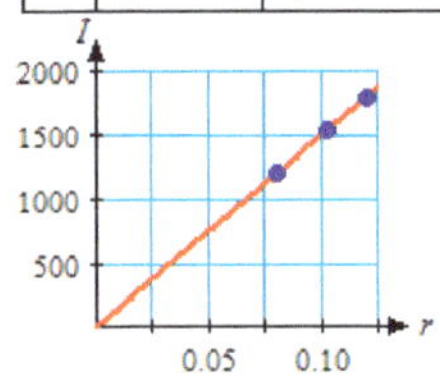

25. a. $S = \frac{540}{n}$

b.

n	20	25	30
S	27	21.6	18

c. 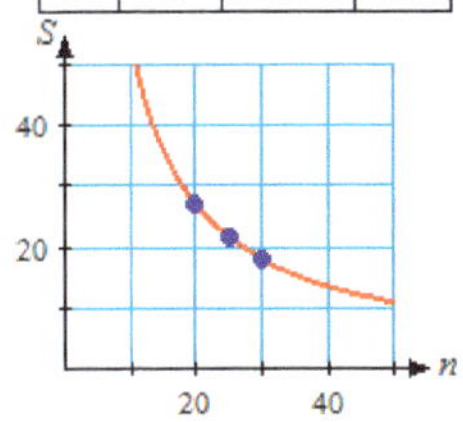

b. $B + 32 = 157$ c. 125 pounds

25. a. $w =$ Hourly wage
b. $20w = 136$ c. \$6.80

26. a. $P =$ Total Profit
b. $\frac{P}{8} = 64$ c. \$512

27. $89 = p - 26$
Her mother paid \$115.

28. $360 = 0.40I$
Emily makes \$900 per month.

29. $1978 = B - 2378$
There wase \$4356 in her account.

30.a. price of new car : n; price of used car : u;
$u = n - 3400$
b. \$11,100 c. \$12,600

31. a. number of games played: p;
number of games won: w; $w = 0.60p$
b. 72 c. 160

32.a. selling price : s; profit : P;
$P = 0.18s$
b. \$10.80 c. \$40

Homework 1.4

1.

t	q
2	11
4	13
6	15
9	18
21	30
30	39

To find a value for q, add 9 to the value of t. To find a value for t, subtract 9 from the value of q.

2.

n	p
0	0
2	10
4	20
5	25
7	35
11	55

To find a value for p, multiply the value of n by 5. To find a value for n, divide the value of p by 5.

3. a. yes b. no c. yes
4. $x = 14$ 5. $y = 2.8$ 6. $y = 36$
7. $b = 12$ 8. $a = 3.9$ 9. $x = 4$
10. $x = \frac{106}{17}$ 11. $z = \frac{10}{3}$ 12. $k = 0$
13. a. $I = Prt$ becomes $75 = P \cdot 0.03 \cdot 1$
b. $P = 2500$; Clive loaned his brother \$2500.
14. a. $A = \frac{S}{n}$ becomes $38.25 = \frac{S}{8}$
b. $S = 306$; Andy had 306 points.
15. a. $d = rt$: $234 = 13t$
b. $t = 18$; It takes 18 hours.
16. a. $A = lw$: $400 = l \cdot 16$
b. $l = 25$; The roll is 25 feet long.
17. $x + 7 = 26$ 18. $x + 7 = 26$
19. $\frac{x}{7} = 26$ 20. $7x = 26$
21. $\frac{x}{26} = 7$ 22. $x - 7 = 26$
23. a. $B =$ Amount before Craft Fair
b. $B - 24 = 39$ c. \$63
24. a. $B =$ Brenda's weight

Homework 1.5

1. 14 2. 3 3. 72 4. 18
5. $\frac{3}{2}$ 6. 23 7. 0 8. 9
9. 4 10. 9 11. 3 12. 1
13. 0.859 14. 0.729
15. 42.705 16. 2204.533
17.a. 18 b. 50 18a. 3 b. 18
19a. 2 b. 8 20a. 42 b. 12
21a. 2 b. 18 22a. 125 b. 5

23.

z	$5z$	$5z - 3$
2	10	7
4	20	17
5	25	22

24.

Q	$12 + Q$	$2(12 + Q)$
0	12	24
4	16	32
8	20	40

25. 26 26. 53 27. $\frac{7}{12}$
28. $\frac{19}{18}$ 29. $\frac{15}{2}$ 30. $\frac{13}{10}$
31. a. $(20 - 2) \cdot 8 + 1 = 145$
b. $20 - 2 \cdot (8 + 1) = 2$
c. $20 - 2 \cdot 8 + 1 = 5$
32. $4(5 + 7) \div (10 - 8)$
33.a. $\dfrac{20 - 8}{6 + 4 + 10}$ b. $(20 - 8) \div (6 + 4 + 10)$
34. First multiply 32 and 12, then divide the result by 4, then subtract the result from 825, and finally add the result to 2.
35. $\dfrac{12}{m - 3}$ 36. $\dfrac{m}{12} - 3$
37. $\dfrac{m - 3}{12}$ 38. $\dfrac{12}{m} - 3$
39. Subtracting $9 - 3$ is incorrect. Should be

$$9 - 3(2) = 9 - 6 = 3$$

40. $(2.3 + 5.7)6 - (1.2 + 3.3)2 = 39$ sq cm

Chapter 1 Review

1.

Gallons	3	5	8	10	11	12
Miles	66	110	176	220	242	264

b. Multiply 22 by the number of gallons.
c. $m = 22g$

d.

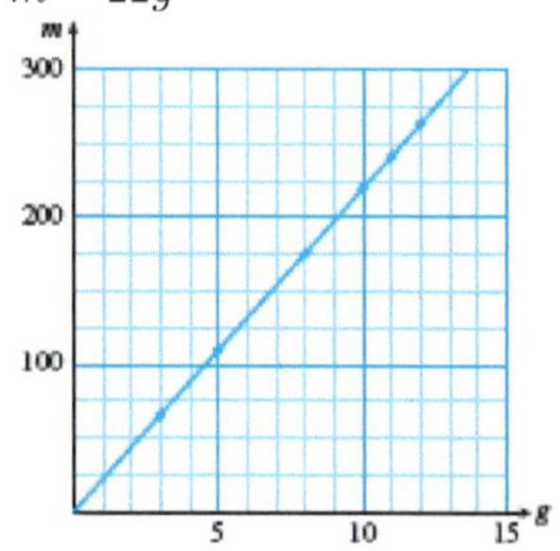

3.

x	.5	1	1.5	2	4	6	7.5
y	.125	.25	.375	.5	1	1.5	1.875

$y = \frac{x}{4}$

5. $z + 5$ 7. $f - 60$
9. $y + 6$ 11. $0.08b$
13a. $d = 88t$ b. 440 ft, 44 ft, 2640 ft
15a. $I = 35t$ b. \$35, \$70, \$175
17a. $n = b - 60$

b.

b	100	120	150	180	200
n	40	60	90	120	140

c.

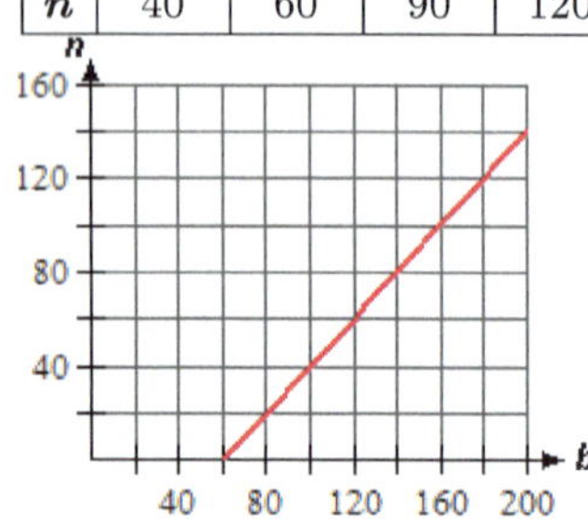

19a. $n = 2s$

b.

s	2	3	5	8
n	4	6	10	16

21. a, c, d
23a. yes   b. no
25. 9   27. 0.4   29. 0
31a. 32   b. 24
33. $171 = 0.095P$; \$1800
35. $106,000 = 0.53W$; 200,000
37. $\frac{w}{85} = 0.7$; 59.5 lb
39. 3   41. 6
43. $\frac{1}{2}d + 4$ 45. $3(l + 5.6)$
47. 4   49. 5
51. \$1300

Homework 2.1

1a. 2   b. -2 c. -8 2a. -35 b. -5 c. 5
3a. -25 b. 4.1   c. $\frac{-1}{6}$ 4a. -49 b. 2.4   c. $\frac{-9}{8}$
5a. $4 + (-8) = -4$ b. $3 + 9 = 12$ c. $-8 + 6 = -2$
d. $-6 + (-5) = -11$
6a. 6   b. 10   c. -10 d. 2
7a. -11 b. -9 c. 0   d. -8
8a. 34   b. 6   c. -2 d. -10
9a. 10   b. 10   c. 10
10a. -8 b. -8 c. -8
11a. -5 b. -9 c. 5   d. 9   e. 5
12a. 8   b. -18 c. -8 d. 18   e. -8
13a. -6 b. -18 14a. -13 b. -34
15a. -2 b. 12   c. -39 16a. 4   b. -6 c. -9
17a. 32   b. -3 c. 4
18a. 18   b. undefined   c. 0
19a. 0   b. -8 c. undefined
20a. $\frac{-3}{8}$ b. 2.6   c. $\frac{2}{3}$
21a. -16 b. 48   c. -8 d. -16 e. -3
22a. -6 b. 9   c. 0   d. -6 e. 1
23a. $5, -5$ b. $5, 5$ c. $0, 5$ d. $0, 5$
24a. $5, \frac{1}{5}$ b. $-5, \frac{1}{5}$ c. $5, 5$ d. $5, -5$
25. $x = 5$ 26. $z = \frac{-4}{3}$ 27. $a = -32$
28. $x = 6$ 29. $t = -13$ 30. $b = \frac{9}{2}$
31a. $x = 3$ b. $x = -6$ 32. 9
33. \$500   34. 40 feet   35. $-6°$
36. $-$ \$695   37. 14,776 feet   38. $-$\$35.40
39. -48 meters   40. $-28.75°$ per hour
41. $n + 12 = 5$ 42. $-12 + n = -5$
43. $-12 + n = -5$ 44. $-12 + n = 5$

Homework 2.2

1. 29.8 meters   2. 24 square cm   3. \$5400
4. $37°\,C$ 5. -18 6. 2
7. -15 8. -15
9a. -24 b. 24   c. -24 d. 48   e. -48 f. 48
10a. negative   b. positive
11a. 2   b. -2 12a. 7   b. -22
13a. -6 b. -2 14a. 12.96   b. -19.2
15a. -60 b. 7   c. 27   d. 6   e. -23 f. 35
16a. 2   b. -3 c. -6 d. 27   e. 2   f. -8
17. 13   18. -12 19. -18 20. $\frac{1}{13}$
21. $-\frac{1}{9}$ 22. $\frac{7}{6}$ 23a. i. $-\frac{15}{4}$ ii. $-\frac{15}{4}$ iii. -3.75
b. No   24. a, d   25. $2x + 10$ 26. $7x + 42$
27a. \$4400 ;\$2600 b. Multiply the number of weeks by \$200, then subtract from \$5000.

c.

weeks	2	4	5	6	10	15	20
left	4600	4200	4000	3800	3000	2000	1000

d. $5000 - 200w$ e. $S = 5000 - 200w$
28.a. \$720; \$960   b. Subtract \$2000 from her income, then multiply the result by 0.12
c.

I	5000	7,000	12,000	15,000	20,000	24,000	30,000
T	360	600	1200	1560	2160	2640	3360

d. $0.12(I - 2000)$ e. $T = 0.12(I - 2000)$
29a. $w =$ width (inches): $2w - 3$ b. 23 inches
30a. $P =$ principle (\$): $0.40P + 20$ b. \$220
31a. $t =$ number of childrens tickets; $8(150 - t)$
b. \$536
32a. $p =$ profit; $\frac{1}{3}(p - 50)$ b. \$150
33. $65 + 30t$ 34. $800 + \dfrac{a}{2}$ 35. $\dfrac{80 - M}{4}$
36. $50 + \frac{1}{3}I$ 37. $200 - 15w$ 38. $600 + 80t$
39. $\dfrac{c - 50}{10}$ 40. $35(x + 3)$
41a. $\dfrac{x}{4 + x}$ b. Answers may vary
42a. $(a + 5)(a - 5)$ b. Answers may vary

Homework 2.3

1a. No b. Yes c. No d. Yes

2a. Yes b. Yes c. No d. No

3a. $(-6, 2)$ b. $(-5, -7)$ c. $(8, -4)$
d. $(1, 7)$ e. $(0, -9)$ f. $(-2, 0)$

4a. $(-10, 0)$ b. $(-2, -4)$ c. $(0, -5)$
d. $(2, -6)$ e. $(10, -10)$

5a. $(0, -600)$ b. $(10, -200)$
c. $(15, 0)$ d. $(20, 200)$

6a. $P = 100 - 6d$

b.

d	0	2	5	10	15
P	100	88	70	40	10

c.

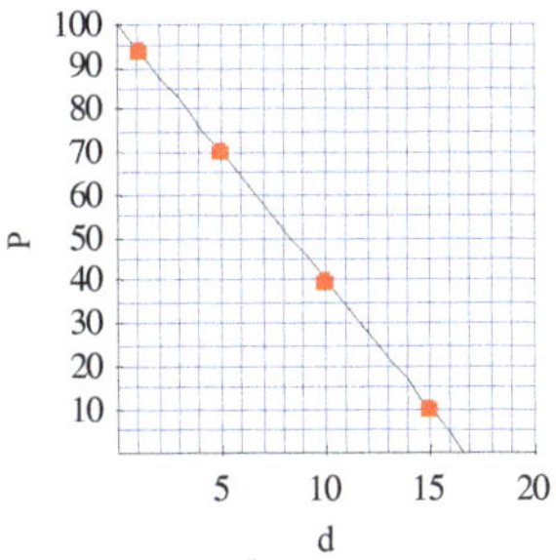

7a. $M = 200 + \frac{1}{5}U$

b.

U	20	40	80	100	200
M	204	208	216	220	240

c.

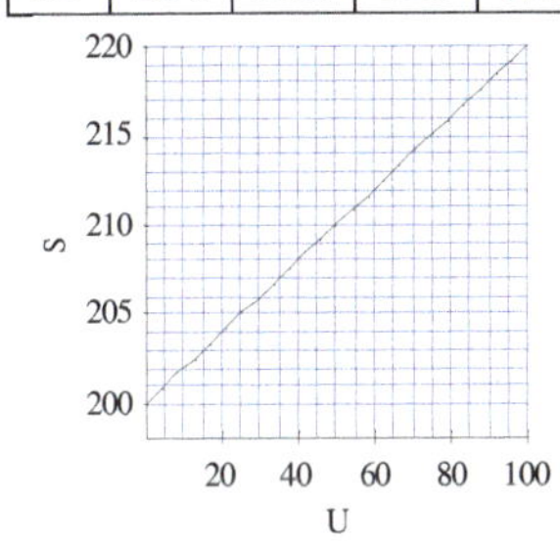

8a. $B = -100d + 2000$

b.

d	B
-15	3500
-5	2500
0	2000
5	1500
15	500
20	0

c.

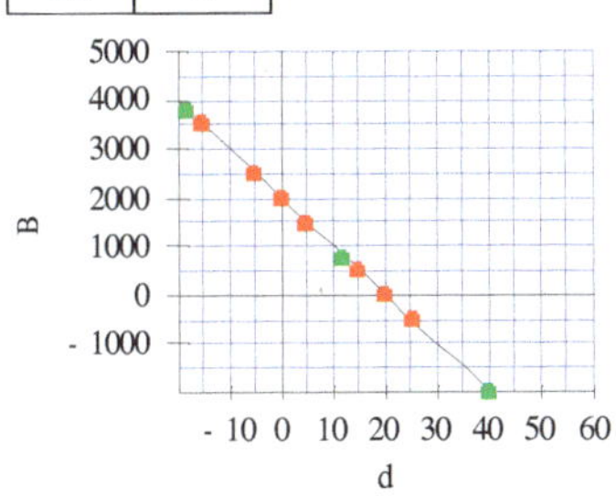

d. \$3500 e. In 15 days

9a. \$540 b. \$30 c. \$360 d. 18

e. $B = 540 - 30m$

10a. $s = 100 - 5x$

b.

x	2	5	6	12
s	90	75	70	40

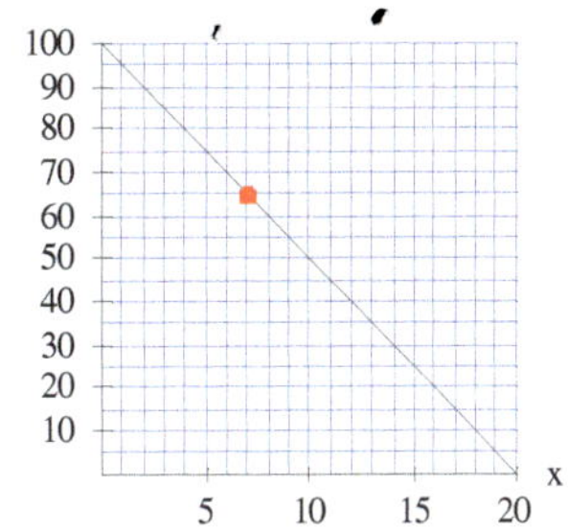

c. $65 = 100 - 5x$; 7

11a. $d = 120 - \frac{1}{2}w$

b.

w	24	36	84	120
d	108	102	78	60

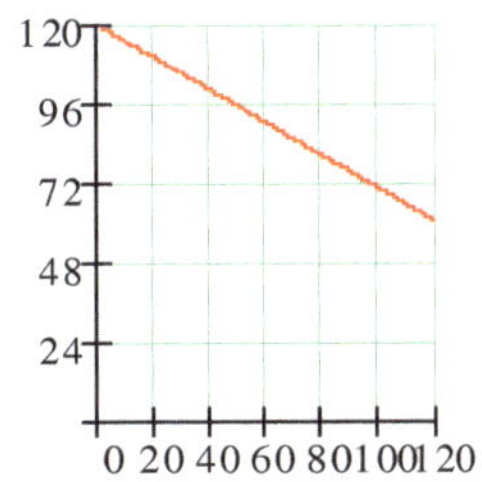

c. 48 weeks; $96 = 120 - \frac{1}{2}w$

12.

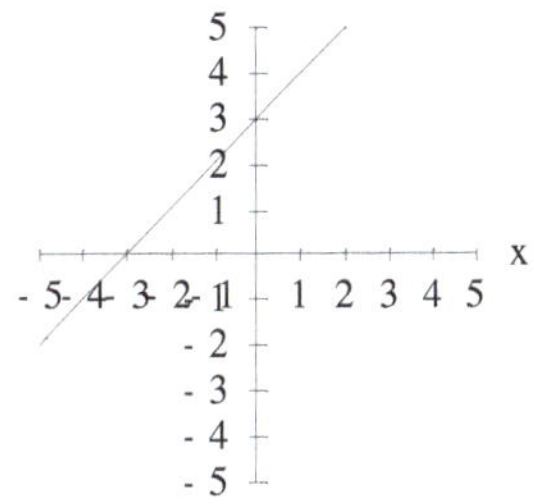

13.

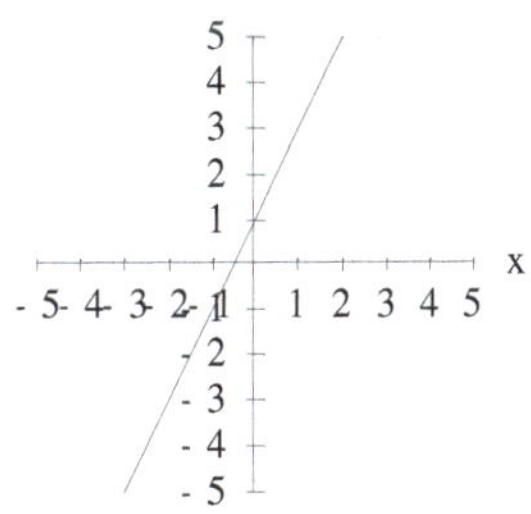

14.

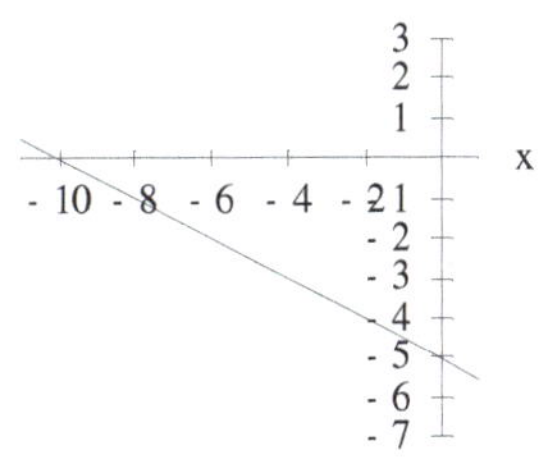

15.

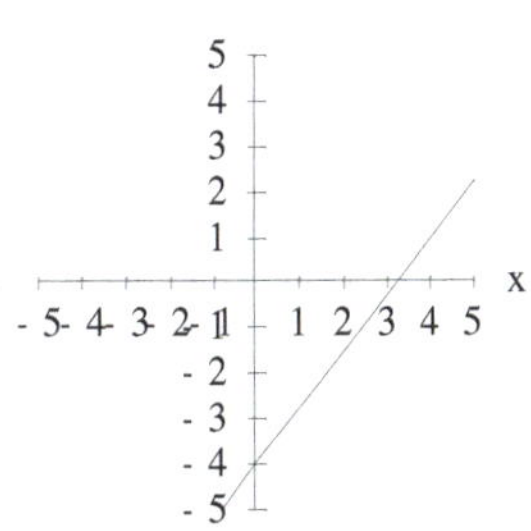

16a. -4 b. -5 c. $-4 = 2(-5) + 6$
d. 1 e. -4.9, -4, etc.
17a. $x = -16$ b. $x = -40$
18a. $x = 6$ b. $x = 12$
19. $t \approx 1.5$ 20. $x \approx -2.5$

Homework 2.4

1.

x	$2x$	$2x+4$
3	**6**	10
6	**12**	16
5	**10**	14
8	**16**	20

2.

q	$q-3$	$5(q-3)$
3	**0**	**0**
5	**2**	10
4	**1**	**5**
7	**4**	20

3. $x = 3$ 4. $a = 20$ 5. $x = 4$
6. $x = 0$ 7. $p = 13$ 8. $z = 0$
9. $k = 41$ 10. $x = 18$ 11. $b = 6$
12. $c = -2$ 13. $t = 1$ 14. $h = 5$
15. $b = 18$ 16. $y = 10$ 17. $x = -4$
18. $w = 13.4$ 19. $x = 3.4$ 20a. $x \le -15$
b. 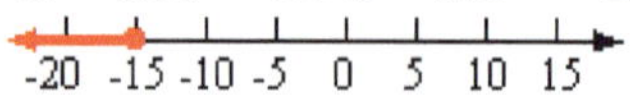

c. $x = -15.1$ is a solution, $x = -14.9$ is not.
21a. $y > -5$
b. 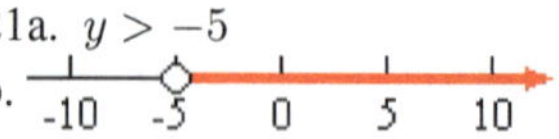

c. $y = -4.9$ is a solution, $y = -5$ is not.
22a. $x \le 12$
b. 

c. $x = 12$ is a solution, $x = 13$ is not.
23. $x > 2$ 24. $x \ge -3$ 25. $x < -6$
26. $-1 \le x \le 4$ 27. $-7 < b \le -2$
28. $-7 \le w < -3$ 29. 7
30. 6 31. -3
32a. 8.6 lbs b. 36 weeks c. $w = 3.8 + 0.6t$
d. $26 = 3.8 + 0.6t$; $t = 37$ weeks
33a. $P \approx 20$ b. $m \approx 80$ minutes
c. $P = 13 + 0.15m$
d. $25 = 13 + 0.15m$; $80 = m$
34a. m = each installment ($)
b. $10200 = 1200 + 36m$
c. Each installment is $250.
35a. d = assigned distance
b. $5(d + 0.5) = 22.5$
c. She is assigned 4 miles.
36a. p = number of planted seeds
b. $112 = 0.6p - 38$
c. 250 seeds were planted.
37a. c = average temperature change per day
b. $26 = -6 + 4c$
c. The average change was 8 degrees per day.
38a. w = time it takes (weeks) b. $162 = 196 - 4w$
c. It will take 8.5 weeks.
39. 175 yards 40. 6 meters
41. $4x + 8 = 36$; $x = 7$
42. $\frac{1}{2}(2)(3 + x) = 20$; $x = 17$

Homework 2.5

1. $-4x$ 2. $-12.8a$ 3. t 4. $-bc$
5. $12y - 4$ 6. $-8st + 9s - 2$ 7. $9x - 2$
8. $6y - 10$ 9. $2a$ 10. $-5b + 2$ 11. 4
12. $\frac{-9}{2}$ 13. 0 14. $x > -2$ 15. $\frac{8}{3}$
16. $y \le -4$ 17. $w \ge 21$ 18. $p = -1$
19. a. -21 b. $-6y - 6$ c. -21 20. -170
21. $x = 0$: $2 + 7x = 2 + 7(0) = 2$;
$9x = 9(0) = 0$
22. $a = 0$: $-(a - 3) = -(0 - 3) = 3$;
$-a - 3 = -0 - 3 = -3$
23. $6x + 8y$ 24. $13a + 7$
25a. $d + 8$ b. $2d + 8$ c. There are 16 dogs.
26a. $3x$ b. $4x$ c. Nine are smokers.
27a. $9.8t$ b. 98 feet c. 15 seconds
28a. $0.08x$ b. $1.08x$ c. $860
29a. $596x - 1355$ b. 30
30. $8b = 5b + 12$; $b = 4$ oz.
31. $3p + 33 = p + 131$; $p = 49$ in
32. $12c + 1.8 = 8c + 5$; $c = \$0.8$
33. $3g - 70 = 2g + 220$; $g = 290$ mg
34. $9s + 15 + 14 = 10s + 23$; $s = 6$ points

Chapter 2 Review

1. f 3. c 5. b
7. $\frac{5}{3}$ 9. 8 11. $V = 2000 - 200t$

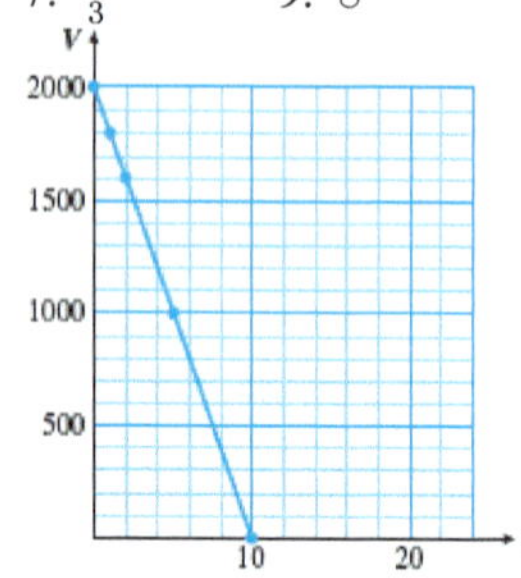

13. $5.75 + 4.50b = 19.25$; 3 bushels
15a. 3 b. -6 c. 6 d. -3
17. $>$ 19. $>$ 21. 20 23. 9
25. 31 27. $\frac{-13}{16}$ 29. -3 31. -1
33. 6°F 35. $460,000 37. -12 yd
39. 6.9, 6, 5 41. 3, 2, 1

43a.

h	-4	-2	0	1	3	5	8
T	30	24	18	15	9	3	-6

b. $T = 18 - 3h$

c. 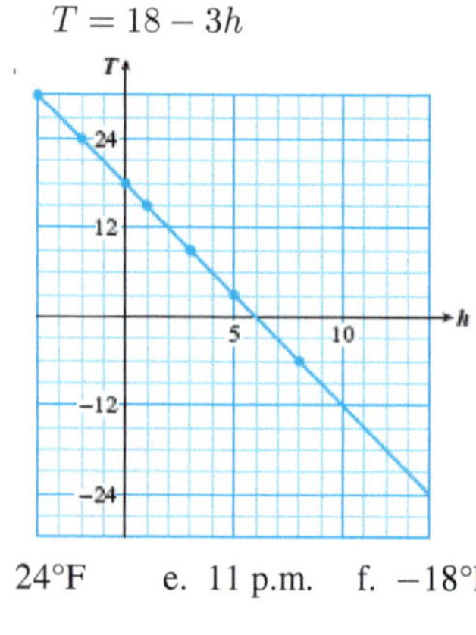

d. 24°F e. 11 p.m. f. -18°F

45. 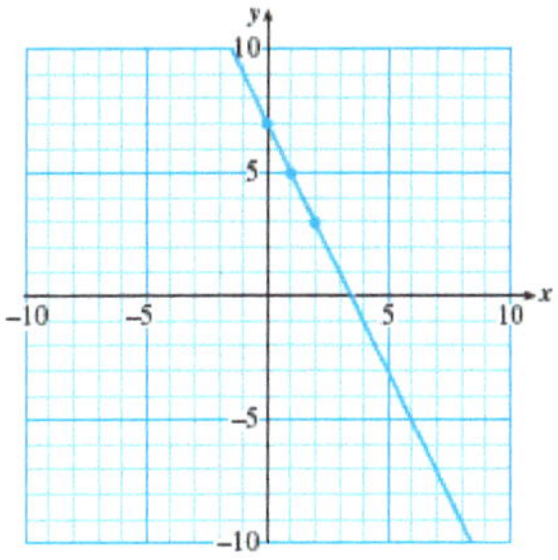

47. $2m + 7n$ 49. $z = -1$
51. $w = -3$ 53. $h = -6$
55. a. -5 b. 2
57. a. $3x$ b. $8x$ c. 6 cm by 18 cm

59. $z \geq 3$ 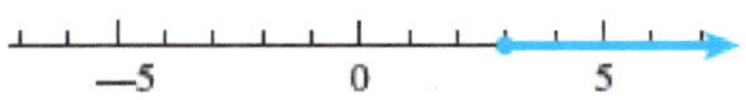

61. $k > -6$ 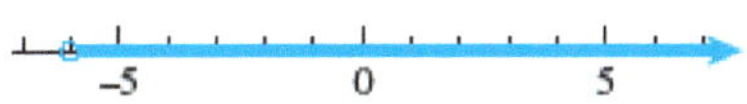

63. $7 > n \geq 3$ 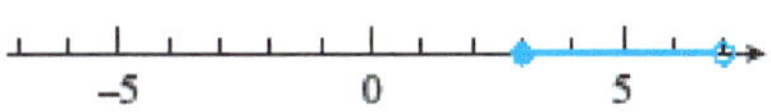

Homework 3.1

1. a. $(0, 2), (4, 0)$ b. 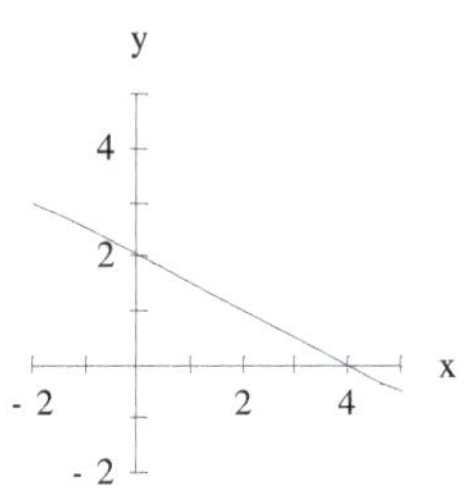

2. a. $(0, -5), (-10, 0)$ b. 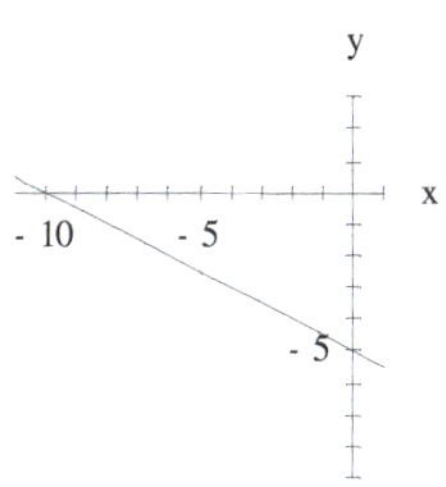

3. a. $(7, 0), (0, -2)$ b. 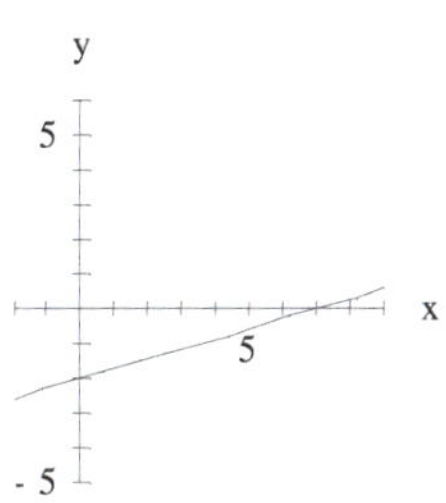

4. a. $(0, 8), (2, 0)$ b. 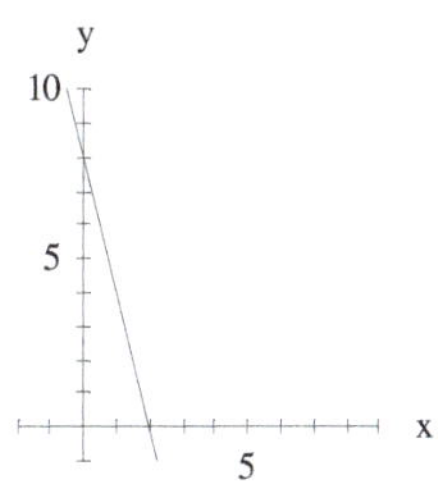

5a. $(20, 0), (0, 30)$ b. 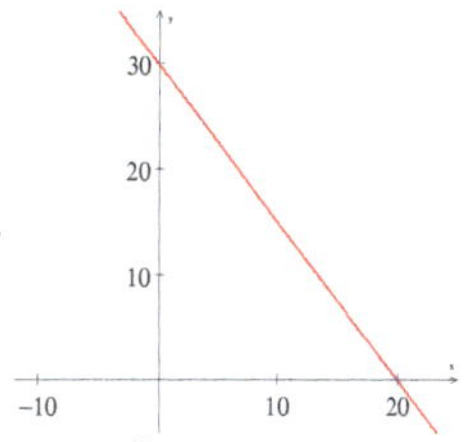

6.a. $(0, -60), (40, 0)$ b. 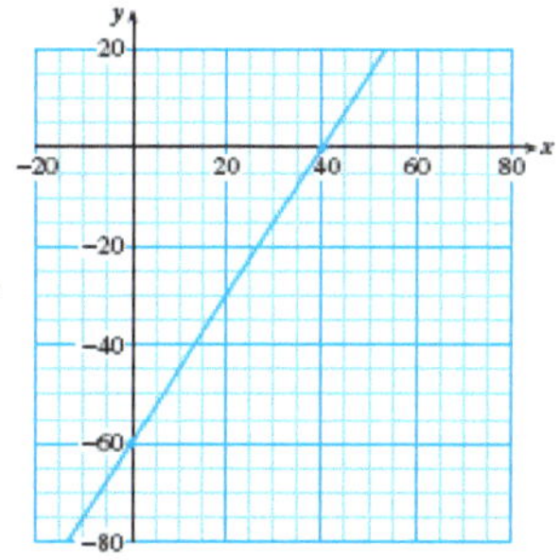

7. d 8. a 9. c 10. b 11. a 12. c
13a. $x = -3$ b. $x = -\frac{k}{2}$
14a. $x = 3$ b. $x = 7 - k$
15a. $x = 14$ b. $x = k + 5$
16a. $x = \dfrac{5}{2}$ b. $x = \dfrac{8 - k}{2}$
17a. $x = 3$ b. $x = \dfrac{k - 15}{-4}$
18a. $x = \dfrac{-10}{3}$ b. $x = \dfrac{-10}{k}$
19a. 360 miles b. 24 hours c. 180 miles
d. 60 miles 20a. $(13\frac{1}{3}, 0), (0, 200)$

b. 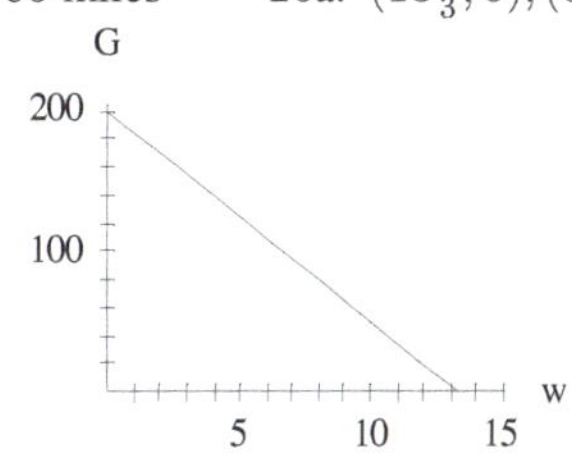

c. $(\frac{40}{3}, 0)$: The fuel will run out after $\frac{40}{3} \approx 13.3$ weeks.
$(0, 200)$: There were 200 gallons in the tank when they turned on the furnace.
21a. $(-9, 0), (0, 225)$

b. 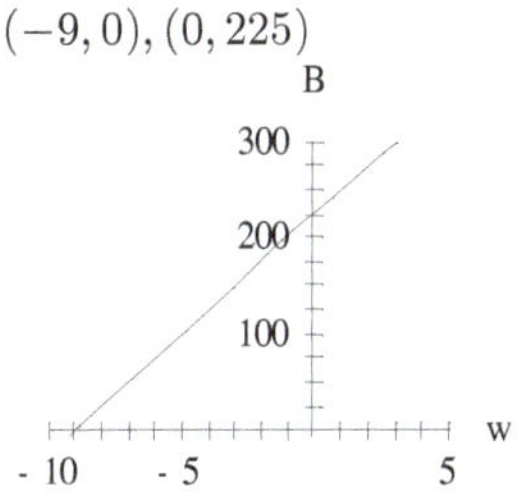

c. $(0, 225)$: She has $225 this week.
$(-9, 0)$: She had no balance 9 weeks ago.
22a. $(15, 0), (0, -600)$

b. 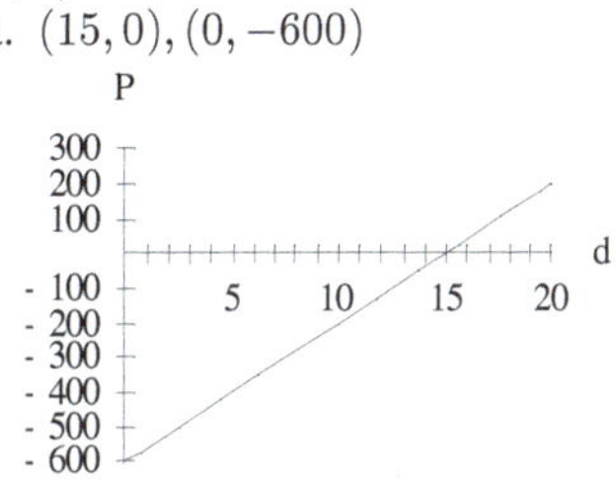

c. $(15, 0)$: Delbert must groom 15 dogs to break even.
$(0, -600)$: Delbert had spent \$600 before grooming the first dog.

23a. $(12.5, 0)$ b. $(0, -5)$ 24. Increasing

Homework 3.2

1. $\frac{5}{7}$ 2. 8.72 3. $\frac{11}{6}$ 4. $\frac{30}{7}$

5. a, b, and d are proportions.

6. $x = 6$ 7. $w = \frac{15}{2}$ 8. $b = \frac{63}{5}$ 9. $x = \frac{7}{2}$

10. yes 11. no 12. no 13. no

14. yes 15. yes 16. no 17. 40 lbs

18. 100 liters 19. 80 kilometers

20. 5753 21. about 13.78 in. 22. 2 cups

23a. $w = 7.5h$

b. 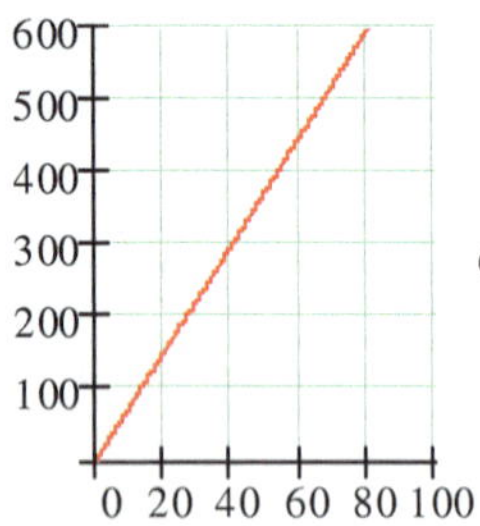

c. 67 hours

24a. $c = 0.03g$

b. 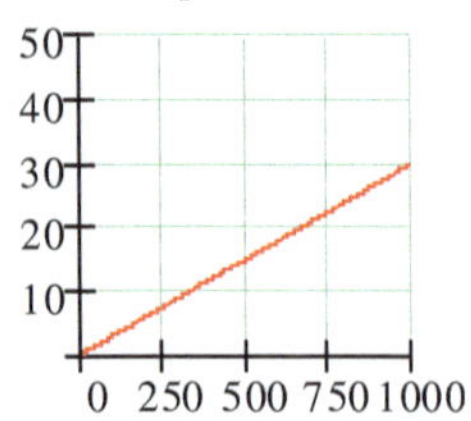

c. 30 grams

25a. $C = 0.04S$ b. \$8000

26a. $d = 8t$

b. 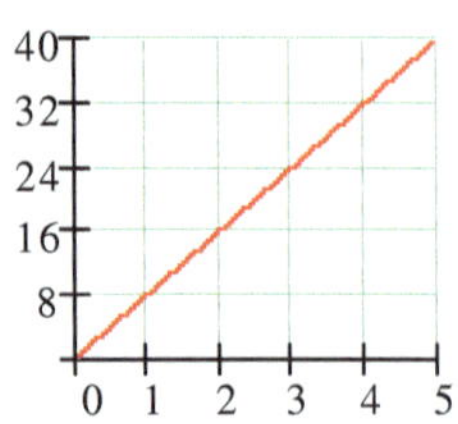

c. The distance doubles also.

27a. $T = 0.06P$

b. 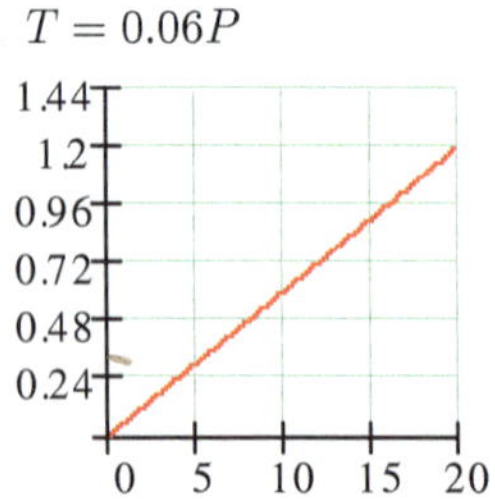

c. The tax is also cut in half.

28a.

Units	Tuition	Units	Tuition
3	620	6	740
5	700	10	900
8	820	16	1140

b. No c. No

29a. $C = 2\pi r$

b.

r	2	5	7	10
C	12.6	31.4	44.0	62.8

c. 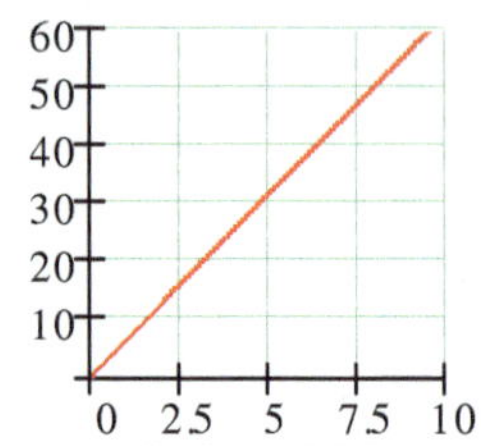

d. Yes

30a. $A = s^2$

b.

s	2	5	6	8
A	4	25	36	64

c. 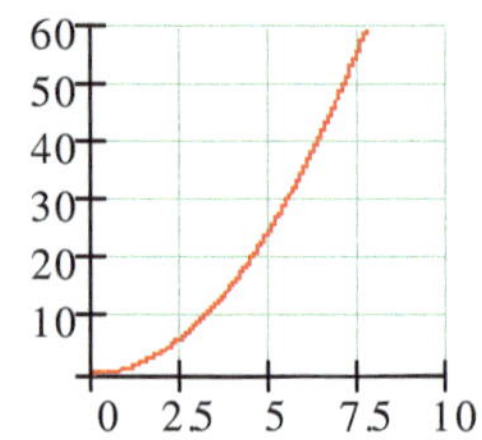

d. No

31a. 3 to 2; $416\frac{2}{3}$ servings; $7\frac{1}{2}$ cans

b. 2.4 oz; 15.5 lb c. 24.75 cans; 90 lb

Homework 3.3

1a. $\frac{3}{2}$ b. -2

2a. $\frac{-1}{4}$ b. 3 3. $\frac{1}{15}$

4. -12

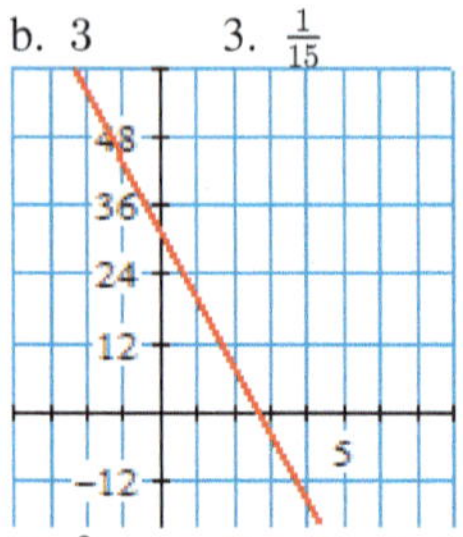

5. $\frac{-1}{2}$ 6. $\frac{3}{2}$ 7a. $(0, 4)$, $(6, 0)$

b. 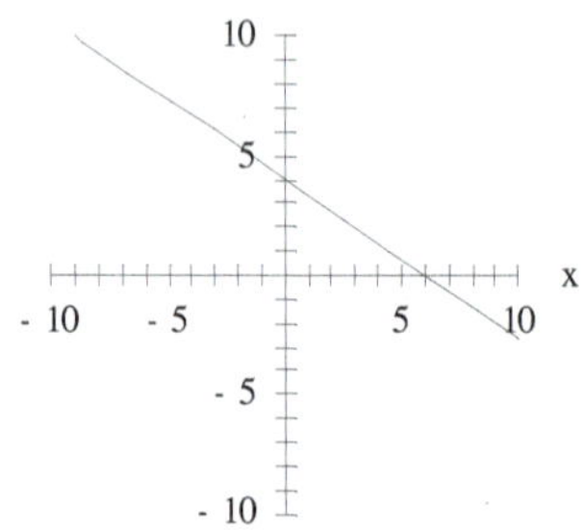

c. $\frac{-2}{3}$

8a. $(0, -5), (2, 0)$

b. 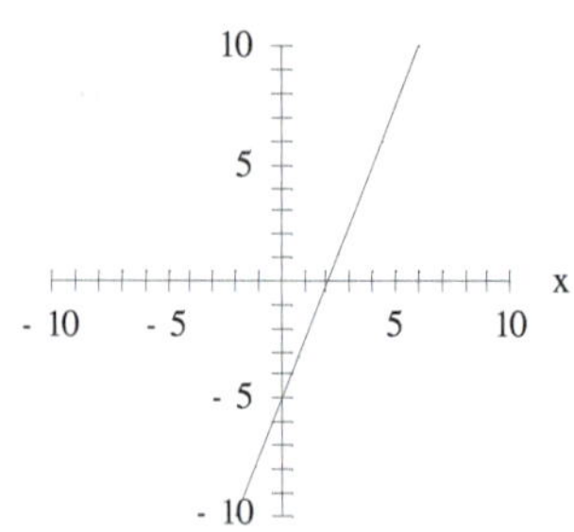

c. $\frac{5}{2}$

9a. $(0,5),(5,0)$

b.

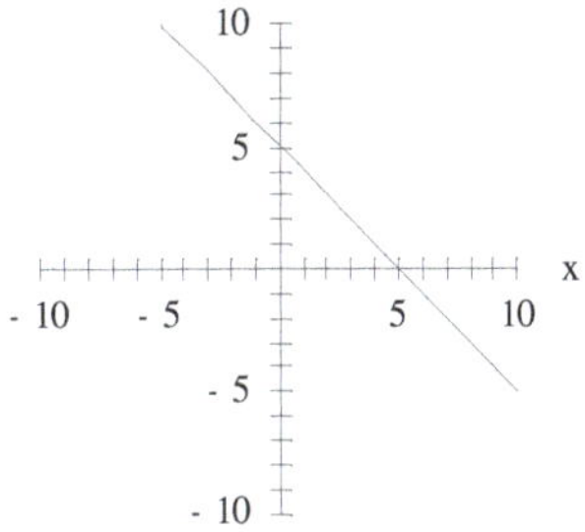

c. -1

10a. $(0,-2),(4,0)$

b.

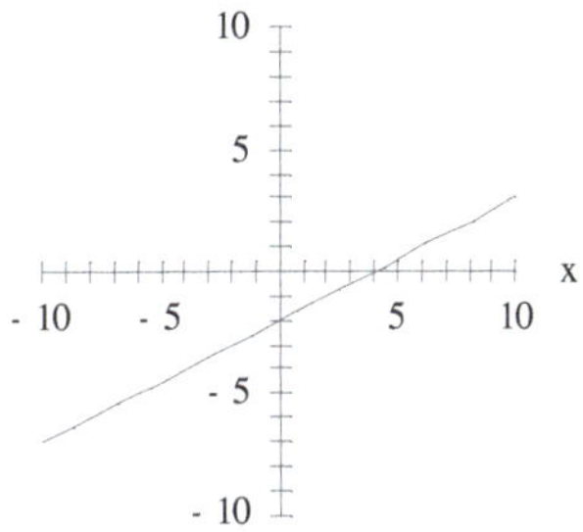

c. $\frac{1}{2}$

11. $\frac{2}{3}$ 12. $\frac{-5}{3}$

13. The line with slope $\frac{5}{3}$ is steeper.

14. The line with slope -2 is decreasing.

15. $\Delta y = 17.5$ 16. $\Delta x = 1.5$

17a. $d = 25g$ b. $\dfrac{\text{25 miles}}{\text{1 gallon}}$; the car gets 25 miles per gallon.

18a. $T = 4p$ b. $\dfrac{\text{4 cents}}{\text{1 dollar}}$; the sales tax is 4 cents per dollar purchased.

19a. \$50 b. $(0,200),(2,300)$, $\dfrac{\Delta A}{\Delta w} = 50$

c. dollars/week. She is saving \$50 per week.

20a. 4 lbs; 3 lbs b. $\frac{1}{2}$pound per week

21. 950.4 ft 22. 16 ft

23a. 0 b. All the y-coordinates on the line are the same, so the change in y is 0.

c. $\dfrac{\Delta y}{\Delta x} = \dfrac{0}{\Delta x} = 0$

24a. undefined b. All the x-coordinates on the line are the same, so the change in x is zero.

c. $\dfrac{\Delta y}{\Delta x} = \dfrac{\Delta y}{0} = 0$

25a. (1) b. (1): $m = 3$; (2): $m = 20$; (2)

26a. $(8,2.4),(16,4.8)$ b. 0.3 dollars/oz

Homework 3.4

1. $y = 3x + 4$; slope 3 and y-intercept $(0,4)$.
2. $y = -2x + \frac{5}{3}$; slope -2 and y-intercept $\left(0,\frac{5}{3}\right)$.
3. $y = \frac{2}{3}x - 2$; slope $\frac{2}{3}$ and y-intercept $(0,-2)$.
4. $y = \frac{5}{4}x$; slope $\frac{5}{4}$ and y-intercept $(0,0)$

5a. $m = 3$; y-intercept $(0,-4)$ b. $y = 3x - 4$

6a. $m = \frac{2}{3}$; y-intercept $(0,-3)$ b. $y = \frac{2}{3}x - 3$

7a. $m = -\frac{8}{5}$; y-intercept $(0,1)$ b. $y = \frac{-8}{5}x + 1$

8a. $m = -\frac{1}{3}$; y-intercept $(0,-1)$ b. $y = -\frac{1}{3}x - 1$

9a.

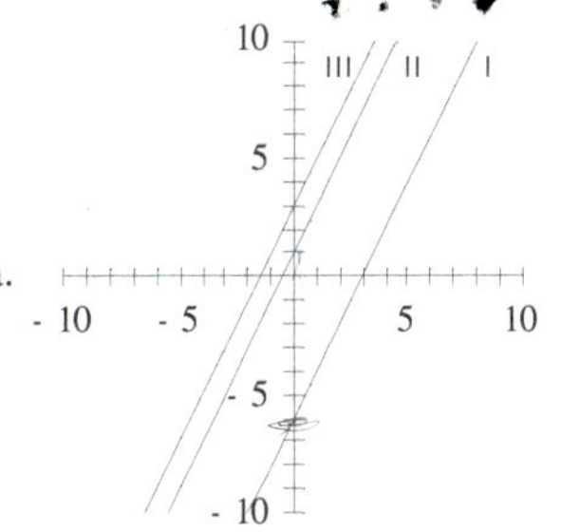

b. The slopes are all 2. c. $(0,-6),(0,1),(0,3)$

10a.

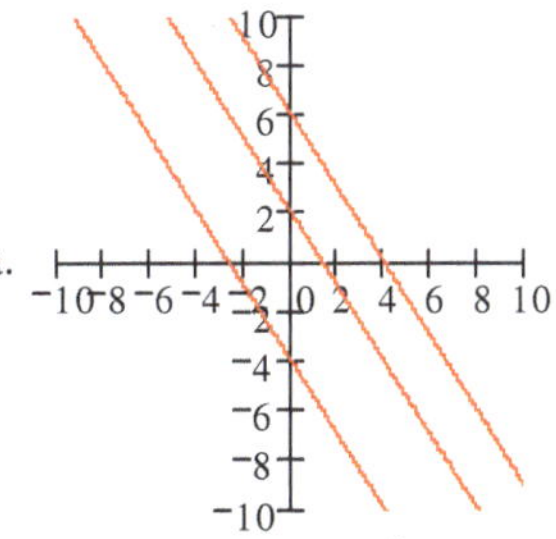

b. The slopes are all $\frac{-3}{2}$. c. $(0,\ -2)$, $(0,\ 4)$, $(0,\ 6)$

11a.

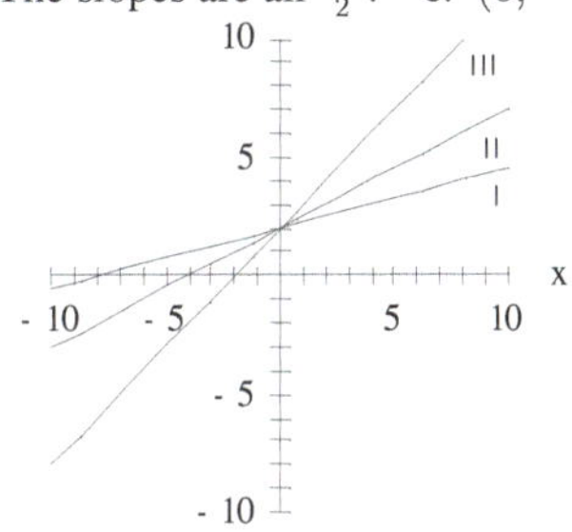

b. I. $\frac{1}{4}$ II. $\frac{1}{2}$ III. 1 c. $(0,2)$

12a.

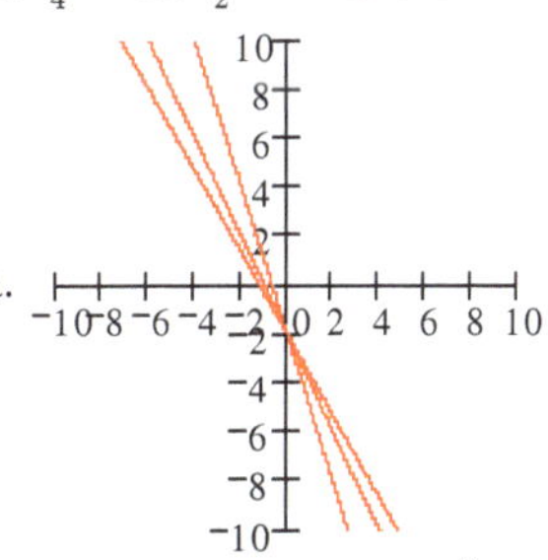

b. I. -3 II. -2 III. $\frac{-5}{3}$ c. $(0,-2)$

13a. $(0,3),(4,0)$

b.

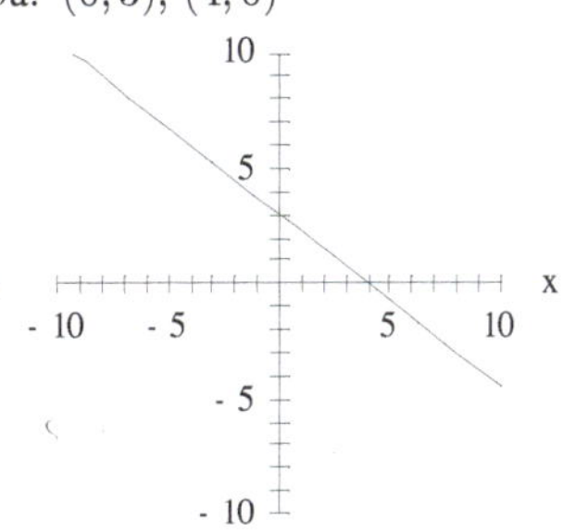

c. $\frac{-3}{4}$ d. $y = \frac{-3}{4}x + 3$

14a. $(0,8), (\frac{8}{3},0)$

b.

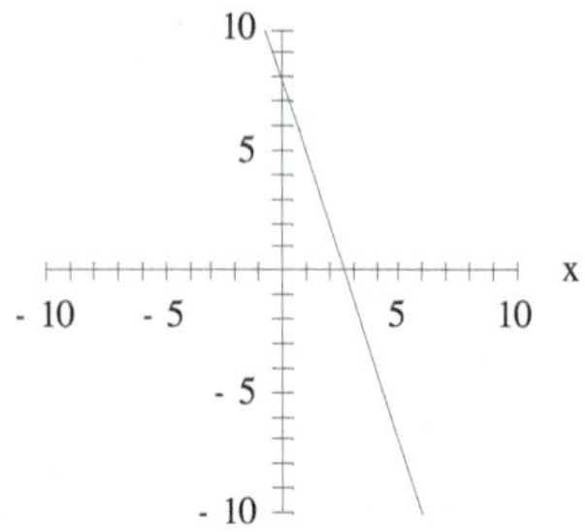

c. -3 d. $y=-3x+8$

15a. $y=\frac{3}{5}x$ b. $(0,0)$, $\frac{3}{5}$

c.

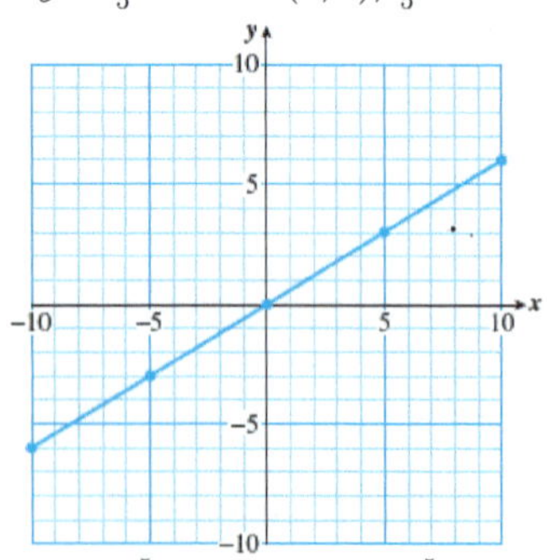

16a. $y=\frac{-5}{4}x$ b. $(0,0)$, $\frac{-5}{4}$

c.

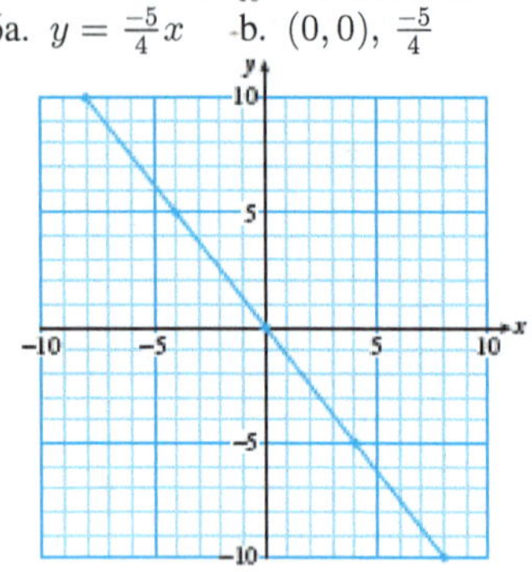

17.

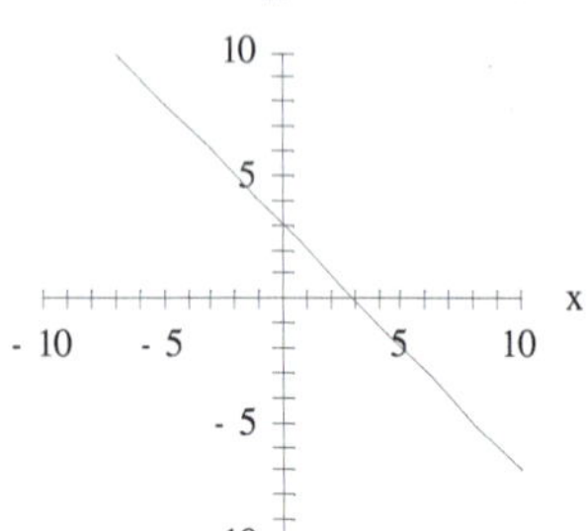

18.

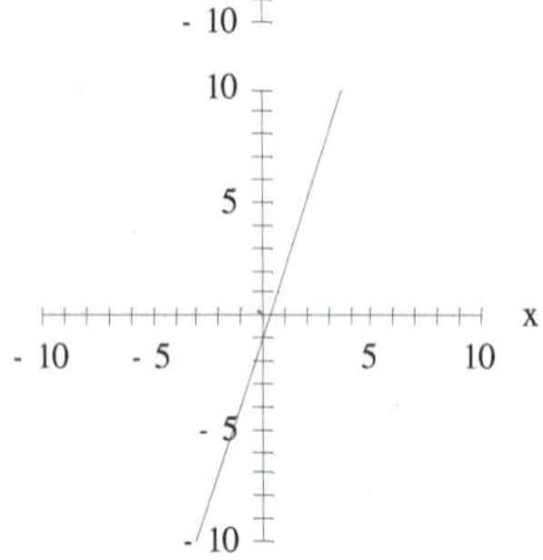

19.

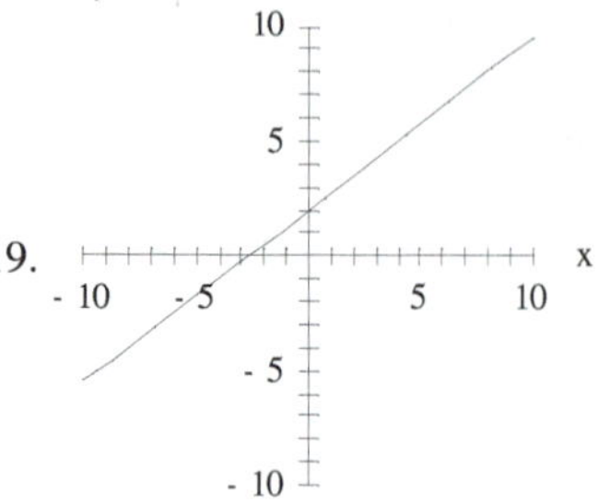

20.

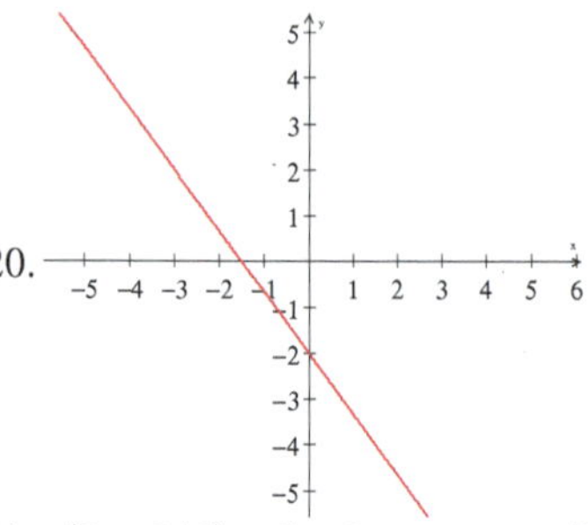

21a. $(0,-240)$ b. 3 c. $P=-240+3x$

d. He spent $240 to open the shop and makes $3 of profit per smoothie.

22a.

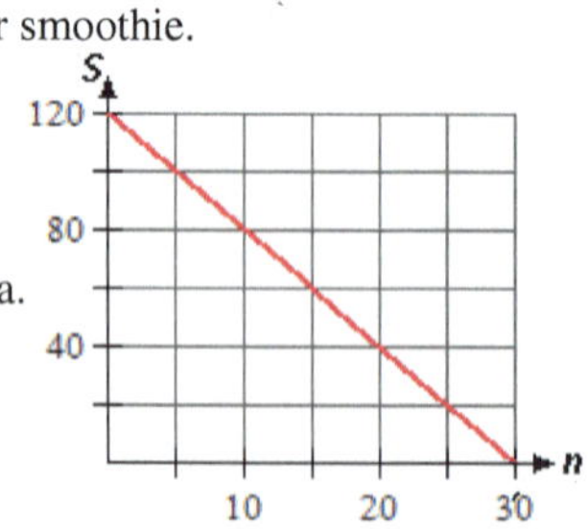

b. The slope means that for each incorrect answer, she loses 4 points. The S-intercept means that if she has zero wrong answers, her score will be 120 points.

23a.

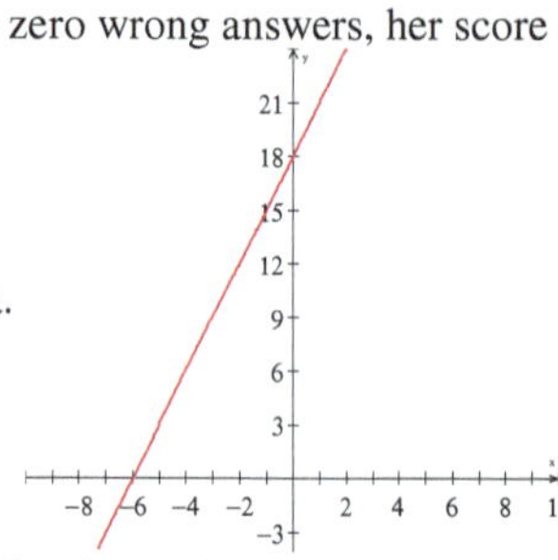

b. The slope gives the weekly rate that the beans are growing: 3 inches per week. The h-intercept, (0,18), means that today, the beans are 18 inces tall.

24a.

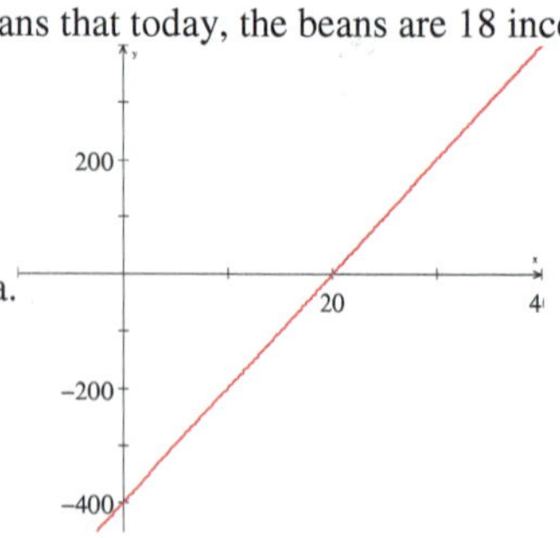

b. The slope, 20 points per question, means that he gets 20 points for each question he gets right. The S-intercept, $(0, -400)$, means that his score at the end of the first round is -400.

Homework 3.5

1. Parallel 2. Neither 3. Neither
4. Parallel: a, g, h; perpendicular: c, f

5a. 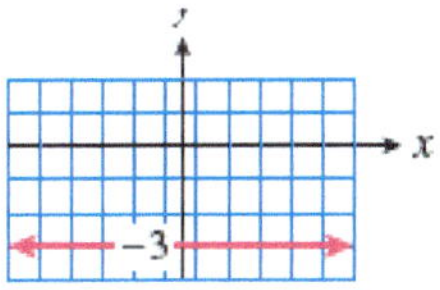

b. $m = 0$

6a. 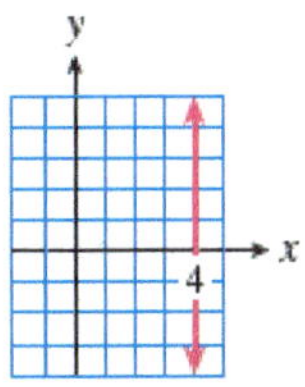

b. m is undefined

7a. 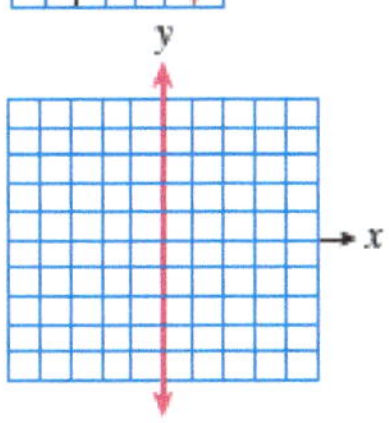

b. m is undefined.

8. $y = -5$ 9. $x = 2$
10. $x = -8$ 11. $y = 0$
12. $x = 4$
13. $m = 0,\ b = -6$ 14. $m = -6,\ b = 0$
15. $m = \frac{-1}{6},\ b = 0$ 16. $m = -1,\ b = 6$
17. $\frac{5}{3}$ 18. -1 19. $\frac{1}{9}$ 20. 0
21a. $(5, 0),\ (0, 7)$ b. $\frac{-7}{5}$
22a. $(-2.4, 0),\ (0, 1.6)$ b. $\frac{2}{3}$
23a. $(\frac{1}{3}, 0),\ (0, \frac{7}{2})$ b. $\frac{-21}{2}$
24a. $(0, -3)$ b. $(-2, -7)$
25a. $(6, -8)$ b. $(-4, -3)$
26a. \$1200/1 year b. The college fund grows at a rate of \$1200 per year.
27a. \$0.08/1 page b. The cost of making copies is 8 cents per page.
28a. -500 feet/1min b. The elevation decreases at an average rate of 500 feet per minute.
29a. $-3°$/1 hour b. The temperature dropped at an average rate of 3° per hour.
30a. cost per pound of peanuts
b. 1.50 dollars per pound c. \$30
31a. -75 miles/1 hour b. Roy's distance from home decreases by 75 miles each hour.
c. 800; Roy boarded the train 800 miles away from home.
d. $10\frac{2}{3}$; Roy will arrive home in about ten hours and 40 minutes.

Chapter 3 Review

1. $(2, 0),\ (0, -3)$ 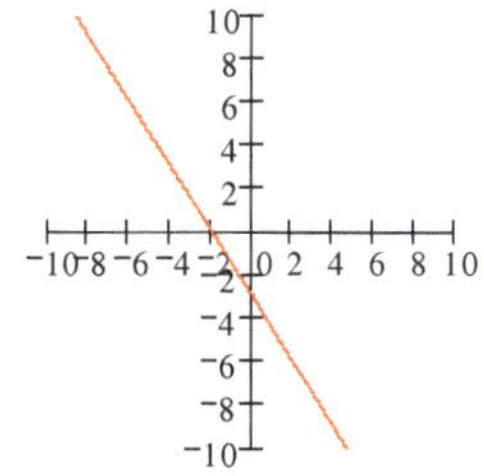

3a. $S = 7800 - 600m$
b. $2400 = 7800 - 600m;\ m = 9$ months

c.

m	S
0	7800
13	0

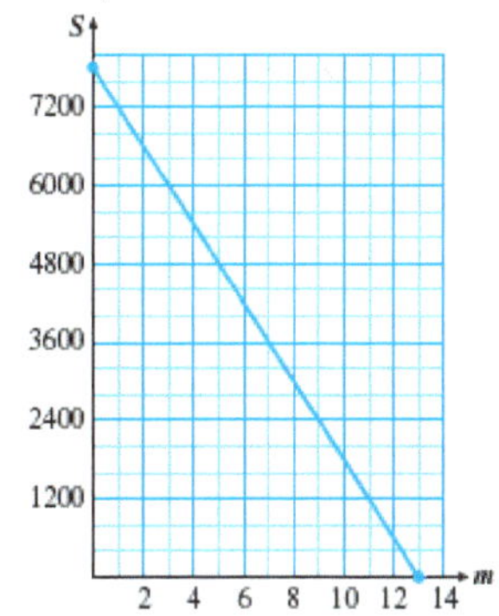

The S-intercept at $S = 7800$ shows that she started with \$7800. The m-intercept at $m = 13$ shows that she will use all the money in 13 months. d. 9 months
5. $\frac{2}{3}$ 7. 119 9. 50 11. 52 hr
13a. 10 mi by 15 mi b. 50 mi; 30 cm c. $\frac{5}{3}$
d. 150 sq mi; 54 sq cm e. $\frac{25}{9}$
15a. No b. Yes
17. $\frac{-3}{2}$ 19. 8515 ft
21a. $(0, -6),\ (3, 0)$

b. 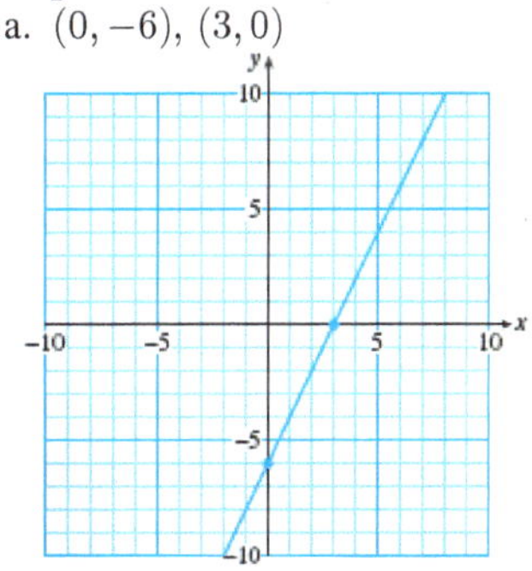

c. $m = 2$

23a. $(2, 0),\ (0, 3)$

b. 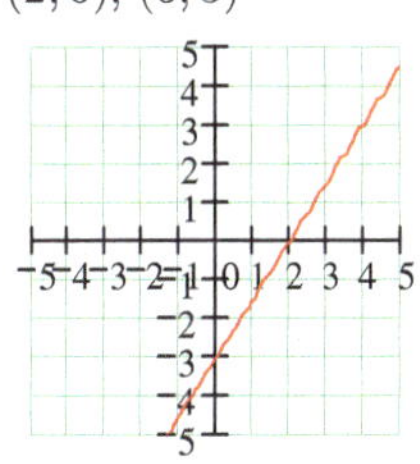

c. $m = \frac{3}{2}$

25a. $m = \frac{-3}{5},\ b = 3$ b. $y = \frac{-3}{5}x + 3$

27a. $m = \frac{-5}{2}$, $b = -5$

b. 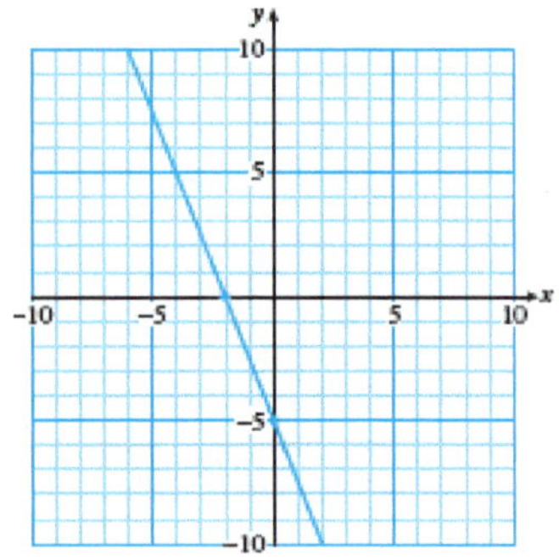

29a. $m = \frac{3}{4}$, $b = 2$

b. 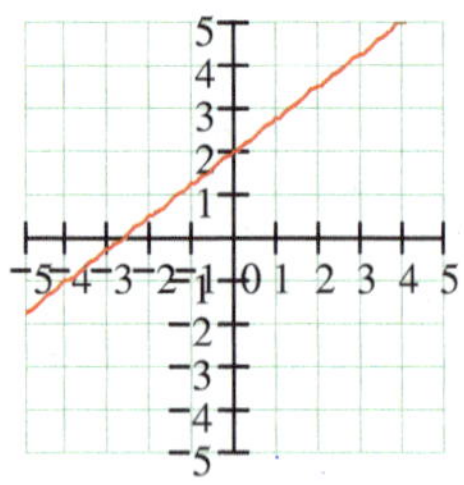

31. $h = 36 + 6t$

33. $b = 500$. At 2 pm, Beryl's altitude was 500 ft. $m = -15$. Beryl is descending at a rate of 15 ft/sec.

35. 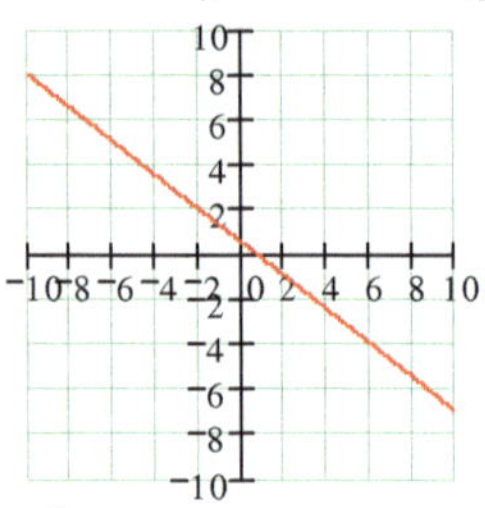

37. $\frac{-3}{2}$ 39. -0.4

41. undefined

43a. 4.2 b. Gas costs \$4.20 per gallon.

45. $y = 5x - 1$ 47. $x = -4$

49a. parallel b. perpendicular

51. 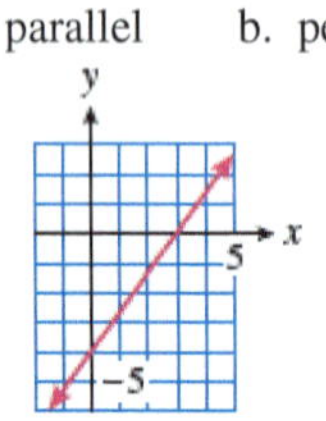

53. 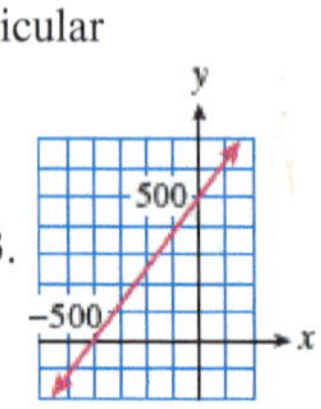

55. 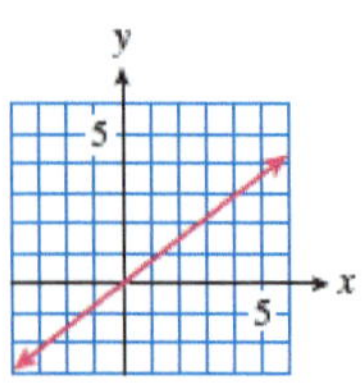

57. 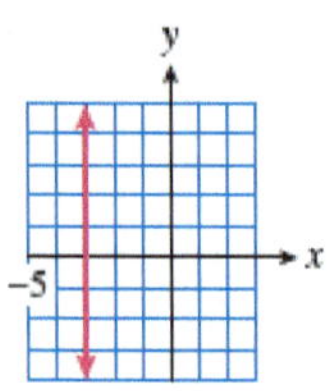

Homework 4.1

1. $10y - 15$ 2. $-8x - 16$ 3. $-5b + 3$

4. $36 - 12t$ 5a. $32c$ b. $32 + 8c$

The distributive law is used in (b).

6a. $-16 - 2t$ b. $-16t$

The distributive law is used in (a).

7. d 8a. $2xy$; 54 b. $2x + 2y$; 24

c. $2 - xy$; -25 d. $-2xy$; -54

9. $-4x - 6$ 10. $x + 23$ 11. $-12z - 15$

12. $-24a + 4$ 13. $y = -2$ 14. $w = 6$

15. $c = 0$ 16. $a \le -4$ 17. $t < 2$

18. $x = 9$ 19. $a = 4$ 20. $b = 3.25$

21a. $x = 0$: $5(x + 3) = 5(0 + 3) = 15$;
$5x + 3 = 5(0) + 3 = 3$

b. $c = 0$ 22a. 48 b. $-12x + 12$ c. 48

23a. $2w - 8$ b. $6w - 16$ c. $2w^2 - 8w$

24a. $108 - c$ b. $216 - 2c$ c. $324 - 2c$

25a. $3s$ b. $3s + 45$ c. 45 d. $6s + 45$

26a. $15 - a$ b. $0.35a$ c. $2.25 - 0.15a$

d. $0.2a + 2.25$ 27a. $8 - x$ b. $0.6x$

c. $2 - 0.25x$ d. $0.35x + 2$ 28a. $10 - x$

b. $0.12x$ c. $5.5 - 0.55x$ d. $5.5 - 0.43x$

29. $5(2x + 3)$; $10x + 15$ 30. $8x(3y + 2)$; $24xy + 16x$

31a. $6w + 12$ b. 5 by 16 yards

32. $\$13,714,385.71$

33a. $260 - a$ b. $780 - a$ c. 120

34a. $47 - x$ b. $10x$ c. $282 - 6x$ d. $4x + 282$

35. $10x + 6(47 - x) = 330$; she sold 12 reserved tickets.

Homework 4.2

1. No 2. No 3. intercept 4. slope-intercept

5. slope-intercept 6. $(1, -3)$ 7. $(-4, -5)$

8. $(4, 0)$ 9. $(-2, 3)$ 10. inconsistent

11. inconsistent 12. The graphs have the same slope but different y-intercepts.

13a. Plan A: $y = 0.03x + 20,000$
Plan B: $y = 0.05x + 15,000$

b.

x	*Earnings under Plan A*	*Earnings under Plan B*
0	20,000	15,000
50	21,500	17,500
100	23,000	20,000
150	24,500	22,500
200	26,000	25,000
250	27,500	27,500
300	29,000	30,000
350	30,500	32,500
400	32,000	35,000

c. 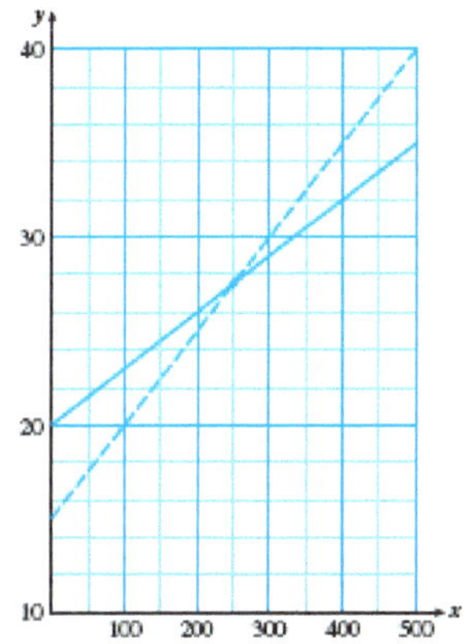

d. $250,000

14a. Cost: $y = 60x + 6000$; Revenue: $y = 80x$

b.

x	Cost	Revenue
0	6,000	0
50	9,000	4,000
100	12,000	8,000
150	15,000	12,000
200	18,000	16,000
250	21,000	20,000
300	24,000	24,000
350	27,000	28,000
400	30,000	32,000

c. 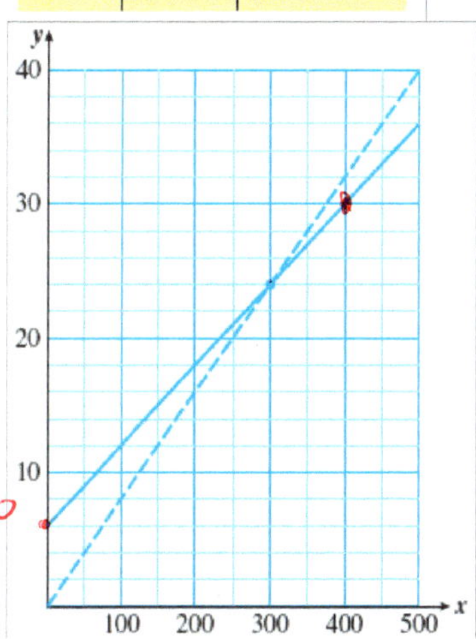

d. 300 clarinets
e. $4000 profit; −$2000 or a loss of $2000
15. rose: $1.50, carnation: $1.00
16. base angles: 65°, vertex angle: 50°

Homework 4.3

1. $(2, 4)$ 2. $(-1, 2)$
3. $(5, 2)$ 4. $(22, -28)$
5. $(2/3, -4/3)$ 6. $(1/3, -2)$
7. $(3, 2)$ 8. $(1, 1)$
9. $(1, 2)$ 10. $(-3, -2)$
11. $(1, -1)$ 12. $\left(\frac{3}{4}, \frac{-1}{4}\right)$
13. inconsistent 14. inconsistent
15. dependent 16. elimination
17. elimination
18. hamburger: 650 cal; shake: 380 cal
19. 12 tables and 48 chairs
20. 17 meters by 4 meters
21. bacon: $2.20; coffee: $5.60

Homework 4.4

1. bonds: $90,000; mutual fund: $60,000
2. $25,000 at 12%; $5000 at 15%
3. student loan: $5000; car loan: $10,000
4c. 26.4% 5c. 40%
6. 15 liters of 12% solution; 30 liters of 30% solution
7. 160 women
8. Delbert: 50 mph; Francine: 45 mph
9. 7 hours; 280 miles 10. 10 hours; 400 miles
11. 40 miles 12a. $g = 0.4t + 43.2$
b. 69.2, 75.2 c,d. 92

Homework 4.5

1a. 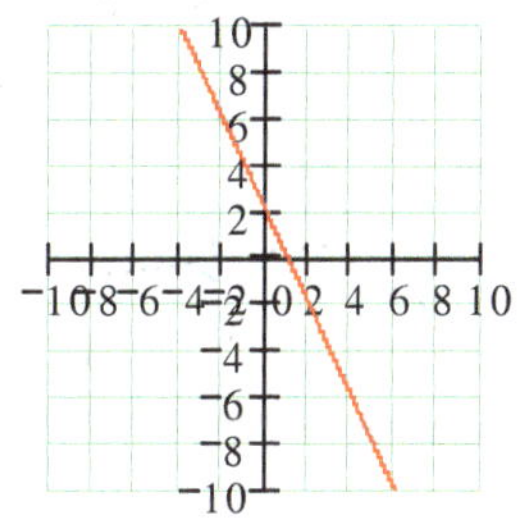

b. $y = -2x + 2$ c. $(1, 0)$

2a. 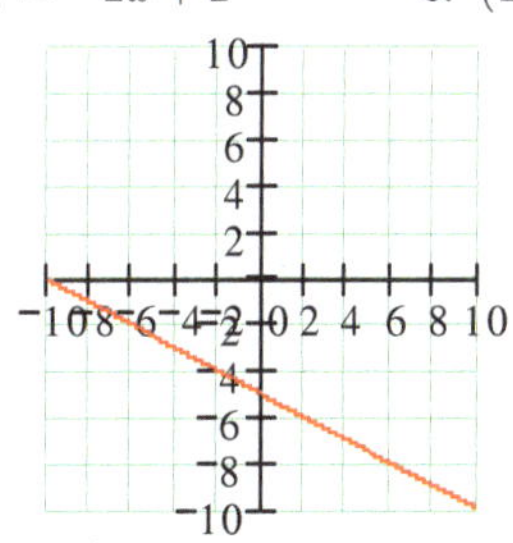

b. $y = -\frac{1}{2}x - 5$ c. $(-10, 0)$

3a. 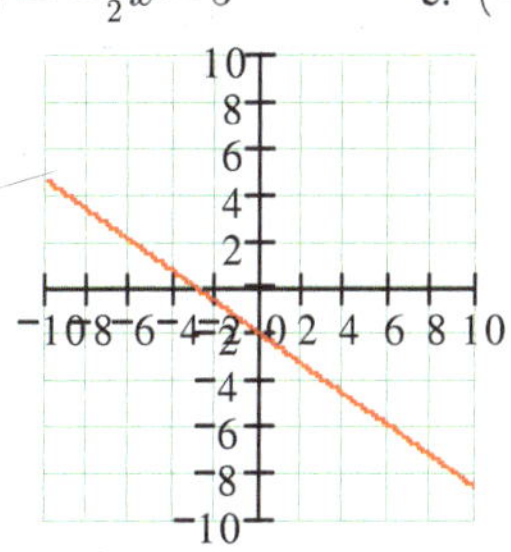

b. $y = \frac{-2}{3}x - 2$ c. $(-3, 0)$

4a. 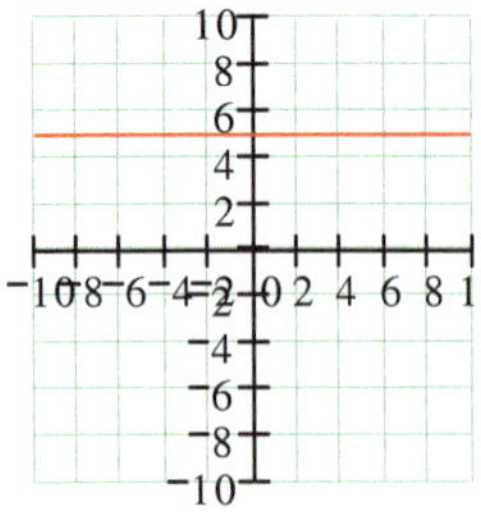

b. $y = 5$ c. no x-intercept

5a. 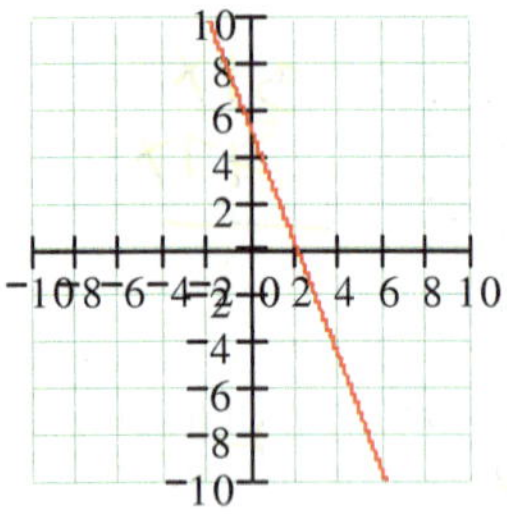

b. $y = -\frac{5}{2}x + 5$ c. $(2, 0)$

6a. 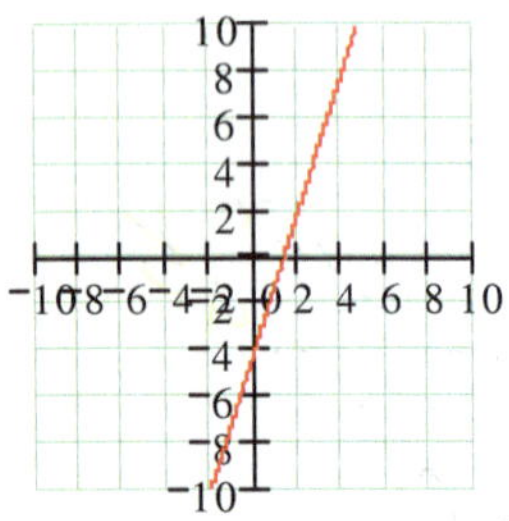

b. $y = 3x - 4$ c. $(\frac{4}{3}, 0)$

7. $m = \frac{3}{5}$, $(0, -7)$ 8. $m = 3$, $(-5, 2)$

9. $m = \frac{4}{3}$, $(0, 0)$ 10. $y = x + 6$

11. $y = \frac{5}{3}x$ 12. $y = \frac{-1}{8}x + \frac{19}{4}$

13. $y = \frac{-5}{7}x + \frac{11}{7}$ 14. $y = \frac{1}{4}x - \frac{1}{4}$

15. $y = \frac{-5}{3}x - \frac{1}{3}$ 16. $y = \frac{10}{7}x - \frac{8}{7}$

17a. $y = \frac{3}{2}x + \frac{5}{2}$ b. $\frac{3}{2}$

c. 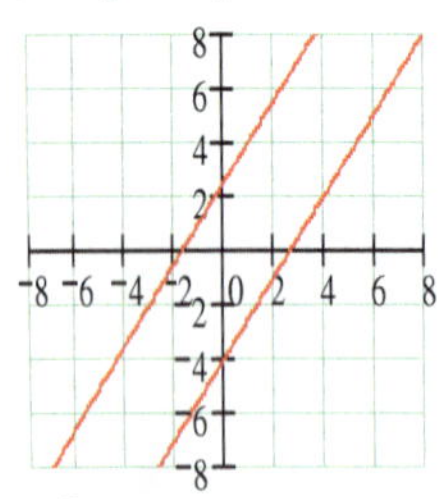

d. $y = \frac{3}{2}x - 4$

18a. $y = \frac{1}{2}x - \frac{5}{2}$ b. -2

c. 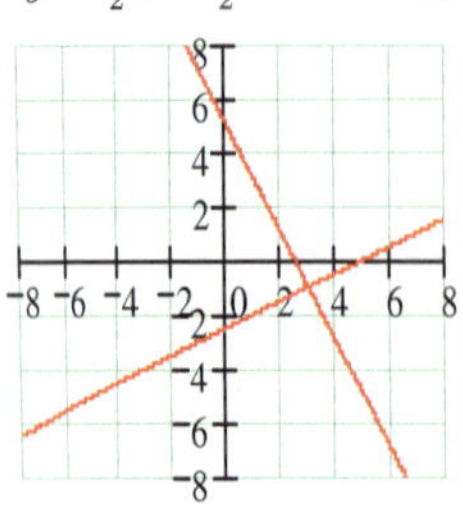

d. $y = -2x + 5$

19a. 2 b. 2 c. $y = 2x + 5$

20a. -3 b. $\frac{1}{3}$ c. $y = \frac{1}{3}x + \frac{8}{3}$

21. $y = -0.0025x + 85$

a. $m = -0.0025$ degrees/ft. The temperature decreases by 0.0025 degrees for every 1 foot increase in altitude. b. 60° c. 85°

22. $y = 200x + 25,000$

a. $m = 200$ dollars/dryer. It costs \$200 to produce each dryer. b. 375 c. \$25,000

23. $C = 65h + 125$

a. 65 dollars/hr. The hourly rate for lessons is 65 dollars per hour. b. \$775 c. \$125

24. $p = 2.2k$

a. 2.2 lbs/kg. One kilogram is equivalent to 2.2 pounds. b. 22.73 kg

c. Zero kg equals zero lb.

Chapter 4 Review

1a. $30m$ b. Requires distributive law: $-30 + 5m$

3. $2m + 7n$ 5. $-15w + 26$

7. $p = 25$ 9. $a < 1$

11a. $3x$ b. $8x$ c. width: 6 cm; length: 18 cm

13a. $30 - x$ b. $1200x$

c. $800(30 - x)$ d. $400x + 24,000$

e. 12 with speakers and 18 without speakers

15. No 17. $(3, 2)$

19. $(-3, -5)$ 21. $(1, -3)$ 23. $(\frac{1}{2}, \frac{7}{2})$

25. $(12, 0)$ 27. Consistent 29. Dependent

31. 12 lb of cereal and 18 lbof dried fruit

33. \$500 in the first account and \$700 in the second

35. 2 lb of the 60% alloy and 6 lb of the 20% alloy

37. Alida: 62 mph, Steve: 31 mph

39. 26 41. \$3181.82 at 8%, \$1818.18 at 13.5%

43a. 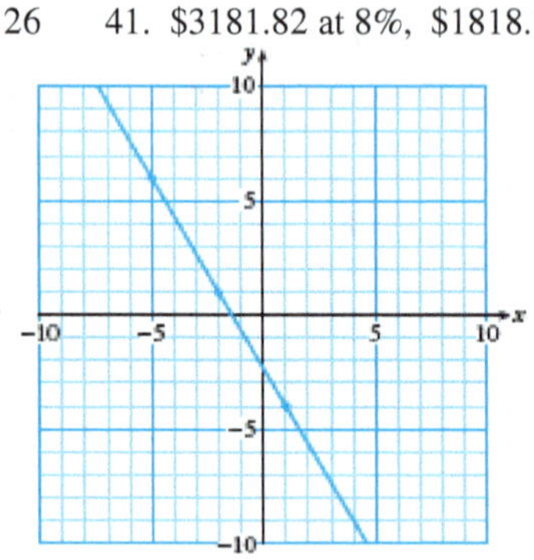

b. $y = \frac{-5}{3}x - \frac{7}{3}$ 45. $y = \frac{-9}{5}x + \frac{2}{5}$

47a. $F = 500 + 0.10C$

b. 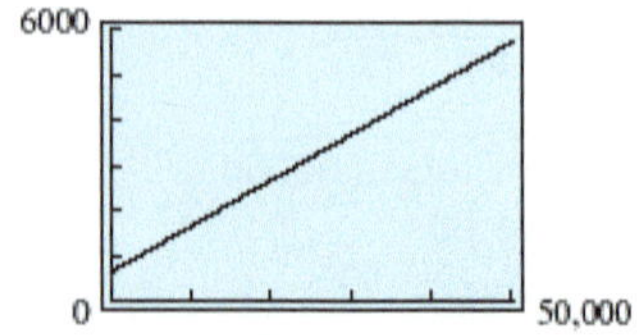

c. $m = 0.10$. The decorator charges 10% of the cost of the job (plus a \$500 flat fee.)

49.

t	0	15
P	4800	6780

b. $P = 4800 + 132t$

c. $m = 132$ people per year gives the rate of population growth

51a. $y = \frac{5}{2}x + b$ b. $y = \frac{-2}{5}x + b$
53. $y = \frac{-2}{3}x + \frac{14}{3}$

Homework 5.1

1a. 64 b. 125 c. 625
2a. $\frac{16}{81}$ b. $\frac{64}{125}$ c. $\frac{121}{81}$
3a. 29.79 b. 45.70 c. 0.41
4a. -25 b. -125 c. 25 d. -125
5a. -4 b. 8 c. -12 d. -36
6a. -40 b. 20 c. 1 d. 13
7a. 192 b. 61 c. 361 d. -84
8a. $3x$ b. x^3 9a. $25a^2$ b. $10a$
10a. $-3q$ b. $-q^3$ 11a. $-6m$ b. $9m^2$
12. $x + x = 2x$ 13. $x \cdot x = x^2$
14. $x^2 + x^2 = 2x^2$ 15. $x \cdot x^2 = x^3$
16. $-2a^2$ 17. cannot be simplified.
18. $-2m^2$ 19. $12k^2$ 20. cannot be simplified
21. $7k^2$ 22. $12b^3$ 23. $2y^3 + 2y^2 + y + 1$
24. $3x^3 + 3x^2 + 11$
25. The exponents can't be combined; $14w^3$
26. The terms are not like terms, so they can't be combined; $6 + 3x^2$
27. $7 - (-5) = 12$; $t^2 + 12$
28. The terms are not like terms, so they can't be combined; $5b^2 - 3b$
29. 49 30. $5 + x^3$ 31. $0.25h^3$
32a. $-0.3x^2 + 9x - 50$ b. \$10, \$17.50, \$10
33a. $0.3h$ b. $50 - h$
c. $0.25(50 - h)$ d. $0.3h + 0.25(50 - h)$
34a. For $z = 2$, $3z^2 = 3(2)^2 = 3 \cdot 4 = 12$, but $(3z)^2 = (3 \cdot 2)^2 = 6^2 = 36$
b. $(3z)^2 = (3z)(3z) = 9z^2$, so $(3z)^2 = 9z^2$
35a. $x = 0$, $x = 2$ b. Any values except $x = 0, x = 2$
c. No
36a. 350 b. 3500 c. 35,000
37a. 7.4 b. 74 c. 740
38a. 2400 b. 891,000 c. 30
39a. 39 b. 234 c. 3456
40. $12x$ 41. $60b$
42. area: $63v^2$, perimeter: $32v$

Homework 5.2

1a. -12 b. -2 2a. 9 b. 3
3a. $19/2$ b. $-2/3$ 4. 8.660 5. -3.055
6. 1.899 7. -15.544 8. 1.293 9. -1.512
10. b 11. c 12. b 13a. 85
b. 168 c. 4082 14a. 3, 5, 8, 12 b. a
c. 4082 15a. 16 b. 29 c. x
16a. 7 b. 20 c. 4 17a. $\sqrt{6}$
b. $-\sqrt{15}$ c. $2\sqrt{m}$ 18a. $2b$ b. $9a$
c. $8b$ 19. $\frac{5}{4}$, 2, 2.3, $\sqrt{8}$
20. $\sqrt{6}$, 3, $2\sqrt{3}$, $\frac{23}{6}$ 21. $\sqrt{5} \approx 2.236$; 1; 0
22. -1; $2 - \sqrt{7} \approx -0.646$; $10 - \sqrt{103} \approx -0.149$
23. 30; $4 - 4\sqrt{2} \approx -1.657$; $\frac{-7}{8}$
24. ≈ 193.81 miles 25. ≈ 9.61 sec
26. A rational number is one that can be expressed as a ratio of integers, such as 5, $\frac{3}{4}$, and $\frac{-11}{2}$. An irrational number cannot be expressed as a ratio of integers. $\sqrt{3}$, $-\sqrt{5}$, and π are irrational numbers.
27. $\sqrt{4}$, $\sqrt{9}$, $\sqrt{64}$. $\sqrt{2}$, $\sqrt{6}$, $\sqrt{11}$
28. Rational: b, c, e, f
29. No

a	b	$a+b$	$(a+b)^2$	a^2	b^2	a^2+b^2
2	3	5	25	4	9	13
3	4	7	49	9	16	25
1	5	6	36	1	25	26
-2	6	4	16	4	36	40

30. No

a	b	$a+b$	a^2	b^2	a^2+b^2	$\sqrt{a^2+b^2}$
3	4	7	9	16	25	5
2	5	7	4	25	29	$\sqrt{29} \approx 5.4$
1	6	7	1	36	37	$\sqrt{37} \approx 6.1$
-2	-3	-5	4	9	13	$\sqrt{13} \approx 3.6$

31. No

a	b	$a+b$	$\sqrt{a+b}$	$\sqrt{a}$	$\sqrt{b}$	$\sqrt{a}+\sqrt{b}$
2	7	9	3	$\sqrt{2}$	$\sqrt{7}$	$\sqrt{2}+\sqrt{7} \approx 4.1$
4	9	13	$\sqrt{13} \approx 3.6$	2	3	5
1	5	6	$\sqrt{6} \approx 2.4$	1	$\sqrt{5}$	$1+\sqrt{5} \approx 3.2$
9	16	25	5	3	4	7

32. No

a	b	$a+b$	$\sqrt{a}$	$\sqrt{b}$	$\sqrt{a}+\sqrt{b}$	$\left(\sqrt{a}+\sqrt{b}\right)^2$
4	9	13	2	3	5	25
1	4	5	1	2	3	9
3	5	8	$\sqrt{3}$	$\sqrt{5}$	$\sqrt{3}+\sqrt{5}$	≈ 15.7
6	10	16	$\sqrt{6}$	$\sqrt{10}$	$\sqrt{6}+\sqrt{10}$	≈ 31.5

Homework 5.3

1. $16w^2 - 36t^2$ 2. $4\pi w^2 - 25$
3. $\dfrac{32\pi h^3}{3}$ 4. $42c^3$
5. $22a^2$ 6. $4\sqrt{3}m^2 + 4\sqrt{3}m + 6m$
7. 8 8. 37 9. $\sqrt{68}$ 10. $\sqrt{15}$
11. $k = 5\sqrt{2}$, $3k = 15\sqrt{2}$
12. No 13. Yes 14. No
15. Not a right triangle
16. Sides should be squared
17. The hypotenuse is x, not 11
18. Sides should be squared
19. $w = \dfrac{v}{lh}$ 20. $m = \dfrac{2E}{v^2}$
21. $h = \dfrac{2A}{b+c}$ 22. $C = \dfrac{5}{9}(F-32)$
23. $h = \dfrac{A - 2\pi r^2}{\pi r}$ 24. $x = a - \dfrac{ay}{b}$
25a. $10^2 + 24^2 = x^2$ b. 26 ft
26a. $700^2 + h^2 = 722^2$ b. 177 ft
27a. $90^2 + 90^2 = x^2$ b. 127.3 ft
28a. $5^2 + 7^2 = x^2$ b. 8.6 mi
29. 79.2 m 30. $\sqrt{2}$ in, $\sqrt{3}$ in, 2 in, $\sqrt{5}$ in, $\sqrt{6}$ in, $\sqrt{7}$ in, $\sqrt{8}$ in, 3 in, $\sqrt{10}$ in
31. Area; $49\pi \approx 153.94$ square inches
32. Circumference; 23 inches
33. $\approx 105{,}904.14$ cubic feet
34. $144\pi \approx 452.40$ square feet

Chapter 5

PROBLEM SET 5.1

1. 60 in. 3. 120 in. 5. 6 ft 7. 162 in. 9. $2\frac{1}{4}$ 11. 13,200 ft 13. $1\frac{1}{3}$ yd 15. 1,800 cm 17. 4,800 m 19. 50 cm 21. 0.248 km 23. 670 mm 25. 34.98 m 27. 6.34 dm 29. 20 yd 31. 80 in. 33. 244 cm 35. 65 mm 37. 2,960 chains 39. 120,000 μm 41. 7,920 ft 43. 80.7 ft/sec 45. 19.5 mi/hr 47. 1,023 mi/hr 49. $18,216 51. $157.50 53. 3,965,280 ft 55. 179,352 in. 57. 2.7 mi 59. 18,094,560 ft 61. 144 63. 8 65. 1,000 67. 3,267,000 69. 6 71. 0.4 73. 405 75. 450 77. 45 79. 2,200 81. 607.5

PROBLEM SET 5.2

1. 432 in^2 3. 2 ft^2 5. 1,306,800 ft^2 7. 1,280 acres 9. 3 mi^2 11. 108 ft^2 13. 1,700 mm^2 15. 28,000 cm^2 17. 0.0012 m^2 19. 500 m^2 21. 700 a 23. 3.42 ha

25. NFL: A = $5{,}333\frac{1}{3}$ sq yd = 1.1 acres; Canadian: A = 7,150 sq yd = 1.48 acres; Arena: A = $1{,}416\frac{2}{3}$ sq yd = 0.29 acres

27. 30 a 29. 5,500 bricks 31. 135 ft^2 33. 48 fl oz 35. 8 qt 37. 20 pt 39. 480 fl oz 41. 8 gal 43. 6 qt 45. 9 yd^3 47. 5,000 mL 49. 0.127 L 51. 4,000,000 mL 53. 14,920 L 55. 16 cups 57. 34,560 in^3 59. 48 glasses 61. 20,288,000 acres 63. 3,230.93 mi^2 65. 23.35 gal 67. 21,492 ft^3 69. 285,795,000 ft^3 71. 192 73. 6,000 75. 300,000 77. 12,500 79. 12.5

PROBLEM SET 5.3

1. 128 oz 3. 4,000 lb 5. 12 lb 7. 0.9 T 9. 32,000 oz 11. 56 oz 13. 13,000 lb 15. 2,000 g 17. 40 mg 19. 200,000 cg 21. 508 cg 23. 4.5 g 25. 47.895 cg 27. 1.578 g 29. 0.42 kg 31. 48 g 33. 4 g 35. 9.72 g 37. 120 g 39. 3 L 41. 1.5 L 43. 20.32 45. 6.36 47. 50 49. 56.8 51. 122 53. 248 55. 38.9

PROBLEM SET 5.4

1. 15.24 cm 3. 13.12 ft 5. 6.56 yd 7. 32,200 m 9. 5.98 yd^2 11. 24.7 acres 13. 8,195 mL 15. 2.12 qt 17. 75.8 L 19. 339.6 g 21. 33 lb 23. 365°F 25. 30°C 27. 3.94 in. 29. 7.62 m 31. 46.23 L 33. 17.67 oz 35. Answers will vary. 37. 91.46 m 39. 20.90 m^2 41. 88.55 km/hr 43. 2.03 m 45. 38.3°C 47. 75 49. 82 51. 3.25 53. 22 55. 41 57. 48 59. 195 61. 3.27 63. $90.00

PROBLEM SET 5.5

1. **a.** 270 min **b.** 4.5 hr 3. **a.** 320 min **b.** 5.33 hr 5. **a.** 390 sec **b.** 6.5 min 7. **a.** 320 sec **b.** 5.33 min 9. **a.** 40 oz **b.** 2.5 lb 11. **a.** 76 oz **b.** 4.75 lb 13. **a.** 54 in. **b.** 4.5 ft 15. **a.** 69 in. **b.** 5.75 ft 17. **a.** 9 qt **b.** 2.25 gal 19. 11 hr 21. 22 ft 4 in. 23. 11 lb 25. 5 hr 40 min 27. 3 hr 47 min 29. 52 min

31.

Triathlete	Swim Time (Hr:Min:Sec)	Bike Time (Hr:Min:Sec)	Run Time (Hr:Min:Sec)	Total Time (Hr:Min:Sec)
Peter Reid	0:50:36	4:40:04	2:47:38	8:18:18
Lori Bowden	0:56:51	5:09:00	3:02:10	9:08:01

33. 00:06:15 35. $104 37. 10 hr 39. $150 41. $6

CHAPTER 5 TEST

1. 3,600 in. 2. .025 m 3. 3.65 dm 4. 144 in^2 5. 1,000,000 mm^2 6. 450 gal. 7. 192 oz. 8. 300,000 cg 9. 50 mi 10. 122 in^3 11. 38.9 12. a. 195 min b. 3.25 hr 13. 7 hr 41 min

Chapter 6

Problem Set 6.1

1. 3 **3.** 6 **5.** 16,194 **7.** 7.5 **9.** 11 **11.** 50 **13.** 900
15. 3.2 **17.** 18 **19.** 87 **21.** 1 **23.** 22 **25.** 2.9 **27.** 38
29. Mean: 79.5; Median: 83 **31.** Both are $31,000
33. a. 78.5 **b.** 76 **c.** 76 **35. a.** size 10 **b.** size 10
37. Mean: $19,985; Median: $19,788 **39.** 47 **41.** 10
43. $1,530 **45. a.** 3,055 **b.** Mean **47.** 7

Problem Set 6.2

1. a.

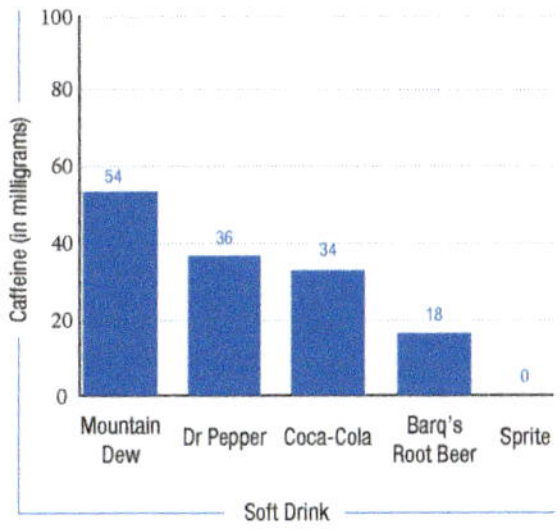

b. 28.4 mg
c. 34 mg
d. 54 mg

3. a.

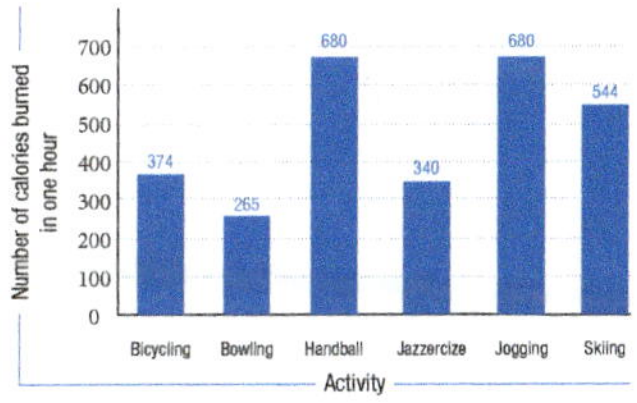

b. 480.5 Cal
c. 680 Cal
d. 415 Cal

5. 5.6 seconds

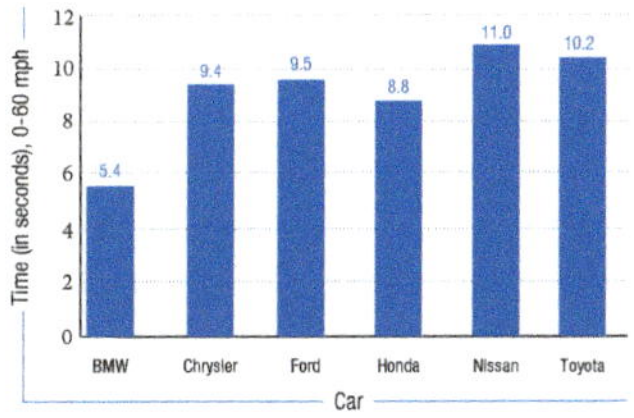

7. a.

Mountain	Planet/Moon	Height (km)
Mauna Kea	Earth	10
Maat Mons	Venus	8
Olympus Mons	Mars	25
Doom Mons	Saturn's Moon, Titan	1.5
Caloris Montes	Mercury	2
Boosaule Montes	Jupiter's Moon, Io	18
Mons Huygens	Earth's Moon	6

b. 15 km **c.** 150%

9.

Concentration of Antidepressant

Days Since Discontinuing	Fraction Remaining in Patient's System
0	1
5	$\frac{1}{2}$
10	$\frac{1}{4}$
15	$\frac{1}{8}$
20	$\frac{1}{16}$

11. a.

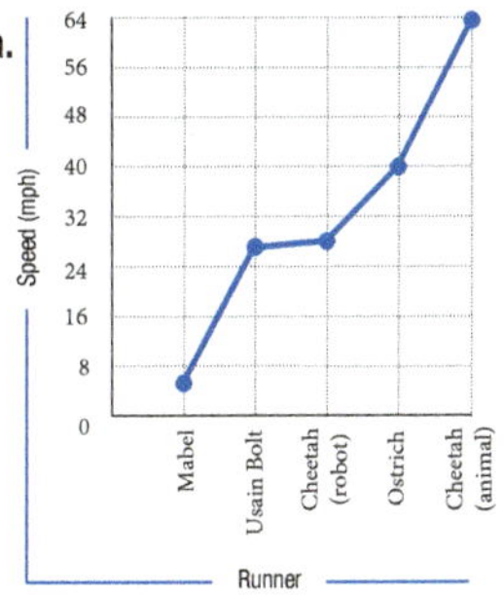

b. 35.7 mph
c. 21.5 mph
d. 1.87%

13. a.

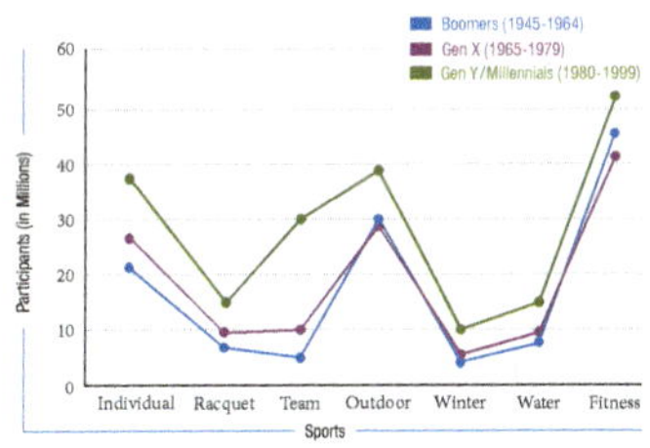

b. Gen Y has greater participation than the other generations.
c. $\frac{5}{30} = \frac{1}{6}$
d. $\frac{11}{26}$

15. a.

Social Network	Number of Worldwide Active Users (in millions; as of January 2014)
Facebook	1,310
QZone	632
Google+	300
Linkedin	259
Twitter	232
Tumblr	230

b. 44.2%; 21.3%

c.

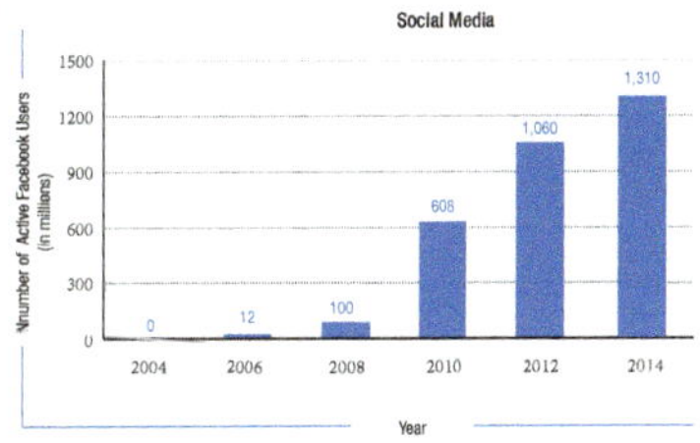

d. $\frac{608}{1310}$; User growth nearly doubled from 2010 to 2014.
e. Increase **f.** 131 users per year

17. a.

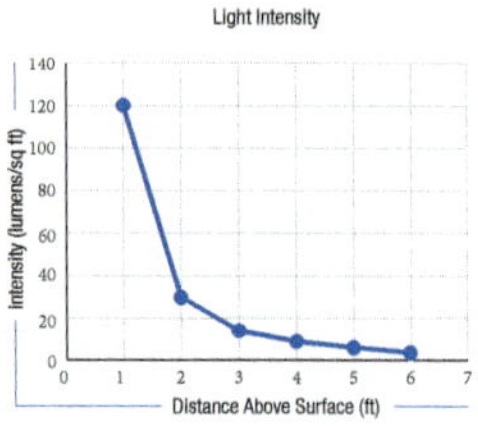

b. 29.8 Lumens/sq ft.
c. 116.7 Lumens/sq ft.

19.

Interval	Frequency
47-56	1
57-66	3
67-76	10
77-86	5
87-96	1

21.

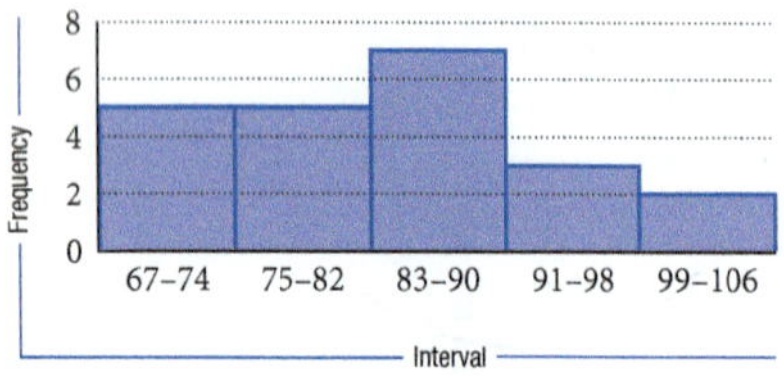

23. a.

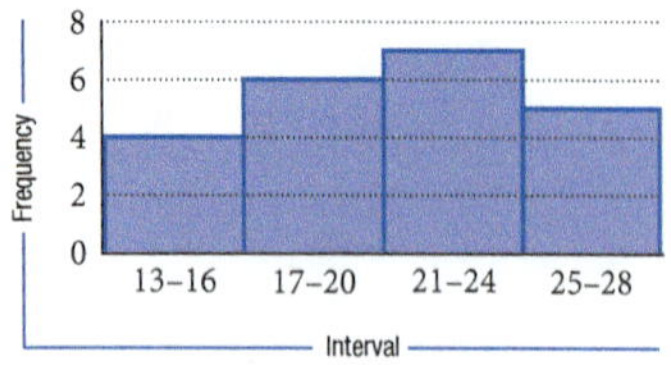

25. a. 87,209 students **b.** 587 million
27. a. ≈ 10.2% **b.** ≈ 9.6%
c. Internet users in the U.S.: 80.3%; Internet users in the world: 35.0%. The percentage of internet users in the U.S. is greater than the percentage of internet users in the world.
29. 60% **31.** $8\frac{1}{3}$% **33.** 144 **35.** 54 **37.** 9 **39.** 1

Problem Set 6.3

1. a. $\frac{103}{200}$ **b.** $\frac{12}{103}$ **c.** $\frac{47}{50}$ **d.** $\frac{47}{3}$
3. a. 64% **b.** 15% **c.** 27% **d.** 36%
5. a. 540 people **b.** 570 people **c.** 840 people
d. 1,890 people

7.

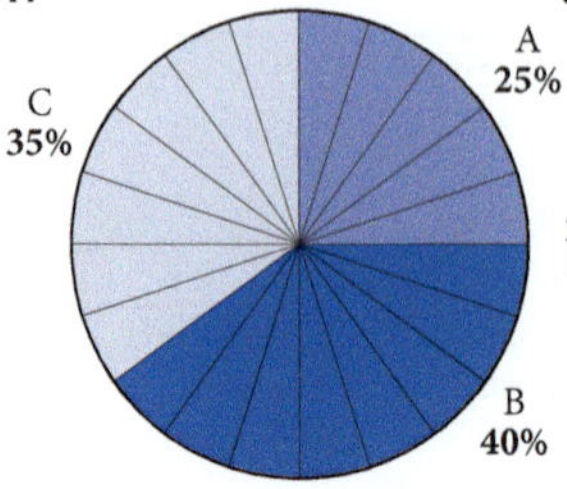

9.

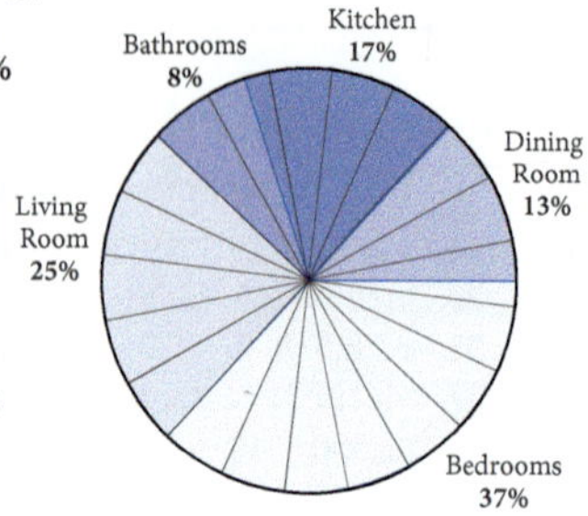

11. 0 **13.** $\frac{1}{13}$ **15.** 0.09

Chapter 6 Test

1. a. 28 min **b.** 3,388 min **c.** 3,388 min
2. a. 174.9 ft **b.** 169 ft **c.** 169 ft **d.** 43 ft
3. a.

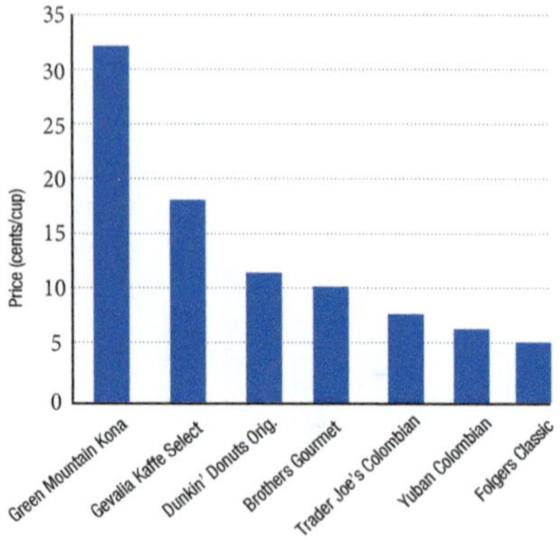

b. 12.6 cents/cup **c.** 10 cents/cup **d.** 27 cents/cup
4. a. 6.6 sec **b.** 11.4 sec **c.** 10.8 sec
d.

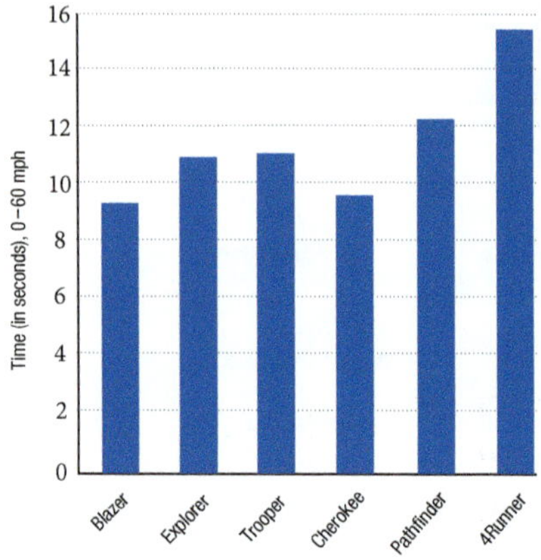

5. a.

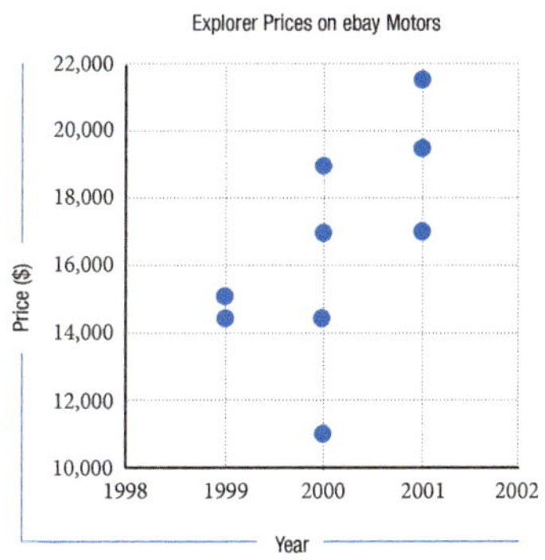

b. \$16,900 **c.** \$10,450

6. a.

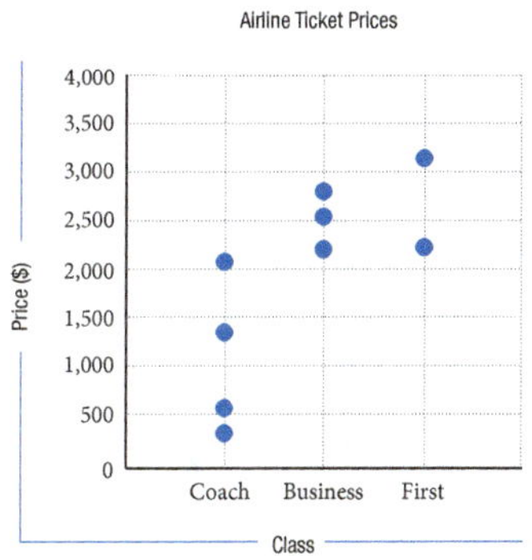

b. \$2,251 **c.** \$2,816

7.

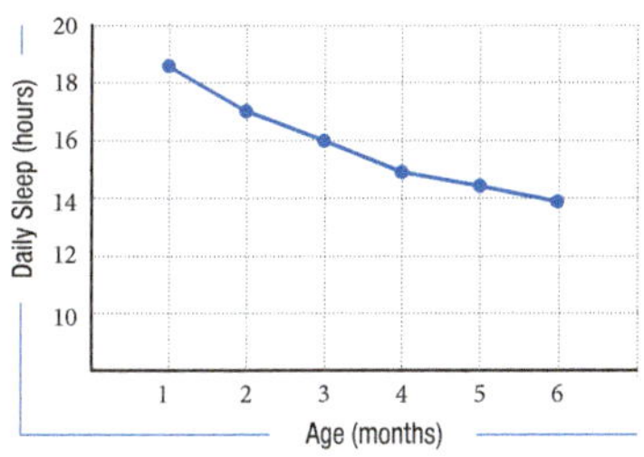

8. a.

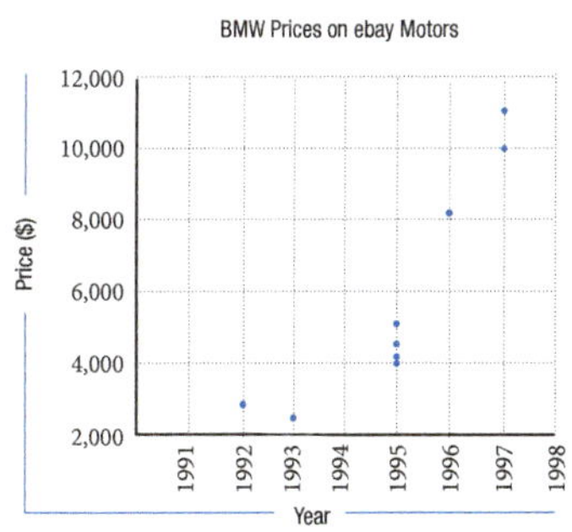

b. \$5,767 **c.** \$4,450 **d.** \$8,900

9. a. \$2,036 **b.** \$2,443 **c.** \$2,443 **d.** \$3,564

e.

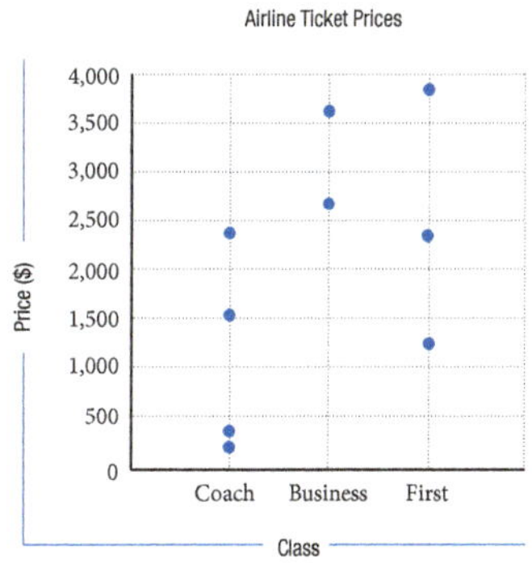

10.

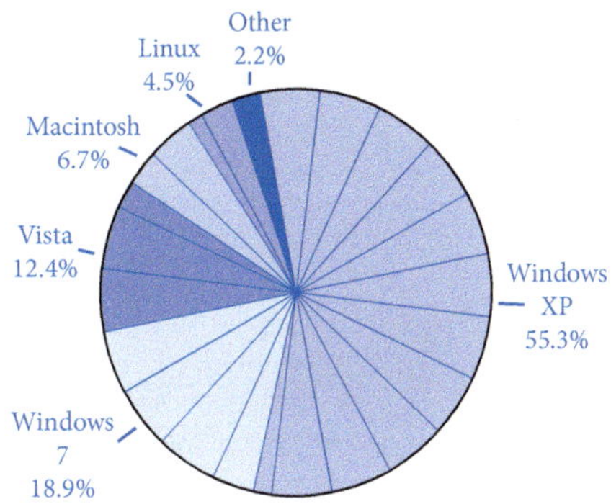

Pie Chart Template Each slice is 5% of the area of the circle.

11.

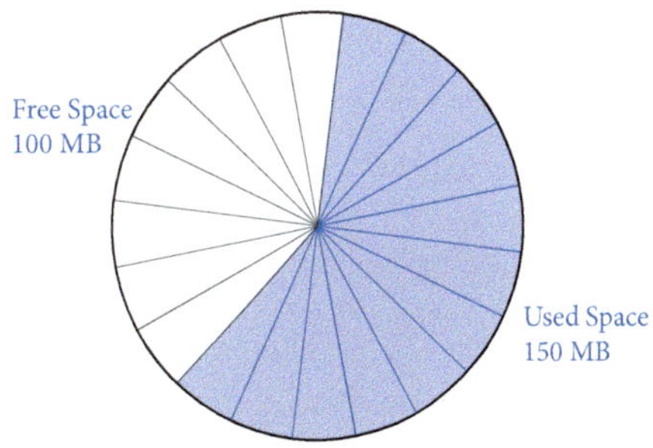

Pie Chart Template Each slice is 5% of the area of the circle.

12.

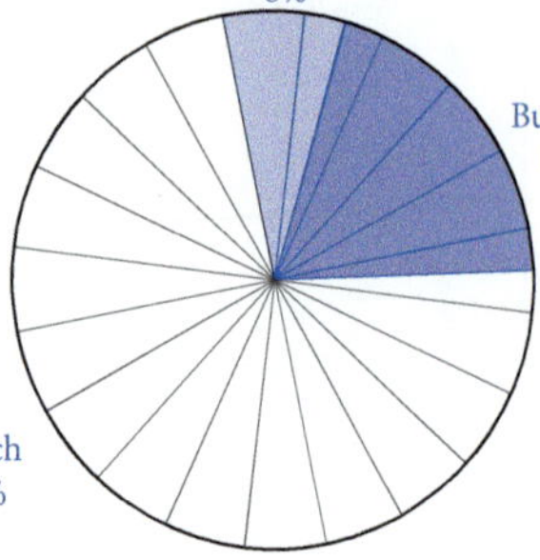

Pie Chart Template Each slice is 5% of the area of the circle.

13. $\frac{13}{52} = \frac{1}{4}$ **14.** $\frac{16}{52} = \frac{4}{13}$ **15.** $\frac{5}{6}$ **16.** $\frac{1}{2}$ **17.** $\frac{25}{80} = \frac{5}{16}$
18. $\frac{27}{80}$